Rhizosphere Engineering and Stress Resilience in Plants

"In the hidden realm beneath our feet, a complex network of interactions unfolds between plant roots and the surrounding soil."

The rhizosphere is a zone of intense biological and chemical activity, which plays a critical role in shaping plant growth, nutrient uptake, and ecosystem functioning. Recent advances in our understanding of root signaling and rhizosphere engineering have revealed the vast potential for manipulating these interactions to enhance crop productivity, improve soil health, and mitigate environmental stresses. It is worthwhile to decipher the molecular signals and nutrient exchanges between plants and microbes, which can in turn facilitate the design of an innovative strategy to optimize rhizosphere functioning.

It is imperative to bring about the development of more climate-resilient crops, improve soil carbon sequestration, and reduce the need for synthetic fertilizers. This field also offers opportunities for novel biotechnological applications, including the development of microbe-based biofertilizers and biopesticides.

As we continue to face the challenges of a growing global population and environmental degradation, rhizosphere engineering offers a cutting-edge solution to crop productivity. It enables harnessing the power of plant–microbe interactions, unlocking new avenues for sustainable agriculture, and ecosystem restoration. This preface sets the stage for exploring the latest advances and future directions in rhizosphere engineering, highlighting its potential to revolutionize our understanding of soil ecosystems and promote a more sustainable future.

This book explores the cutting-edge science and applications of rhizosphere engineering and root signaling, bringing together insights from plant biology, soil science, microbiology, and ecology. Unraveling the intricate communication networks between roots and soil can unveil new strategies for sustainable agriculture, ecosystem restoration, and environmental resilience.

Rhizosphere Engineering and Stress Resilience in Plants

Concepts and Applications

Edited by

Piyush Mathur and Soumya Mukherjee

CRC Press
Taylor & Francis Group
Boca Raton London New York

CRC Press is an imprint of the
Taylor & Francis Group, an **informa** business

Designed cover image: Shutterstock

First edition published 2025
by CRC Press
2385 NW Executive Center Drive, Suite 320, Boca Raton FL 33431

and by CRC Press
4 Park Square, Milton Park, Abingdon, Oxon, OX14 4RN

CRC Press is an imprint of Taylor & Francis Group, LLC

ISBN: 9781032406480 (hbk)
ISBN: 9781032406497 (pbk)
ISBN: 9781003354109 (ebk)

DOI: 10.1201/9781003354109

Typeset in Times
by Newgen Publishing UK

Contents

Preface

The rhizosphere is a dynamic region of the root-soil interface, which encounters a plethora of biomolecular interactions. The rhizospheric region is considered to be the hot spot for the microbial inhabitants. Due to ample secretions from plant roots into the surrounding soil, considerable numbers of microbes are attracted towards the rhizosphere. These microbes can be beneficial for the plants or they may cause detrimental effects. Rhizosphere engineering deals with the tailoring of these beneficial microbes into plant roots under suitable soil and climatic conditions. The book will highlight theengineering of these components for augmentation of plant growth, productivity, and alleviation of stress. The book will discuss the advanced techniques used to study root microbiomes and their identification. With regard to this, use of omics will be highlighted and very importantly application of metagenomics for identification of rhizosphere microbes from varied habitats will be deliberated. The positive effects of plant growth-promoting rhizobacteria (PGPRs) as an inoculant for improving plant health and stress tolerance is a promising area in the current trend of invesitgations. It is important to understand the mechanisms of PGPRs in bioremediation and xenobiotics. Arbuscular mycorrhizal fungi (AMF) have been widely used for biofertilizers for augmenting phosphorus uptake and provide resistance to plants against a number of biotic and abiotic stresses. Furthermore, it is important to understand the affirmative role endophytes in sustaining growth and development of crop plants under stressed and unstressed conditions. Endophytes play a vital role in the production of bioactive secondary metabolites that have wide application in industries, agriculture, and pharmaceuticals. The book will provide fruitful information related to tailoring of plant roots in vivo with beneficial microbes under appropriate soil composition.

About the Authors/Editors

Piyush Mathur is presently working as an Assistant Professor in the Department of Botany, Institute of Science, Banaras Hindu University, Varanasi, Uttar Pradesh, India. He has been teaching for the past 6 years and has taught both at undergraduate and post-graduate level. He completed his Ph.D. in botany with a specialization in mycology and plant pathology at Delhi University. Presently, he is supervising four Ph.D. research scholars who are working on microbial diversity, bioprospecting, and plant disease management. He has published a substantial number of papers in well-known international journals of good repute. He has also contributed articles as book chapters in both international and national books. He is serving as a member of the editorial board of *BMC Plant Biology*. He has previously edited three books with Springer and is working on other books too. Dr. Mathur is an active member of the Society of Rapeseed Mustard Research (SRMR), Bharatpur, and Delhi University Botanical Society (DUBS) Delhi, India and India Phytopathological Society, New Delhi, India

Soumya Mukherjee accomplished his Ph.D in plant physiology from the Department of Botany, University of Delhi, India. He is an ex-faculty member of Ramjas College, University of Delhi. Currently he is employed as a faculty member at the Department of Botany, Jangipur College, University of Kalyani, West Bengal, India. He has published both research and review articles in various peer-reviewed international journals (SCOPUS and SCI indexed). He is serving as an associate editor in the editorial board of *Plant Signalling and Behaviour* (Taylor and Francis) and *Journal of Plant Growth Regulation* (Springer). He has contributed various chapters (plant physiology, abiotic stress signaling) and popular articles to various books published from national and international publishers. He has authored an e-learning module in Plant Physiology and Biochemistry published as an effort of the National Mission on Education through ICT (MHRD Project undertaken by University of Delhi). He has published edited volumes in *Plant Signaling*, and *Communication* (Springer), *Rhizobiology* (Springer), *Phytomelatonin* (Springer), *Crop sustainability and IPR* (Apple Academic Press), and *Strigolactones, Alkamides and Karrikins* (CRC Press).

Contributors

Malay Kumar Adak, Department of Botany, University of Kalyani, Kalyani, West Bengal, India

Nahid Akhtar, School of Bioengineering and Biosciences, Lovely Professional University, Jalandhar, Punjab, India

Fatama S.I. Ali Agricultural Botany Department, Faculty of Agriculture, Ain Shams University, Cairo, Egypt

Mahmoud Adel Ahmed Ali, Department of Horticulture, Faculty of Agriculture, Ain Shams University, Cairo, Egypt

Anshul Arya, Subject Matter Specialist (Plant Protection), Krishi Vigyan Kendra (GBPUAT) Jakhdhar, Rudraprayag, Uttarakhand, India

Abhinobo Chakraborty, Department of Plant Nutrition, Scientist-R&D, Indofil Industries Ltd. Thane, Maharashtra, India

Arka Pratim Chakraborty, Department of Botany, Raiganj University, Raiganj, West Bengal, India

Nilakshi Chakraborty, Department of Botany, University of Kalyani, Kalyani, West Bengal, India

Papan Chowhan, Department of Botany, Raiganj University, Raiganj, West Bengal, India

Subash Chudali, Sharda School of Agricultural Sciences, Sharda University, Greater Noida, Uttar Pradesh, India

Abir Das, Department of Botany, University of Kalyani, Kalyani, West Bengal, India

Abhijit Dey, Department of Life Sciences, Presidency University, Kolkata, West Bengal, India

Dolly Wattal Dhar, Sharda School of Agricultural Sciences, Sharda University, Greater Noida, Uttar Pradesh, India.

Hany G. Abd El-Gawad, Department of Horticulture, Faculty of Agriculture, Ain Shams University, Cairo, Egypt

Mahmoud A.M. Abd El-Hady, Horticulture Department, Faculty of Agriculture, Damietta University, Damietta, Egypt

Ahmed Abou El-Yazied, Department of Horticulture, Faculty of Agriculture, Ain Shams University, Cairo, Egypt

Abhrajyoti Ghosh, Department of Biochemistry, Centenary Campus, Bose Institute, Kolkata, India

Raja Ghosh, Plant Biochemistry Laboratory, Department of Botany, University of North Bengal, Rajarammohunpur, Dist. Darjeeling, West Bengal, India

Trisha Ghosh, Department of Botany, University of Gour Banga, Malda, West Bengal, India

Astha Gupta, Department of Biotechnology, Sharda School of Engineering and Technology, Sharda University, Greater Noida, Uttar Pradesh, India

Santanu Gupta, Department of Botany, University of Gour Banga, Malda, West Bengal, India

Saran Kumar Gupta, Department of Botany, Kalimpong College, West Bengal, India

Shyamalina Haldar, Department of Biochemistry, Asutosh College, Kolkata, India

Soma Halder (Paul), Department of Botany, Asutosh College, Kolkata, India

Mohamed F.M. Ibrahim, Department of Agricultural Botany, Faculty of Agriculture, Ain Shams University, Cairo, Egypt

Suchisree Jha, Department of Plant Nutrition, Scientist-R&D, Indofil Industries Ltd., Thane, Maharashtra, India

Aarshi Kongbrailatpam, Sharda School of Agricultural Sciences, Sharda University, Greater Noida, Uttar Pradesh, India

Meghna Laiphangbam, Sharda School of Agricultural Sciences, Sharda University, Greater Noida, Uttar Pradesh, India

Garima Malik, Department of Botany, University of Delhi, New Delhi, India

Suthar Shubhamkumar Manojkumar, Sharda School of Agricultural Sciences, Sharda University, Greater Noida, Uttar Pradesh, India

Piyush Mathur, Microbiology Laboratory, Department of Botany, University of North Bengal, Rajarammohunpur, Dist. Darjeeling, West Bengal, India; Department of Botany, Institute of Science, Banaras Hindu University, Varanasi, India

Sohair A. Nasr, Agricultural Microbiology Department, Faculty of Agriculture, Ain Shams University, Cairo, Egypt

Praveen Kumar Oraon, Department of Botany, University of Delhi, New Delhi, India

Sayan Pal, Plant Physiology and Molecular Biology Research Unit, Department of Botany, University of Kalyani, Kalyani-741235, West Bengal, India

Debamalla Pathak, Microbiology Laboratory, Department of Botany, University of North Bengal, Rajarammohunpur, Dist. Darjeeling, West Bengal, India

Kawthar A.E. Rabie, Agricultural Botany Department, Faculty of Agriculture, Ain Shams University, Cairo, Egypt

Farida Rahayu, Research Center for Applied Microbiology, National Research and Innovation Agency, Bogor, Indonesia

Swarnendu Roy, Plant Biochemistry Laboratory, Department of Botany, University of North Bengal, Rajarammohunpu, Dist. Darjeeling, West Bengal, India

Dasputre Yash Sanjaykumar, Sharda School of Agricultural Sciences, Sharda University, Greater Noida, Uttar Pradesh, India

Bipul Sarkar, Department of Botany, University of Kalyani, Kalyani, West Bengal, India

Nidhi Selwal, School of Bioengineering and Biosciences, Lovely Professional University, Jalandhar, Punjab, India

Tripti Sharma, Sharda School of Agricultural Sciences, Sharda University, Greater Noida, Uttar Pradesh, India

Sudipta Kumar Sil, Department of Botany, University of Gour Banga, Malda, West Bengal, India

Kumari Prabha Singh, Sharda School of Agricultural Sciences, Sharda University, Greater Noida, Uttar Pradesh, India

Prasun Kumar Singh, Sharda School of Agricultural Sciences, Sharda University, Greater Noida, Uttar Pradesh, India

Surendra Pratap Singh, Plant Molecular Biology Laboratory, Department of Botany, Dayanand Anglo-Vedic (PG) College, Chhatrapati Shahu Ji Maharaj University, Kanpur, India

Vandana Singh, Sharda School of Allied Health Sciences, Sharda University, Greater Noida, Uttar Pradesh, India

Surbhi, Sharda School of Agricultural Sciences, Sharda University, Greater Noida, Uttar Pradesh, India

Sapna Tiwari, Department of Entomology, G. B. Pant University of Agriculture and Technology, Pantnagar, Uttarakhand, India

Prabha Toppo, Microbiology Laboratory, Department of Botany, University of North Bengal, Rajarammohunpur, Dist. Darjeeling, West Bengal, India

Ab Waheed Wani, Department of Horticulture, School of Agriculture, Lovely Professional University, Jalandhar, Punjab, India

Atif Khurshid Wani, School of Bioengineering and Biosciences, Lovely Professional University, Jalandhar, Punjab, India

Priyanka Yadav, Department of Botany, University of Delhi, New Delhi, India

Sujata Singh Yadav, Department of Plant Protection, CSIR-Central Institute of Medicinal and Aromatic Plants, Lucknow, Uttar Pradesh, India

Taring Yakum, Sharda School of Agricultural Sciences, Sharda University, Greater Noida, Uttar Pradesh, India

1 Plant Growth-Regulating Substances in the Rhizosphere

Kawthar A.E. Rabie, Sohair A. Nasr, Fatama S.I. Ali and Mohamed F.M. Ibrahim

1.1 INTRODUCTION

Plant physiologists have recognized for many years that soil microorganisms can produce plant growth-regulating substances (PGRs). The root systems of all higher plants are related to a community of soil microorganisms capable of synthesizing plant hormones. Many of these microorganisms have alternate relationships with plant roots, being highly dependent on their survival on preformed substrates, such as amino acids, vitamins and other growth substances, excreted in root exudates.

Microbial communities in the soil, particularly the rhizosphere, have great potency to produce a wide range of metabolites (biologically active substances) that may affect plant growth and development directly after being taken up by the plant, or indirectly by modifying the soil environment (Frankenberger and Arshad, 1995).

The major focus of this chapter is biologically active substances, particularly phytohormones, their effects on plant growth and development, and recommendation regarding future research needs.

1.2 BIOLOGICALLY ACTIVE SUBSTANCES (BIOTIC SUBSTANCES)

Biologically active substances have an ecological effectiveness on living organisms. Microorganisms are considered the primary source of biologically active substances in soil, although plants may also participate in the soil pool through root exudation (Figure 1.1). Secondary metabolites produced by soil microorganisms may affect the growth and development of the plants (through plant growth regulators, vitamins and inhibitors) and their health and physiological condition (through phytotoxins and pathogenicity).

Plant responses to inoculation with microbial isolates have been attributed indirectly to the ability of inocula to excrete organic substances in the rhizosphere. The term plant growth regulators has been used to include some of the defined biologically active substances that regulate plant growth, and rhizosphere microbiota are considered the potential source of these PGRs. A term equivalent to PGRs, plant growth-promoting microorganisms (PGPM) are rhizosphere microorganisms that colonize the root environment. Some of them are beneficial rhizobacteria while others are fungi that efficiently colonize roots and rhizosphere (ALKahtani et al., 2020; Asghar et al., 2020). These microorganisms can improve agriculture production and can also be used as biofertilizers under stressful environmental conditions. Plant growth-promoting rhizobacteria (PGPR) has been introduced by Kloepper and Schroth (1978) to define free-living, beneficial, root-colonizing bacteria. Production of PGRs by PGPR is one of the suggested mechanisms through which these bacteria affect plant growth and development.

These PGPR promote plant growth through either direct or indirect mechanism of action (Nadeem et al., 2014). Direct growth promotion takes place in different ways like by providing

DOI: 10.1201/9781003354109-1

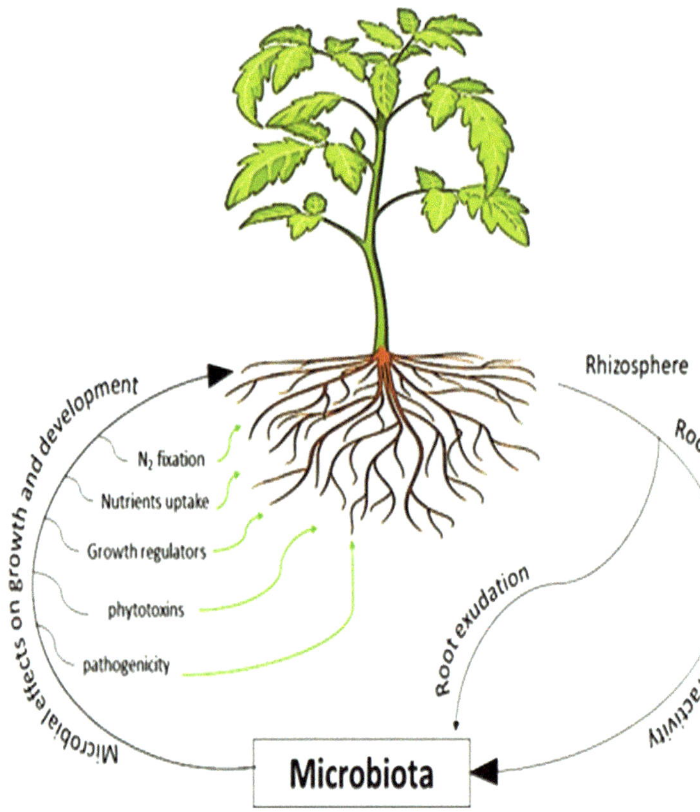

FIGURE 1.1 Plant–microbe interactions.

beneficial compounds to the host plant synthesized bacterium and/or by facilitating nutrient uptake from soil environment (Nadeem et al., 2014; Cordero et al., 2018). While indirect growth may occur when microbes prevent or reduce the virulence of pathogenic microbes by producing antimicrobial products or by increasing resistance in plants against pathogens (Maksimov et al., 2011).

The well-known PGPR include bacteria belonging to the genera *Azotobacter, Azospirillium, Klebsiella, Bacillus, Pseudomonas, Arthrobacter, Enterobacter*, and *Serratia*.

Young et al. (1991) screened a PGPR set for production of PGRs and found a good correlation between the ability of a rhizobacterial strain to create root elongation or promote germination and emergence and significant production of PGRs. They suggested that screening a PGPR for PGRs may improve the competence of selection for effective PGPR strains.

Kloepper (1994) has just discussed the significance of PGPR in agriculture. *Azotobacter and Azospirillium* are the most quietly investigated PGPR. A commercial azotobacterial preparation (Azotobakterin) has been used extensively as a biofertilizer on a commercial scale in various countries.

N_2-fixation by *Azotobacter* spp. or *Azospirillium* spp. was initially considered as the mechanism responsible for increasing the crop yields (Döbereiner and Day, 1975; Subba Rao, 1984). Allison (1947) proposed additional mechanisms of action for *Azotobacter* on plant growth: production of PGRs and antagonism against pathogens. Numerous studies have demonstrated the ability of *Azotobacter* spp. to synthesize plant hormones (auxins, gibberellins, and cytokinins). Hence, the production of PGRs by *Azotobacter* spp. in the rhizosphere and their uptake by plant roots is considered a more credible and alternative clarification for the useful effects of *Azotobacter* spp.

on plant growth. On the other hand, the literature indicated that *Azospirillium* is also a potential producer of auxins, gibberellins, and cytokinins in culture media and, in some cases, growth alterations caused by *Azospirillium* inoculation are similar to those caused by plant hormones. A critical discussion on the mechanism of action of *Azospirillium* spp. on plant growth has been reported by Hubbell and Gaskins (1984).

1.3 PHYTOHORMONES AND MICROBIAL–PLANT INTERACTION

Plant growth regulators are organic substances that affect physiological processes of plants at very low concentrations. When produced endogenously by plants, they are preferably referred to as phytohormones (plant hormones), where the term PGRs includes a large number of synthetic and naturally occurring compounds.

Exogenous supplementation of PGRs to plant roots is necessary to maximize crop yield. Plants themselves synthesize auxins, gibberellins, cytokinins, ethylene, and abscisic acid. Plants may not have the capacity to synthesize sufficient endogenous phytohormones for optimal growth and development under suboptimal climate and environmental conditions. Scientists have shown that plant growth can be improved by using microbial strains "inoculate" seeds or roots of agricultural crops due to microbial production of PGRs. Microbiota capable of producing phytohormones in vitro are present in appreciable numbers in the rhizosphere of the plants. However, microbial production of assured plant hormones (Figure 1.2) has been investigated much more thoroughly than any other biologically active substance of microbial origin; thus, the established phytohormones are the subject of this chapter.

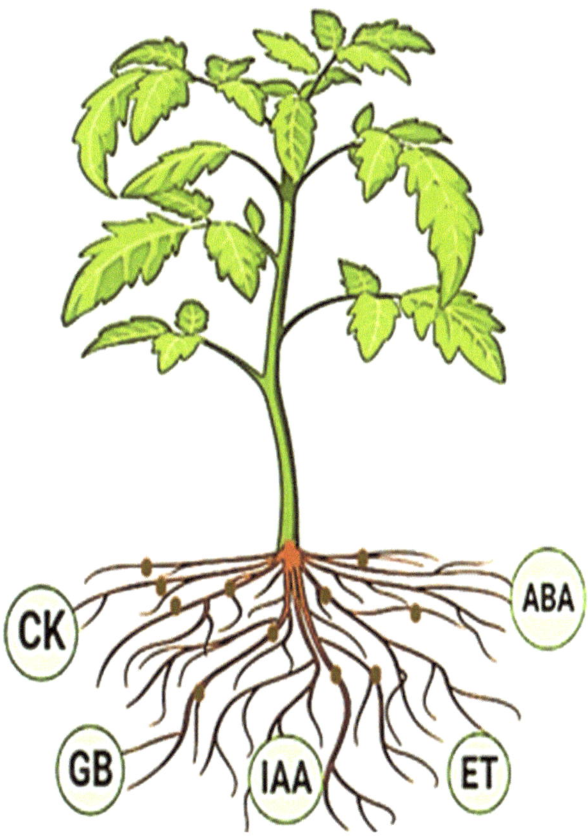

FIGURE 1.2 Microbial production of assured plant hormones.

Scientists have reported that the physiological responses as a result of microbial–plant interaction may involve microbially derived phytohormones. In addition to symbiotic associations (nodulation, mycorrhizae).

1.4 PRECURSOR–INOCULUM INTERACTIONS

Phytohormones of microbial origin in the rhizosphere may modify plant growth as a result of their contact with the plant cells. The lack of understanding of the mechanisms by which a specific inoculum promotes plant growth limits the inoculation technology. Little is known about how the soil type, host plant genotype, and environmental conditions affect the ability of the inocula to colonize and persist on plant roots.

To enhance the ability to reproduce the beneficial effects of inocula by PGR production, Frankenberger and Arshad (1991a, 1991b) and Martens and Frankenberger (1994) reported the success of precursor–inoculum interactions, in which physiological precursors of PGRs alone or precursors plus microbial inocula are added to the rhizosphere. However, the later combined application is often better than the precursors alone (Nieto and Frankenberger, 1991).

1.5 PHYTOHORMONES OF MICROBIAL ORIGIN

1.5.1 Auxins

A various group of soil microorganisms are qualified for producing physiologically active auxins that may have evident effects on plant growth and development. Microbial isolates from the rhizosphere of various crops appear to have a greater potential to synthesize and release auxins as secondary metabolites because of the rich supply of substrates (Strzelczyk and Pokojska-Burdziej, 1984). Production of auxins by microbial isolates varies greatly between different species and strains of the same species and is also affected by culture conditions, growth stage, and availability of the substrates. Therefore, they also follow different biosynthetic pathways for auxin production. Nasr (2002) tested some bacterial and yeast strains for auxin biosynthesis. The highest levels of auxin were achieved by *Pseudomonas flurescens*, *Bacillus cereus*, *Candida pelliculosa*, and *Saccharomycopsis lipolytica* grown on an orbital shaker as a batch culture.

1.5.2 Biosynthesis

Microorganisms use tryptophan (TRP) as an auxin precursor, various intermediates were found in a TRP-supplemented medium revealed that auxin production by microorganisms proceeds via more than one pathway, and even strains of the same microbial species may use different pathways.

There is increasing evidence that various microbes can produce indole-3-acetic acid (IAA) from TRP (Figure 1.3). Tryptophan-derived 1AA production through the indole-3-pyruvic acid (IPyA) pathway (Ruckdäschel and Klingmüller, 1992). Many workers believe that IPyA is converted into 1AA through indole-3-acetaldehyde (IAALd) (Norberg, 1968; Marchal and Rigaud, 1970), but IPyA may also be transformed into indole-3-lactic acid (ILA) (Zimmer et al., 1988). Labeled ILA was identified as one metabolite of labeled TRP added to the soil, indicating the ability of soil microflora to carry out this transformation (Martens and Frankenberger, 1993a).

Another side reaction of IAA pathway is the conversion of 1AALd into indole-3-ethanol (TOL), a transformation considered to be a regulatory reaction in IAA synthesis (Ruckdäschel and Klingmüller, 1992).

Tryptamine (TAM) has been identified as an intermediate in TRP-dependent IAA formation by several microorganisms (Sridhar et al., 1968), and it is metabolized into IAALd by an amine oxidase.

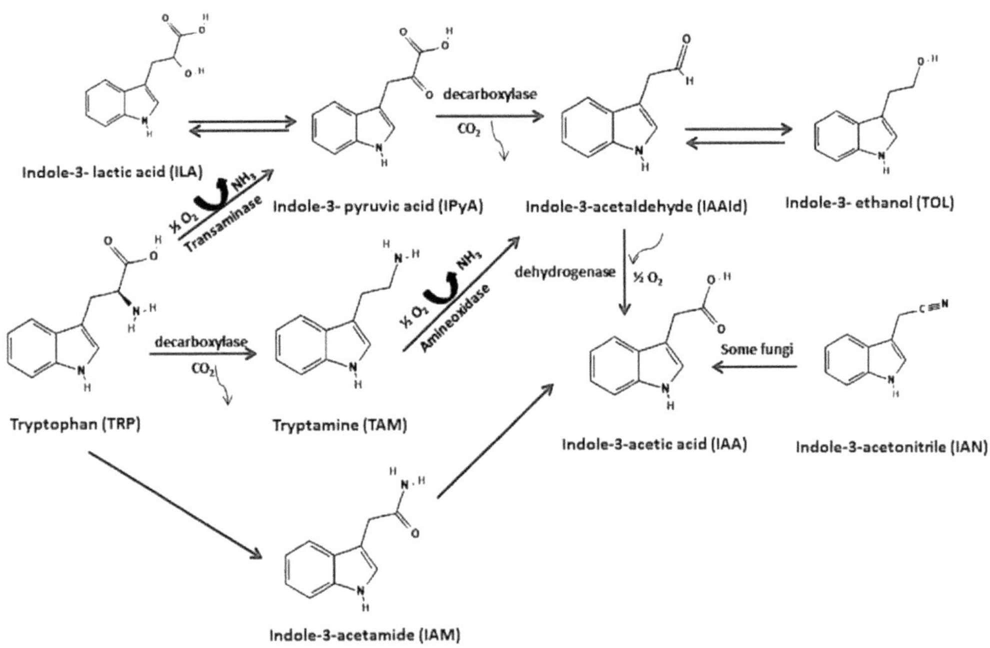

FIGURE 1.3 Pathways of IAA synthesis by microorganisms.

Frankenberger and Brunner (1983) proposed that TRP-derived IAA synthesis in soil occurs by the indole-3-acetamide (IAM) pathway, in addition to the IPyA pathway. When labeled TRP was added to the soil, its catabolites including IAM, ILA, IAA, TOL, and IAALd were identified (Martens and Frankenberger, 1993a), which further confirms the presence of the IAM pathway for IAA formation by soil microflora. The conversion of indol-3-acetonitrile (IAN) to IAA by some fungi also occurs (Magie et al., 1963). However, microbial metabolism of IAN to IAA does not appear to be widespread (Thimann and Mahadevan, 1964).

1.5.3 Microbial Metabolism of 1AA

Indole-3-acetic acid is metabolized readily by microorganisms to several products. The reactions leading to these products include oxidation, conjugation, and hydroxylation of IAA.

1.5.3.1 Oxidation

Oxidation of IAA led to various indole derivatives (Figure 1.4) and is believed to play a vital role in detoxification of high levels of IAA. IAA oxidation was induced by a different oxidase system. Tsubokura et al. (1961) isolated the bacterium from air capable of metabolizing IAA oxidatively into anthranilic acid. Indole-3-acetic acid can be oxidized by other broad-spectrum enzymes, including tyrosinase and laccase, which are produced by a large number of soil bacteria and fungi (Mahadevan, 1982).

1.5.3.2 Conjugation

The formation of IAA conjugates (ester and amide) by microorganisms is also common (Figure 1.5). *Bacillus megatherium* is active in the glycosylation of IAA (Mahadevan, 1984). There are no reports on the formation of IAA-aspartate conjugate by microbial cultures, a common conjugate in plants, except for its formation in root nodules of peas when labeled IAA was exogenously

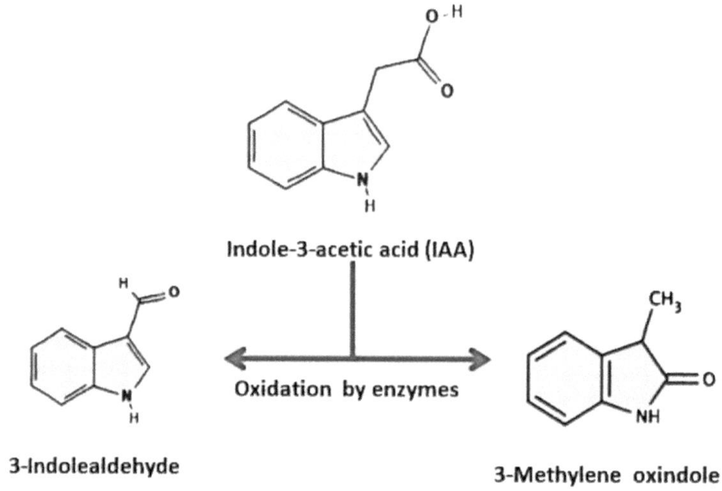

FIGURE 1.4 Oxidation of indole-3-acetic acid (IAA) (from Mahadevan, 1984).

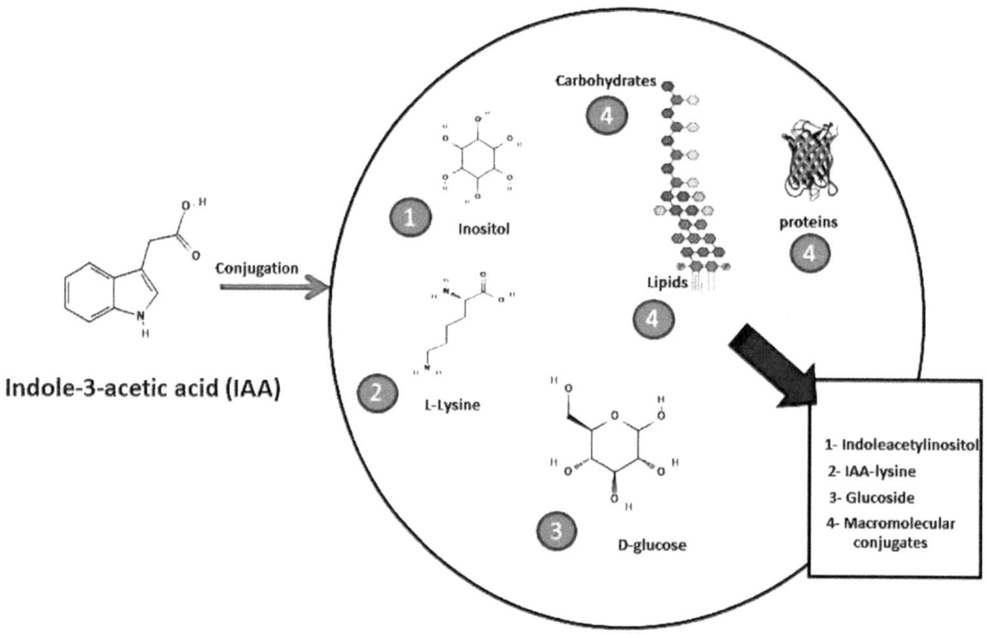

FIGURE 1.5 Conjugation of indole-3-acetic acid into inactive forms.

applied (Badenoch-Jones et al., 1983). However, it is unclear whether the source of this conjugate is *Rhizobium* or the plant root.

1.5.3.3 Hydroxylation

Hydroxylation of IAA results in the loss of auxin activity. *Aspergillus niger* and *A. sydowii* produced 5-OH-IAA from TAM through IAA (Figure 1.6) (Dvornikova et al., 1970). Martens and Frankenberger (1993a) found 5-OH-IAA among the catabolites of TRP, TAM, and IAALd added to

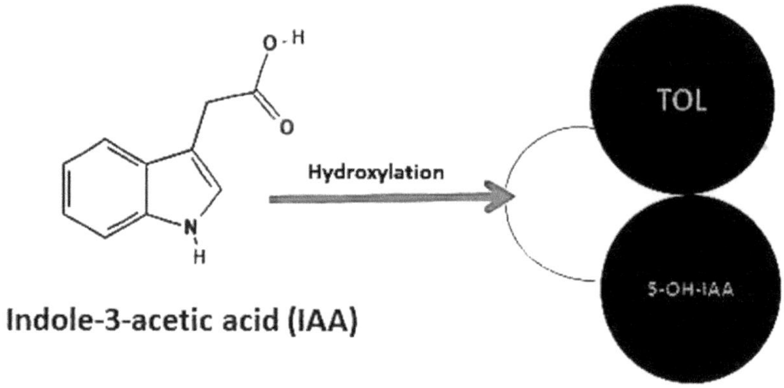

FIGURE 1.6 Hydroxylation of IAA.

the soil. It is highly likely that 5-OH- IAA might be the metabolic product of 1AA derived by the soil-indigenous microflora from added precursors.

1.5.4 Plant Response to Auxins of Microbial Origin

Many soil microorganisms (e.g., *Rhizobium*, *Azotobacter*, *Azospirillum*, *Pseudomonas*, mycorrhizal fungi, and others) that significantly affect plant growth and development are also able to produce auxins and other plant hormones. Auxins of microbial origin are ecologically safe. Microbially released auxins provide a continuous supply to plant root, which may be better than a one-time application of synthetic auxins. High concentrations of auxins led to inhibition of root growth; therefore, a continuous release of auxins at low concentrations as the outcome of microbial activity in the rhizosphere may be more beneficial.

Numerous studies have shown an improvement in plant growth and development in response to seed or root inoculation with various microbial inoculants. Loper and Schroth (1986) observed a significant linear relationship between IAA accumulation of the rhizobacterial strains and decreased root elongation and increased shoot/root ratios of the sugar beet seedlings. Holl et al. (1988) showed that production of plant growth-promoting compounds, like IAA, by the bacterium were stimuli an increment in plant productivity. Müller et al. (1989) reported that morphogenetic effects caused by rhizosphere bacteria may be a result of different ratios of plant hormones produced by roots as well as rhizosphere bacteria.

Tryptophan is considered the physiological precursor of auxin for both plant and soil microorganisms. Its addition to soil has been shown to increase auxin content and may have physiological effects on plant growth through microbially derived auxins. Low availability of TRP could be the most limiting factor in auxin production in the rhizosphere of plants. The only natural source of TRP for rhizosphere microflora is the root exudate. Detectable amounts of TRP in the root exudate of some, but not in all varieties of wheat have been reported (Martens and Frankenberger, 1993b). This indicates that not all the plants can release adequate quantities of TRP into the rhizosphere for microbial production of auxins, and auxin production in the root zone is limited by the genetic and physiological properties of both the microorganisms and the plants.

1.6 GIBBERELLINS

Gibberellins are tetracyclic diterpenoid acid, about more than 136 different chemical structures have been characterized as naturally occurring gibberellins. In addition to an unknown number of glucose-conjugate forms (Schneider, 1983) and likely intermediates yet to be identified in biosynthetic and catabolic pathways (Sponsel and Hedden, 2004).

However, gibberellins are produced not only by higher plants and fungi (MacMillan, 2002) but also by bacteria (Janzen et al., 1992; Gutiérrez-Mañero et al., 2001; MacMillan, 2002). In fungi and bacteria there is no known role for gibberellins, rather they seem to be secondary metabolites that may play a role as signaling factors towards the host plant. However, evidence is available that GA-producing microorganisms often induce reactions in host plants, which are beneficial to their growth (Rademacher, 1994).

Apart from *Azospirillum* sp. and *Rhizobium* sp., production of gibberellins has been confirmed in *Acetobacter diazotrophicus*, *Herbaspirillum seropedicae* (Bastian et al., 1998), and *Bacillus* sp. (Gutiérrez-Mañero et al., 2001).

In fungi, the general pathway is similar to that of higher plants, although the genes and enzymes involved differ. Thus, the information on the enzymes and their genetic control involved in the metabolism of gibberellins in higher plants and fungi is abundant, evidence for bacterial biosynthesis is scarce.

1.6.1 Putative Biosynthetic Pathway for Gibberellins in *Azospirillum* sp.

Piccoli et al. (1996, 1997) demonstrated that *Azospirillum* sp. can metabolize gibberellins in vitro as well as in vivo, i.e., in association with higher plants (Cassán et al. 2001a, 2001b). Piccoli and Bottini (1994) found that the bacteria produce GA_9, and GA_{19} in chemically defined culture may indicate the existence of two branches for biosynthetic pathway in *Azospirillum* sp. (Figure 1.7), i.e.,

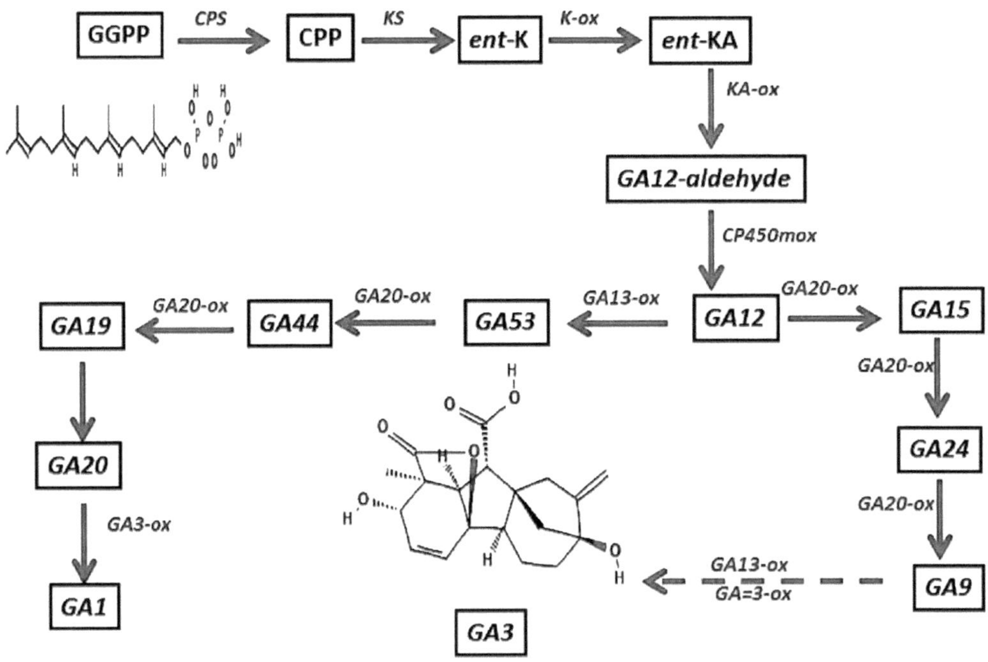

FIGURE 1.7 Putative biosynthetic pathway for gibberellins in *Azospirillum* sp. based on well-established steps in vascular plant and fungi, and data available on gibberellin metabolism studies with this bacterium (Piccoli and Bottini, 1994). (geranylgeranyldiphosphate (GGPP); copalyl diphosphate (CPP); ent-kaurene (ent-k); ent-kaurenoic acid (ent-KA); cytochromeP450monooxygenases (CP450 mox); 2- oxoglutarate-dependent soluble dioxygenases (GA_{20}-OX); 13-hydroxylase (GA_{13}-OX); 3β-hydroxylases (GA_3-OX)).

early 13-hydroxylation involving the metabolism of GA_{19} (and its metabolite, GA_{20}) to GA_1, and an early nonhydroxylation branch where GA_9 is (presumably) the precursor of GA_3.

Piccoli and Bottini (1996), found that the influence of blue light on *A. Lipoferum* cultures, led to two- to three-fold increase in the amount of GA_3, relative to GA_1. Other environmental factors, such as N supply (Piccoli and Bottini, 1994), O_2 availability and osmotic potential (Piccoli et al., 1999), can influence both the quantity and type of gibberellin produced by *Azospirillum* cultures. The reverse was true for high concentrations of NH_4CL, which reduced the quantity of GA_3 produced (Piccoli and Bottini, 1994).

From the results of the previous research, as in higher plants, early steps of the gibberellin biosynthetic pathway in the bacterium may be regulated by membrane-related cytochrome P450 monooxygenases, and the late hydroxylate steps by soluble 2-oxoglutarate-dependent dioxygenases.

There are a number of interconversions between gibberellins, there is also evidence that bound gibberellins exist as glucosides; however, it is still not clear whether they are for inactivation or as a storage source.

1.6.2 Plant Response to Gibberellins Produced by Microorganisms

The widespread ability of microorganisms to produce more than one plant hormone, which makes it very difficult to assess the role of microbially produced GAs alone in inducing a physiological response in plants (Frankenberger and Arshad, 1995). Many facts showed that GAs of microbial origin, particularly those synthesized by the microbial community within the rhizosphere or by microsymbionts, can affect plant growth.

Williams and de Mallorca (1982) showed that rhizobial–plant interactions, most likely involving GA_S, in modifying plant growth. Similarly, Allen et al. (1982) proposed that GAs may be involved in host–mycorrhizal interactions. Yakushkina and Tarasenko (1975) reported that infection of sterilized seeds with epiphytic microorganisms led to an increase in endogenous hormones (including GAs) and increased growth rate.

The increases in shoot length induced by suitable bacterial strains are relatively small and amount to around 25% (Dobert et al., 1992). Cassán et al. (2003) suggested that the microorganism modifies hormonal levels in the nodules, either by affecting plant cell metabolism or by affecting gibberellin production by the bacterium. Yanni et al. (2001) noted that the Rhizobium–rice combination promotes root and shoot growth, thereby improving seedling vigor and increasing grain yield.

Kucey (1988) found an increment of ^{15}N uptake after inoculation of wheat roots with *Azospirillum* sp. and *Bacillus* sp.

Furthermore, reversal of dwarfism, both genetic and induced by inhibitors of gibberellin biosynthesis, was demonstrated in both rice and maize seedlings that had been inoculated with *Azospirillum* sp. and showed the endophytic presence of the bacteria (Lucangeli and Bottini, 1996, 1997).

Cohen et al. (2001) showed that prior inoculation with *A. lipoferum* promoted growth of both maize roots and shoots under drought. Probanza et al. (2002) reported that inoculation with *Bacillus licheniformis* and *B. pumilus* enhanced growth of *Pinus pinea* plants, probable by bacterial gibberellin production.

1.7 CYTOKININS

Cytokinins (CKs) are a group of plant hormones that, together with other plant hormones, promote cytokinesis and affect many other aspects of plant development (Mok, 1994). Despite the classification of plant hormones, CKs are evolutionally older than plants, constituting a class of highly preserved low-molecular mass molecules present in many organisms, including bacteria, amoebae, filamentous fungi, algae, nematodes, insects, and humans (Frébortová and Frébort, 2021)

Naturally occurring CKs are adenine derivatives, with either isoprenoid or aromatic side-chain attached to the exocyclic nitrogen atom. These include N6-(2-isopentenyl) adenine (iP), trans-zeatin (tZ), cis-zeatin (cZ) and dihydrozeatin (DHZ) in the group of isoprenoid CKs. Cytokinin molecules extend to various conjugates with sugars, sugar phosphates, and amino acids (such as ribosides, ribonucleotides, N-glucosides, O-glucosides, O-xylosides, Lupinic acid, or discadenine), or other compounds substituted either at the C2 position of adenine ring (2-methylthiolated CKs) or on the isopentenyl side-chain (mono and dimethylated iP) (Radhika et al., 2015).

Both plants and microorganisms are known to produce cytokinins, but little is known about the ecological significance of microbially synthesized cytokinins. It has been known for several decades that bacteria can synthesize CKs (Stirk and Van Staden, 2010), but most of them only in a form bound to tRNA. Notably, CKs are produced by plant-associated bacteria, both pathogenic and beneficial (Akhtar et al., 2020).

1.7.1 Biosynthesis

Biosynthesis of free cytokinins has been studied in cell-free systems of both plants and microorganisms. Adenine is considered the common precursor for cytokinin production in plant tissues (Dickinson et al., 1986) and some microorganisms like bacteria, *Corynebacterium fascians* (Murai, 1980) and *Azotobacter* spp. (Nieto and Frankenberger, 1989) (Figure 1.8). Other possible precursors include adenine, adenosine-5′-monophosphate (AMP), isopentyl alcohol and isoprene served as precursors of cytokinin synthesis by *Azotobacter chroococcum* (Nieto and Frankenberger, 1989).

Free cytokinins occur in plants by the turnover of cytokinin-containing tRNA or by de novo synthesis, or by the combination of both mechanisms. Adenine is incorporated into tRNA with subsequent hydrolysis and the release of free cytokinin in the 5′-monophosphate form (Klemen and Klämbt, 1974). The relatively constant ratios of Zeatin riboside [9R] Z and 2-Methylthiozeatin riboside [2MeS 9R]Z in tRNA and their derived bases found in *C. fascians* led to the view that tRNA may serve as the source of cytokinins for the bacterium (Murai, 1980).

The survey of the spectrum of CK metabolites produced by plants and various microorganisms suggests that not all pathways involved in plant CK metabolism function in microorganisms (Figure 1.9). For example, there is no experimental evidence for the presence of aromatic CKs in bacteria. In addition to the missing N9-glucosylation pathway of CK deactivation, most bacteria lack CK dehydrogenase (CKX) to degrade CKs (Wang et al., 2020). In contrast, the only characterized CK species exclusive to bacteria are derivatives of iP with mono- and dimethylated isopentenyl sidechain (1-Me iP and 2-Me iP) (Radhika et al., 2015). On the other hand, the origin of cis-Zeatin

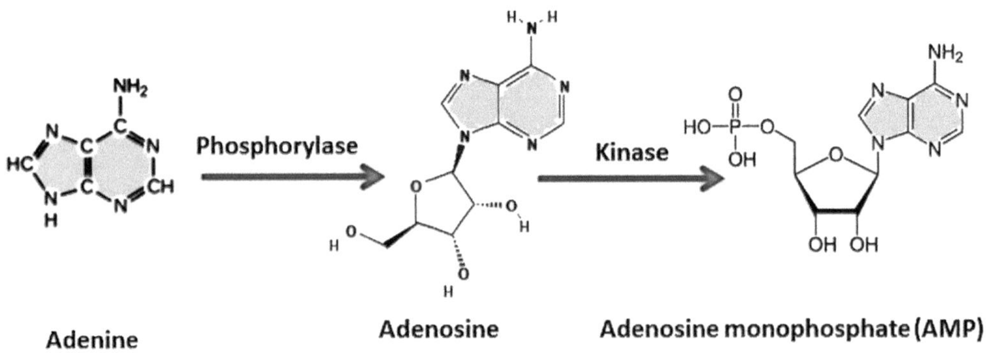

FIGURE 1.8 5′-adenosine monophosphate (AMP).

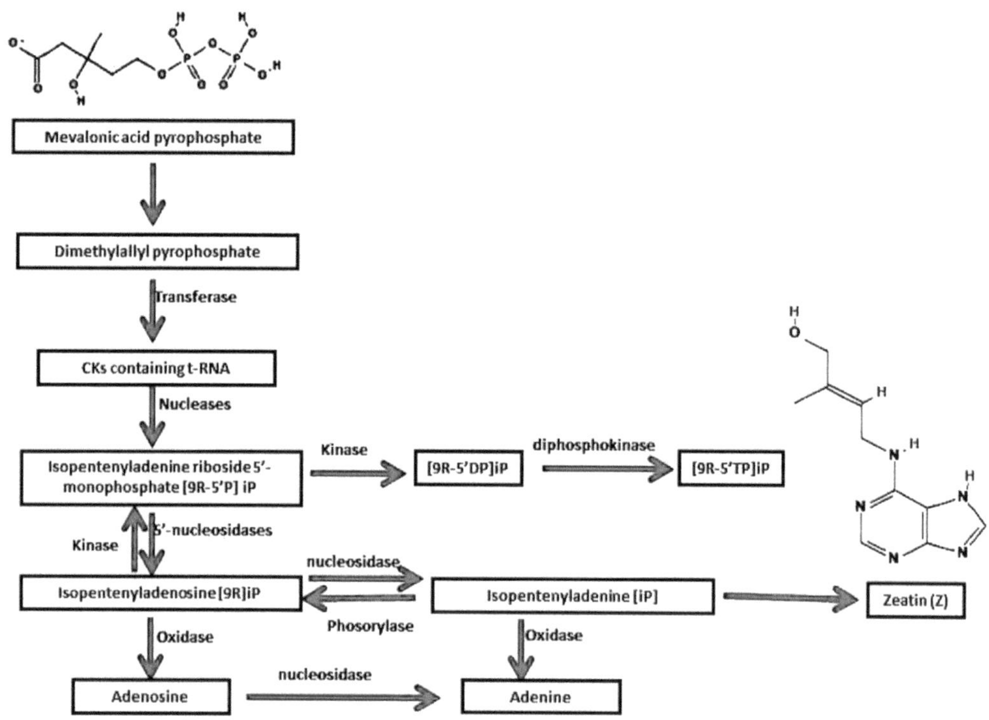

FIGURE 1.9 Biosynthesis of cytokinins (adapted from Chen et al., 1982).

riboside (cZR) and 2-methylthio-cis-zeatin riboside (2 Me ScZR) in plants must differ from that in bacteria due to the absence of the *miaE* gene in eukaryotes (Frébort et al., 2011).

1.7.2 Metabolism

According to McGaw (1987), Cytokinin metabolism can be categorized under four reactions as follows:

1.7.2.1 Conjugation

Conjugation may lead to the formation of less active cytokinins, with decreased biological activity. In some cases, conjugation may lead to detoxification of very high cytokinin levels. Various kinds of conjugates have been induced, including ribosides and ribotides. Ribosyl conjugation is confined to position 9 of the purine ring. The ribosides and their aglycones (i.e., Z and iP) are highly active in cytokinin bioassays (Letham et al., 1983). Whether or not these compounds are the "active" molecule is unknown because of their intensive interconversion (McGaw, 1987). The enzymes responsible for these interconversions include adenosine phosphorylase found in *Escherichia coli* (Sivadjian et al., 1969), which catalyzes iP to [9R] iP.

1.7.2.2 Hydrolysis

The resultant conjugates may be subjected to hydrolysis to yield free bases or other active cytokinins. The enzymatic hydrolysis of these conjugates plays an important role in maintaining the desired cytokinin activity in plant tissues.

1.7.2.3 Reduction

The enzymatic conversion of Z into (di H)Z and its ribosides and ribotides has not been characterized (Entsch et al., 1980). The interconversion of Z (di H) Z may be important in the maintenance of cytokinin activity levels in an oxidized environment.

1.7.2.4 Oxidation

The oxidative N^6 side-chain cleavage of cytokinin leads to the irreversible destruction of cytokinin activity. The mechanism of oxidative cleavage is not fully understood, but 3-methyl-2-butenal (Brownlee et al., 1975) and adenine (McGaw and Horgan, 1983) have been identified by GC-MS as the products of cytokinin oxidase action on iP.

1.7.3 Plant Response to Cytokinins Produced by Microorganisms

Microbially produced cytokinins may play an important role in stimulating plant growth and development. Barea et al. (1976) found that 90% of the bacteria isolated from the rhizosphere of various crop plants were potential producers of cytokinin-like compounds. As a result of the intimate relation of rhizosphere microorganisms with the root surface, the exogenous production of cytokinins by these organisms may have a profound effect on plant growth. This exogenous supply may provide supplemental quantities to the plant's endogenous level or alter the plant's synthesis and metabolism of other plant growth regulators. Nieto and Frankenberger (1991) demonstrated that microbially derived cytokinins can be of great value in enhancing crop production.

Many rhizospheric and epiphytic plant growth-promoting bacteria belonging to the diverse genera, including *Bacillus*, *Azospirillum*, *Pseudomonas*, *Rhizobium*, or *Methylobacterium*, can produce CKs, which affect the growth and productivity of plants, increase tolerance to environmental stress, or enhance disease resistance (Numan et al., 2018).

1.8 ETHYLENE

Among the compounds with growth-regulating characteristics, ethylene (C_2H_4) is the only phytohormone with simple structure and the only plant growth regulator that is a gas under physiological conditions of temperature and pressure. Ethylene, a natural product of plant metabolism produced by almost all plant tissues and organs has been shown to be involved in numerous aspects of plant growth and development from seed germination through senescence and death of the plant. It has numerous but complex interactions with other plant hormones, particularly with auxins.

A diverse group of microbiota, including both pathogens and nonpathogens, are very active in producing C_2H_4. As a result of microbial activity, C_2H_4 accumulates in the soil atmosphere in concentrations enough to be biologically active. According to Primrose (1979), C_2H_4 is synthesized by many species of bacteria and fungi, but is oxidized by only a limited number of microorganisms.

Because of the ability of microorganisms to derive C_2H_4 from a variety of compounds, the biochemistry of microbial production of C_2H_4 is very complex. A wide range of compounds have been reported as possible precursors for C_2H_4 generation by different microbial isolates, including alcohol, sugars, organic acids, amino acids, Krebs cycle acids, methionine analogues, and humic phenolic compounds (Goto et al., 1985; Arshad and Frankenberger, 1989). However, usually L-methionine (L-MET) has been the most favorable substrate.

1.8.1 Biosynthetic Pathways

The ability of microbial isolates to generate C_2H_4 from a variety of structurally unrelated compounds indicates that more than one pathway may be involved in the microbial biosynthesis of C_2H_4. Although most soil microorganisms can synthesize C_2H_4 from MET, the pathway does not appear

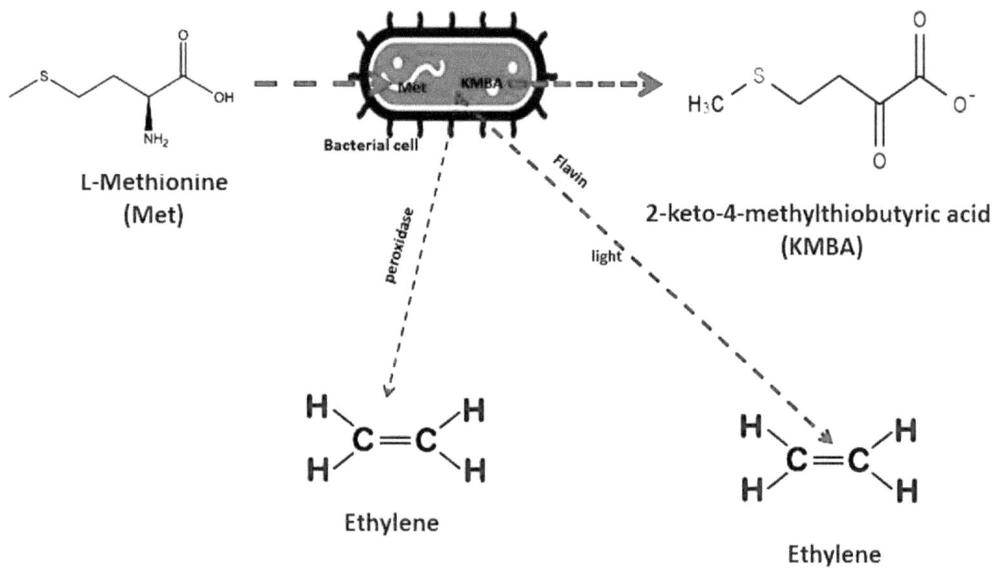

FIGURE 1.10 Proposed scheme for microbial biosynthesis of ethylene (from Primrose, 1977).

to be the same as that of higher plants [i.e., methionine (MET) S-adenosylmethionine (SAM) 1-aminocyclo- propane-1-carboxylic acid (ACC) ethylene (C_2H_4)].

1. The proposed pathway reveals that L-MET is deaminated to KMBA, which is then degraded to C_2H_4 either by a peroxidase or photochemically in the presence of flavin (Figure 1.10) (Primrose, 1977). The acceptance of KMBA as an intermediate also explains the need for sulfur atoms in all C_2H_4 precursors.

   Escherichia coli produced C_2H_4 from amino acids containing sulfur atom and MET analogues. Primrose (1977, 1979) suggested that the MET analogues are converted to MET before C_2H_4, formation.

   The addition of catalase, a known inhibitor of peroxidase, suppressed C_2H_4 formation by *E. coli* (Primrose, 1977), suggesting that peroxidases are involved in microbial C_2H_4 synthesis in the absence of light.
2. The enzymatic conversion of substrate 2-oxoglutarate (=α-ketoglutarate; AKG) into ethylene by an ethylene-forming enzyme (EFE; 2-oxoglutarate-dependent ethylene/succinate-forming enzyme). This pathway was discovered in bacterium, *Pseudomonas syringae*, but is also known for additional bacterial species and several fungi (Figure 1.11) (Weingart and Volksch, 1997).
3. The nonenzymatic conversion of 2-keto-4-methylthio-butyric acid (KMBA) into ethylene (Figure 1.12) (Ladygina et al., 2006). KMBA is the transaminated derivative of methionine that is formed by an NADH: Fe (111) EDTA oxidoreductase, which is enhanced under limited ammonia (Shipston and Bunch, 1989). This pathway is known to occur in *Escherichia coli* and other bacteria, and the fungus *Cryptococcus albidus* (Eckert et al., 2014).

1.8.2 Sources of Ethylene in Soil

Accumulation of C_2H_4 in the soil is of great ecological significance because a very minute amount of C_2H_4 proximity of roots can have an enormous effect on plant growth and development (Arshad and Frankenberger, 1991a, 1991b).

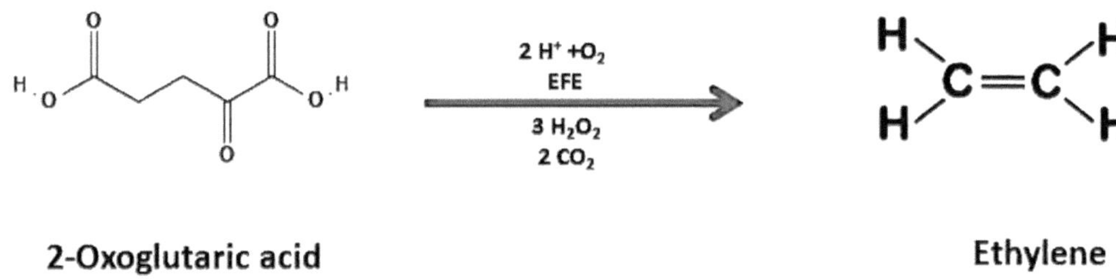

FIGURE 1.11 The bacterial and fungal pathway using 2-oxoglutarate as the substrate and the ethylene-forming enzyme (EFE).

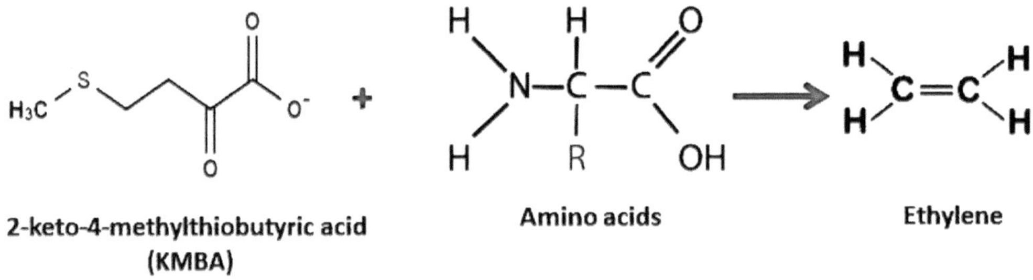

FIGURE 1.12 The nonenzymatic bacterial pathway using 2-keto-4-methylthio-butyric acid (KMBA).

Both biological and chemical processes seem to contribute to C_2H_4 accumulation in soil. However, C_2H_4 production as a result of microbial activity is the primary source of soil C_2H_4. Some compounds present in native soil organic matter can yield C_2H_4 through nonenzymatic transformations (Arshad and Frankenberger, 1991b). The same authors showed that nonsterilized soils amended with various organic compounds, release C_2H_4.

Although it is well accepted that under natural conditions microorganisms are the primary producers of C_2H_4 in soil, a small fraction may also be contributed by plant roots. In flooded soils, water acts as a barrier to the escape of C_2H_4 produced in roots (Schneider and Musgrave, 1992). Several studies have demonstrated that various soil conditions and environmental factors significantly affect C_2H_4 accumulation in soil.

1.8.3 Ecological Significance of Soil Ethylene

Ethylene produced by soil microorganisms, released by plant roots, and formed abiotically from soil organic matter contributes to the C_2H_4 pool of soil. Since C_2H_4 is biologically active within an extremely low concentration range, it is most likely that C_2H_4 present in the soil atmosphere surrounding the roots can affect plant growth and crop productivity (Arshad and Frankenberger, 1988).

Under certain conditions, C_2H_4 can accumulate at relatively high concentration in the soil atmosphere. Waterlogging of soil tends to promote accumulation of C_2H_4. Many of the plant responses to flooding have been attributed to enhanced endogenous levels of C_2H_4 (Jackson, 1985). It has been suggested that these flooding-caused effects on plant growth could be the result not only of increased endogenous levels of C_2H_4, but also of elevated levels of C_2H_4 present within the rhizosphere. Lu and Kirkham (1991) reported that the drought-resistant genotype of maize evolved more C_2H_4 than the drought-sensitive genotype when exposed to metal (Cd, Cu, Zn) toxicity. They suggested that genotypes that evolve more C_2H_4 may be more resistant to heavy metals than those that release less C_2H_4.

1.9 ABSCISIC ACID

Abscisic acid (ABA) is a phytohormone that is found in all photosynthetic organisms (Hunter, 2007). In higher plants, ABA is known as a stress-related hormone and plays a critical role in the regulation of various stress responses, including cold, salt, and drought stresses (Tijero et al., 2016; Jiang et al., 2019). It is also involved in the dormancy, immunity and other developmental processes (Cao et al., 2011). ABA also seems to play an important role in compatible mutualistic interactions such as mycorrhiza and rhizosphere bacteria with plants, and possibly also the animal gut microbiome. The frequent use of ABA in interspecies communication could be a possible reason for the wide distribution and reinvention of ABA as a signaling molecule in different organisms (Lievens et al., 2017).

Strikingly, this sesquiterpene could also be produced by a small number of fungi including the phytopathogenic species *Botrytis cinerea* (Marumo et al., 1982; Kettner and Dörffling, 1995), *Cercospora pini-densiflorae*, *C. rosicola*, and *C. cruenta* (Oritani and Kiyota, 2003), bacteria, cyanobacteria, bryophytes, lichens, and algae (Hartung, 2010). It is believed that ABA contents in algae are usually lower than in higher plants, but it plays a more important role in physiological activity (Huang et al., 2010). The ability to synthesize ABA was found in several PGPRs (Sgroy et al., 2009). Fungi produce relatively large amounts of ABA that could be released into the environment (Crocoll et al., 1991). Although little is known about the role of ABA in the metabolism of fungi. There is much evidence that ABA production is a significant contributor to the virulence of fungal phytopathogens (Spence and Bias, 2015).

1.9.1 Biosynthesis

However, information about the biochemistry of ABA biosynthesis and metabolism is very limited (Lievens et al., 2017; Olds et al., 2018). Different organisms have developed different biosynthesis and signal transduction pathways related to ABA. Nevertheless, there are results of preliminary studies that indicate the possible existence of more than one pathway of ABA catabolism in bacteria.

However, the ABA biosynthetic pathway in fungi differs from the carotenoid pathway described in plants. Efforts have been made to elucidate the ABA biosynthetic pathway in three related fungi: namely, *C. rosicola*, *C. cruenta*, and *C. pini-densiflorae.* Although these fungi use acetate and mevalonate (MVA) as substrates for ABA biosynthesis, there is some evidence that variation, particularly at latter stages occurs in their biosynthetic pathways. Fungi are also able to produce this sesquiterpene, but they use a distinct pathway (i.e., *B. cinerea*) that starts with the cyclization of farnesyl diphosphate (FPP) into 2Z, 4E-α-ionylideneethane, which is then subjected to several oxidation steps to form ABA (Figure 1.13) (Inomata et al., 2004).

A few studies have demonstrated the isolation of ABA from green and brown algae, including *Ascophyllum nodosum*, *Enteromorpha compressa*, and *Stigeoclonium* (Boyer and Dougherty, 1988; Tietz et al., 1989).

Yang et al. (2022) suggested that ABA is synthesized via the carotenoid pathway in *Neopyropia yezoensis* in a similar manner to terrestrial plants. They also suggested that both the MVA and methylerythritol phosphate (MEP) pathways likely exist simultaneously due to the inhibition of ABA synthesis under the corresponding specific inhibitor treatments.

Now, it is generally accepted that there are two ABA synthesis pathways, including the carotenoid pathway (indirect pathway), which is present in higher plants, and the direct pathway, which is mainly found in two fungi, *Cercospora rosicola* and *C. cruenta* (Takino et al., 2019). In both pathways, ABA is derived from the five-carbon precursor, isoprenyl pyrophosphate (IPP) (Nambara and Marion-Poll, 2005). IPP can be biosynthesized via two pathways: (1) the mevalonate (MVA) pathway, where it is synthesized in the cytoplasm using acetyl CoA as a substrate, and (2) the methylerythritol phosphate (MEP) pathways where it is synthesized in the plastids using pyruvate and glyceraldehyde-3-phosphate (GA-3P) as substrates (Yang et al., 2012). The MEP pathway is

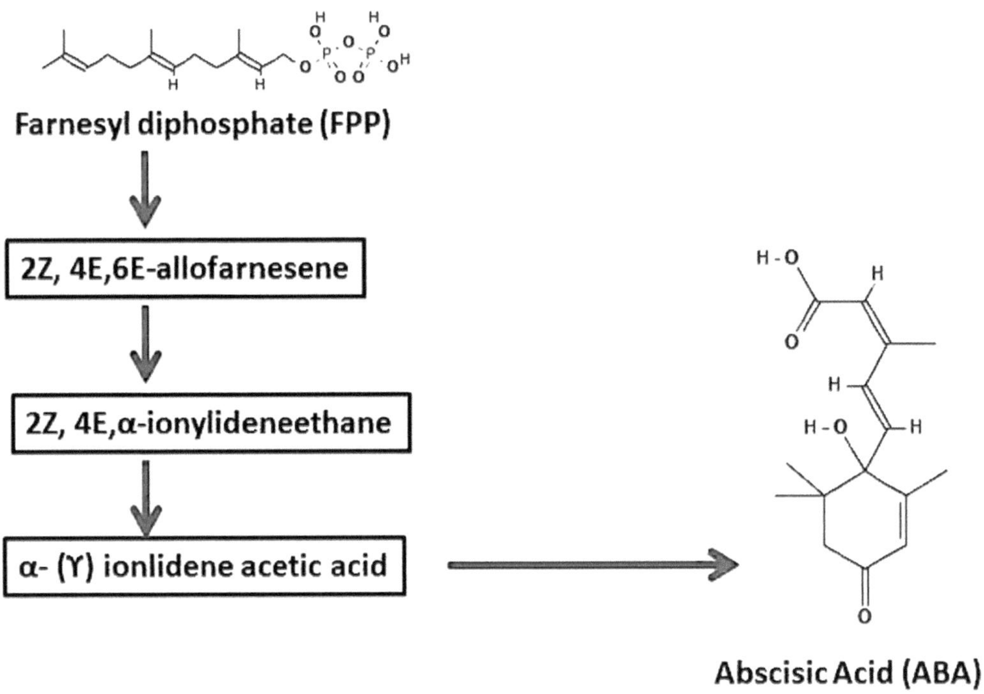

FIGURE 1.13 Structures of some of the sesquiterpenes produced by *B. cinerea* (Inomata et al., 2004).

the main synthesis pathway for the IPP in higher plants and operates in bacteria and green algae (Hunter, 2007; Phillips et al., 2008). However, the MVA pathway, mainly found in the cytoplasm, is the only synthetic pathway of 1PP in many fungi and animals. Studies have suggested that some green algae, such as *C. reinhardtii* and *Scenedesmus obliquus*, only have an MEP pathway (Disch et al., 1998). However, there are different opinions about red algae (Rhodophyta) (Grauvogel and Petersen, 2007), especially *N. yezoensis*.

1.9.2 Metabolism

The catabolism of ABA by microorganisms has received less attention than its synthesis. Abscisic acid is catabolized by soil microorganisms, but apart from its reduction to the 1,4-diol by a mixed soil culture, no bacterial breakdown products have been identified (Milborrow, 1983). Several polar, labeled products are formed by bacterial metabolism, but none coincide with the products of higher plant metabolism. It is most likely that different microflora metabolizes ABA by different routes.

1.9.3 Abscisic Acid in Soil

There are a few studies on the accumulation of ABA in soil. Müller et al. (1989) showed that ABA and other phytohormones were detected in maize rhizosphere and adjacent bulk soil. They found a high concentration gradient of ABA from the rhizoplane to rhizosphere and bulk soil. This gradient decreased as the distance from the rhizosphere increased. They also reported that ABA in the rhizosphere soil changed during the growth period and was correlated with the growth stage or an external influence. Since ABA appears during a period of severe water stress, the roots could have possibly released ABA to the rhizosphere soil. However, the root exudates could have possibly released some precursors for ABA synthesis by natural soil microbiota.

1.9.4 Ecological Significance of ABA of Microbial Origin

Soil microbiota could serve as an exogenous potential source of ABA for uptake by plant roots. Müller et al. (1989) demonstrated the vital role of ABA and other phytohormones present in the rhizosphere and in plant–microbe interactions affecting plant growth and development. They showed that ABA of microbial origin in the rhizosphere may exert a significant effect on plant growth.

Abscisic acid, particularly in combination with other phytohormones, can be used for improvement of plant growth under a water-stressed environment, such as found in arid and semiarid climates. The stability of ABA in soil should be determined, particularly in relation to wetting and dry of soil. Dodd et al. (2015) showed that ABA accumulation during water stress may play a role in maintaining plant water status and growth in drying soil. In addition, the accumulation of ABA in soil can have a variety of effects on the development, resistance, and fitness of plants.

The negative effects of ABA accumulation should probably be offset by processes of its conversion. Since it is supposed that plants can share to, but do not regulate, the ABA concentration in soil, the plant-associated microorganisms can demand this role. This assumption is supported by the evidence that PGPR influences the rhizosphere concentration of other phytohormones, such as auxin (Spence and Bias, 2015) and ethylene (Belimov et al., 2015), through the utilization of hormones or their precursors as carbon and/or nitrogen sources.

1.10 ROOT-ASSOCIATED PHYTOHORMONE-PRODUCING MICROBES

Plant hormones (phytohormones) are naturally occurring small organic molecules that affect numerous aspects of growth and differentiation in plants and are involved in alleviating different biotic and abiotic stresses (Davies, 2010). So far, nine categories of phytohormones have been identified. These include auxins, cytokinins (CK), gibberellins (GA), abscisic acid (ABA), ethylene (ET), brassinosteroids (BR), salicylates (SA), jasmonates (JA), and strigolactones (SL) (Su et al., 2017). These hormones are not only produced by plants but are also produced by beneficial and pathogenic microorganism (e.g., bacteria, fungi, insects, microalgae, etc.) that may modulate plant growth, physiology, and immunity. Among phytohormones, ET, SA, and JA are particularly known for regulating defense response in plants against pathogens and are called as the immunity hormones. However, the role of cytokinin in inducing immunity in plant upon plant–pathogenic interaction has also been recognized (Grosskinsky et al., 2016; Dowd et al., 2017).

Root-associated microbes, including free living, symbiotic or endophytic microbes, can produce various type of phytohormones and belong to different genera and species (Sgroy et al., 2009). For example, Piccoli et al. (2011) isolated the endophytic diazotrophic bacterium *Arthrobacter koreensis*, which produce ABA, IAA, GA_3, and Jasmonic acid from the roots of the halophyte shrub *Prosopis strombulifera*. Mishra et al. (2017) isolated bacteria with IAA production ability from extreme environments, which were identified as *Pseudomonas* spp. and *Ochrobactrum* spp. IAA-producing *Mycobacterium* species was found in the rhizosphere of orchid (Tsavkelova et al., 2007). Also, *Klebsiella* and *Enterobacter* isolated from sugar cane synthesize IAA (de Santi Ferrara et al., 2012).

Naz et al. (2009) also observed cytokinin-producing species, such as *Arthrobacter*, *Bacillus*, *Azospirillum*, and *Pseudomonas*, that stimulated the root development of plants. ABA was also detected in root-associated microbes from various plants. Among IAA-producing bacteria associated with plant grown under saline soil, Rhizobia have also been shown to synthesize auxins, cytokinins and abscisic acid, increase plant growth and development, and improve the yield of agricultural crops (Hayat et al. 2008). Actinobacteria have also been found to produce IAA, CK, GA-like substances (Vijayabharathi et al., 2016). Ruanpanun et al. (2010) found high IAA-producing nematophagous actinomycete and fungal isolates, such as *Aspergillus* and *Streptomyces*. Müller et al. (1989) reported that growth effects of bacteria on various maize root parameters could be described by the combined effect of IAA and ABA. They have shown that this elevated level of ABA in maize rhizosphere could influence root development, but it is also possible that ABA is taken up by the root system and translocated by xylem to other parts of the plant.

Soils are sources of diverse organisms, including fungi, bacteria, and plants (Mendes et al., 2013). Plant roots are heavily colonized with microorganisms because of the rich nutrient component of root exudates (Hashem et al., 2016), which attract microbes that utilize the various nutrients released by the root. In turn, the microbes synthesize biologically active compounds, including phytohormones (auxin, cytokinins, gibberellins, and ABA), antifungal compounds, enzymes, and compatible solutes. These microbial metabolites play a vital role in plant growth, nutrition, and development (Egamberdieva et al., 2017). They can stimulate plant growth and development, provide resistance to various abiotic and biotic stress factors, improve nutrient acquisition and protect plants from various soil-borne pathogens (Cho et al., 2015).

The microbes relieve stress responses by regulating the nutritional and hormonal balance in plants and inducing systematic tolerance to stress. One of the mechanisms of improvement of plant growth and stress tolerance by microbes is their phytohormone synthesizing ability in the rhizosphere or root tissue (Etesami et al., 2015). Microbial phytohormones affect the metabolism of endogenous growth regulators in plant tissue (Sorty et al., 2016) and play a key role in changing root morphology upon exposure to drought, salinity, extreme temperature, and heavy metal toxicity (Khan et al., 2011). El-haddad et al. (1986) and Warembourg et al. (1987) have indicated that the beneficial effect on plant growth cannot be ascribed only to the nitrogen feeding of plants but also to the role of growth-promoting substances produced by azospirilla.

Kawthar et al. (2002) investigated the effect of mineral nitrogen and/or a biofertilizer on the yield and hormone content of the edible part of potato plant (*Solanum tuberosoum* cv. Diamond). The highest GA_3 content was achieved in the case of Bio. + half N dose. On the other hand, mineral nitrogen fertilization led to an increment of IAA content as compared to biofertilization only or the combination between mineral and biofertilization. The same trend was obtained for ABA content of tubers at harvest stage. Moreover, biofertilization singly induced a clear increase of zeatin content to nearly ten folds as compared to nitrogen fertilization only.

1.11 PROSPECTS

The beneficial effects on plants mediated by microbes, such as the stimulation of plant growth, tolerance to abiotic stresses, and resistance to pathogens, are based on the microbe's ability to produce auxins, gibberellins, salicylic acid, abscisic acid, and cytokinins in plant tissues. Thus, plant-associated microbes hold the potential to modulate hormone levels and metabolism in plant tissue, especially in biochemical processes that can prevent the damaging effects on the plant. Thus, future studies require an intensive effort to determine the practical significance of this approach in terms of enhancing crop yields. Future research should focus on the following aspects, including the mode of action of microbial phytohormones origin and their interactions with plants.

- Screening economical physiological precursors to be used as soil amendments to promote GA synthesis in the rhizosphere.
- Numerous important aspects concerning phytohormones of microbial origin in evoking physiological responses in plants.
- Catabolism of phytohormones by microorganisms.
- Evaluating precursor–inoculum interactions to provide an exogenous source of C_2H_4 to plant roots for improving plant growth and development.
- Studying the interactions of ABA with other phytohormones in the rhizosphere on plant growth.

REFERENCES

Akhtar, S. S.; Mekureyaw, M. F.; Pandey, C. and Roitsch, T. (2020). Role of cytokinins for interactions of plants with microbial pathogens and pest insects. *Frontiers in Plant Science*, 10, 1777.

ALKahtani, M. D. F.; Fouda, A.; Attia, K. A.; Al-Otaibi, F.; Eid, A. M.; Ewais, E. E.-D.; Hijri, M.; St-Arnaud, M.; Hassan, S. E.-D.; Khan, N., et al. (2020). Isolation and characterization of plant growth promoting endophytic bacteria from desert plant and their application as bioinoculants for sustainable agriculture. *Agronomy*, 10, 1325.

Allen, M. F., Moore, T. S. Jr. and Christensen, M. (1982). Phytohormone changes in *Bouteloua gracilis* infected by vesicular-arbuscular mycorrhizae. II. Altered levels of gibberellin-like substances and abscisic acid in the host plant. *Candian Journal of Botany,* 60, 468–471.

Allison, F. E. (1947). *Azotobacter* inoculation of crops-I-Historical. *Soil Science,* 64, 413–429.

Arshad, M., and Frankenberger, W. T. Jr. (1988). Influence of ethylene produced by soil microorganisms on etiolated pea seedlings. *Applied and Environmental Microbiology*, 54(11), 2728–2732.

Arshad, M. and Frankenberger, W. T. Jr. (1989). Biosynthesis of ethylene by *Acremonium falciforme. Soil Biology and Biochemistry,* 21, 633–638.

Arshad, M., and Frankenberger, W. T. Jr. (1991a). Microbial production of plant hormones. *Plant and Soil,* 133, 1–8.

Arshad, M., and Frankenberger, W. T. Jr. (1991b). Effects of soil properties and trace elements on ethylene production in soils. *Soil Science,* 151, 377–386.

Asghar, W.; Kondo, S.; Igudhi, R.; Mahmood, A. and Kataoka, R. (2020). Agricultural utilization of unused resources: Liquid food waste material as a new source of plant growth- promoting microbes. *Agronomy*, 10, 954.

Badenoch-Jones, J., Rolfe, B. G., and Letham, D. S. (1983). Phytohormones, *Rhizobium* mutants, and nodulation in legumes: III. Auxin metabolism in effective and ineffective pea root nodules. *Plant Physiology*, 73(2), 347–352.

Barea, J. M.; Navarro, E. and Montoya, E. (1976). Production of plant growth regulators by rhizosphere phosphate-solubilizing bacteria. *Journal of Applied Bacteriology,* 40, 129–134.

Bastian, F.; Cohen, A.; Piccoli, P.; Luna, V.; Baraldi, R. and Bottini, R. (1998). Production of indole-3-acetic acid and gibberellins A_1 and A_3 by *Acetobacter diazotrophicus* and *Herbaspirillum seropedicae* in chemically defined culture media. *Plant Growth Regulators,* 24, 7–11.

Belimov, A. A.; Dodd, I. C.; Safronova, V. I.; Shaposhnikov, A. I.; Azarova, T. S.; Makarova, N. M.; Davies, W. J. and Tikhonovich, I. A. (2015). Rhizobacteria that produce auxins and contain ACC deaminase decrease amino acid concentrations in the rhizosphere and improve growth and yield of well-watered and water-limited potato (*Solanum tuberosum*). *Annals of Applied Biology,* 167, 11–25.

Boyer, G. L., and Dougherty, S. S. (1988). Identification of abscisic acid in the seaweed *Ascophyllum nodosum. Phytochemistry*, 27(5), 1521–1522.

Brownlee, B. G.; Hall, R. H. and Whitty, C. D. (1975). 3-Methyl-2-butenal: An enzymatic degradation product of the cytokinin, N^6 – (2-isopentenyl) adenine. *Candian Journal of Biochemistry,* 53, 37–41.

Cao, F. Y.; Yoshioka, K., and Desveaux, D. (2011). The roles of ABA in plant pathogen interactions. *Journal of Plant Research,* 124, 489–*499.*

Cassán, F.; Bottini, R.; Schneider, G., and Piccoli, P. (2001a). *Azospirillum brasilense* and *Azospirillum lipoferum* hydrolyze conjugates of GA_{20} and metabolize the resultant aglycones to GA_1 in seedlings of rice dwarf mutants. *Plant Physiology*, 125(4), 2053–2058.

Cassán, F.; Lucangeli, C.; Bottini, R. and Piccoli, P. (2001b). *Azospirillum spp.* Metabolize [17,17-$^2H_2$] Gibberellin A_{20} to [17,17-$^2H_2$] Gibberellin A_1 in vivo in *dy* rice mutant seedlings. *Plant and Cell Physiology,* 42, 763–767.

Cassán, F.; Piccoli, P. and Bottini, R. (2003). Promocion del crecimiento vegetal por *Azospirillum sp*. A traves de la produccion de giberelinas. Un modelo alternativo para incrementar la produccion Agricola. In *Microbiologia Agricola. Un aporte de la investigacion en Argentina para la Sociedad (*A. Albanesi, C. Kunst, A. Anriquez, S. Luna and R. Ledesma, eds.*)* pp. 1–16. Universidad Nacional de Santiago del Estero, Ssntiago.

Chen, C. M.; Melitz, D. K. and Clough, F. W. (1982). Metabolism of cytokinin: phosphoribosylation of cytokinin bases by adenine phosphoribosyl transferase from wheat germ. *Archives of Biochemistry and Biophysics*, 214(2), 634–641.

Cho, S. T.; Chang, H. H.; Egamberdieva, D.; Kamilova, F.; Lugtenberg, B. and Kuo, C. H. (2015). Genome analysis of *Pseudomonas fluorescens* PCL1751: a rhizobacterium that controls root diseases and alleviates salt stress for its plant host. *PLOS ONE,* 10, e0140231.
Cohen, A.; Travaglia, C.; Reinoso, H.; Piccoli, P. and Bottini, R. (2001). *Azospirillum* inoculation and inhibition of gibberellin and ABA synthesis in maize seedlings under drought. *Proceedings of Plant Growth Regulations Society of America,* 28, 88–93.
Cordero, I.; Balaguer, L.; Rincon, A. and Pueyo, J. J. (2018). Inoculation of tomato plants with selected PGPR represents a feasible alternative to chemical fertilization under stress. *Journal of Plant Nutrition and Soil Science,* 181, 694–703.
Crocoll, C.; Kettner, J. and Dorffling, K. (1991). Abscisic acid in saprophytic and parasitic species of fungi. *Phytochemistry*, 30, 1059–1060.
Davies, P. J. (2010). *The plant hormones: their nature, occurrence, and functions. Plant hormones*. P. J. Davies Ed. Springer, Dordrecht, pp. 1–15.
de Santi Ferrara, F. I.; Oliveira, Z. M.; Gonzales, H. H. S.; Floh, E. I. S. and Barbosa, H. R. (2012). Endophytic and rhizospheric enterobacteria isolated from sugar cane have different potentials for producing plant growth-promoting substances. *Plant and Soil,* 353, 409–417.
Dickinson, J. R.; Forsyth, C. and Van Staden, J. (1986). The role of adenine in the synthesis of cytokinins in tomato plants and in cell-free root extracts. *Plant Growth Regulation*, 4, 325–334.53
Disch, A.; Schwender, J.; Müller, C.; Lichtenthaler, H. K. and Rohmer, M. (1998). Distribution of the mevalonate and glyceraldehyde phosphate/pyruvate pathways for isoprenoid biosynthesis in unicellular algae and the *Cyanobacterium synechocystis* PCC 6714. *Biochemical Journal*, 333(2), 381–388.
Dobereiner, J. and Day, J. M. (1975). Nitrogen fixation in the rhizosphere of tropical grasses. In *Nitrogen Fixation by Free -Living Microorganisms (*W. D. P. Stewart, ed.*)* pp. 39–56. Cambridge, London.
Dobert, R. C.; Rood, S. B.; Zanewich, K. and Blevins, D. G. (1992). Gibberellins and the legume-Rhizobium symbiosis. III. Quantification of gibberellins from stems and nodules of lima bean and cowpea. *Plant Physiology,* 100, 1994–2001.
Dodd, I. C.; Puértolas, J.; Huber, K.; Pérez-Pérez, J. G.; Wright, H. R. and Blackwell, M. S. (2015). The importance of soil drying and re-wetting in crop phytohormonal and nutritional responses to deficit irrigation. *Journal of Experimental Botany,* 66, 2239–2252.
Dowd, C. D.; Chronis, D.; Radakovic, Z. S.; Siddique, S.; Schmulling, T.; Werner, T., et al. (2017). Divergent expression of cytokinin biosynthesis, signaling and catabolism genes underlying difference in feeding sites induced by cyst and root-knot nematodes. *Plant Journal,* 92, 211–228.
Dvornikova, T. P.; Skryabin, G. K. and Suvorov, N. N. (1970). Enzymatic transformation of tryptamine by fungi. *Microbiology,* 39, 32–35.
Eckert, C.; Xu, W.; Xiong, W.; Lynch, S.; Ungerer, J.; Tao, L.; Gill, R.; Maness, P-C. and Yu, J. (2014). Ethylene forming enzyme and bioethylene production. *Biotechnology Biofuels,* 7, 33.
Egamberdieva, D.; Wirth, S.; Behrendt, U.; Parvaiz, A. and Berg, G. (2017). Antimicrobial activity of medicinal plants correlates with the proportion of antagonistic endophytes. *Frontiers in Microbiology,* 8, 199.
El-haddad, M. E.; Ishac, Y. Z.; Borollosy, M. A. El.; Wedad, E. E. and Girgis, M. E. (1986). Studies Endophytic and rhizospheric enterobacteria isolated from sugar cane have different potentials for producing plant growth-promoting substances. *Plant and Soil,* 353, 409–417.
Entsch, B.; Letham, D. S.; Parker, C. W.; Summons, R. E. and Gillnow, B. I. (1980). Metabolites of cytokinins. In *Plant Growth Substances* (F. Skoog, ed.) pp. 109–118. Springer-Verlag, Berlin.
Etesami, H.; Alikhani, H. A. and Hosseini, H. M. (2015). Indole-3-acetic acid (IAA) production trait, a useful screening to select endophytic and rhizosphere competent bacteria for rice growth promoting agents. *MethodsX,* 2, 72–78.
Frankenberger, W. T. Jr. and Arshad, M. (1991a). Yield response of watermelon and muskmelon to L-tryptophan applied to soil. *Horticulture Science,* 26(1), 35–37.
Frankenberger, W. T. Jr. and Arshad, M. (1991b). *Yield response of Capsicum annuum to auxin precursor, L-tryptophan applied to soil. PGRSA Quarterly,* 19, 231–240.
Frankenberger, W. T. Jr. and Arshad, M. (1995). *Phytohormones in soil: microbial production and function.* Marcel Dekker, Inc. New York, pp. 76–85 and 233–235..
Frankenberger, W. T. Jr. and Brunner, W. (1983). Method of detection of auxin-indole-3-acetic acid in soils by high performance liquid chromatography. *Soil Science Society of America Journal,* 47(2), 237–241.

Frébort, I.; Kowalska, M.; Hluska, T.; Frébortová, J. and Galuszka, P. (2011). Evolution of cytokinin biosynthesis and degradation. *Journal of Experimental Botany*, 62(8), 2431–2452.

Frébortová, J. and Frébort, I. (2021). Biochemical and Structural Aspects of Cytokinin Biosynthesis and Degradation in Bacteria. *Microorganisms,* 9, 1314.

Goto, M.; Ishida, Y.; Takikawa, Y. and Hyodo, H. (1985). *Ethylene production by the Kudzu strain of Pseudomonas syringae pv. Phaseolicola causing halo blight in Pueraria lobata (Wild)* Ohwi. *Plant and Cell Physiology,* 26, 141–150.

Grauvogel, C. and Petersen, J. (2007). Isoprenoid biosynthesis authenticates the classification of the green alga Mesostigma *viride* as an ancient streptophyte. *Gene,* 396, 125–133.

Grosskinsky O. K.; Tafner, R.; Moreno, M. V.; Stenglein, S. A.; de Salamone, I. E. G.; Nelson, L. M., et al. (2016). Cytokinin production by *Pseudomonas fluorescens* G 20 - 18 determines biocontrol activity against *Pseudomonas syringae* in Arabidopsis. *Scientific Report,* 6, 23310.

Gutiérrez-Mañero, F. J.; Ramos-Solano, B.; Probanza, A. N.; Mehouachi, J.; R. Tadeo, F. R. and Talon, M. (2001). The plant--growth-promoting rhizobacteria *Bacillus pumilus* and *Bacillus licheniformis* produce high amounts of physiologically active gibberellins. *Physiologia Plantarum,* 111(2), 206–211.

Hartung, W. (2010). The evolution of abscisic acid (ABA) and ABA function in lower plants, fungi, and lichen. *Functional Plant Biology*, 37(9), 806–812.

Hashem, A.; Abd_Allah, E. F.; Alqarawi, A.; Al-Huqail, A. A.; Wirth, S. and Egamberdieva, D. (2016). The interaction between arbuscular mycorrhizal fungi and endophytic bacteria enhances plant growth of *Acacia gerrardii* under salt stress. *Frontiers of Plant Science,* 7, 1089.

Hayat, S.; Hasan, S. A.; Fariduddin, Q. and Ahmad, A. (2008). Growth of tomato (*Lycopersicon esculentum*) in response to salicylic acid under water stress. *Journal of Plant Interactions*, *3*(4), 297–304.

Holl, F. B.; Chanway, C. P.; Turkington, R. and Radley, R. A. (1988). Response of crested wheatgrass (*Agropyron cristatum* L.), perennial ryegrass (*Lolium perenne* and white clover (*Trifolium repens* L.) to inoculation with *Bacillus polymyxa*. *Soil Biology and Biochemistry*, 20(1), 19–24.

Huang, X.; Ke, C. H. and Wang, W. X. (2010). Cadmium and copper accumulation and toxicity in the macroalga *Gracilaria tenuistipitata. Aquatic Biology,* 11, 17–26.

Hubbell, D. H. and Gaskins, M. H. (1984). *Associative N2 fixation with Azospirillum. In Biological Nitrogen Fixation: Ecology, Technology and Physiology (*Martin Alexander, ed.*)*, Springer US. 201–224.

Hunter, W. N. (2007). The non-mevalonate pathway of isoprenoid precursor biosynthesis. *Journal of Biological Chemistry,* 282, 21573–21577.

Inomata, M.; Hirai, N.; Yoshida, R. and Ohigashi, H. (2004). The biosynthetic pathway to abscisic acid via ionylideneethane in the fungus *Botrytis cinerea. Phytochemistry*, 65(19), 2667–2678.

Jackson, M. B. (1985). Ethylene and responses of plants to soil waterlogging and submergence. *Annual Review of Plant Physiology,* 36, 145–174.

Janzen, R.; Rood, S.; Dormar, J. and Mc Gill, W. (1992). *Azospirillum brasilense* produces gibberellins in pure culture and chemically- medium and in co- culture on straw. *Soil Biology and Biochemistry*, 24, 1061–1064.

Jiang, D.; Zhou, L.; Chen, W.; Ye, N.; Xia, J. and Zhuang, C. (2019). Overexpression of a microRNA-targeted NAC transcription factor improves drought and salt tolerance in rice via ABA-mediated pathways. *Rice,* 12, 76.

Kawthar, A. E. R.; Selim, S. M. and Nasr, S. A. (2002). *Nitrate and nitrite accumulation in potato tubers in relation to mineral nitrogen and biofertilization. Annals Agricultural Science, Ain Shams University,* 47(1), 107–122.

Kettner, J. and Dörffling, K. (1995). Biosynthesis and metabolism of abscisic acid in tomato leaves infected with *Botrytis cinerea*. *Planta,* 196, 627–634.

Khan, A. L.; Hamayun, M.; Kim, Y.-H.; Kang, S. M.; Lee, J. H. and Lee, I. N. (2011). Gibberellins producing endophytic *Aspergillus fumigatus* sp. LH02 influenced endogenous phytohormonal levels, isoflavonoids production and plant growth in salinity stress. *Process Biochemistry,* 46, 440–447.

Klemen, F. and Klämbt, D. (1974). Half-life of sRNA from primary roots of *Zea mays*. A contribution to the cytokinin production. *Physiologia Plantarum*, 31(3), 186–188.

Kloepper, J. W. (1994). *Plant growth promoting rhizobacteria*. In: *Azospirillum/Plant Association (Y.Okon* ed.) pp. 137–166. CRC Press, Boca Raton.

Kloepper, J. W. and Schroth, M. N. (1978). *Plant growth promoting rhizobacteria on radishes. Proceedings of International Conference of Plant Pathology Bacteria (Angers, France),* 2, 879–882.

Kucey, R. M. N. (1988). Plant growth- altering effects of *Azospirillum brasilense* and *Bacillus* C-11-25 on wheat cultivars. *Journal of Applied Bacteriology,* 64, 187–196.
Ladygina, N.; Dedyukhina, E. G. and Vainshtein, M. B. (2006). A review on microbial synthesis of hydrocarbons. *Process Biochemistry*, 41(5), 1001–1014.
Letham, D. S.; Palni, L. M. S.; Tao, G. Q.; Gollnow, B. I. and Bates, C. M. (1983). The activities of cytokinin glucosides and alanine conjugates in cytokinin bioassays. *Journal of Plant Growth Regulation,* 2, 103–115.
Lievens, L.; Pollier, J.; Goossens, A.; Beyaert, R. and Staal, J. (2017). Abscisic acid as pathogen effector and immune regulator. *Frontiers in Plant Science*, 8, 587.
Loper, J. E. and Schroth, M. N. (1986). Influence of bacterial sources of indole-3-acetic acid on root elongation of sugar beet. *Phytopathology*, 76(4), 386–389.
Lu, W. P. and Kirkham, M. B. (1991). Genotypic tolerance to metals as indicated by ethylene production. *Water, Air and Soil Pollution,* 57–58, 605–615.
Lucangeli, C. and Bottini, R. (1996). Reversion of dwarfism in *dwarf-1* maize (*Zea mays* L.) and *dwarf-x* rice (*Oryza sativa* L.) mutants by endophytic *Azospirillum* spp. *Biocell,* 20, 223–228.
Lucangeli, C. and Bottini, R. (1997). Effects of *Azospirillum* spp. on endogenous gibberellin content and growth of maize (*Zea mays* L.) treated with uniconazole. *Symbiosis,* 23, 63–72.
MacMillan, J. (2002). Occurrence of gibberellins in vascular plants, fungi, and bacteria. *Journal of Plant Growth Regulation,* 20, 387–442.
Magie, A. R.; Wilson, E. E. and Kosuge, T. (1963). Indole acetamide as an intermediate in the synthesis of indoleacetic acid in *Pseudomonas savastanoi. Science*, 141, 1281–1282.
Mahadevan, A. (1982). *Biochemical Aspects of Plant Disease Resistance, Part I. Preformed Inhibitory Substances "Prohibitions."* Today and Tomorrow, New Delhi.
Mahadevan, A., ed. (1984). *Growth Regulators, Microorganisms and Diseased Plants.* Oxford and IBH Publishing, New Delhi, India, p. 466.
Maksimov, I. V.; Abizgil'Dina, R. R. and Pusenkova, L. I. (2011). Plant growth promoting rhizobacteria as alternative to chemical crop protectors from pathogens. *Applied Biochemistry and Microbiology,* 47, 333–345.
Marchal, P. and Rigaud, J. (1970). Sur la presence et le metabolisme de l'indolyl-3-acetaldehyde chez *Nectria galligena* Bres. *var. major Wr, C. R. Academic Science,* 271D, 479–481.
Martens, D. A. and Frankenberger, W. T. Jr. (1993a). Stability of microbial- produced auxins derived from L-tryptophan added to soil. *Soil Science,* 155, 263–271.
Martens, D. A. and Frankenberger, W. T. Jr. (1993b). Metabolism of tryptophan in soil. *Soil Biology and Biochemistry*, 25(12), 1679–1687.
Martens, D. A. and Frankenberger, W. T. Jr. (1994). Assimilation of exogenous 2′-^{14}C-indole-3-acetic acid and 3′-^{14}C-tryptophan exposed to the roots of three wheat varieties. *Plant and Soil,* 166, 281–290.
Marumo, S.; Katayama, M.; Komori, E.; Ozaki, Y.; Natsume, M. and Kondo, S. (1982). Microbial production of abscisic acid by *Botrytis cinerea. Agricultural and Biological Chemistry*, 46(7), 1967–1968.
McGaw, B. A. (1987). Cytokinin biosynthesis and metabolism. In *Plant Hormones and Their Role in Plant Growth and Development* (P. J. Davies, ed) pp. 76–92. Martinus Nijhoff, Dordrecht.
McGaw, B. A. and Horgan, R. (1983). Cytokinin catabolism and cytokinin oxidase. *Phytochemistry,* 22, 1103–1105.
Mendes, R.; Garbeva, P. and Raaijmakers, J. M. (2013). The rhizosphere microbiome: significance of plant beneficial, plant pathogenic, and human pathogenic microorganisms. *FEMS Microbiology Reviews*, 37(5), 634–663.
Milborrow, B. V. (1983). Pathways to and from abscisic acid. In *Abscisic Acid* (F. T. Addicott, ed.) pp.79–111. Praeger Publisher, New York.
Mishra, S. K.; Khan, M. H.; Misra, S.; Dixit, K. V.; Khare, P.; Srivastava, S. and Chauhan, P. S. (2017). Characterization of *Pseudomonas spp. and Ochrobactrum sp. isolated from volcanic soil. Antonie Van Leeuwenhoek*, 110, 253–270.
Mok, M. C. (1994).Cytokinins and plant development, An overview. In *Cytokinins Chemistry, Activity, and Function*, 1st ed. (D.W.S. Mok and M. C. Mok, eds.) pp. 155–166. CRC Press, Boca Raton, FL, USA.
Müller, M.; Deigele, C. and Ziegler, H. (1989). Hormonal interactions in the rhizosphere of maize (*Zea mays* L.) and their effects on plant development. *Z. Pflanzen. Bodenk,* 152, 247–254.

Murai, N. (1980). Cytokinin biosynthesis and its relationship to the presence of plasmids in strains of *Corynebacterium fascians*. In *Metabolism and Molecular Activities of Cytokinins* (J. Guern and C. Peaud-Lenoel, eds.) pp. 17–26. Springer-Verlag, Berlin.

Nadeem, S. M.; Ahmad, M.; Zahir, Z. A.; Javaid, A. and Ashraf, M. (2014). The role of mycorrhizae and plant growth promoting rhizobacteria (PGPR) in improving crop productivity under stressful environments. *Biotechnology Advances*, 32(2), 429–448.

Nambara, E. and Marion-Poll, A. (2005). Abscisic acid biosynthesis and catabolism. *Annual Review Plant Biology,* 56, 165–185.

Nasr, S. A. (2002). Stimulation of auxin biosynthesis by some bacterial and yeast strains. *Arab University Journal of Agricultural Science Ain Shams University,* 10(1), 89–107.

Naz, I.; Bano, A. and Ul-Hassan, T. (2009). Isolation of phytohormones producing plant growth promoting rhizobacteria from weeds growing in Khewra salt range, Pakistan, and their implication in providing salt tolerance to *Glycine max* L. *African Journal of Biotechnology,* 8, 5762–5766.

Nieto, K. F. and Frankenberger, W. T. Jr. (1989). Biosynthesis of cytokinins by *Azotobacter chroococcum. Soil Biology and Biochemistry*, 21(7), 967–972.

Nieto, K. F. and Frankenberger, W. T. Jr. (1991). Influence of adenine, isopentyl alcohol and *Azotobacter chroococcum* on the vegetative growth of *Zea mays. Plant and Soil,* 135, 213–221.

Norberg, S. O. (1968). Studies in the production of auxins and other growth stimulating substances by *Exobasidium. Symbolae Botanicae Upsalienses,* 19, 1–117.

Numan, M.; Bashira, S.; Khan, Y.; Mumtaz, R.; Shinwaric, Z. K.; Khan, A. L.; Khan, A. and AL-Harrasib, A. (2018). Plant growth promoting bacteria as an alternative strategy for salt tolerance in plants: A review. *Microbiological Research*, 209, 21–32.

Olds, C. L.; Glennon, E. K. K. and Luckhart, S. (2018). Abscisic acid: New perspectives on an ancient universal stress signaling molecule. *Microbes Infection,* 20, 484–492.

Oritani, T. and Kiyota, H. (2003). Biosynthesis and metabolism of abscisic acid and related compounds. *Natural Product Reports*, 20(4), 414–425.

Phillips, M. A.; Leon, P.; Boronat, A. and Rodriguez-Concepcion, M. (2008). The plastidial MEP pathway: unified nomenclature and resources. *Trends Plant Science,* 13, 619–623.

Piccoli, P. and Bottini, R. (1994). Effects of C/N relationships, N content, pH, and time of culture on growth and gibberellin production by *Azospirillum lipoferum* cultures. *Symbiosis,* 17, 229–236.

Piccoli, P. and Bottini, R. (1996). Light enhancement of gibberellin production by *Azospirillum lipoferum* cultures. *Biocell,* 20, 200–207.

Piccoli, P.; Masciarelli, O. and Bottini, R. (1996). Metabolism of 17,17 [$^2H_2$]-Gibberellins A_4, A_9, and A_{20} by *Azospirillum lipoferum* in chemically defined culture medium. *Symbiosis,* 21, 167–178.

Piccoli, P.; Lucangeli, D.; Schneider, G. and Bottini, R. (1997). Hydrolysis of [17,17- $^2H_2$] Gibberellin A_{20}-Glucoside and [17,17-$^2H_2$] Gibberellin A_{20}- glucosyl ester by *Azospirillum lipoferum* cultured in a nitrogen-free biotin-based chemically defined medium. *Plant Growth Regulation,* 23, 179–182.

Piccoli, P.; Masciarelli, O. and Bottini, R. (1999). Gibberellin Production by *Azospirillum lipoferum* cultured in chemically defined medium as affected by oxygen availability and water status. *Symbiosis,* 27, 135–146.

Piccoli, P.; Travaglia, C.; Cohen, A.; Sosa, L.; Cornejo, P.; Masuelli, R. and Bottini, R. (2011). An endophytic bacterium isolated from roots of the halophyte *Prosopis strombulifera* produces ABA, IAA, gibberellins A_1 and A_3 and jasmonic acid in chemically defined culture medium. *Plant Growth Regulation*, 64, 207–210.

Primrose, S. B. (1977). Evaluation of the role of methional, 2-keto-4-methyl-thiobutyric acid and peroxidase in ethylene formation by *Escherichia coli. Journal of General Microbiology*, 98(2), 519–528.

Primrose, S. B. (1979). Ethylene and agriculture: the role of microbes. *Journal of Applied Bacteriology,* 46, 1–25.

Probanza, A.; Garcıa, J. A. L.; Palomino, M. R.; Ramos, B. and Mañero, F. J. G. (2002). *Pinus pinea* L. seedling growth and bacterial rhizosphere structure after inoculation with PGPR *Bacillus* (*B. licheniformis* CECT 5106 and *B. pumilus* CECT 5105). *Applied Soil Ecology*, 20(2), 75–84.

Rademacher, W. (1994). Gibberellin formation in microorganisms. *Plant Growth Regulation,* 15, 303–314.

Radhika, V.; Ueda, N.; Tsuboi, Y.; Kojima, M.; Kikuchi, J.; Kudo, T. and Sakakibara, H. (2015). Methylated cytokinins from the phytopathogen *Rhodococcus fascians* mimic plant hormone activity. *Plant Physiology*, 169(2), 1118–1126.

Ruanpanun, P.; Tangchitsomkid, N.; Hyde, K. D. and Lumyong, S. (2010). Actinomycetes and fungi isolated from plant-parasitic nematode infested soils: screening of the effective biocontrol potential, indole-3-acetic acid and siderophore production. *World Journal of Microbiology and Biotechnology,* 26, 1569–1578.

Ruckdäschel, E. and Klingmüller, W. (1992). Analysis of IAA biosynthesis in Azospirillum lipoferum and Tn5 induced IAA mutants. *Symbiosis,* 13, 123–131.

Schneider, G. (1983). Gibberellin conjugates. In *The Biochemistry and Physiology of Gibberellins* (A. Crozier ed.) *vol. 1,* pp. 389–456. Praeger, New York.

Schneider, R. W. and Musgrave, M. E. (1992). The soil atmosphere. In *Methods for Research on Soil Borne Phytopathogenic Fungi* (L. L. Singleton, J. D. Mihail and C. M. Rush, eds.) Amer Phytopathological Society; Second Printing edition (January 1, 1993) St. Paul, MN 55121 USA. pp. 215–224.

Sgroy, V.; Cassán, F.; Masciarelli, O.; Del Papa, M. F.; Lagares, A. and Luna, V. (2009). Isolation and characterization of endophytic plant growth-promoting (PGPB) or stress homeostasis-regulating (PSHB) bacteria associated to the halophyte *Prosopis strombulifera. Applied Microbiology and Biotechnology,* 85, 371–381.

Shipston, N. and Bunch, A. W. (1989). The physiology of L-methionine catabolism to the secondary metabolite ethylene *Escherichia coli. Journal of General Microbiology,* 135, 1489–1497.

Sivadjian, A.; Sadorge, P.; Gawer, M.; Terrine, C. and Guern, J. (1969). Enzymic synthesis of some N^6-substituted adenine nucleosides. *Physiology Vegatable,* 7, 31–42.

Sorty, A. M.; Meena, K. K.; Choudhary, K.; Bitla, U. M.; Minhas, P. S. and Krishnani, K. K. (2016). Effect of plant growth promoting bacteria associated with halophytic weed (*Psoralea corylifolia* L) on germination and seedling growth of wheat under saline conditions. *Applied Biochemistry and Biotechnology,* 180, 872–882.

Spence, C. and Bias, H. (2015). Role of plant growth regulators as chemical signals in plant-microbe interactions: A double-edged sword. *Current Opinion in Plant Biology,* 27, 52–58.

Sponsel, V. M. and Hedden, P. (2004). Gibberellin biosynthesis and catabolism. In *Plant Hormones: Biosynthesis, Signal Transduction, Action!* (P. J. Davies ed.). Kluwer, Dordrecht.

Sridhar, R.; Chandramahan, D. and Mahadevan, A. (1968). The role of tryptamine in indole acetic acid synthesis by *Fusarium oxysporum* f. vasinfectum and *Pyricularia oryzae. Indian Journal of Experimental Biology,* 6, 59–60.

Stirk, W. A. and Van Staden, J. (2010). Flow of cytokinins through the environment. *Plant Growth Regulation,* 62, 101–116.

Strzelczyk, E. and Pokojska-Burdziej, A. (1984). Production of auxin and gibberellin-like substances by mycorrhizal fungi, bacteria and actinomycetes isolated from soil and the mycorrhizosphere of pine (*Pinus silvestris* L.). *Plant and Soil,* 81, 185–194.

Su, Y. I.; Xia, S.; Wang, R. and Xiao, L. (2017). Phytohormonal quantification based on biological principles. *Hormone Metabolism and Signaling in Plants.* 13, 431–470.

Subba Rao, N. S., ed. (1984). *Biofertilizers in Agriculture.* A. A. Balkema, Rotterdam, pp. 186.

Takino, J.; Kozaki, T.; Ozaki, T.; Liu, C.; Minami, A. and Oikawa, H. (2019). Elucidation of biosynthetic pathway of a plant hormone abscisic acid in phytopathogenic fungi. *Bioscience, Biotechnology, and Biochemistry,* 83, 1642–1649.

Thimann, K. V. and Mahadevan, S. (1964). Nitrilase: I. Occurrence, preparation, and general properties of the enzyme. *Archives of Biochemistry and Biophysics,* 105(1), 133–141.

Tietz, A.; Ruttkowski, U.; Kohler, R. and Kasprik, W. (1989). Further investigations on the occurrence and the effects of abscisic acid in algae. *Biochemistry and Physiology Pflanz,* 184, 259–266.

Tijero, V.; Teribia, N.; Munoz, P. and Munne-Bosch, S. (2016). Implication of abscisic acid on ripening and quality in sweet cherries: differential effects during pre- and post-harvest. *Frontiers in Plant Science,* 7, 602.

Tsavkelova, E. A.; Cherdyntseva, T. A.; Botina, S. G. and Netrusov, A. I. (2007). Bacteria associated with orchid roots and microbial production of auxin. *Microbiological Research,* 162(1), 69–76.

Tsubokura, S.; Sakamoto, Y. and Ichihara, K. (1961). The bacterial decomposition of indoleacetic acid. *The Journal of Biochemistry,* 49(1), 38–42.

Vijayabharathi, R.; Sathya, A. and Gopalakrishnan, S. (2016). A renaissance in plant growth-promoting and biocontrol agents by endophytes. In *Microbial Inoculants in Sustainable Agricultural Productivity* (Dhananjaya Pratap Singh, Harikesh Bahadur Singh, Ratna Prabha, eds.): Vol. 1: *Research Perspectives*, Springer, New Delhi, pp. 37–61.

Wang, X.; Ding, J.; Lin, S.; Liu, D.;Gu, T.; Wu, H.; Trigiano, R. N.; McAvoy, R.; Huang, J. and Li, Y. (2020). Evolution and roles of cytokinin genes in angiosperms 2: Do ancient CKX_S play housekeeping roles while non-ancient CKX_S play regulatory roles? *Horticulture. Result*, 7, 29.

Warembourg, F. R.; Dressen, R.; Vlassak, K. and Lafont, F. L. (1987). The peculiar effect of *Azospirillum* inoculation on growth and nitrogen balance of winter wheat (*Triticum aestivum*). *Biology and Fertility of Soils*, 4, 55–59.

Weingart, H. and Volksch, B. (1997). Ethylene production by *Pseudomonas syringae* pathovars in vitro and in planta. *Applied and Environmental Microbiology*, 63(1), 156–161.

Williams, P. M. and de Mallorca, M. S. (1982). Abscisic acid and gibberellin-like substances in roots and root nodules of *Glycine max*. *Plant and Soil*, 65, 19–26.

Yakushkina, N. I. and Tarasenko, A. A. (1975). Comparative study of hormone content and growth of seedlings cultivated under sterile and unsterile conditions. *Soviet Plant Physiology*, 22, 129–133.

Yang, D.; Du, X.; Liang, X.; Han, R.; Liang, Z.; Liu, Y. and Zhao, J. (2012). Different roles of the mevalonate and methylerythritol phosphate pathways in cell growth and tanshinone production of *Salvia miltiorrhiza* hairy roots. *PloS One*, 7(11), 46797.

Yang, J.; Gu, W.; Feng, Z.; Yu, B.; Niu, J. and Wang, G. (2022). Synthesis of Abscisic Acid in *Neopyropia yezoensis* and Its Regulation of Antioxidase Genes Expressions Under Hypersaline Stress. *Frontiers in Microbiology*, 12, 775–710.

Yanni, Y. G.; Rizk, R. Y.; Abd El-Fattah, F. K.; Squartini, A.; Corich, V.; Giacomini, A. and Dazzo, F. B. (2001). The beneficial plant growth-promoting association of *Rhizobium leguminosarum* bv. trifolii with rice roots. *Functional Plant Biology*, 28(9), 845–870.

Young, S.; R. P. Pharis ; D. Reid ; M. S. Reddy ; R. Lifshitz and G. Brown. (1991). PGPR: Is there a relationship between plant growth regulators and the stimulation of plant growth or biological control activity? The Second International Workshop on Plant Growth Promoting Rhizobacteria. (C. Keel, B. Koller and G. D e'faso, eds) pp. 102–103, *Bull. Srop. Interlaken, Switzerland*, October 14–19, 1990.

Zimmer, W.; Roeben, K. and Bothe, H. (1988). An alternative explanation for plant growth promotion by bacteria of the genus *Azospirillum*. *Planta*, 176, 333–342.

2 Role of Root Exudates and Rhizodepositions in the Management of the Rhizosphere Microbial Community

Soma Halder (Paul)

2.1 INTRODUCTION

Development of agriculture is an evolutionary process where the wild plant species has been replaced by the genetically improved cultivars of crop species, and soil is the foundation of agriculture and microbes play a vital role in sustaining natural ecosystems. The rhizospheric soil is different from the bulk soil by its various properties and composition, which imparts some effects on the soil microflora, i.e., different types of microorganisms (rhizobacteria, phytopathogenic bacteria, other organic matter decomposing fungi, mineralizing bacteria, mycorrhiza) and nematodes of that region. Different abiotic and biotic stress condition changes the plant growth pattern as well as metabolism pattern, which also has some impact on root exudates' composition. Genetic diversity of the crop plants as well as different modern agricultural practices (mixed cropping, intercropping) has also some impact on the root exudates' pattern and composition due to differences in growth and metabolic pattern, which also alters the plant–microbe interaction. How do these root exudates and rhizodeposits influence these microorganisms that need to be studied? The use of artificial root exudates and a multi-omic approach, such as metabolomic, metagenomic, isotope labeling, imaging, and multifluidic technology may help in finding out the result and progressing the relation between root exudates and rhizosphere-microbiome function (Sun et al. 2021). These interactions ultimately influence the plant growth and productivity. Here in this chapter, some discussions are given about the different kinds of interactions between the rhizospheric microbiome and rhizodeposits and root exudates.

2.2 RHIZOSPHERE

The term rhizosphere defines the zone of soil that includes the zone of soil, which includes the plant roots along with various kinds of microorganisms, such as mycorrhiza, rhizobia, plant growth-promoting rhizobacteria (PGPR), and also nematodes (Morgan et al. 2005; Pinton et al. 2001). The rhizosphereic zone can be divided into three distinct zones like the rhizoplane, ectorhizosphere, and endorhizosphere (Lynch 1987). The endorhizosphere is the zone of endodermis and the cortex of the root tissue, whereas the ectorhizosphere consists of the soil particle outside the root surface with all root exudates or rhizodeposits (Figure 2.1). The microbes can colonize within the endorhizosphere so it has been termed as the internal root colonization zone rather than the

DOI: 10.1201/9781003354109-2

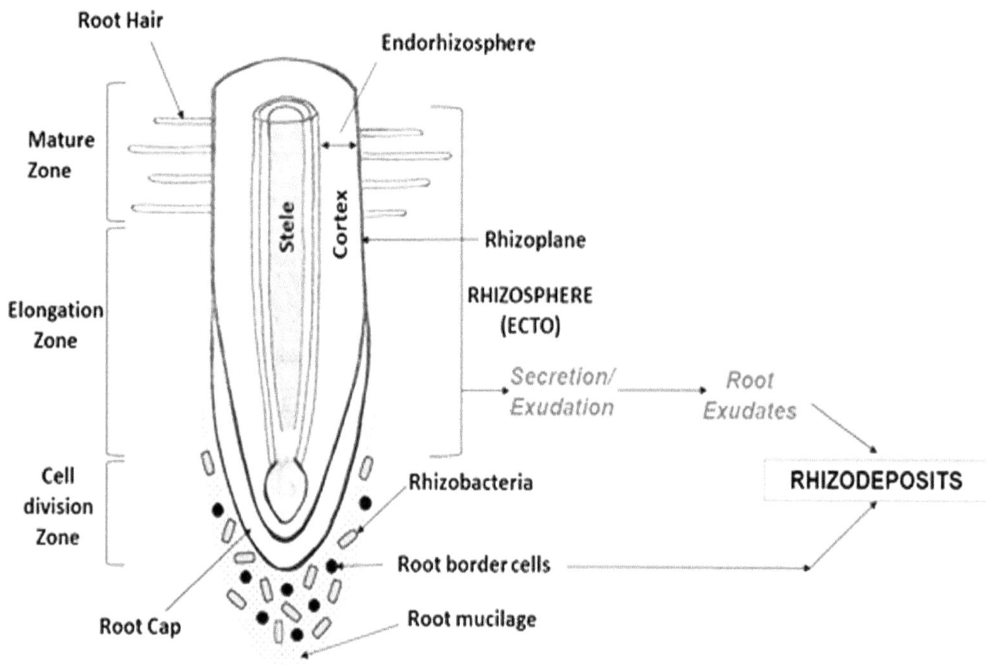

FIGURE 2.1 Different zones and components around the plant root.

rhizospheric zone. In the rhizosphere, the root exudates and the plant rhizodeposits play a major role in determining the physical and chemical properties of soil, which influence the microbial and other organism community of soil directly or indirectly (Badri et al. 2009b; Chaparro et al. 2013). Compared to bulk soil, the rhizospheric soil contains a greater amount of microorganisms as supported by the rhizodeposits. The interaction between plant and microbes in the soil depends on the criteria of rhizodeposits.

2.3 ROOT EXUDATES

The root exudates are mainly composed of carbon and nitrogen, which may be of low molecular weight, such as amino acids, sugars, organic acids, phenolics, and fatty acids, which are mainly primary metabolites; there are also some secondary metabolites, high molecular weight compounds, such as some sterols, growth factors, and enzymes. There are also flavonoids, nucleotides/purines, and also some polysachharides and peptides (Bais et al. 2006; Badri and Vivanco 2009). It has been observed that about 11% of the fixed carbon or 27% of total carbon in the root are exuded through the root, though the composition of root exudates varies greatly depending on the plant species, age or growth stage of the plant, metabolic condition of the cell, nutritional status of the plant, etc. Most of the studies have been performed on herbaceous plants whilst a few works have been carried out on tree plants also. The different phytochemicals present in the root exudates regulate much of the oil biogeochemical cycle, alter the oil structure, and have the potential to increase the rhizospheric microbe. Some of them have allelopathic properties, or chelating properties, and can bind with toxic chemicals, sometimes having growth regulatory properties, which are helpful for the growth of other neighboring plants (Table 2.1).

TABLE 2.1
Different Phytochemicals Present in Root Exudates Help in the Plant Defence Mechanism

Name of the Compound	Activities	Reference
Benzoxazinoid	Alter the rhizopheric microbiome Inhibit/decrease the plant growth of herbivore, helpful for next generation crop plant	Hu et al. 2018
Coumarin	Inhibit soil borne fungal pathogen	Tringli et al. 2018
Jamonic acid	Enrichment of beneficial microbe through biological control	Ruan et al. 2019
Malic acid	Activate plant immune system with the help of Bacillus	Yuan et al. 2018
Romarinic acid	Mimic the quorum sensing molecule, promote plant defene mechanism	Corral-lugo et al. 2016
Alicylic acid	Act a elective filter for rhizopheric microbiome, Inhibit Actinobacteria and Proteobacteria	Arah et al. 2015
Triterpene	Plant defence through antimicrobial activities	Huang et al. 2019

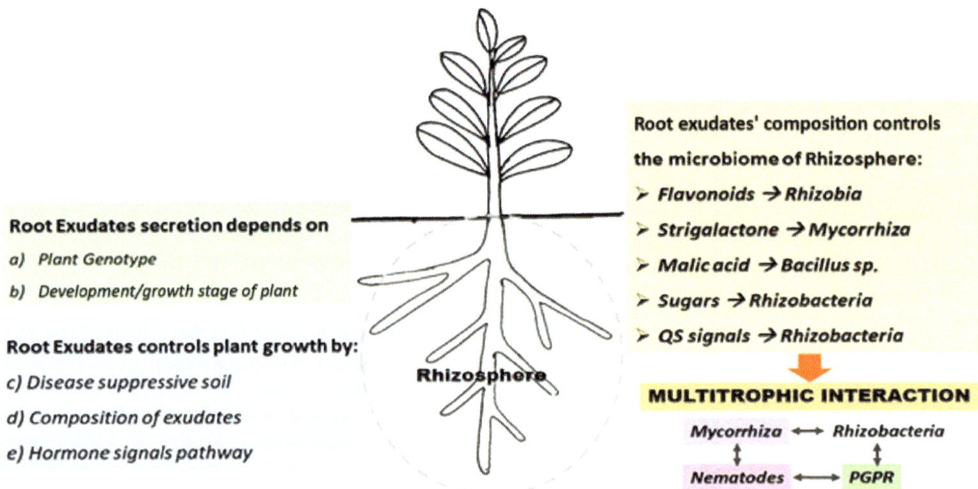

FIGURE 2.2 Interaction in between root exudates and rhizosphere.

2.4 RHIZODEPOSITS

Rhizodeposits are actually the secreated materials from plant root cells, which are the root exudates along with the root border cells, which are naturally slaughed off from the root tips of healthy plants. The root border cells become separated from the root tip due to the enzymatic activity of several pectin-degrading enzymes, such as methylesterases, polygalactouronases, etc. These root border cells also remain viable according to the availability of the nutrient in the rhizosphere region; they also play some role on the plant growth process by having a positive effect on beneficial microbes and a negative effect by inhibiting the plant pathogenic microbes.

Rhizodeposition takes place in different root zones, as the root possesses mature, elongation, and divisional zones, exudation occurs from the elongation zone, root border cells are released from the root cap region. The rhizospheric microbes colonize both in the endo- and ecto-rhizospheric zone where the root exudates provide more nutrients to the rhizospheric microbes (Figure 2.2). Thus, the microbial community in the rhizosphere is controlled by the rhizodeposits, root exudates, and root border cells, which ultimately influence the plant growth.

2.5 MECHANISM OF ROOT EXUDATES' SECRETION IN SOIL

The plant possesses various types of mechanisms for secretion of these various categories of root exudates in the rhizosphere, low molecular weight compounds may be released through diffusion or through ionic channels or vesicles, i.e., through passive transport. But secondary metabolites, suhc as polysaccharides and proteins or others are secreated or released through active transport, i.e., through ATP binding cassette (ABC) or through extrusion transporters like the multidrug and toxic compound extrusion family (MATE) (Vives-Peris et al. 2020)

In the case of small compounds, the passive transport through the process of diffusion takes place due to the gradient created between the root cell cytoplasm and the rhizosphere. Two different types of ionic channels are there: slow anion channels and quick anion channels. Passive transport also includes vesicle transport (Dreyer et al. 2012) where the higher molecular weight metabolites are stored and pass through (Badri and Vivanco 2009)

In the case of active transport of root exudates secreated by the plant root, ABC transporters are the primary transporters, which use the ATP as the energy source and are found mostly in all plants (Weston et al. 2012). Another is MATE family transporters, which have been studied mostly in *Arabidopsis* where citrate exudation occurs and in *Oryza* where the phenolics' secreation is regulated (Zhou et al. 2013; Takanashi et al. 2014).

2.6 FACTORS EFFECTING THE ROOT EXUDATES' COMPOSITION

Different physical, chemical, and biological factors play some roles in root exudation pattern either qualitatively or quantitatively. How the chemical and physical factors influence the root exudates' secretion can be studied in in vitro conditions by changing the parameters. There are several kinds of stress conditions that the plants may face, e.g., abiotic stress like droughts, high salinity, flooding, extreme temperatures, or nutrient starvation. All of these stresses have some effects on plant metabolism and growth so they also play some role in root exudates' composition as this process of exudation depends on plant metabolism (Table 2.2).

Nutrient availability in soil and CO_2 in atmosphere are the most important factors for plant growth; it has been observed that due to the low availability of nitrogen, phosphorus, potassium, and iron in the root exudates' composition are changed (amino acid, carbohydrate, organic acid secretion). Simliar to nitrogen starvation lessening the amino acid and increasing the organic acid secretion, in the case of phosphorus deficiency, the root exudates contain more organic acids so that it can solubilize soil phosphorus by acidification (Khorassani et al. 2011) and also contains more sugars, whereas potassium starvation inhibits it; iron deficiency increases the glutamate, glucose, ribitol, or citrate secretion.

Plant biomass production is effected by CO_2 starvation as it is directly related with the total CO_2 fixed by plants. Water stress also effects the composition of root exudates likely both deficiency and excess of water increase carbon content, with a higher content of succinic, fumaric or malic acids or with higher secretion of oxalic or malonic acid (Henry et al. 2007). Salinity stress may be osmotic or ion toxicity, both of these effects the root exudates, e.g., flavonoid secretion may become increased due to salinity as observed in *Glycine max* and *Phaseolus vulgaris* (Dardanelli et al. 2010, 2012).

Environmental temperature plays an important role on different plant metabolic processes like photosynthesis, respiration, translocation, etc., also it has been observed by Pramanik et al. (2000) that root exudate composition also changes in *Cucumis sativa*, temperature changes from 30/25 °C to 25/20 °C (day/night temperature) promotes the secretion of organic acids like benzoic, phthalic, palmitic acid in root exudates when grown in hydroponics.

Other abiotic stress conditions like the presence of toxic ions or heavy metals change the composition of root exudates as studied in *Arabidopsis* under aluminium toxicity, in *Sedum alfredii* under cadmium and lead toxicity (Luo et al. 2014, 2017).

TABLE 2.2
Microorganisms Associated with Roots of Different Plant Species Under Different Stress Conditions in Rhizosphere Soil

Name of the Plant	Type of Stress	Mycorrhiza	PGPR
Wheat *(Triticum aestivum)*	Salinity and heat	*Rhizophagus irregularis* (Zhu et al. 2016)	*Enterobactor* sp. (Singh & Jha 2016) *Pseudomonas putida* (Ali et al. 2011)
Maize (*Zea mays*)	Drought	*Rhizophagus irregularis* (Quiroga et al. 2017)	
Cucumber (*Cucumis sativus)*	Cold/ heat	*Glomus mosseae* (Haghighi et al. 2015)	
	Salinity & Drought		*Burkholdera cepacia* *Promicromonospora* *Acinetobacter calcoaceticus* (Kang et al. 2014)
Tomato *(Lycopersicon esculentum)*	Flooding & cold	*Rhizophagus irregularis* (Calvo-Polanco et al. 2014) *Funneliformis mosseae* (Liu et al. 2016)	*Enterobacter cloacae* *Pseudomonas putida* (Grichko & Glick 2001)
Bell pepper (*Capsicum annum)*	Drought		*Bacillus licheniformis* (Lim & Kim 2013)
	Osmotic stress		*Arthrobacter* sp., *Bacillus* sp. (Sziderics et al. 2007)
Sorghum (*Sorghum bicolor)*	Drought		*Bacillus* sp. (Grover et al. 2014)

Light also indirectly influences the root exudate secretion, as CO_2 fixed by plant ultimately controls the composition of root exudates, the light wavelength as well as photoperiod both effect the exudation process as studied by several workers; as in *Cucumis sativus* (Pramanik et al. 2020), *Alnus glutinosa* (Hughes et al. 1999), *Centaurea stoebe* (Tharayil and Triebwasser 2010) in Seagrass (Martin et al. 2018). In all studies the composition of root exudates got changed.

Biological factors also have some influence on the root exudation process, such as genotype of plant, growth, and age of plant etc. As it has been observed that some of the cultivars produce more root exudates than others, it has also been observed that genetic changes in *Arabidopsis* could change the root exudate production as revealed by a change in the rhizosphere microbial community. Growth stage of a plant also determines the composition of root exudates as it has been studied in *Arabidopsis*, which also effects the microbial community in the rhizosphere according to the composition of root exudates (Chaparro et al. 2013), in potato (Inceoğlu et al. 2010).

According to soil nutrient availability, the plant growth is regulated, which also effects the root exudate composition and changes the nutrient availability to the plant, controlling the microflora in the soil. Sometimes the root exudates play a role in controlling other insects in the soil, such as some arthropods, nematodes, fungi, and other bacteria either positively or negatively, i.e., their presence or absence in the soil gets modulated by root exudate composition from different plant species (Schrey et al. 2015, Dessaux et al. 2018).

Positive and negative interaction is also observed among the plants of the same species or of the different species due to different composition of root exudates from different plants. The mechanism of negative interaction includes allelopathy and production of phytotoxins, which are mostly aromatic compounds of hydroxyl or ketone groups, flavonoids, quinones, quinolines, and hydroxamic acids (Bais et al. 2006) (Figure 2.2).

2.6 FACTORS AFFECTING THE RHIZOSPHERE MICROBIOME

As the rhizospheric soil is enriched in all types of nutrients from different components of rhizodepositions so the microbial community is more abundant than the bulk soil as it has been observed by many studies. It has been detected that the proteobacteria, Actinobacteria, Firmicutes are predominating taxa in the rhizosphere through the assembly of different types of microbes is governed by different abiotic factors and also the biotic factors in that particular area (Xun et al. 2016; Lekberg et al. 2021; Thiergart et al. 2020).

Different physico-chemical properties of soil, i.e., soil pH, availability of nutrient, moisture content, temperature, and soil type influence the microbial community in the bulk soil of a particular area. But the rhizospheric of the same place differs due to the presence of different plant species. The presence of different plant species of different genotypes strongly affects the microbial community as the rhizodeposition or root exudate pattern differs. Also the genotypic differences of plants are likely to attract different types of microorganisms and trigger their growth, so the species-species rhizosphere nitches are formed (Stopnisek and Shade 2021; Zhalnina et al. 2018; Huang et al. 2019; Lebeis et al. 2015).

The plant immune system also plays a role to govern the distribution and abundance of plant rhizospheric microbes. Several immunogenic microbes are associated molecular patterns (MAMPs), which are found both in plant pathogens and beneficial microbes, which trigger the immunogenic response in the host plant, i.e., MAMP-triggerd immunity (MTI) (Teixeira et al. 2019).

Interdependence between plant roots and rhizospheric microbes exists due to the exchange of nutrients on which the relationships may be associative, symbiotic neutral, or parasitic. Root exudates sometimes contain phytochemicals, which attracts some specific microorganisms that in turn helps the plant roots for assimilation of nitrogen, phosphorus, iron, etc., or sometimes they get induced to produce the plant growth regulators, such as auxin, gibberellin, cytokinin, and ethylene-like substances.

Quorum sensing is another important phenomenon among the rhizospheric microbiome. The root exudates sometimes contain various components, which mimics QS signals to attract the associated or adjacent bacteria. As has been observed, the QS compounds in root exudates regulate the metabolic and behavioral activities of the bacterial community. This type of interaction occurs through chemical signals called autoinducers, autoinducingpeptides synthesized by the bacteria. Bacterial quorum sensing occurs for various complex processes depending on the species density (Reinhold-Hurek et al. 2015). The plant rhizospheric region has a higher amount of QS molecules, which leads to the threshold level of the PGPR, which strongly induces plant performance (Rodrignez et al. 2020) (Figure 2.3).

2.7 ROOT EXUDATE–RHIZOSPHERE INTERACTION

As there are different kinds of plant root secreted root exudates, which can mediate a number of interactions between plant to plant, plant to microbes, and also plant with other organisms. These interactions may be positive or negative, in the case of positive interactions with PGPR, symbiotic rhizobacteria, mycorrhiza, and other endophytes some of these exudates helps in more colonization and are growth elicitors, act as vectors for symbioses with soil fauna. The exudates may be allelopathic, antibacterial, QS mimics, phytotoxins, or nematicidal or insecticidal compound (Bais et al. 2006) (Table 2.3).

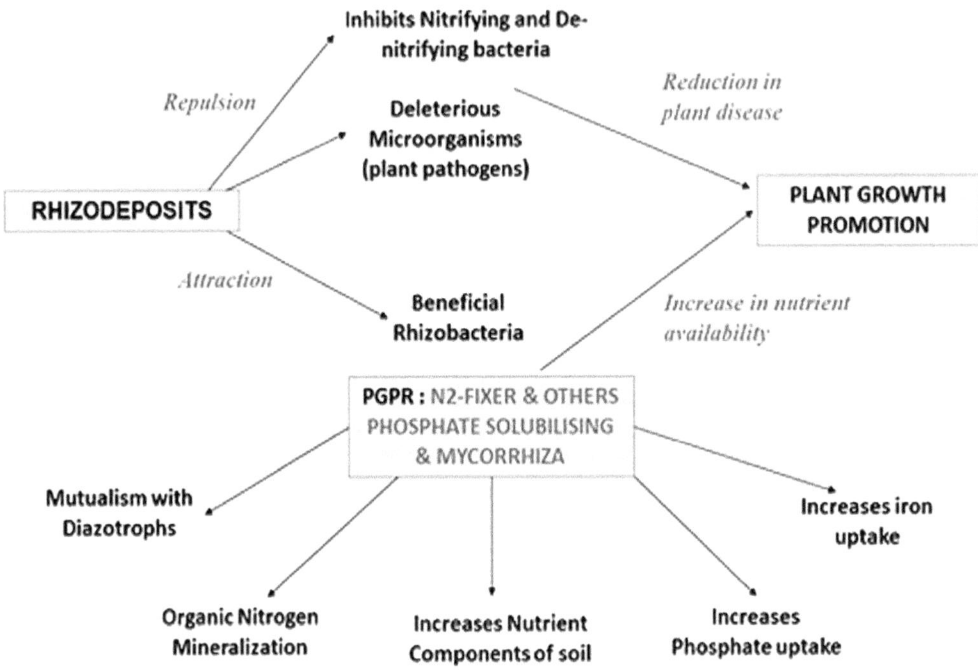

FIGURE 2.3 Role of PGPR and rhizodeposits in plant growth promotion.

TABLE 2.3
Components of Root Exudates of Different Crop Plants, the PGPR Recruited in Rhizosphere and their Activity

Name of the Plant	Components of Root Exudates	PGPR	Activity
Rice (*Oryza sativa*)	Carbohydrates, amino acids, like glycine, alanine, proline, valine histidine	*Azospirillum brazilience* (Bacilo-Jimez et al. 2003)	Attracts endophytic bacteria like *Corynebacterium* sp. *Bacillus sp.* due to chemoattractant nature
	Salicylic acid	*Pseudomonas chlororaphis* (Bowya and Balachendar 2020)	Enhances the available form of nitrogen, phosphorus, potassium, zinc in the rhizosphere
	Flavonoids and Hydroxycinnamic acid	*Azospirillum* sp. (Charmam et al. 2013)	Enhances metabolite activities and plant growth
Wheat (*Triticum aestivum)*	2-4-diacetyl phloroglucinol (DAPG)	*Pseudomonas fluorosence* (Bonsal et al. 1997)	Act as biocontrol agent
	Organic acids likeacetic acid, oxalic acid, succinic acid and tartaric acid	*Arthrobacter* sp. *Bacillus* sp. (Chen et al. 2019)	Enhances the available form of nitrogen, phosphorus, potassium, zinc in the rhizosphere
Maize (*Zea mays*)	Humic acid	*Herbaspirillum seropedacae* (Canellas & Oliveres 2017)	Enhances the production of root border cells, signalling and colonization of nitrogen fixer

TABLE 2.3 (Continued)
Components of Root Exudates of Different Crop Plants, the PGPR Recruited in Rhizosphere and their Activity

Name of the Plant	Components of Root Exudates	PGPR	Activity
	Amino acids(proline), soluble sugars, and endopolysaccharides	*Bacillus* sp. (Vardharajula et al. 2011)	Enhances seedling and plant growth
Soyabean (*Glycine max)*	Isoflavonoid	*Rhizobium* sp. (White et al. 2017)	Helps in plant defence and facilitates symbiotic interaction within rhizosphere microbiome
Tomato *(Lycopersicon esculentum)*	Organic acids likeCitric, Lactic, malic, succinic etc.	*Pseudomonas fluorosence* (Kamilova et al. 2006b)	Act as biocontrol agent and helps to increase the biomass production
Ground nut (*Arachis hypogea)*	Nanogenin, oleic acid, citric acid lactic acid threonine and glycolic oxime	*Bradyrhizobium* sp. *Azospirillum* sp. (Cesari et al. 2019) *Pseudomonas aeruginosa* (Arkail & Podle 2019)	Enhances growth of groundnut, enhances the exudation process and PGPR interaction

2.8 ROLE OF ROOT EXUDATES ON RHIZOSPHERIC MICROBIOME

The rhizobacteria in the rhizosphere communicate with the root exudates through quorum sensing, cross talk, and nutrient management (Tashiro et al. 2013, Singh et al. 2017, Keswani et al. 2020a). Increase in population of plant growth-promoting bacteria like nitrogen fixing bacteria or the mycorrhiza are due to the positive plant–microbe interaction due to the presence of some kind of root exudates, whereas negative interaction is also found in the case of plant pathogenic microorganisms (Halder and Sen Gupta 2018, Compant et al. 2019, Bashir et al. 2021). The nodule formation in legume is triggered by a plant root secreted flavonoid compound, which mediates the signals for nodulation. Some of the plant pathogenic microorganisms are negatively controlled by the presence of other rhizobacteria, which indirectly help in plant growth by reducing the disease occurrence. So it can be concluded that the plant growth-promoting microorganisms control the plant growth-promoting microorganism, controlling the plant growth directly or indirectly through biofertilizers (Mahmud et al. 2021) or by phytostimulators (Spaepen et al. 2007) or by bio-control activity (BCA).

2.8.1 Root Exudates Control the Bio-Availability of Nitrogen

Bioavailability of soil nitrogen depends on three factors, such as fixation of atmospheric nitrogen to soil nitrogen by symbiotic or nonsymbiotic N_2 fixing rhizobacteria, degradation of organic nitrogen into soil nitrogen by reducing the rate of nitrification and denitrification. Nitrogen fixation in soil by N_2 fixing bacteria happens to be by symbiotic association of several species of *Rhizobium*, *Bradyrhizobium*, *Sinorhizobium*, *Mesorhizobium*, *Azorhizobium*, etc. (Lugtenberg 2015). Symbiotic fixation needs a nodulation process in the roots where the flavonoid type of root exudates help to trigger the process through nod-factor production by the nodulation gene. In the case of nonsymbiotic N_2 fixing bacteria it has been found that some chemoattractant secreated as plant root exudates, which can enhance the population of the same, such as mucilage secreated by maize aerial roots containing arabinose, fucose, and galactose can be used by diazotrophs as energy sources (Van Deynze et al. 2018).

An endophytic fungi *Phomopsis liquidambris* has been found to increase the production of phenolic and flavonoid compound in pea plants, which could increase the expression of nodulation gene in *Bradyrhizobium* and as a result the crop yield increased (Xie et al. 2019).

Root exudates can play a major role to enhance the mineralization of soil organic nitrogen through microbes. Ammonifying bacteria play the major role in converting the organic nitrogenous matter and this is the major source of nitrogen to the plant root. Root-derived carbon compound can also help in this regard as it has been shown to increase the rhizosphere microbial growth and microbial demand of nitrogen and mineralization of organic nitrogen was promoted (Meier et al. 2017).

The root exudates also known to contain biological nitrogen inhibitors (BNIs), i.e., which can reduce the loss of conserved nitrogen in soil. Only a few BNIs have been studied that could inhibit the growth of nitrifying bacteria, which in turn restricts the loss of nitrogen from ammonia. Denitrification is another microbial process for nitrogen loss from soil. Although there is not much evidence that root exudates can play a role, it has been found in an investigation that sometimes invasive plants can reduce the rate of nitrogen loss through denitrification and some biological denitrification inhibitor (BDI) was identified. Bardon et al. (2014, 2016) studied a polycyanidin type of compound, and found the possible mechanism may act as an allosteric inhibitor (Bardon et al. 2016).

Dissimilatory nitrate reduction to ammonia (DNRA) is another process where mobile nitrate is transformed to ammoniacal compounds by microbes, and it has been observed that this process is more in rhizospheric soil than in bulk soil, which may be due to the availability of carbon from root exudates playing a role for a gretaer amount of microbes in this process (Li et al. 2014).

2.8.2 Root Exudates Control Phosphorus and Iron Uptake

Except nitrogen, other important elements required by plants are phosphorus and iron, which are not very easily available to plants, sometimes the availability of these elements increased in the presence of root exudates as it has been shown that root exudates of potato, tomato, onion help to increase the phosphorus availability (Khasi U Rahman et al. 2021); citrate and phylase present in tobacco root exudates can have an influence in the phosphorus acquisition of tobacco (Giles et al. 2018).

The same is true for iron availability also, several studies have revealed that root exudates can play a role in improving the Fe acquisition by a reduction-based strategy or using a chelation-based strategy (Lo et al. 2016, Chen et al. 2017). Riboflavin or phenolic compounds or coumarins present in root exudates and phytosiderophore production influence the Fe acquisition either by reduction or by chelation (Chen et al. 2017, Tsai and Schmidt 2017).

2.8.3 Root Exudates Control the Plant Defense Mechanism

With the help of metabolomic study, a number of phytochemicals have been identified in root exudates, which can be helpful for the plant defense process. The plant roots are always prone to infection by phytopathogens existing in the soil. The susceptible plant gets infected but it has been studied that some of the root exudates help in colonizing some of the helpful nonpathogenic bacteria, which can reduce the pathogenic population or may be inhibitory to other pathogenic microbes, also these beneficial microbes may be helpful to activate the plant defense signaling pathways or inhibit the growth of the pathogen by secreting antibiotic-like compounds. Here a few examples are given, some of the facts are studied in some plant systems. A number of studies have also been performed to find out the regulatory network of plant gene for different plant metabolites biosynthesis, which are involved in the plant immune system. Some of these metabolites have the antibacterial activities through biological control, through QS signal or through selective association with other beneficial microbes. These play the role for reduction in plant disease occurrence.

2.9 CONCLUSION

For sustainable agricultural practices multiple approaches are necessary and exploring the rhizosphere microbiome is also required for increased crop production. The rhizosphere microbiome is regulated by plant root exudates and rhizodeposits, so it can be manipulated to support more crop growth and production through different types of agricultural management practices. In conventional agricultural practices, we use different kinds of chemical fertilizers, which imbalance the soil properties and has indirect effects on the rhizosphere microbiome, as well as the use of different kinds of fungicides or pesticides, inhibiting the pathogenic microbes but also imbalancing the dynamic process on microbial population in the rhizosphere.

Modern agricultural practices towards crop diversity (i.e., crop rotation, intercropping, etc.) require the complex interactions among all of these to be understood, i.e., rhizosphere microbiome, rhizodeposits, and plant root exudates are very important, so that we can modulate the strategies of agricultural methods for different crop production.

Wild crops have different kinds of genetic patterns that control the root exudate components, which indirectly controls the rhizospheric microbiome. Now the modern crop varieties have been developed by newer breeding techniques as well as through a transgenic approach, these newer genetic constituents might have some indirect effect on the secretion or exudation pattern of the roots or its constituents. As these components have some indirect effect on the rhizobacteria assimilation, there will be a change in the rhizosphere microbiome pattern. Without complete knowledge of the role of different root exudate components on the microbial population, it is not possible to develop newer strategies for sustainable agricultural practices. A combined approach with the help of transcriptomics and proteomics may help to reveal the interaction patterns between plant and rhizosphere microbiome.

As in the reverse way it has been studied that rhizospheric microbes also have some effects on plant growth and development by their own activities, those can also be manipulated accordingly for better production of crops. For this purpose, more studies are required to understand the integrated interaction pattern among all these three, i.e., root exudates, rhizodeposits, and rhizosphere microbiome at molecular and functional levels. There is more information available on how the root exudates controls the microbial population and plant growth but much more information is required on how a specific microbe plays a role to change the associated microbial population through QS signals, which ultimately changes the rhizospheric soil components and indirectly or directly affects the plant growth and development. Further research is also required to understand how plant root exudates' pattern and components are regulated by the rhizospheric microbiome.

REFERENCES

Ali SZ, Sandhya V, Grover M, Linga VR, Bandi V (2011) Effect of inoculation with a thermotolerant plant growth promoting Pseu domonas putida strain AKMP7 on growth of wheat (Triticum spp.) under heat stress. *J Plant Interact* 6:239–246. https://doi.org/10.1080/17429145.2010.545147

Ankati, S., and Podile, A. R. (2019). Metabolites in the root exudates of groundnut changes during interection with plant growth promoting rhizobacteria in a strain-specific manner. *J. Plant Physiol.* 24:153057. https://doi.org/10.1016/j.jplph.2019.153057

Bacilio-Jimenez, M., Aguilar-Flores, S., Ventura-Zapata, E., Perez-Campos, E., Bouquelet, S., and Zenteno, E. (2003). Chemical characterization of root exudates from rice (Oryza sativa) and their effects on the chemotactic response of endophytic bacteria. *Plant Soil* 249:271–277. https://doi.org/10.1023/A:1022888900465

Badri DV, Vivanco JM. (2009). Regulation and function of root exudates. *Plant Cell Environ* 32(6):666–681. https://doi.org/10.1111/j.1365-3040.2009.01926.x. PMID: 19143988.

Badri DV, Weir TL, van der Lelie D, Vivanco, JM. (2009b). Rhizosphere chemical dialogues: plant–microbe interactions. *Curr Opin Biotechnol* 20(6):642–650. https://doi.org/10.1016/j.copbio.2009.09.014. PMID: 19875278.

Bais HP, Weir TL, Perry LG, Gilroy S, Vivanco JM. (2006). The role of root exudates in rhizosphere interactions with plants and other organisms. *Annu Rev Plant Biol* 57: 233–266. https://doi.org/10.1146/annurev.arplant.57.032905.105159

Bardon C, Piola F, Bellvert F, Haichar FZ, Comte G, Meiffren G, Pommier T, Puijalon S, Tsafack N, Poly F. (2014). Evidence for biological denitrification inhibition (BDI) by plant secondary metabolites. *New Phytol* 204:620–630.

Bardon C, Poly F, Piola F, Pancton M, Comte G, Meiffren G, Haichar FZ. (2016). Mechanism of biological denitrification inhibition: procyanidins induce an allosteric transition of the membranebound nitrate reductase through membrane alteration. *FEMS Microbiol Ecol* 92 :fiw034.

Bashir S, Iqbal A, Hasnain S, White JF. (2021). Screening of sunflower associated bacteria as biocontrol agents for plant growth promotion. *Arch Microbiol* 203: 4901–4912. https://doi.org/10.1007/s00203-021-02463-8

Bonsall, R. F., Weller, D. M., and Thomashow, L. S. (1997). Quantification of 2,4-diacetylphloroglucinol produced by fluorescent Pseudomonas spp. in vitro and in the rhizosphere of wheat. *Appl. Environ. Microbiol.* 63:951–955. https://doi.org/10.1128/aem.63.3.951-955.1997

Bowya, T., and Balachandar, D. (2020). Rhizosphere engineering through exogenous growth-regulating small molecules improves the colonizing efficiency of a plant growth-promoting rhizobacterium in rice. 3 *Biotech* 10:277. https://doi.org/10.1007/s13205-020-02275-5

Calvo-Polanco M, Molina S, Zamarreño AM, García-Mina JM, Aroca R (2014) The symbiosis with the arbuscular mycorrhizal fungus Rhizophagus irregularis drives root water transport in flooded tomato plants. *Plant Cell Physiol* 55:1017–1029. https://doi.org/10.1093/pcp/pcu035

Canellas, L. P., and Olivares, F. L. (2017). Production of border cells and colonization of maize root tips by Herbaspirillumseropedicae are modulated by humic acid. *Plant Soil.* 417:403–413. https://doi.org/10.1007/s11104-017 3267-0

Cesari, A., Paulucci, N., Lopez-Gomez, M., Hidalgo-Castellanos, J., Pla, C. L., and Dardanelli, M. S. (2019). Restrictive water condition modifies the root exudates composition during peanut-PGPR interaction and conditions early events, reversing the negative effects on plant growth Plant. *Physiol. Biochem.* 142:519–527. https://doi.org/10.1016/j.plaphy.2019.08.015

Chamam, A., Sanguin, H., Belllvert, F., Meiffren, G., Comte, G., Wisniewski Dye, F., et al. (2013). Plant secondary metabolite profiling evidences strain dependent effect in the Azospirillum- Oryza sativa association. *Phytochemistry* 87:65–77. https://doi.org/10.1016/j.phytochem.2012.11.009

Chaparro JM, Badri DV, Bakker MG, Sugiyama A, Manter DK, Vivanco JM. (2013). Root exudation of phytochemicals in arabidopsis follows specifc patterns that are developmentally programmed and correlate with soil microbial functions. *PLoS One* 8:e55731. https://doi.org/10.1371/journal.pone.0055731

Chen, S., Waghmode, T. R., Sun, R., Kuramae, E. E., Hu, C., and Liu, B. (2019). Root-associated microbiomes of wheat under the combined effect of plant development and nitrogen fertilization. *Microbiome* 7:136. https://doi.org/10.1186/s40168-019-0750-2

Chen YT, Wang Y, Yeh KC. (2017). Role of root exudates in metal acquisition and tolerance. *Curr Opin Plant Biol* 39:66–72. https://doi.org/10.1016/j.pbi.2017.06.004

Chen Y, Ding Q, Chao Y, Wei X, Wang S, Qiu R. (2018). Structural development and assembly patterns of the root-associated microbiomes during phytoremediation. *Sci Total Environ* 644:1591–1601. https://doi.org/10.1016/j.scitotenv.2018.07.095

Compant, S., Samad, A., Faist, H., and Sessitsch, A. (2019). A review on the plant microbiome: ecology, functions, and emerging trends in microbial application. *J. Adv. Res.* 19:29–37. https://doi.org/10.1016/j.jare.2019.03.004

Corral-Lugo A, Daddaoua A, Ortega A et al (2016). Rosmarinic acid is a homoserinelactone mimic produced by plants that activates a bacterial quorum-sensing regulator. *Sci Signal* 9:ra1

Dardanelli MS, Manyani H, González-Barroso S, Rodríguez-Carvajal MA, Gil-Serrano AM, Espuny MR, López-Baena J, Bellogín RA, Megías M, Ollero FJ. (2010). Efect of the presence of the plant growth promoting rhizobacterium (PGPR) *Chryseobacterium balustinum* Aur9 and salt stress in the pattern of flavonoids exuded by soybean roots. *Plant Soil* 328:483–493. https://doi.org/10.1007/s11104-009-0127-6

Dardanelli MS, de Córdoba FJF, Estévez J, Contreras R, Cubo MT, Rodríguez-Carvajal MA, Gil-Serrano AM, López-Baena FJ, Bellogín R, Manayni H, Ollero FJ, Megías M. (2012). Changes in favonoids secreted by *Phaseolus vulgaris* roots in the presence of salt and the plant growth-promoting rhizobacterium *Chryseobacterium balustinum*. *Appl Soil Ecol* 57 :31–38.

Dessaux Y, Grandclément C, Faure D. (2018). Engineering the rhizosphere. *Trends Plant Sci* 21(3):266–278. https://doi.org/10.1016/j.tplants.2016.01.002

Dreyer I, Gomez-Porras JL, Riaño-Pachón DM, Hedrich R, Geiger D. (2012). Molecular evolution of slow and quick anion channels (SLACs and QUACs/ALMTs). *Front Plant Sci* 3:263. https://doi.org/10.3389/fpls.2012.00263

Giles CD, Richardson AE, Cade-Menun BJ, Mezeli MM, Brown LK, Menezes-Blackburn D, Darch T, Blackwell MSA, Shand CA, Stutter MI, Wendler R, Cooper P, Lumsdon DG, Wearing C, Zhang H, Haygarth PM, George TS. (2018). Phosphorus acquisition by citrate- and phytase-exuding Nicotiana tabacum plant mixtures depends on soil phosphorus availability and root intermingling. *Physiol Plant* 163:356–371. https://doi.org/10.1111/ppl.12718

Grichko VP, Glick BR (2001). Amelioration of flooding stress by ACC deaminase-containing plant growth-promoting bacteria. *Plant Physiol Biochem* 39:11–17. https://doi.org/10.1016/S0981 -9428(00)01212 -2

Grover M, Madhubala R, Ali SZ, Yadav SK, Venkateswarlu B (2014). Influence of Bacillus spp. strains on seedling growth and physi ological parameters of sorghum under moisture stress condi tions. *J Basic Microbiol* 54:951–961. https://doi.org/10.1002/ jobm.201300250

Gupta, V., Kumar, G. N., and Buch, A. (2020). Colonization by multi potential Pseudomonas aeruginosa P4 stimulates (Arachis hypogaea L.) growth, defence physiology and root system functioning to benefit the root rhizobacterial interface. *J. Plant Physiol.* 248:153144. https://doi.org/10.1016/j.jplph.2020. 153144

Haghighi M, Mozafariyan M, Abdolahipour B (2015). Effect of cucum ber mycorrhiza inoculation under low and high root temperature grown on hydroponic conditions. *J Crop Sci Biotechnol* 18:89 96. https://doi.org/10.1007/s12892-014-0083-4

Haldar S, Sengupta S. (2015). Plant-microbe cross-talk in the rhizosphere: insight and biotechnological potential. *Open Microbiol J* 9: 1–7. https://doi.org/10.2174/1874285801509010001

Henry A, Doucette W, Norton J, Bugbee B. (2007). Changes in crested wheatgrass root exudation caused by food, drought, and nutrient stress. *J Environ Qual* 36:904–912. https://doi.org/10.2134/jeq2006.0425sc

Hu L, Robert CAM, Cadot S, Zhang X, Ye M, Li B, Manzo D, Chervet N, Steinger T, van der Heijden MGA, Schlaeppi K, Erb M (2018). Root exudate metabolites drive plant-soil feedbacks on growth and defense by shaping the rhizosphere microbiota. *Nat Commun* 9:2738. https://doi.org/10.1038/s41467-018-05122-7

Huang AC, Jiang T, Liu Y-X, Bai Y-C, Reed J, Qu B. (2019). A specialized metabolic network selectively modulates *Arabidopsis* root microbiota. *Science* 364(6440):eaau6389. https://doi.org/10.1126/science:aau6389

Hughes M, Donnelly C, Crozier A, Wheeler CT. (1999). Efects of the exposure of roots of *Alnus glutinosa* to light on flavonoids and nodulation. *Can J Bot* 77:1311–1315. https://doi.org/10.1139/b99-077

Inceoğlu Ö, Salles JF, van Overbeek L, van Elsas JD. (2010). Efects of plant genotype and growth stage on the betaproteobacterial communities associated with diferent potato cultivars in two felds. *Appl Environ Microbiol* 76:3675–3684. https://doi.org/10.1128/AEM.00040-10

Kamilova, F., Kravchenko, L. V., Shaposhnikov, A. I., Azarova, T., Makarova, N., and Lugtenberg, B. (2006b). Organic acids, sugars, and l-tryptophane in exudates of vegetables growing on stonewool and their effects on activities of rhizosphere. *Bacter. Mol. Plant Microbe Interact.* 19:250–256. https://doi.org/10.1094/MPMI-19-0250

Kang SM, Khan AL, Waqas M, You YH, Kim JH, Kim JG, Hamayun M, Lee IJ (2014). Plant growth-promoting rhizobacteria reduce adverse effects of salinity and osmotic stress by regulating phy tohormones and antioxidants in Cucumis sativus. *J Plant Interact* 9:673–682. https://doi.org/10.1080/17429145.2014.894587

Keswani C, Singh HB, García-Estrada C, Caradus J, He YW, MezaacheAichour S, et al. (2020a). Antimicrobial secondary metabolites fromagriculturally important bacteria as next-generation pesticides. *Appl Microbiol Biotechnol* 104: 1013–1034. https://doi.org/10.1007/s00253-019-10300-8

Khashi u Rahman M, Wang X, Gao D et al (2021). Root exudates increase phosphorus availability in the tomato/potato onion intercropping system. Plant Soil 464:45–62. https://doi.org/10.1007/s11104-021-04935-8

Khorassani R, Hettwer U, Ratzinger A, Steingrobe B, Karlovsky P, Claassen N. (2011). Citramalic acid and salicylic acid in sugar beet root exudates solubilize soil phosphorus. *BMC Plant Biol* 11:121. https://doi.org/10.1186/1471-2229-11-121

Lekberg Y, Arnillas CA, Borer ET, Bullington LS, Fierer N, Kennedy PG. (2021). Nitrogen and phosphorus fertilization consistently favor pathogenic over mutualistic fungi in grassland soils. *Nat Commun* 12:3484. https://doi.org/10.1038/s41467-021-23605-y

Lebeis SL, Paredes SH, Lundberg DS, Breakfield N, Gehring J, McDonald M. (2015). Salicylic acid modulates colonization of the root microbiome by specific bacterial taxa. *Science* 349(6250):860–864. https://doi.org/10.1126/science:aaa8764

Li X, Rui J, Xiong J, Li J, He Z, Zhou J, Yannarell AC, Mackie RI. (2014). Functional potential of soil microbial communities in the maize rhizosphere. *PLoS One* 9:e112609.

Lim JH, Kim SD (2013). Induction of drought stress resistance by multi-functional PGPR Bacillus licheniformis K11 in pep per. *Plant Pathol J* 29:201–208. https ://doi.org/10.5423/PPJ. SI.02.2013.0021

Liu A, Chen S, Wang M, Liu D, Chang R, Wang Z, Lin X, Bai B, Ahammed GJ (2016). Arbuscular mycorrhizal fungus alleviates chilling stress by boosting redox poise and antioxidant potential of tomato seedlings. *J Plant Growth Regul* 35:109–120. https:// doi.org/10.1007/s00344-015-9511-z

Lugtenberg B. (2015). *Principles of plant-microbe interactions*. Springer International Publishing, Cham.

Luo Q, Sun L, Hu X, Zhou R (2014). The variation of root exudates from the hyperaccumulator Sedum alfredii under cadmium stress: metabonomics analysis. *PLoS ONE* 9:e115581. https://doi.org/10.1371/journal.pone.0115581

Luo Q, Wang S, Sun L, Wang H (2017). Metabolic profiling of root exu dates from two ecotypes of Sedum alfredii treated with Pb based on GC-MS. *Sci Rep* 7:39878. https://doi.org/10.1038/srep3 9878

Lynch JM. (1987). *The rhizosphere*. Wiley Interscience, Chichester, U.K

Mahmud AA, Upadhyay, SK, Srivastava AK, Bhojiya AA. (2021). Biofertilizers: a nexus between soil fertility and crop productivity under abiotic stress. *Curr Res Environ Sust* 3: 100063. https://doi.org/10.1016/j.crsust.2021.100063

Martin BC, Gleeson D, Statton J, Siebers AR, Grierson P, Ryan MH, Kendrick GA. (2018). Low light availabilityalters root exudation and reduces putative benefcial microorganisms in seagrass roots. *Front Microbiol* 8:2667. https://doi.org/10.3389/fmicb.2017.02667

Meier IC, Finzi AC, Phillips RP. (2017). Root exudates increase N availability by stimulating microbial turnover of fast-cycling N pools. *Soil Biol Biochem* 106:119–128. https://doi.org/10.1016/j.soilbio.2016.12.004

Morgan JAW, Bending GD, White PJ. (2005). Biological costs and benefits to plant–microbe interactions in the rhizosphere. *J Exp Bot* 56(417):1729–1739. https://doi.org/10.1093/jxb/eri205. PMID:15911554.

Pinton R, Varanini Z, Nannipieri P. (Editors). (2001). *The rhizosphere: biochemistry and organic substances at the soil–plant interface*. Marcel Dekker, New York.

Pramanik MHR, Nagai M, Asao T, Matsui Y (2000). Effects of temperature and photoperiod on phytotoxic root exudates of cucumber (Cucumis sativus) in hydroponic culture. *J Chem Ecol* 26:1953–1967. https://doi.org/10.1023/A:1005509110317

Quiroga G, Erice G, Aroca R, Chaumont F, Ruiz-Lozano JM (2017). Enhanced drought stress tolerance by the arbuscular mycorrhizal symbiosis in a drought-sensitive maize cultivar is related to a broader and differential regulation of host plant aquaporins than in a drought-tolerant cultivar. *Front Plant Sci* 8:1056. https://doi.org/10.3389/fpls.2017.01056

Reinhold-Hurek B, Bunger W, Burbano CS, Sabale M, Hurek T. (2015). Roots shaping their microbiome: global hotspots for microbial activity. *Annu Rev Phytopathol* 53: 403–424. https://doi.org/10.1146/annurev-phyto-082712-102342

Rodriguez, M.,Torres, M.,Blanco, L., Bejar, V., Sampedro, I., and Llamas, I. (2020). Plant growth-promoting activity and quorum quenching mediated biocontrol of bacterial phytopathogens by Pseudomonassegetis strain P6. *Sci. Rep.* 10, 4121. https://doi.org/10.1038/s41598-020-61084-1

Ruan J, Zhou Y, Zhou M, Yan J, Khurshid M, Weng W, Cheng J, Zhang K (2019). Jasmonic Acid Signaling Pathway in Plants. *Int J Mol Sci* 20. https://doi.org/10.3390/ijms20102479

Sarah LL, Paredes SH, Lundberg DS et al (2015). Salicylic acid modulates colonization of the root microbiome by specific bacterial taxa. *Science* 349:860–864.

Schrey SD, Hartmann A, Hampp R. (2015). Rhizosphere interactions. In: Krauss GJ, Nies DH (eds) *Ecological biochemistry: environmental and interspecies interactions*, vol. 12. Wiley-VCH, Weinheim, pp. 292–311. https://doi.org/10.1002/9783527686063.ch15

Singh RP, Jha PN (2016) Mitigation of salt stress in wheat plant (Triti cum aestivum) by ACC deaminase bacterium Enterobacter sp. SBP-6 isolated from Sorghum bicolor. *Acta Physiol Plant* 38:2–12. https://doi.org/10.1007/s11738-016-2123-9

Singh R, Singh J, Deval R, Upadhyay SK, Kumar D. (2017). The potentiality of selected strain of PGPR: azotobacter, for sustainable agriculture in India. *G J Environ Sci Technol* 4: 49–53.

Spaepen S, Vanderleyden J, Remans R. (2007). Indole-3-acetic acid in microbial and microorganism-plant signaling. *FEMS Microbiol Rev* 31: 425–448. https://doi.org/10.1111/j.1574-6976.2007.00072.x

Stopnisek N, Shade A. (2021). Persistent microbiome members in the common bean rhizosphere: an integrated analysis of space, time, and plant genotype. *ISME J* 15:2708–2722. 10.1038/s41396-021-00955-5

Stringlis IA, Yu K, Feussner K, de Jonge R, van Bentum S, van Verk MC, Berendsen RL, Bakker PAHM, Feussner I, Pieterse CMJ (2018). MYB72-dependent coumarin exudation shapes root microbiome assembly to promote plant health. *Proc Natl Acad Sci U S A* 115:E5213–E5222. https://doi.org/10.1073/pnas.1722335115

Sun, Haishu & Jiang, Shanxue & Jiang, Cancan & Wu, Chuanfu & Gao, Ming & Wang, Qunhui. (2021). A review of root exudates and rhizosphere microbiome for crop production. *Environmental Science and Pollution Research.* 28:1–14. https://doi.org/10.1007/s11356-021-15838-7

Sziderics AH, Rasche F, Trognitz F, Sessitsch A, Wilhelm E (2007). Bacterial endophytes contribute to abiotic stress adaptation in pepper plants (Capsicum annuum L.). *Can J Microbiol* 53:1195–1202. https://doi.org/10.1139/W07-082

Tashiro Y, Yawata Y, Toyofuku M, Uchiyama H, Nomura N. (2013). Interspecies interaction between Pseudomonas aeruginosa and other microorganisms. *Microbes Environ* 28: 13–24. https://doi.org/10.1264/jsme2.ME12167

Takanashi K, Shitan N, Yazaki K. (2014). The multidrug and toxic compound extrusion (MATE) family in plants. *Plant Biotechnol* 31:417–430. https://doi.org/10.5511/plantbiotechnology.14.0904a

Teixeira PJP, Colaianni NR, Fitzpatrick CR, Dangl JL. (2019). Beyond pathogens: microbiota interactions with the plant immune system. *Curr Opin Microbiol* 49:7–17. https://doi.org/10.1016/j.mib.2019.08.003

Tharayil N, Triebwasser DJ. (2010). Elucidation of a diurnal pattern of catechin exudation by *Centaurea stoebe. J Chem Ecol* 36:200–204. https://doi.org/10.1007/s10886-010-9749-7

Thiergart T, Durán P, Ellis T, Vannier N, Garrido-Oter R, Kemen E. (2020). Root microbiota assembly and adaptive differentiation among European *Arabidopsis* populations. *Nat Ecol Evol* 4(1):122–131. https://doi.org/10.1038/s41559-019-1063-3

Tsai HH, Schmidt W. (2017). Mobilization of Iron by Plant-Borne Coumarins. *Trends Plant Sci* 22:538–548. https://doi.org/10.1016/j.tplants.2017.03.008

Van Deynze A, Zamora P, Delaux P-M, Heitmann C, Jayaraman D, Rajasekar S, et al. (2018). Nitrogen fixation in a landrace of maize is supported by a mucilage-associated diazotrophic microbiota. *PLoS Biol* 16(8): e2006352. doi:10.1371/journal.pbio.2006352

Vardharajula, S., Alia, S. Z., Grover, M., Reddy, G., and Bandi, V. (2011). Drought tolerant plant growth promoting Bacillus spp. effect on growth, osmolytes and antioxidant status of maize under drought stress. *J. Plant Interact.* 6:1–14. https://doi.org/10.1080/17429145.2010.535178

Vives-Peris V, de Ollas C, Gomez-Cadenas A, Perez-Clemente RM (2020) Root exudates: from plant to rhizosphere and beyond. *Plant Cell Rep* 39:3–17. https://doi.org/10.1007/s00299-019-02447-5

Weston, L. A., Ryan, P. R., and Watt, M. (2012). Mechanisms for cellular transport and release of allelochemicals from plant roots into the rhizosphere. *J. Exp. Bot.* 63:3445–3454. https://doi.org/10.1093/jxb/ers054

White, L. J., Ge, X., Brozel, V. S., and Subramanian, S. (2017). Root isoflavonoids and hairy root transformation influence key bacterial taxa in the soybean rhizosphere. *Environ. Microbiol* 19:1391–1406. https://doi.org/10.1111/1462-2920.13602

Xie X-G, Zhang F-M, Yang T, et al. (2019). Endophytic fungus drives nodulation and N2 fixation attributable to specific root exudates. *MBio* 10:e00728–e00719.

Xun W, Zhao J, Xue C, Zhang G, Ran W, Wang B. (2016). Significant alteration of soil bacterial communities and organic carbon decomposition by different long-term fertilization management conditions of extremely low-productivity arable soil in South China. *Environ Microbiol* 18(6):1907–1917. https://doi.org/10.1111/1462-2920.13098

Yuan J, Zhao J, Wen T, Zhao M, Li R, Goossens P, Huang Q, Bai Y, Vivanco JM, Kowalchuk GA, Berendsen RL, Shen Q (2018). Root exudates drive the soil-borne legacy of aboveground pathogen infection. *Microbiome* 6:156. https://doi.org/10.1186/s40168-018-0537-x

Zhalnina K, Louie KB, Hao Z, Mansoori N, da Rocha UN, Shi S. (2018). Dynamic root exudate chemistry and microbial substrate preferences drive patterns in rhizosphere microbial community assembly. *Nat Microbiol* 3(4):470–480. https://doi.org/10.1038/s41564-018-0129-3

Zhou G, Delhaize E, Zhou M, Ryan PR. (2013). The barley *MATE* gene, *HvAACT1*, increases citrate efux and Al3+ tolerance when expressed in wheat and barley. *Ann Bot* 112:603–612. https://doi.org/10.1093/aob/mct135

Zhu X, Song F, Liu S, Liu F (2016). Role of arbuscular mycorrhiza in alleviating salinity stress in wheat (Triticum aestivum L.) grown under ambient and elevated CO2. *J Agron Crop Sci* 202:486–496. https://doi.org/10.1111/jac.12175

3 An Insight into the Role of the Rhizosphere Microbiome in the Sustainability of Plant Health

Shyamalina Haldar and Abhrajyoti Ghosh

3.1 INTRODUCTION

The simple but elegant mode of flourishing and thriving of microorganisms in the soil makes it a rich repository for diverse microbial communities. Although the earliest studies showed the existence of 10,000 different bacterial species only in one gram of soil, later observations demonstrated that one million genomes can be identified from such a small amount of soil (Torsvik et al. 2002; Gans et al. 2005). This fascinating observation led to the growing interest in understanding the microbial diversity of the soil and the factors contributing to the building up and modulation of this enormously diverse ecosystem. With the increasing accumulation of knowledge in this field, researchers also began to understand that the evaluation of microbial functions in the soil environments is of utmost necessity for the accurate preservation of natural ecosystems and the management of human society (Araujo 2022).

Among various factors of soil that affect the structure of the soil microbial community, the presence of plants is known to contribute the most, stimulating an uneven microbial distribution (Pii et al. 2016). The initial studies on bacterial groups associated with plant roots displayed a substantial upsurge in the microbial genetic pool along with an increase in soil functions inclusive of nutrient cycling, fertility of the soil, production of crops, and decomposition of litter by terrestrial ecosystems (Mendes et al. 2013). This is mostly attributed to the prevailing conditions and nutrient availability, such as carbon substrates in the soil areas adjacent to the plants (Pii et al. 2016). The plant microbiome refers to this diverse yet taxonomically organized microbial communities, that are associated along the different parts of a plant, conferring innumerable beneficial effects upon the host plants, including increased bioavailability of nutrients, mobilization and assimilation of nutrients, changing the plant physiology, production of plant growth factors' resilience to biotic and abiotic stresses, alleviating the toxic metals present in the soil, suppressing the soil pathogens and mediating immunity of plants (Igiehon and Babalola 2018; Noman et al. 2021). Due to the dominance of the microbial community in the soil, the transfer of matter and energy between soil and the plants through this nutrient assimilation is a critical function of the resident microbial species that ultimately fosters the establishment of stable plant–microbiome consortia, benefiting the host-plant fitness and ultimately plant-productivity (Araujo et al. 2020).

An exhaustive literature exists on the factors contributing to plant–microbe beneficial associations (Vandenkoornhuyse et al. 2015; Trivedi et al. 2020; Simon et al. 2019). The stability of the plant–microbe ecosystem, referred to as "holobiont" is largely evolutionary-selected (Hamonts et al. 2018). This coevolution of plant–microbe interaction, which has been occurring since 400 million

DOI: 10.1201/9781003354109-3

years ago, is significantly affected by the carry-over effect of the rhizosphere microbiota composition by which the specific species can be transferred from generation to generation by the host plant itself and is responsible for the maintenance of fitness in plants (Pantigoso 2022).

The biotic factors, including genotype and/cultivar and the developmental stage of the plant, along with the abiotic factors, such as geographical conditions of the regions, climate change, and anthropogenic activities, cumulatively determine the dynamicity of the holobiont (Edwards et al. 2018). The recruitment and the assemblage of the rhizobiome depend on the beneficial attributes demonstrated by the microbes and requirements of the host plant at any particular point in time (Torres-Cortés et al. 2018; Wei et al. 2019; Cregger et al. 2018). Therefore, an in-depth evaluation of the factors and mechanisms of microbial colonization in a different niche of the host plants will underpin the beneficial microbial-linked traits while managing the processes of microbial colonization will aid in dealing with the functional evaluation of the same (Noman et al. 2021).

Recently, more new insights have emerged through current advances in culture-independent high-throughput sequencing about the potential of ecologically and economically critical microorganisms (Bergelson et al. 2019). This data can be manipulated through microbial engineering to enhance the efficiency of indigenous microbial consortia, referred to as "core-microbiome," in the present exigent agricultural settings with ever-changing climatic and environmental scenarios (Roman-Reyna et al. 2019). The manipulation of rhizobiome is possible through extensive complex and extremely scrutinized cell-to-cell interaction and by investigating the signaling molecules to screen the habitats of the microbes and altering their functional attributes (Igiehon and Babalola 2018). This knowledge is harbored towards the development of prospective techniques for the commercialization of eco-friendly bioformulants to improve sustainable agriculture (Noman et al. 2021). This will help to overcome the challenges of reduction of agricultural landmasses, increasing environmental pollution and devastation of the environment, and the outbreak of diseases imposed due to the extensive use of commercial agrochemicals (Isobe and Ohte 2014).

Therefore, this chapter focuses on the current knowledge of the fundamental mechanisms of the development of beneficial plant–microbe interactions and multiple facilitating ways engaged by the rhizosphere microbiome in plant growth promotion and stress resistance with a target to throw light upon the deployment of the rhizosphere microbiome towards better productivity of crops in the form of amendments, inoculants, synthetic microbial consortia, prebiotics from root exudates, rhizobiome engineering and crop breeding.

3.2 MICROBIOME IN THE RHIZOSPHERE

The microbes cohabiting with plants are widely distributed among a large group of microorganisms, including bacteria, fungi, protozoa, and protists (Compant et al. 2019). The most dominant inhabitants of the rhizosphere include *Acidobacteria*, *Actinobacteria*, *Bacteroidetes*, *Firmicutes*, and *Proteobacteria* while the fungal phyla include *Ascomycota* and *Basidisomycota*, (Mohanram and Kumar 2019). These are grouped as *beneficial* and *non-beneficial microbes*. Beneficial microbes, referred to as collectively plant growth-promoting microbes (PGPM) participate directly or indirectly in the growth and development of plants (Figure 3.1). PGPM can be classified into five core groups depending on their functions, such as (a) regulation of soil properties and fertility; (b) formation of mycorrhizal structures with plant roots; (c) degradation, fixation, and transformation of soil nutrients; (d) induction of systemic resistance (ISR) of plants; (e) inhibition of the pathogens and decomposition of plant and animal residues and toxins in the soil (Genre et al. 2020; Vejan et al. 2016; Khan et al. 2019).

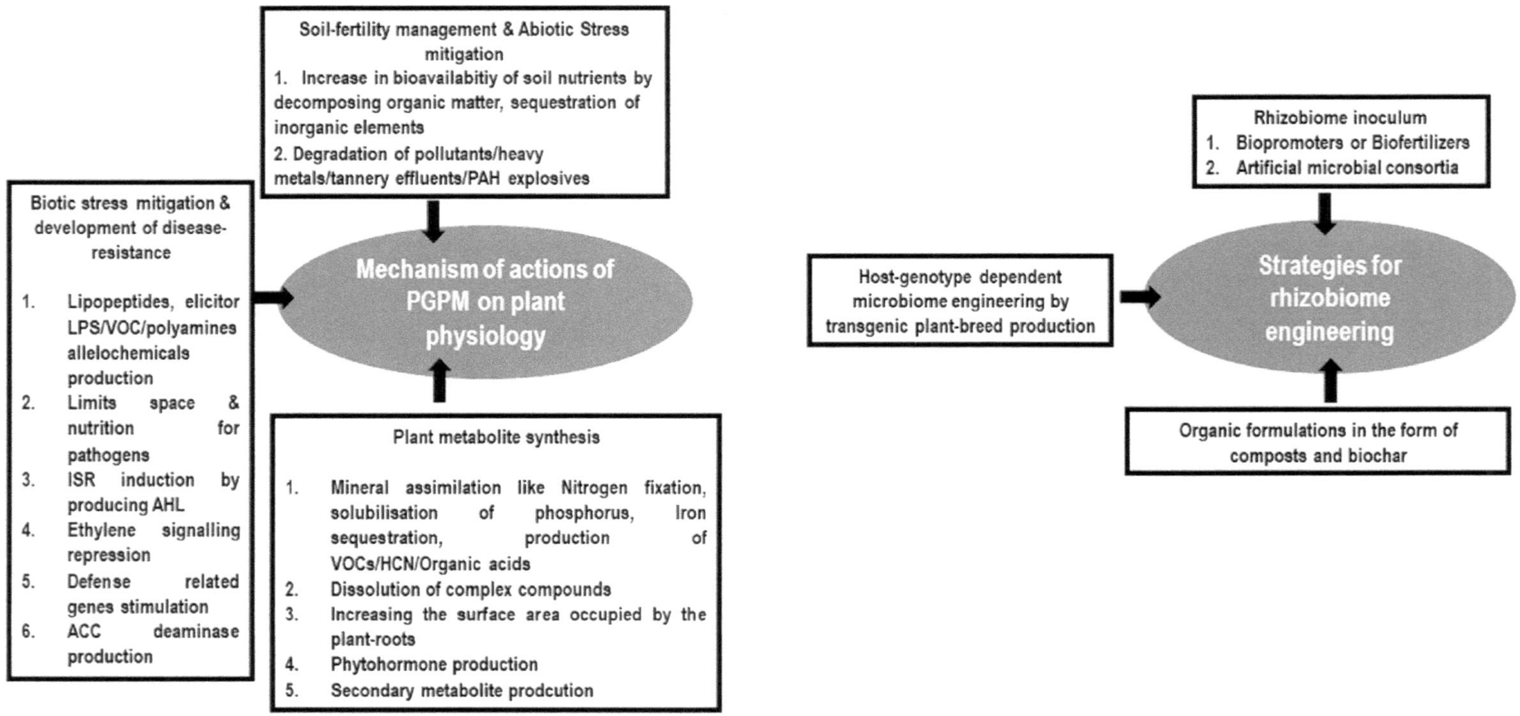

A.

B.

FIGURE 3.1 Mechanisms of actions of (A) PGPM and (B) strategies for rhizobiome engineering.

3.2.1 Regulation of Soil Properties and Fertility

The properties and fertility of the soil with alleviation of nutrient imbalance and deficiency are mediated by the PGPM through any of the following mechanisms: activation of organic and inorganic nutrients and minerals and increasing their bio-availability; fixation of atmospheric nitrogen to improve soil fertility through the transformation of nitrogen elements (ammonia, nitrates, nitrites); extracellular secretion of organic substances to improve the stability of soil aggregates; enhancement of soil organic matter content through decomposition of plant and animal residues (Estrada et al. 2013; Majeed et al. 2018; Liang et al. 2019). The *Azospirillum*, *Burkholderia*, *Klebsiella*, *Nitrosomonas*, *Nitrococcus*, *Nitrobacter*, *Paenibacillus*, *Rhizobium*, and *Pseudomonas* sp. convert atmospheric nitrogen to ammonia, nitrate, and nitrite. In this regard, the most widely established legume-Rhizobia (*Rhizobium*, *Bradyrhizobium*, *Azorhizobium*, and *Mesorhizobium*) symbiosis that can reduce gaseous N_2 to use ammonia for plants being driven by a complex amino acid cycle (Defez et al. 2017; Liang et al. 2019; Shahrajabian et al. 2021; Saeed et al. 2021; Alemneh et al. 2020; Lindstrom et al. 2020; Tewari et al. 2020; Adnan et al. 2017; Guo et al. 2021). The activation of other soil nutrients including solubilization of phosphate and potassium by *Enterobacter*, *Brevibacillus*, *Mortierella*, *Trichoderma* and *Phyllobacterium*, *Paenibacillus*, *Agrobacterium*, *Acinetobacter*, and *Bacillus* sp. plays a crucial role in plant growth and development.

Additionally, the release of organic acids like tartaric acid, citric acid, lactic acid, malic acid, oxalic acid and gluconic acid, and EPS (exopolysaccharides) resulting in a reduction of soil-pH promotes mineral-solubilization by *Trichoderma asperellum*, *Pantoea agglomerans*, *Rahnella aquatilis* and *Pseudomonas orientali* and *Serratia* sp. (Sarikhani et al. 2018; Sun et al. 2020; Yaghoubi et al. 2017). Extracellular polymeric substances (EPS) such as polyuronic, polysaccharides, and amino acids secreted by microbes (hyphae of fungi and actinomycetes) bind soil-particles to increase interparticle cohesion, thereby alleviating erosion and enhancing the internal storage of carbon and water. The necromass of microbes produced by the decomposition of plant and animal residues also adds to soil organic carbon (Kallenbach et al. 2016; Zheng et al. 2016; Rashid et al. 2016).

3.2.2 Enhancement of Environmental Stress Resistance and Plant Growth Promotion

The optimization of the acquisition of nutrients by the rhizosphere microbes is the critical gateway for overall healthy plant growth, particularly in challenging environments. For example, in arid and semi-arid deserts and saline zones, *Bacillus subtilis* enhanced oil production in *Foeniculum vulgare* seeds (Mishra et al. 2016). In droughts, the synergistic actions of phosphate solubilizer and biofertilizer, *Pseudomonas fluorescence,* and arbuscular mycorrhiza fungi (AMF) in *Echinacea purpure* promoted the absorption of water and mobilization of phosphate from roots to leaves. Similarly, *Bacillus amyloliquefaciens* enriched the plant biomass by increasing the utilization of potassium in the rhizosphere. Even DNA methylation and epigenetic modifications were highly promoted in plants induced by the rhizosphere microbes (Chen et al. 2022). The decrease of biomass due to accumulated ethylene in the plants under environmental stresses in drought, salinity, flood, high or low temperature, insect predation, and radiation can be altered by 1-aminocyclopropane-1-carboxylate (ACC) deaminase synthesizing microbes from rhizosphere (*Bacillus*, *Staphylococcus*, *Arthrobacter*, *Oceanobacillus*, *Exiguobacterium*, *Halobacillus* and *Zobellella* species) (Iqbal et al. 2017; Orhan et al. 2021; Ravanbakhsh et al. 2017; Glick et al. 2021; Sadeghi et al. 2017; Brunetti et al. 2021). The plant growth is enhanced by direct manipulation of the architecture, growth, and exudation of roots by the rhizobiome, resulting in an evolutionary advantage in the host plants. The PGPM secreting root-growth-factor auxin and/or stimulation of the roots for better transport of the hormones results in better nutrient acquisition and growth of plants (Freschet et al. 2021). The amino acid efflux is increased by phenazine and zearalenone synthesized by PGPB, which also increases plant biomass (Klein et al. 2022). The stress responses are allayed by the rhizosphere microbes through nutritional and hormonal regulation in plants by synthesizing phytohormones such as auxins, indole-acetic acid (IAA), gibberellins, abscisic acid, salicylic acid, and others (Pantigoso 2022).

3.2.3 Accumulation of Active Ingredients for Secondary Metabolites of Plants

The signaling cascade for secondary metabolites like flavonoids, terpenoids, alkaloids, tannins, and phenolic acids with their simultaneous gene activation in plants are stimulated by the elicitors like oligosaccharides, proteins, and polyunsaturated fatty acids (PUFAs), produced by rhizosphere microbes like *Bacillus cereus*, *Pseudomonas fluorescence*, *Aspergillus niger*, *Coriolus versicolor*, and *Ganoderma lucidum* (Chamkhi et al. 2021; Ho et al. 2018; Wang et al. 2022). The accumulation of nutrients (N and P) by the rhizosphere microbes also facilitates the production of secondary metabolites. Amino acid synthesis is catalyzed by AMF through N-absorption. *Stenotrophomonas* sp. helps in increasing astragaloside content while *Phyllobacterium* and *Inquilinus* help to increase calycosin content in the roots of medicinal plants (Wang et al. 2022).

3.2.4 Removal of Heavy Metals, Pesticides, and Pollutants from Soil

The rich repository of metal-antagonistic and metallo-tolerant microorganisms in the rhizospheres alleviates the phytotoxicity imposed due to the pollutants and heavy metals and thereby enhances the normal growth of plants (Wang et al. 2022). The fungal species like *Aspergillus fumigatus* and *Aspergillus terreus* were found to remove heavy metals while *Aeromonas*, *Bacillus*, and *Pseudomonas* sp. are active in heavy metal-contaminated soil where they accumulate heavy metals like lead, arsenic, copper, and mercury from the soil by biosorption (Khalid et al. 2021; Fakhar et al. 2022). The AMF immobilizes metals by producing phytochelatins, glutathione, metallothionein, and glomalin; thereby aiding the host plants to adapt to the stressful environment (Wang et al. 2022). Even the restriction of transport systems for the uptake of metals in plant roots accelerates the efflux of accumulated heavy metals from plants as found in AMF colonization in arsenic (As)-contaminated soils (Chen et al. 2007). The rhizobiome promotes the phytoremediation of heavy-metal-contaminated soil by decreasing the phytotoxicity of heavy metals and enhancing the process of uptake of the metals by the plants. The organic acids of root exudates hasten the metal-solubilization and increase their bioavailability towards the microbes, which then can degrade them (Khalid et al. 2021). A synergistic action of *Kocuria rhizophila* and secreted citric acid from plants increased the total biomass of the plants followed by accumulation of various amounts of cadmium (Cd), chromium (Cr), copper (Cu) and nickel (Ni) (Hussain et al. 2019). *Bacillus cereus* and *Bacillus thuringiensis* promoted the absorption of manganese (Mn) in *Broussonetia papyrifera.* In addition, the *Bacillus subtilis* fostered redox reactions by which the valency of the heavy metal (Cr) was changed, which enhanced plant growth and photosynthetic activity in plants (Khan et al. 2019; Altowayti et al. 2019; Jin et al. 2018). Besides these, the rhizosphere microbes scavenge reactive oxygen species (ROS) and thereby prevent the heavy-metal-induced oxidative stress and also stabilize the content of malondialdehyde (MDA), promoting gene expression of catalase, peroxidase, and superoxide dismutase (Abdelhameed et al. 2019).

The hydrolysis of xenobiotics by a wide range of microbial groups from rhizobium, including *Agrobacterium*, *Alcaligenes Acinetobacter*, *Bacillus*, *Burkholderia*, *Enterobacter*, *Klebsiella Paenibacillus*, *Pseudomonas*, *Rhizobium*, *Sphingomonas*, *Serratia*, *Stenotrophomonas*, *Streptomyces*, *Trichoderma*, and *Xanthomonas* species convert the refractory organic substances to water and carbon-dioxide and thereby provide the plants with a healthy soil environment (Cunliffe and Kertesz 2006; Masakorala et al. 2013). The organic pollutants like naphthalene, pyrene, and phenanthrene, chlorinated organophosphate-based pesticides, and poly aromatic hydrocarbons (PAHs) are degraded by *Klebsiella*, *Serratia*, *Sphingobium*, *Trichoderma*, and *Pseudomonas* sp. (Smułek et al. 2020; Ping et al. 2014). The presence of specific enzymes for metabolism and through co-metabolism the rhizosphere microbes can potentially transform the pollutants and thus activate the remediation of the contaminated soil, elevating the plant fitness (Gu et al. 2022).

3.2.5 Disease Resistance in Plants

The biocontrol property, referring to the power of the plant microbiome to alleviate the effects of pathogens by directly interacting with the pathogens or indirectly by eliciting immune responses in plants, is an important emerging field of investigation for the improvement of plant health (Ritpitakphong et al. 2016). The origin of disease-suppressive soil is based on the functional soil microbial community when the soil-borne pathogens that are incapable or have a low ability to cause and sustain diseases and/or infections in the plants. This is an outcome of the interplay of that particular soil's diverse microbial residents, which impart a host of direct and indirect mechanisms for soil-pathogen attenuation. However, it is to be borne in mind that not only does the population composition of the microbes influence disease suppression, but the genetic makeup of the plant and the pathogens, along with the environmental factors, determine the outcome of the process (Duran et al. 2018). The pathogen deactivation is the culminating output of collaborative associations of the rhizobiome instead of any specific microbial activity.

Rhizobacteria impart resistance to pests and disease-causing pathogenic microbes by promoting the defense mechanisms of plants and activating ISR and/or systemic acquired resistance (SAR) through the salicylic acid, abscisic acid, jasmonic acid, ethylene, and hormonal signaling routes (Wang et al. 2022). The potential of *Stenotrophomonas* sp., *Sphingobacterium* sp., *Pseudomonas* sp., *Pantoea* sp., *Microbacterium* sp., *Leclercia* sp., *Enterobacter* sp., *Comamonas* sp., *Curtobacterium* sp., *Bacillus* sp., and *Achromobacter* sp. in eliciting plant-defense responses under various pathogen attacks in different host plants have been well studied making them potent biocontrol agents (Sahu et al. 2020). The presence of flagella and polysaccharides on the cell surface are the important characteristics of these biocontrol agents to elicit the ISR/SAR mechanisms in plants. The trickiest mechanism for the biocontrol microbes is to by-pass the immediate immune attacks of the plants, establish a well-coordinated symbiotic relation and then impart ISR/SAR.

In addition, the rhizobacteria enhance the colonization of the microbes with antibiotic properties that can produce lytic enzymes and lipopeptides to destroy the pathogenic cell-membranes. This is possible because the root exudates are modified in the presence of pathogenic species, which in turn accelerates the pathogen-inhibiting microbes (Fira et al. 2018). The other microbial antagonizing properties like antibiosis [production of 2,4-diacetylphloroglucinol (DAPG), butanediol, phenazines (PHZ), and cyclic lipopeptide surfactants] and siderophores synthesis by the rhizobacteria help in the suppression of pathogens and rise in ISR (Beneduzi et al. 2012). This protects plants from destruction by bacterial and fungal pathogens, root-knot nematodes, blue mold, damping off, and systemic viruses (Beneduzi et al. 2012). The beneficial soil fungus *Mortierella verticillate*, *Trichoderma virens*, *Rhodopseudomonas palustris*, *Alcaligenes faecalis*, *Bacillaceae*, *Streptomyces*, *Enterobacter*, and *Burkholderiaceae* have been reported to be potential biocontrol agents residing in the rhizosphere (Buttner et al. 2021; Su et al. 2017; Patel et al. 2019; Kim et al. 2021; Zhong et al. 2022). The rhizosphere of AMF has been studied to have pesticidal traits, which might abolish the adverse effects of chemical pesticides promoting indirectly the biomass of the associated hot plants (Babikova et al. 2014). The siderophore-sequestration of iron from the soil by the PGPM results in the diminishing of iron resources in the soil, thereby preventing the growth of pathogenic fungi, which need iron from optimum spreading and thus act as a biocontrol agent for the host plants (Traxler et al. 2012; Bal et al. 2013). Production of volatile organic compounds (VOCs), ammonia, and hydrogen cyanide (HCN) have been very early-established and well-documented mechanisms of pathogen suppression in the rhizosphere, particularly through stress tolerance, biofilm formation, and coordination of various gene expression required for overall healthy plant growth (Gomez Expósito et al. 2017). Besides this, the competition for ecological niches and acquisition of nutrients, and predate and parasitic interventions play pivotal roles in eradication or reduction of soil-borne diseases in plants (Afridi et al. 2022). This demands the development of robust synthetic microbial communities with abilities to prevent soil-borne diseases by harnessing these isolated species from disease-suppressive soil with screened activities for disease suppression.

3.3 INTERACTIONS IN THE RHIZOSPHERE

The rhizosphere interactions are complex and ever-changing. Hence decoding the mechanisms stimulating the gathering, the existence, and the sensitivity of the microbes along with the dynamic forces functioning at the rhizosphere is important to highlight the functional correlation of the microbes with the better living of the host plants.

3.3.1 Plant–Microbe Interactions

Among the myriad of factors playing in defining a specific rhizobiome environment, the most significant effect is being played by the host compartment on the microbiome assemblage (Gao et al. 2021). Although the soil-bed first determines the composition of the soil microbes, the alteration of the rhizobiome and the individual and/or community-level interactions in the rhizobiome is dependent solely on the root exudates and the immune system of the plants that together progressively modify the rhizosphere structure to establish a specific rhizobiome (Lareen et al. 2016). This establishment of microbial colonization on the root surface requires the perception of signals of the root exudates, mobilization of the microbes, and finally attachment and/or adherence of the microbes on the root surfaces forming a biofilm. The biofilm formation is mediated through the secretions of amino acids like arginine and polyamines like putrescine, followed by modulation of root architecture through signaling and/or nutrient-deficient inhibition of root functioning at the plant–microbe interface. Hence, the study of the microenvironment of the rhizosphere has revealed the presence of multiple genetic elements for mobility, biofilm-formation, flagella assembly, secretions, substrate-transporters, and chemotaxis (Noman et al. 2021). It is also noteworthy to mention here that the rhizodeposits are strategically regulated by the plant age, the photosynthetic abilities, the endogenous evolutionary factors, and the presence of soil pathogens (Olanrewaju et al. 2019).

The root exudates in combination with polymers, border cells, and volatile substances secreted out into the soil, by the process of rhizodeposition are composed of high-molecular-weight (HMW) compounds (proteins, mucilage, enzymes, fatty acids, hormones, nucleotides, steroids, tannins, vitamins, and flavonoids) and low-molecular-weight (LMW) chemicals (amino acids, phenolics, organic acids, sugars, secondary metabolites) (Pérez-Jaramillo et al. 2016). Besides this, the release of antifungal compounds, VOCs, defense-related proteins, biotic stress-responsive proteins, and benzoxazinoids are stimulated by compatible associations of the respective host plants with the microbes (Olanrewaju et al. 2019). There is a strong linkage between the rhizobiome and the stimulation of root exudation, the phenomenon being labeled as systemically induced root exudation of metabolites (SIREM), which indicates that specific systemic variations are being introduced in the rhizosphere by the rhizobiome and minute change in the rhizosphere composition alters the plant-phenotype. Even the differential spatial distribution of metabolites with high concentration at the root tips than anywhere else is also a manipulated action of the rhizobiome for their benefit (Klein et al. 2022) Therefore, the root microbes positively or negatively affect the host plants, which in turn influences the rhizosphere interactions (Bever et al. 2012). However, the environmental stresses induce changes in the root-exudate composition in such a way that the plants can recruit the rhizobiome flora as per their needs. In nutrient-starved soils, specific chemicals, for example, coumarins in iron-starved soils, are secreted to gather siderophore-synthesizing microbes. Similarly, amino acids, shikimic acid, and quinic acid are enhanced during phosphate stress to be absorbed by the soil microbes and positively influence plant growth (Pantigoso 2022).

Although the complete characterization of the root exudates is challenging due to multiple controlling factors, there are few patterns in the distribution of rhizodeposits. The abundance of polysaccharides is high in root-tip, root-cap, and mucilage while organic acids, amino acids, and sugars are mostly allocated in the root-elongation zones. Of these, organic acids are the fundamental chemicals that help to establish microbial colonization, as proved by the widespread distribution of the acids in most of the terrestrial rhizospheres (Noman et al. 2021).

The microbes, on the other hand, bind to the exudates with a phosphotransferase system and periplasmic binding protein, followed by receptor-binding on the plant cells (Noman et al. 2021). However, the responses of the microbes in the rhizosphere vary according to the nature of the metabolites of the root exudates. Flavonoids attract *Bradyrhizobium* sp. and *Phytophthora* sp. in leguminous plants; naringenin triggers spore-germination and branching of hyphae in mycorrhiza and catechin stimulates bacterial quorum sensing in *Combretum albiflorum* (Hassan et al. 2012; Pérez-Montaño et al. 2011). Pyrrolizidine alkaloids, the defense molecules lead to an upsurge of microorganisms that can use these chemicals (Igiehon and Babalola 2018). It is noteworthy to mention that genetic analyses in tomato plants have unveiled three distinct quantitative trait loci involved in symbiotic associations with disease-suppressive *Bacillus cereus* (Igiehon and Babalola 2018). The ABC-transporter mutant *Arabidopsis* sp. secreting high phenols and low sugars as compared to the normal variety and the addition of a single component of exogenous glucosinolates in the root exudates of transgenic *Arabidopsis* sp. affected both α-proteobacteria and fungal communities in the rhizosphere (Badri et al. 2009). The attraction of PGPM by the plants is dependent on the root exudates. The positive chemotactic response was elucidated in *Pseudomonas syringae* and *Bacillus* sp. by citric and malic acids, respectively. 2,4-dihydroxy-7-methoxy-2H-1,4-benzoxazin-3(4H)-one (DIMBOA) from maize roots, showed positive effects towards root colonization of *Pseudomonas putida*. The infection of Arabidopsis leaves with *Pseudomonas syringae* and treatment of foliage by microbe-associated molecular patterns (MAMPs) induced malic acid secretion from roots which in turn attracted *Bacillus* sp. (Pérez-Jaramillo et al. 2016). The *Arabidopsis-Pseudomonas* symbiosis is so much evolutionary linked that the plant-antagonistic bacterium *Pseudomonas protegens* is found to transit to a rhizosphere mutualist within six generations of the plant and promotes the overall fitness through better competitiveness for root exudates and boosted tolerance to antimicrobial compound, scopoletin secreted by the plants (Li et al. 2021). Similar rapid adaptation was observed in *Ensifer meliloti* in *Medicago truncatula* through mutations in symbiotic signaling genes. The exudates induce the expression of antifungal biosynthetic genes in PGPM as in *Pseudomonas fluorescens* CHA0, the phlA and pltA genes, involved in the synthesis of 2,4-diacetylphloroglucinol (DAPG) and pyoluteorin (PLT), both being important antifungal chemicals were highly expressed in the presence of phenolics and pectin (Pérez-Jaramillo et al. 2016). In addition, the amino acids (alanine, tryptophan, proline, arginine), chitinase, oxalic acid, succinic acid, and nicotine are a few of the components from rhizodeposits of different plants (tobacco, rice, tomato, cucumber, wheat, maize) that elicit a response from the various PGPM, resulting in recruitment of these PGPM, which in turn affects plant health status (reviewed in Noman et al. 2021).

The mechanism of perception of signals from microbes by plants is a complex phenomenon and has recently been highlighted in the research on plant immunity development (Olanrewaju et al. 2019). Pieces of evidence depict that the pattern recognition receptors (PRRs) present in plants play a crucial role in identifying the MAMPs produced by the microbes in the rhizosphere. The plant-elicitor peptides (PEPs) and damage-associated molecular patterns (DAMPs) in the plants and the PRPR-MAMPs complexes altogether lead to the immunity development in the plants and activate the host immunity. In this regard, the activation of lipochitooligosaccharides (LCOs) in Rhizobia and the formation of LCOs-lysine receptor-like kinase (LysM) stimulated by the secreted flavonoids in legumes during symbiosis, triggering accretion of cytokinins, calcium spiking, facilitation of curling of root hairs and the subsequent formation of nodules during rhizobia infections have been well studied (Cao et al. 2017).

3.3.2 Microbe–Microbe Interactions

The coordination and competitive interactions in the microbial assemblage of the rhizosphere are the most crucial determining factors for the establishment of a stable rhizosphere ecosystem (Duran et al. 2018; van der Heijden et al. 2016). The simulation based on theoretical data on extensive network analyses of probable interconnectedness among the microbial species of rhizosphere has confirmed that the stable rhizobiome structure is possible in a system of higher modality, less positive but high negative correlations among the community participants (Hernandez et al. 2021; Coyte et al. 2015; Grilli et al. 2016).

A myriad of controlling factors including the survival rate of the microorganisms in the plant-root interface, competitive fitness, responsive capabilities towards both plant and environmental signaling molecules, and protist–prey interactions as a whole define the foundation of stable microbial association on the rhizosphere. The eco-evolutionary processes including niche-shifting by the microbes, ecological drifts, and the selection of the fittest species followed by diversification in genetic elements are the areas of emphasis during the plant–microbe assembly establishments. The association which begins with an arbitrary adaptation of the microbes in the rhizosphere from the soil pool gradually becomes specific depending on the health and necessity of the plants. However, it is noteworthy to mention that despite the sturdiness of plant–microbe interactions, the plants can house new organisms in subsequent phases without any major change in the prevailing community structure (Carlström et al. 2019).

The competition and/or cooperative association between the microorganisms in the plant rhizosphere occurs through the activation of distinct yet diverse genetic clusters, associated with the synthesis of natural substances, of the microbial flora. Although the knowledge is limiting about the processes of this synthesis and microbe–microbe interplay, the biosynthetic genes responsible for producing anti-microbial products like antibiotics, anti-fungal compounds, proteins, polyketides, terpenes, chitinases, ribosomal, and non-ribosomal peptide synthetases (RPS, NRPS) and antibiotic-resistance genes (ARGs) are quite copiously activated among the rhizobiota. This might be targeted to control the overgrowth and spreading of other non-specific or pathogenic microorganisms at the plant-rhizosphere interphase (Noman et al. 2021).

The communication of microorganisms with each other is mediated through the phenomenon of quorum sensing (QS), activated through the signaling molecules like homoserine lactone (HSL), acyl-homoserine lactones (AHL), and terpenes, which contribute to the interfering and/or associative functions between the neighboring microbes. *Enterobacter* sp. produces AHL to attract the pathogen, *Fusarium graminearum* but prevents its entry and traps it instead, by producing biofilms on the root surface. Microorganisms like *Bacillus* and *Pseudomonas* sp. residing at the periphery of the roots also stimulate the roots for exudation of specific molecules using long-distance communication signaling, which will help to establish the colonization of other specific microbes in and around the rhizosphere (Hida et al. 2020). The QS molecules like alcohols, pheromones, and diffusible signal factors (DSF) are immune eliciting and plant growth-modulating factors, thereby indicating the importance of microbe–microbe interactions in the rhizosphere towards the overall sustenance of the plant (Olanrewaju et al. 2019).

Since, QS is a tactic involved in the pathogenesis development by the root pathogens, the interruption of QS by quorum quenching (QQ) by PGPM (*Pseudomonas segetis*) is an important phenomenon by which the QS molecules produced by the pathogens are enzymatically degraded without killing the pathogen. This results in the survival of the pathogen on the root surface without causing any plant disease or infections. Additionally, the microbes can also modulate the root exudates towards the secretion of QQ molecules as a mechanism of pathogen suppression, alternatively maintaining a good health status for the plants (*Medicago truncatula* and edible plants like cinnamon, grapefruit) (Liu et al. 2019). The activity of microbial type VI secretion system is another mechanism of the rhizosphere microbes to produce anti-microbial proteins to outcompete the pathogens and/or competitive species within the rhizosphere niche, as observed in wheat, barley, cucumber, and citrus fruit plants (Noman et al. 2021). It is interesting to note that if the pathogen is successful in invading the plant roots by evading the first line of defense incurred by the microbiome, then the endophytes recruit microorganisms with efficient biocontrol mechanisms like *Actinobacteria* and *Firmicutes* through the signaling and communication with the soil-residents (Trivedi et al. 2017). This results in increasing the abundance of microbes with anti-predator traits, which later become useful for activities towards boosting plant immunity. These modified microbes are highly proficient in producing defense molecules, antimicrobial peptides, and phytohormones (Hünninghaus et al. 2017).

The recent findings have shown that the dynamicity of outbreak of diseases in plants (bacterial wilt) is defined by the inherent microbial interactions of the rhizosphere, thereby indicating that the initial minute variations in the rhizobiome community can predict the infection status of the plants (Gu et al. 2022). The same cultivar of rice shifted to disease-susceptible and resistant varieties with a similar seed-borne pathogen due to the variation in resistance toward a disease brought about by the seed microbiome (Matsumoto et al. 2021). This might be attributed to the fact that root exudates of young plants decrease competitive pressures directed towards more stochastic phenomena in microbe–microbe interactions as well as the interactions between the microbiome of the two different parts of the plants (seed and roots) (Badri et al. 2013). Therefore, from these observations it can be concluded that the rhizosphere microbial interactions determine the status of plant health, giving rise to binary outcomes depending on the initial deviation in the microbial composition that later may give rise to different abilities of disease suppression even among the same plants grown in homogeneous conditions (Gu et al. 2022). This might help predict whether a plant will be affected by a pathogen or will remain unaffected highlighting the importance of diagnostic profiling of rhizobiome for chances of disease occurrence and the healthy-plant enriched distinct microbial groups may be used to formulate bioinoculants to produce disease-resistant crops (Gu et al. 2022). The interactions of the competing microbial residents of the rhizosphere are beneficial for the plant fitness. *Rhizoctonia solani*, causing root-rot in mung bean seedlings are infected by *Trichoderma virens* or the hyperparasite *Ampelomyces* decreases the growth and inhibits the spread of mycelia in the powdery mildew pathogen *Podosphaera plantaginis* (Klein et al. 2022). However, it is noteworthy to mention that the consequences of the microbe–microbe interactions in the rhizosphere upon the host plant's health are context-dependent. These microbial interactions in the rhizosphere are classified as interspecific competition and interspecific facilitation. The interspecific competition changes the rhizosphere microbial association and establishes a dominance of the microbial consortia on the rhizosphere while in the interspecific facilitation, the entry of one induces the colonization of another microbial species. The presence of AMF in *Brassica* sp. has negative effects on growth promotion until it is inoculated simultaneously with *Trichoderma* sp., which then promotes plant growth (Klein et al. 2022).

3.4 RHIZOBIOME ENGINEERING

The greatest challenge faced by human society in recent eras is to supply food to the ever-growing population. Therefore, the target of agriculture is to increase the yield of crops worldwide. However, a large number of natural constraints including the availability of arable soil and nutrient makes it an enormously tedious task for the farmers and the governments of the countries. To overcome this, exhaustive research is going on with a target to qualitatively and quantitatively improve the agroecosystems and considerable labors have been undertaken to engineer the microbiome associated with the plants following diverse approaches. In recent years, there has been an upsurge in the multifaceted strategic applications of biological control agents (BCA) in agriculture for better production of healthy crops. The process of harnessing the rhizobiome towards the suppression of phytopathogen activities in the agricultural fields and overall stimulating the vigor of the plants may be categorized as engineering of rhizosphere microbes using bacterial competitiveness; inoculation of plants with genetically modified organisms (GMO) forming the synthetic microbiome; indigenous microbiome manipulation in situ, use of artificially engineered seed microbiome and development of beneficial plant–fungal interactions by plant-mycobiome management (Figure 1B; Afridi et al. 2022). Although plant–microbe engineering has been reported to be highly fascinating, the greatest hindrance to plant–microbiome engineering is imposed by abiotic factors in field-based studies. The omics studies, systems biology analyses, and investigations into both microbial and plant genomes have thrown light upon the interplaying molecular mechanisms in plant–microbiome consortia, however, it is still a question of how the plants differentiate between pathogens and PGPB, as they are closely related in activities and signaling-responses (Ke et al. 2021; Kaul et al. 2021).

3.4.1 Rhizobiome Inoculants as an Alternative Resource for Agrochemicals

Agrochemicals have been used extensively worldwide for an enormous period, however at the expense of negative impacts on the environment and human health (Dhananjayan et al. 2020). In this regard, nitrogen and phosphorus are mostly the limiting minerals for plant growth and therefore, the supplementation of agricultural fields with nitrogen and/or phosphorus-rich manures is an immediate solution. However, research shows that only 40% is used up by the plants while the rest, 60%, is wasted (Tkacz et al. 2015).

An alternative approach to the use of agrochemicals is the application of biopromoters or fertilizers composed of PGPM (Araujo et al. 2017). However, the greatest challenge, in this case, is to manage the proper proportion and assimilation of the species to be used and to evaluate their functional characteristics under field conditions because few of the microbial species may be unfavorable in such conditions. Another challenge in this respect is the presence of the iron-chelating microorganism in the mixed culture which will hijack the iron for plant growth and thereby will leave the other PGPB in deprivation of the mineral, reducing their overall colonization (Traxler et al. 2012; Bal et al. 2013). Therefore, the study of the effects of the application of biopromoters in the present devastated agro-systems and improving the yield of plants is realistically unachievable in the recent coming years. Rather, tackling plant diseases, which are the causes of a huge loss of crops annually, might be an alternative approach. In this regard, the management of plant microbiomes is helpful as it is a predictor for plant growth and health status, crop productivity, soil quality assessment, and microbial networking. However, the most challenging effort is to choose the appropriate and specific species among the host of disease-suppressing microorganisms (Araujo 2022).

The beneficial microorganisms are inoculated directly to the soil, plants, seeds, and seed coats or are sprayed as formulations upon flowers to be incorporated within the progeny seeds (Sessitsch et al. 2019). The use of rhizosphere inoculants in the form of economically and ecologically sustainable biofertilizers are alternative solutions to chemical fertilizers for prolonged use to increase soil fertility. A better nutritional and fungal-resistance level in the soil can be maintained by the use of biofertilizers with the accretion of quercetrol, coumaric acid, ferulic acid, and isoquercitrin and inducing the accumulation of PGPM like *Bacillus* sp. and *Trichoderma* sp. (Afridi et al. 2022). But the greatest challenge is the outcompete of the inoculated microbes by the indigenous microflora, resulting in the complete disappearance of the species. Therefore, the most critical task is the strain selection, which will not hinder the existing plant and/or soil microbiome and will be able to compete and interact with them. To date, *Pseudomonas* sp. with *Streptomyces* sp. for *Aspergillus* sp., *Pythium* sp., and *Fusarium oxysporum* and *Pseudomonas* sp. with *Trichoderma* sp. for *Rhizoctonia solani* has been well studied for such inoculum experiments. The application of different combined strains of *Rhizobia* sp. in tomato and maize reduced their dependency on chemical fertilizers by 20% with an additional growth increase of 12–18%. (Araujo 2022). Similarly, *Stenotrophomonas* sp. and *Bacillus* sp. inoculants boosted growth in wheat and cotton plants through enhanced plant hormones and disease-suppression signaling (Qin et al. 2021; Liu et al. 2021). However, such inoculant-induced growth has been observed in several crops, including cereals, cucumber, pepper, and cabbage (reviewed in Araujo 2022). The microbial inoculum used in amalgamation with organic amendments has been beneficial to crop production and yield to a much higher extent than inoculum alone (Bibi et al. 2022).

The solution to the challenges bourn by the applied biofertilizers such as out competition, compatibility issues, and compromised performance in the field conditions is the application of artificial microbial consortia (AMC). For developing AMC, the particular microbial consortium of interest needs to be identified and artificially grown in the laboratory, with successive identification of the dominantly active species, refinement of the interactions between the microbes and finally assessing the efficiency of the consortium (Kong et al. 2018). Co-inoculation of AM-fungus with *Pseudomonas* sp., *Bacillus* sp. with *Acinetobacter* sp., *Pseudomonas* sp. with *Stenotrophomonas* sp. or *Trichoderma* sp. are effective in treating soil-borne fungal pathogens (reviewed in Noma et al. 2021). However, the continuous application of AMC in field trials is the only way to establish its stability through generations and to achieve effective sustainability.

3.4.2 Host-Genotype-Dependent Microbiome Engineering

The modification of the genetic makeup of plants to breed for microbe-friendly host plants is a method for sustainable agricultural production and improvement of crop yields. In this case, the transgenic breeds are stimulated to secrete root exudates promoting the association of PGPB only. This will prevent the unnecessary wastage of energy and resources in selecting a host of microbes, which do not have any beneficial interactions with the host plants. Although such studies have shown differences in microbial assemblages between natural and genetically modified cultivars (tomato, beans, wheat), the maintenance of the functional microbial consortia under field conditions is challenging due to other intervening biotic and abiotic factors (Noman et al. 2021). Also, due to the large size of the plant genome, there is a dearth of reference genomes for most plants (Olanrewaju et al. 2019).

3.4.3 Organic Formulations as a Conventional Approach

The organic formulations consist of organic residues or wastes, peat, compost, and biochar are applied in agricultural fields targeting the propagation of PGPM. Organic amendments (OA) like in composts and biochar are an effective treatment in nutrient-deficient and disease-inducing soil where the microbial consortia can be targeted in a slow but sustainable way in contrast to chemical fertilizers. OA encourages the activities of PGPB like *Actinobacteria*, *Proteobacteria*, *Firmicutes*, and TM7 in the soil, enriches the edaphic factors of soil, and changes the microbial community composition by high accumulation of rutoside (Bonanomi et al. 2016). The chitin and exudates from the bark of plants containing biocidal components like terpenoids, essential oils, alcohols, and phenolic derivatives present in OA and its decomposition to VOCs with nematicides/nematotoxic/nematostatic properties (methyl-isothiocyanate, MITC) promote the overall growth of plants by increasing the nematode-antagonistic microbial activities leading to the suppression of parasitic nematodes in the soil (Castaño et al. 2011). In composting, bacterial species like *Bacillus* sp., *Pseudomonas* sp., *Streptomyces* sp., *Enterobacter* sp., *Trichoderma* sp., *Burkholderia* sp., *Pantoea* sp., and *Lysobacter* sp. also function similarly by changing the soil microbiome structure, producing plant growth-promoting hormones and anti-metabolites (tannins, siderophores, phenols) for the suppression of soil-borne pathogens and ultimately enrich the fertility and nutritional status of the soil and the crops. Hence, OA has been used to degrade the organic components to humus and proved to be successful for years in treating fungal diseases (De Corato et al. 2019).

The char co-product (biochar) is produced from the thermochemical processing of biomass and is utilized as an amendment or carbon-sequestration agent in disease-inducing soil. Biochar manages the overall soil health and fertility, availability of nutrients to the plants by inhibiting the leaching of nutrients, conservation of soil water, and eradication of detrimental compounds from the soil (Rafique et al. 2019). The microbial inoculum of *Azotobacter* sp., *Rhizobium* sp., *Pseudomonas* sp., and *Azospirilium* sp. can be added to biochar to make it a more potent driver for increasing microbial diversity and function in the soil like that of denitrifying rhizobials. The porous nature combined with the large surface area make the biochar an appropriate microbe-inhabiting and proliferating site, ultimately profiting the plants for their healthy growth (Awasthi et al. 2020). Biofumigation is another strategy for soil conditioning by using components from plants and natural compounds that over the phases of degradation produce glucosinolates that subsequently catabolize into nematotoxic isothiocyanates and thus inhibit the growth of pathogenic fungi (Bonanomi et al. 2018).

3.5 SUMMARY

It is very well accepted that the harnessing of the plant microbiome is the solution for an economically feasible and ecologically viable approach for sustainable agriculture to ensure food security worldwide. However, this approach is not free from challenges imposed by the complexity and diversity of the plant microbiome and its interactions. Although the recent advent in technologies has unveiled various aspects of plant microbiome, basic questions like how the plants would differentiate between disease-suppressing and disease-causing agents and how the barriers hindering the effectiveness of the engineered rhizobiome formulations in the fields can be overcome to translate the laboratory findings to reality needs to be answered.

Although this chapter focuses on microbial interactions in the rhizosphere, the monitoring of micro-macro fauna cannot be overlooked during rhizobiome engineering. Among many other environmental factors, the changing and unpredictable climatic scenarios due to global warming impose the greatest threat to agriculture at present. One of the cost-effective ways to combat this state is to increase plantations that will remove CO_2 and will aid in soil organic carbon restoration. Later, the forests could be harvested for biochar at their climax stage.

To address these problems, it is necessary to decode the molecular mechanisms of signaling dynamicity and the evolutionary pressures involved in plant–microbe and microbe–microbe interactions. An integrated systemic study combining the fields of experimental biology, bioinformatics, chemistry, and genomics needs to be used to decipher the pattern in the complex root-exudate profiles under various environmental circumstances, along the growth phases and nutritional states of the plants. These in combination would help to formulate techniques to assess the correlation between microbial functionality and plant fitness, determine the stability of the communities, and eventually the identified issues can be targeted during rhizobiome engineering.

3.6 ACKNOWLEDGEMENTS

The authors would like to thank the Department of Biochemistry, Asutosh College and Bose Institute, Centenary campus, Kolkata for financial and infrastructure facilities.

REFERENCES

Abdelhameed RE, Metwally RA. (2019). Alleviation of cadmium stress by arbuscular mycorrhizal symbiosis. *International Journal of Phytoremediation,* 21, 663–671.

Adnan M, Shah Z, Fahad S, Arif M, Alam M, Khan IA, Mian IA, Basir A, Ullah H, Arshad M, et al. (2017). Phosphate-solubilizing bacteria nullify the antagonistic effect of soil calcification on bioavailability of phosphorus in alkaline soils. *Scientific Reports*, 7, 16131.

Alemneh AA, Zhou Y, Ryder MH, Denton MD. (2020). Mechanisms in plant growth-promoting rhizobacteria that enhance legume-rhizobial symbioses. *Journal of Applied Microbiology,* 129, 1133–1156.

Altowayti W, Algaifi HA, Bakar SA, Shahir S. (2019). The adsorptive removal of As (III) using biomass of arsenic resistant Bacillus thuringiensis strain WS3, Characteristics and modelling studies. *Ecotoxicology and Environmental Safety,* 172, 176–185.

Afridi MS, Ali S, Salam A, César Terra W, Hafeez A, Sumaira, Ali B, S. AlTami M, Ameen F, Ercisli S, Marc RA, Medeiros FHV, Karunakaran R. (2022). Plant microbiome engineering, hopes or hypes. *Biology,* 11(12), 1782. https//doi.org/10.3390/biology11121782.

Araujo R. (2022). Advances in Soil Engineering, Sustainable Strategies for Rhizosphere and Bulk Soil Microbiome Enrichment. *Frontiers in Bioscience (Landmark Ed)*, 27(6), 195. Doi: 10.31083/j.fbl2706195.

Araujo R, Dunlap C, Franco C. (2020). Analogous wheat root rhizosphere microbial successions in field and greenhouse trials in presence of biocontrol agents *Paenibacillus peoriae* SP9 and *Streptomyces fulvissimus* FU14. *Molecular Plant Pathology,* 2020 , 1–14.

Araujo R, Kaewkla O, Franco CMM. (2017). Endophytic Actinobacteria, beneficial partners for sustainable agriculture. In Maheshwari DK (ed.) *Endophyte Vol I, Biology and Biotechnology*, (pp. 171–191). Springer International Publishing AG, Cham, Switzerland.

Awasthi M, Duan Y, Liu T, Awasthi S, Zhang Z. (2020). Relevance of biochar to influence the bacterial succession during pig manure composting. *Bioresource Technology,* 304, 122962.

Babikova Z, Johnson D, Bruce T, Pickett J, Gilbert L. (2014). Underground allies, How and why do mycelial networks help plants defend themselves? *BioEssays,* 36, 21–26.

Badri DV, Chaparro JM, Zhang R, Shen Q, Vivanco JM. (2013). Application of natural blends of phytochemicals derived from the root exudates of Arabidopsis to the soil reveal that phenolic-related compounds predominantly modulate the soil microbiome. *Journal of Biological Chemistry,* 288, 4502–12.

Badri DV, Quintana N, El Kassis EG, Kim HK, Choi YH, Sugiyama A, Verpoorte R, Martinoia E, Manter DK, Vivanco JM. (2009). An ABC transporter mutation alters root exudation of phytochemicals that provoke an overhaul of natural soil microbiota. *Plant Physiology,* 151, 2006–2017.

Bakker MG, Manter DK, Sheflin AM, Weir TL, Vivanco JM. (2012). Harnessing the rhizosphere microbiome through plant breeding and agricultural management. *Plant Soil,* 360(1–2), 1–13.

Bal HB, Das S, Dangar TK, Adhya TK. (2013). ACC deaminase and IAA producing growth promoting bacteria from the rhizosphere soil of tropical rice plants. *Journal of Basic Microbiology,* 53, 972–984.

Beneduzi A, Ambrosini A, Passaglia LM. (2012). Plant growth-promoting rhizobacteria (PGPR), their potential as antagonists and biocontrol agents. *Genetics and Molecular Biology*, 35, 1044–1051.

Bergelson J, Mittelstrass J, Horton MW. (2019). Characterizing both bacteria and fungi improves understanding of the Arabidopsis root microbiome. *Scientific Reports*, 9, 24.

Bever JD, Platt TG, Morton ER. (2012). Microbial population and community dynamics on plant roots and their feedbacks on plant communities. *Annual Review of Microbiology,* 66, 265–283.

Bibi F, Ilyas N, Arshad M, Khalid A, Saeed M, Ansar S. (2022). y testing of bio-organic fertilizer produced through solid-state fermentation of agro-waste by *Burkholderia cenocepacia. Chemosphere,* 291, 132762.

Bonanomi G, De Filippis F, Cesarano G, La Storia A, Ercolini D, Scala F. (2016). Organic farming induces changes in soil microbiota that affect agroecosystem functions. *Soil Biology and Biochemistry,* 103, 327–336.

Bonanomi G, Lorito M, Vinale, F, Woo SL. (2018). Organic amendments, beneficial microbes, and soil microbiota, toward a unified framework for disease suppression. *Annual Review of Phytopathology*, 56, 1–20.

Brunetti C, Saleem AR, Della RG, Emiliani G, De Carlo A, Balestrini R, Khalid A, Mahmood T, Centritto M. (2021). Effects of plant growth-promoting rhizobacteria strains producing ACC deaminase on photosynthesis, isoprene emission, ethylene formation and growth of Mucuna pruriens (L.) DC. in response to water deficit. *Journal of Biotechnology,* 331 , 53–62.

Buttner H, Niehs SP, Vandelannoote K, Cseresnyes Z, Dose B, Richter I, Gerst R, Figge, MT, Stinear TP, Pidot SJ, et al. (2021). Bacterial endosymbionts protect beneficial soil fungus from nematode attack. *Proceedings of Natural Academics of Science, USA,* 118, e2110669118.

Cao Y, Halane MK, Gassmann W, Stacey G. (2017). The role of plant innate immunity in the legume-rhizobium symbiosis. *Annual Review of Plant Biology,* 68, 535–561.

Carlström CI, Field CM, Bortfeld-Miller M, Müller B, Sunagawa S, Vorholt JA. (2019). Synthetic microbiota reveal priority effects and keystone strains in the Arabidopsis phyllosphere. *Nature Ecology and Evolution,* 3, 1445–1454.

Castaño R, Borrero C, Avilés M. (2011). Organic Matter Fractions by SP-MAS 13C NMR and Microbial Communities Involved in the Suppression of Fusarium Wilt in Organic Growth Media. *Biological Control*, 58 , 286–293.

Chamkhi I, Benali T, Aanniz T, El Menyiy N, Guaouguaou F, El Omari N, El-Shazly M, Zengin G, Bouyahya A. (2021). Plant-microbial interaction, The mechanism and the application of microbial elicitor induced secondary metabolites biosynthesis in medicinal plants. *Plant Physiology and Biochemistry,* 167, 269–295.

Chen B, Xiao X, Zhu YG, Smith FA, Xie ZM, Smith SE. (2007). The arbuscular mycorrhizal fungus *Glomus mosseae* gives contradictory effects on phosphorus and arsenic acquisition by Medicago sativa Linn. *Science of the Total Environment*, 379, 226–234.

Chen C, Wang M, Zhu J, Tang Y, Zhang H, Zhao Q, Jing M, Chen Y, Xu X, Jiang J, et al. (2022). Long-term effect of epigenetic modification in plant-microbe interactions: modification of DNA methylation induced by plant growth-promoting bacteria mediates promotion process. *Microbiome,* 10, 36.

Compant S, Samad A, Faist H, Sessitsch A. (2019). A review on the plant microbiome, ecology, functions, and emerging trends in microbial application. *Journal of Advanced Research*, 19, 29–37.

Coyte KZ, Schluter J, Foster KR. (2015). The ecology of the microbiome, networks, competition, and stability. *Science,* 350(6261), 663–6. https//doi.org/1 0.1126/science.aad2602.

Cregger M, Veach A, Yang Z, Crouch M, Vilgalys R, Tuskan G, Schadt C. (2018). The Populus holobiont, Dissecting the effects of plant niches and genotype on the microbiome. *Microbiome,* 6, 31.

Cunliffe M, Kertesz MA. (2006). Effect of *Sphingobium yanoikuyae* B1 inoculation on bacterial community dynamics and polycyclic aromatic hydrocarbon degradation in aged and freshly PAH-contaminated soils. *Environmental Pollution,* 144, 228–237.

De Corato U, Patruno L, Avella N, Lacolla G, Cucci G. (2019). Composts from green sources show an increased suppressiveness to soilborne plant pathogenic fungi, relationships between physicochemical properties, disease suppression, and the microbiome. *Crop Protection,* 124, 104870.

de Vries FT, Griffiths RI, Bailey M, Craig H, Girlanda M, Gweon HS, et al. (2018). Soil bacterial networks are less stable under drought than fungal networks. *Nature Communications*, 9(1), 3033. https//doi.org/ 10.1038/s41467-018-05516-7.

Defez R, Andreozzi A, Bianco C. (2017). The overproduction of indole-3-acetic acid (IAA) in endophytes upregulates nitrogen fixation in both bacterial cultures and inoculated rice plants. *Microbial Ecology,* 74, 441–452.

Dhananjayan V, Jayanthi P, Jayakumar S, Ravichandran B. (2020). *Agrochemicals impact on ecosystem and bio-monitoring*. Springer, Singapore. 349–388.

Duran P, Thiergart T, Garrido-Oter R, Agler M, Kemen E, Schulze-Lefert P, et al. (2018). Microbial interkingdom interactions in roots promote Arabidopsis survival. *Cell*, 175(4), 973–83 e14. https//doi.org/10.1016/ j.cell.2018.1 0.020.

Edwards JA, Santos-Medellín CM, Liechty ZS, Nguyen B, Lurie E, Eason S, Phillips G, Sundaresan V. (2018). Compositional shifts in root-associated bacterial and archaeal microbiota track the plant life cycle in field-grown rice. *PLoS Biology,* 16, e2003862.

Estrada GA, Baldani VLD, De Oliveira DM, Urquiaga S, Baldani JI. (2013). Selection of phosphate-solubilizing diazotrophic *Herbaspirillum* and *Burkholderia* strains and their effect on rice crop yield and nutrient uptake. *Plant Soil,* 369,115–129.

Fakhar A, Gul B, Gurmani AR, Khan SM, Ali S, Sultan T, Chaudhary HJ. (2022). Heavy metal remediation and resistance mechanism of *Aeromonas*, *Bacillus*, and *Pseudomonas*, a review. *Critical Review of Environmental Science and Technology,* 52 , 1868–1914.

Fira D, Dimki´c I, Beri´c T, Lozo J, Stankovi´c S. (2018). Biological control of plant pathogens by *Bacillus* species. *Journal of Biotechnology,* 285, 44–55.

Freschet GT, Roumet C, Comas LH, Weemstra M, Bengough AG, Rewald B, Bardgett RD, De Deyn GB, Johnson D, Klimešová J, Lukac M, McCormack ML, et al. (2021). Root traits as drivers of plant and ecosystem functioning, Current understanding, pitfalls and future research needs. *New Phytologist*, 232(3), 1123–1158. https//doi.org/10.1111/nph.17072.

Gans J, Wolinsky M, Dunbar J. (2005). Computational improvements reveal great bacterial diversity and high metal toxicity in soil. *Science,* 309, 1387–1390.

Gao M, Xiong C, Gao C, Tsui CKM, Wang MM, Zhou X, Zhang AM, Cai L. (2021). Disease-induced changes in plant microbiome assembly and functional adaptation. *Microbiome*, 9(1), 187. https//doi.org/10.1186/ s40168-021-01138-2.

Genre A, Lanfranco L, Perotto S, Bonfante P. (2020). Unique and common traits in mycorrhizal symbioses. *Nature Reviews of Microbiology,* 18, 649–660.

Glick BR, Nascimento FX. (2021). Pseudomonas 1-Aminocyclopropane-1-carboxylate (ACC) deaminase and its role in beneficial plant-microbe interactions. *Microorganisms,* 9, 2467.

Gomez Expósito R, de Bruijn I, Postma J, Raaijmakers JM. (2017). Current insights into the role of rhizosphere bacteria in disease suppressive soils. *Frontiers of Microbiology*, 8, 2529.

Grilli J, Rogers T, Allesina S. (2016). Modularity and stability in ecological communities. *Nature Communications*, 7(1), 12031. https//doi.org/10.1038/ ncomms12031.

Gu D, Xiang X, Wu Y, Zeng J, Lin X. (2022). Synergy between fungi and bacteria promotes polycyclic aromatic hydrocarbon cometabolism in lignin-amended soil. *Journal of Hazardous Materials,* 425, 127958.

Gu Y, Banerjee S, Dini-Andreote F, Xu Y, Shen Q, Jousset A, Wei Z. (2022). Small changes in rhizosphere microbiome composition predict disease outcomes earlier than pathogen density variations. *International Society for Microbial Ecology Journal*, 16(10), 2448–2456. https//doi.org/10.1038/s41396-022-01290-z.

Guo S, Feng B, Xiao C, Wang Q, Chi R. (2021). Phosphate-solubilizing microorganisms to enhance phytoremediation of excess phosphorus pollution in phosphate mining wasteland soil. *Bioremediation Journal,* 25, 271–281.

Hamonts K, Trivedi P, Garg A, Janitz C, Grinyer J, Holford P, Botha FC, Anderson IC, Singh BK. (2018). Field study reveals core plant microbiota and relative importance of their drivers. *Environmental Microbiology,* 20, 124–140.

Hassan S, Mathesius U. (2012). The role of flavonoids in root–rhizosphere signalling, Opportunities and challenges for improving plant–microbe interactions. *Journal of Experimental Botany,* 63 , 3429–3444.

Hernandez DJ, David AS, Menges ES, Searcy CA, Afkhami ME. (2021). Environmental stress destabilizes microbial networks. *International Society for Microbial Ecology Journal*, 15(6), 1722–34. https//doi.org/10.1038/s41396-020-00882-x.

Hida A, Oku S, Miura M, Matsuda H, Tajima T, Kato J. (2020). Characterization of methyl-accepting chemotaxis proteins (MCPs) for amino acids in plant-growth-promoting rhizobacterium Pseudomonas protegens CHA0 and enhancement of amino acid chemotaxis by MCP genes overexpression. *Bioscience Biotechnology Biochemistry,* 84, 1948–1957.

Ho TT, Lee JD, Jeong CS, Paek KY, Park SY. (2018). Improvement of biosynthesis and accumulation of bioactive compounds by elicitation in adventitious root cultures of Polygonum multiflorum. *Applied Microbiology Biotechnology*, 102, 199–209.

Hünninghaus M, Koller R, Kramer S, Marhan S, Kandeler E, Bonkowski M. (2017). Changes in Bacterial Community Composition and Soil Respiration Indicate Rapid Successions of Protist Grazers during Mineralization of Maize Crop Residues. *Pedobiologia,* 62, 1–8.

Hussain A, Kamran MA, Javed MT, Hayat K, Farooq MA, Ali N, Ali M, Manghwar H, Jan F, et al. (2019). Individual and combinatorial application of *Kocuria rhizophila* and citric acid on phytoextraction of multi-metal contaminated soils by *Glycine max* L. *Environment and Experimental Botany,* 159, 23–33.

Iqbal N, Khan NA, Ferrante A, Trivellini A, Francini A, Khan M. (2017). Ethylene role in plant growth, development and senescence, Interaction with other phytohormones. *Frontiers in Plant Science*, 8, 475.

Igiehon NO, Babalola OO. (2018). Rhizosphere microbiome modulators, contributions of nitrogen fixing bacteria towards sustainable agriculture. *International Journal of Environmental Research and Public Health*, 15(4), 574. https//doi.org/10.3390/ijerph15040574.

Isobe K, Ohte N. (2014). Ecological perspectives on microbes involved in N-cycling. *Microbes and Environments*, 29(1), 4–16.

Jin Y, Luan Y, Ning Y, Wang L. (2018). Effects and mechanisms of microbial remediation of heavy metals in soil, A critical review. *Applied Science*, 8, 1336.

Kallenbach CM, Frey SD, Grandy AS. (2016). Direct evidence for microbial-derived soil organic matter formation and its ecophysiological controls. *Nature Communications,* 7, 13630.

Kaul S, Choudhary M, Gupta S, Dhar MK. (2021). Engineering host microbiome for crop improvement and sustainable agriculture. *Frontiers in Microbiology*, 12, 635917.

Ke J, Wang B, Yoshikuni Y. (2021). Microbiome engineering, synthetic biology of plant-associated microbiomes in sustainable agriculture. *Trends in Biotechnology,* 39, 244–261.

Khalid M, Ur-Rahman S, Hassani D, Hayat K, Pei Z, Nan H. (2021). Advances in fungal-assisted phytoremediation of heavy metals; a review. *Pedosphere,* 31, 475–495.

Khan I, Aftab M, Shakir S, Ali M, Qayyum S, Rehman MU, Haleem KS, Touseef I. (2019). Mycoremediation of heavy metal (Cd and Cr)-polluted soil through indigenous metallotolerant fungal isolates. *Environmental Monitoring and Assessment,* 191, 585.

Kim DR, Jeon CW, Cho G, Thomashow LS, Weller DM, Paik MJ, Lee YB, Kwak YS. (2021). Glutamic acid reshapes the plant microbiota to protect plants against pathogens. *Microbiome,* 9, 244.

Klein M, Stewart JD, Porter SS, Weedon JT, Kiers ET. (2022). Evolution of manipulative microbial behaviors in the rhizosphere. *Evolutionary Applications*, 15(10), 1521–1536. https//doi.org/10.1111/eva.13333.

Kong Z, Hart M, Liu H. (2018). Paving the way from the lab to the field, Using synthetic microbial consortia to produce high-quality crops. *Frontiers in Plant Science*, 9, 1467.

Lakshmanan V, Selvaraj G, Bais HP. (2014). Functional soil microbiome, belowground solutions to an aboveground problem. *Journal of Plant Physiology*, 166(2), 689–700

Lareen A, Burton F, Schäfer P. (2016). Plant root-microbe communication in shaping root microbiomes. *Plant Molecular Biology*, 90, 575–587.

Li E, de Jonge R, Liu C, Jiang H, Friman VP, Pieterse CMJ, Bakker PAHM, Jousset A. (2021). Rapid evolution of bacterial mutualism in the plant rhizosphere. *Nature Communications,* 12(1), 3829. https//doi.org/10.1038/s41467-021-24005-y.

Liang C, Amelung W, Lehmann J, Kastner M. (2019). Quantitative assessment of microbial necromass contribution to soil organic matter. *Global Changes in Biology,* 25, 3578–3590.

Lindstrom K, Mousavi SA. (2020). Effectiveness of nitrogen fixation in rhizobia. *Microbial Biotechnology,* 13, 1314–1335.

Liu H, Khan MY, Carvalhais LC, Delgado-Baquerizo M, Yan L, Crawford M, Dennis, PG, Singh B, Schenk PM. (2019). Soil amendments with ethylene precursor alleviate negative impacts of salinity on soil microbial properties and productivity. *Scientific Reports*, 9, 6892.

Liu H, Li J, Carvalhais LC, Percy CD, Prakash Verma J, Schenk PM, et al. (2021). Evidence for the plant recruitment of beneficial microbes to suppress soil-borne pathogens. *New Phytologist,* 229, 2873–2885.

Liu X, Etienne RS, Liang M, Wang Y, Yu S. (2015). Experimental evidence for an intraspecific Janzen-Connell effect mediated by soil biota. *Ecology*, 96(3), 662–671.

Majeed A, Muhammad Z, Ahmad H. (2018). Plant growth promoting bacteria, Role in soil improvement, abiotic and biotic stress management of crops. *Plant Cell Reproduction,* 37 , 1599–1609.

Mansfield J, Genin S, Magori S, Citovsky V, Sriariyanum M, Ronald P, Dow M, Verdier V, Beer SV, Machado MA. (2012). Top 10 plant pathogenic bacteria in molecular plant pathology. *Molecular Plant Pathology*, 13(6), 614–629.

Masakorala K, Yao J, Cai M, Chandankere R, Yuan H, Chen H. (2013). Isolation and characterization of a novel phenanthrene (PHE) degrading strain *Psuedomonas* sp. USTB-RU from petroleum contaminated soil. *Journal of Hazardous Materials,* 263 , 2493–2500.

Matsumoto H, Fan X, Wang Y, Kusstatscher P, Duan J, Wu S, et al. (2021). Bacterial seed endophyte shapes disease resistance in rice. *Nature Plants,* 7, 60–72.

Mendes R, Garbeva P, Raaijmakers J. (2013). The rhizosphere microbiome, significance of plant beneficial, plant pathogenic, and human pathogenic microorganisms. *FEMS Microbiology Reviews,* 37, 634–663.

Mishra BK, Meena KK, Dubey PN, Aishwath OP, Kant K, Sorty AM, Bitla U. (2016). Influence on yield and quality of fennel (*Foeniculum vulgare Mill.*) grown under semi-arid saline soil, due to application of native phosphate solubilizing rhizobacterial isolates. *Ecological Engineering,* 97 , 327–333.

Mohanram S, Kumar P. (2019). Rhizosphere microbiome, re-visiting the synergy of plant-microbe interactions. *Annals of Microbiology*, 69(4), 307–320.

Noman M, Ahmed T, Ijaz U, Shahid M , Azizullah, Li D, Manzoor I , Song F. (2021). Plant-Microbiome Crosstalk, Dawning from Composition and Assembly of Microbial Community to Improvement of Disease Resilience in Plants. *International Journal of Molecular Science,* 22(13), 6852. https//doi.org/10.3390/ijms22136852.

Orhan F. (2021). Potential of halophilic/halotolerant bacteria in enhancing plant growth under salt stress. *Current Microbiology,* 78, 3708–3719.

Olanrewaju OS, Ayangbenro AS, Glick BR, Babalola OO. (2019). Plant health, feedback effect of root exudates-rhizobiome interactions. *Applied Microbiology Biotechnology*, 103(3), 1155–1166. https//doi.org/10.1007/s00253-018-9556-6.

Pantigoso HA, Newberger D, Vivanco JM. (2022) The rhizosphere microbiome, Plant-microbial interactions for resource acquisition. *Journal of Applied Microbiology*, 133(5), 2864–2876. https//doi.org/10.1111/jam.15686.

Patel S, Thaker V, Kuvad R, Saraf M. (2019). Role of lipopolysaccharide extracted from *Alcaligenes faecalis* as elicitor for the induction of plant defense against fusarium wilt. *Journal of Plant Pathology*, 102, 351–357.

Pérez-Jaramillo JE, Mendes R, Raaijmakers JM. (2016). Impact of plant domestication on rhizosphere microbiome assembly and functions. *Plant Molecular Biology*, 90(6), 635–44. https//doi.org/10.1007/s11103-015-0337-7.

Pérez-Montaño F, Guasch-Vidal B, González-Barroso S, López-Baena FJ, Cubo T, Ollero FJ, Gil-Serrano AM, Rodríguez-Carvajal MÁ, Bellogín RA, Espuny MR. (2011). Nodulation-gene-inducing flavonoids increase overall production of autoinducers and expression of N-acyl homoserine lactone synthesis genes in rhizobia. *Research in Microbiology*, 162 , 715–723.

Pii Y, Borruso L, Brusetti L, Crecchio C, Cesco S, Mimmo T. (2016). The interaction between iron nutrition, plant species and soil type shapes the rhizosphere microbiome. *Plant Physiology and Biochemistry*, 99, 39–48.

Ping L, Zhang C, Zhang C, Zhu Y, He H, Wu M, Tang T, Li Z, Zhao H. (2014). Isolation and characterization of pyrene and benzo[a]pyrene-degrading *Klebsiella pneumonia* PL1 and its potential use in bioremediation. *Applied Microbiology Biotechnology,* 98, 3819–3828.

Qin L, Tian P, Cui Q, Hu S, Jian W, Xie C, et al. (2021). *Bacillus circulans* GN03 alters the microbiota, promotes cotton seedling growth and disease resistance, and increases the expression of phytohormone synthesis and disease resistance-related genes. *Frontiers in Plant Science,* 12, 644597.

Rafique M, Ortas I, Ahmed IA, Rizwan M, Afridi MS, Sultan T, Chaudhary HJ. (2019). Potential Impact of Biochar Types and Microbial Inoculants on Growth of Onion Plant in Differently Textured and Phosphorus Limited Soils. *Journal of Environmental Management,* 247, 672–680.

Rashid MI, Mujawar LH, Shahzad T, Almeelbi T, Ismail IM, Oves M. (2016). Bacteria and fungi can contribute to nutrients bioavailability and aggregate formation in degraded soils. *Microbiology Research,* 183, 26–41.

Rasmann S, Ali JG, Helder J, van der Putten WH. (2012). Ecology and evolution of soil nematode chemotaxis. *Journal of Chemical Ecology*, 38(6), 615–628.

Ravanbakhsh M, Sasidharan R, Voesenek LA, Kowalchuk GA, Jousset A. (2017). ACC deaminase-producing rhizosphere bacteria modulate plant responses to flooding. *Journal of Ecology,* 105, 979–986.

Ritpitakphong U, Falquet L, Vimoltust A, Berger A, Métraux JP, L'Haridon F (2016).The microbiome of the leaf surface of Arabidopsis protects against a fungal pathogen. *New Phytology*, 210, 1033–1043.

Roman-Reyna V, Pinili D, Borjaa FN, Quibod I, Groen SC, Mulyaningsih ES, Rachmat, A, Slamet-Loedin IH, Alexandrov N, Mauleon R. (2019). The rice leaf microbiome has a conserved community structure controlled by complex host-microbe interactions. Cell Press. http://dx.doi.org/10.2139/ssrn.3382544.

Sadeghi H, Ghanaatiyan K. (2017). Probing the responses of four chicory ecotypes by ethylene accumulation and growth characteristics under drought stress. *Italian Journal of Agronomy,* 12, 773.

Saeed Q, Wang X, Haider FU, Kucerik J, Mumtaz MZ, Holatko J, Naseem M, Kintl A, Ejaz M, Naveed M, et al. (2021). Rhizosphere bacteria in plant growth promotion, biocontrol, and bioremediation of contaminated sites, A comprehensive review of effects and mechanisms. *International Journal of Molecular Science*, 22, 10529.

Sahu KP, Kumar A, Patel A, Kumar M, Gopalakrishnan, S, Prakash G, Rathour R, Gogoi R. (2020). Rice blast lesions, an Unexplored phyllosphere microhabitat for novel antagonistic bacterial species against *Magnaporthe oryzae*. *Microbial Ecology*, 81, 731–745.

Sarikhani MR, Oustan S, Ebrahimi M, Aliasgharzad N. (2018). Isolation and identification of potassium-releasing bacteria in soil and assessment of their ability to release potassium for plants. *European Journal of Soil Science,* 69, 1078–1086.

Sessitsch A, Pfaffenbichler N, Mitter B. (2019). Microbiome applications from lab to field, facing complexity. *Trends in Plant Science,* 24, 194–198.
Shahrajabian MH, Sun W, Cheng Q. (2021). The importance of *Rhizobium*, Agrobacterium, Bradyrhizobium, Herbaspirillum, Sinorhizobium in sustainable agricultural production. *Notulae Botanicae Horti Agrobotanici Cluj-Napoca*, 49, 12183.
Simon JC, Marchesi JR, Mougel C, Selosse MA. (2019). Host-microbiota interactions, from holobiont theory to analysis. *Microbiome,* 7(1), 5. https//doi.org/10.1186/s40168-019-0619-4.
Smułek W, Sydow M, Zabielska-Matejuk J, Kaczorek E. (2020). Bacteria involved in biodegradation of creosote PAH-A case study of long-term contaminated industrial area. *Ecotoxicology and Environmental Safety*, 187, 109843.
Su P, Tan X, Li C, Zhang D, Cheng J, Zhang S, Zhou X, Yan Q, Peng J, Zhang Z, et al. (2017). Photosynthetic bacterium *Rhodopseudomonas palustris* GJ-22 induces systemic resistance against viruses. *Microbial Biotechnology,* 10 , 612–624.
Sun F, Ou Q, Wang N, Guo ZX, Ou Y, Li N, Peng C. (2020). Isolation and identification of potassium-solubilizing bacteria from *Mikania micrantha* rhizospheric soil and their effect on *M. micrantha* plants. *Global Ecology and Conservation,* 23, e01141.
Tanaka S, Han X, Kahmann R. (2015). Microbial effectors target multiple steps in the salicylic acid production and signaling pathway. *Frontiers in Plant Science,* 6, 349.
Tewari S, Pooniya V, Sharma S. (2020). Next generation bioformulation prepared by amalgamating *Bradyrhizobium*, cell free culture supernatant, and exopolysaccharides enhances the indigenous rhizospheric rhizobial population, nodulation, and productivity of pigeon pea. *Applied Soil Ecology*, 147, 103363.
Tkacz A, Poole P. (2015). Role of root microbiota in plant productivity. *Journal of Experimental Botany,* 66, 2167–2175.
Torres-Cortés G, Bonneau S, Bouchez O, Genthon C, Briand M, Jacques MA, Barret M. (2018). Functional microbial features driving community assembly during seed germination and emergence. *Frontiers in Plant Science*, 9, 902.
Torsvik V, Øvreås L. (2002). Microbial diversity and function in soil, from genes to ecosystems. *Current Opinion in Microbiology,* 5, 240–245.
Traxler MF, Seyedsayamdost MR, Clardy J, Kolter R. (2012). Interspecies modulation of bacterial development through iron competition and siderophore piracy. *Molecular Microbiology,* 86, 628–644.
Trivedi P, Delgado-Baquerizo M, Jeffries TC, Trivedi C, Anderson IC, Lai K, McNee M, Flower K, Pal Singh B, Minkey D, Singh BK.(2017.) Soil aggregation and associated microbial communities modify the impact of agricultural management on carbon content. *Environmental Microbiology*, 19, 3070–3086.
Trivedi P, Leach JE, Tringe SG, Sa T, Singh BK. (2020). Plant-microbiome interactions, from community assembly to plant health. *Nature Reviews of Microbiology*, 18(11), 607–21. https//doi.org/10.1038/s41579-020-0412-1.
Vandenkoornhuyse P, Quaiser A, Duhamel M, Le Van A, Dufresne A. (2015). The importance of the microbiome of the plant holobiont. *New Phytology*, 206(4), 1196–206. https//doi.org/10.1111/nph.13312.
van der Heijden MG, de Bruin S, Luckerhoff L, van Logtestijn RS, Schlaeppi K. (2016). A widespread plant-fungal-bacterial symbiosis promotes plant biodiversity, plant nutrition and seedling recruitment. *International Society for Microbial Ecology Journal,* 10(2), 389–99. https//doi.org/10.1038/ISMEj.2015.120.
Vejan P, Abdullah R, Khadiran T, Ismail S, Nasrulhaq BA. (2016). Role of plant growth promoting rhizobacteria in agricultural sustainability—A review. *Molecules,* 21, 573.
Wang G, Ren Y, Bai X, Su Y, Han J. (2022). Contributions of beneficial microorganisms in soil remediation and quality improvement of medicinal plants. *Plants (Basel)*, 11(23), 3200. https//doi.org/10.3390/plants11233200.
Wang JW, Zhang Z, Tan RX. (2001). Stimulation of artemisinin production in Artemisia annua hairy roots by the elicitor from the endophytic *Colletotrichum* sp. *Biotechnology Letters,* 23, 857–860.
Wei Z, Gu Y, Friman VP, Kowalchuk GA, Xu Y, Shen Q, Jousset A. (2019). Initial soil microbiome composition and functioning predetermine future plant health. *Science Advances*, 5, eaaw0759.
Xu J, Zhang Y, Zhang P, Trivedi P, Riera N, Wang Y, Liu X, Fan G, Tang J, Coletta-Filho HD. (2018). The structure and function of the global citrus rhizosphere microbiome. *Nature Communications,* 9, 4894.

Yaghoubi Khanghahi M, Pirdashti H, Rahimian H, Nematzadeh G, Sepanlou MG. (2017). Potassium solubilising bacteria (KSB) isolated from rice paddy soil, From isolation, identification to K use efficiency. *Symbiosis,* 76, 13–23.

Zheng W, Morris EK, Lehmann A, Rillig MC. (2016). Interplay of soil water repellency, soil aggregation and organic carbon, A meta-analysis. *Geoderma,* 283, 39–47.

Zhong Y, Xun W, Wang X, Tian S, Zhang Y, Li D, Zhou Y, Qin Y, Zhang B, Zhao G, et al. (2022). Root-secreted bitter triterpene modulates the rhizosphere microbiota to improve plant fitness. *Nature Plants,* 8, 887–896.

4 Mycorrhiza
A Paradigm and Panacea for Sustainable Agriculture

Saran Kumar Gupta, Suchisree Jha and Abhinobo Chakraborty

4.1 INTRODUCTION

A ubiquitous terrestrial mutualism between plant roots and specific soil fungi is known as mycorrhizae. The term "mycorrhiza" was coined by Frank in 1885, which comes from the Greek word "mykes," which stands for fungus, and "rhiza," which means root. Within a symbiotic relationship, the fungi colonize the host plant roots, establishing a mutually beneficial association. Within this association, the fungus provides increased water uptake along with nutrients, such as phosphorus and nitrogen, while the plant provides photosynthetically derived carbohydrates to the fungus for its growth. Besides the nutrients, mycorrhizae also enhanced the tolerance level of the plant against pathogens. This exchange of nutrients allows both the plant and the fungus to succeed (Dhiman et al., 2022).

There are several types of mycorrhizae (Figure 4.1), which differ in their structure and the types of plants they associate with. Some common types of mycorrhizae are arbuscular mycorrhiza (AM). AM is the most prevalent and widespread mycorrhiza and is symbiotically associated with more than 80% of plant families. The AM fungi produces highly branched structures, known as arbuscules (van der Heijden et al., 2015). Despite AM, there is another type of mycorrhiza known as ectomycorrhiza (ECM). The ectomycorrhiza does not penetrate the root cells of the host plant but it develops a dense hyphal network (fungal filaments) around the host root tip, which helps to absorb nutrients (Figure 4.2), showing the penetration or infection pattern of the mycorrhiza. This type of mycorrhiza involves fungi from several different phyla, including *Basidiomycota* and *Ascomycota*, and are observed in most of the forest trees, namely oaks, pines, and birches.

The ericoid mycorrhiza is mainly found in the plants of *Ericaceae* family, such as blueberries and heather. In ericoid mycorrhiza, the fungus forms a sheath around the root like ectomycorrhiza, the only difference is that it does not penetrate the root. This type of mycorrhiza involves fungi from the *Ascomycota* and *Basidiomycota* phyla. Beside this type of mycorrhiza, orchid mycorrhiza and monotropoid mycorrhiza are another unique type found in orchids and the *Monotropaceae* family (such as Indian pipe), respectively. Orchid mycorrhizal hyphae penetrates the seed of the orchid and helps the seed to germinate and grow. In monotropoid mycorrhiza, the fungus associates with the roots of monotropa plants which is a non-chlorophyllous plants growing under forest trees like *Quercus*, *Fagus*, *Pinus*, and *Salix* as epi-parasites depending on the fungal associate.

Mycorrhizal associations with plant kingdom are highly prevalent. According to estimates, 74% of all plant species are associated with AMs. Approximately, 1% of plants develop have ericoid mycorrhizal association, 9% of plants are associated with orchid mycorrhizas, and 2% of plants form ECM associations. Although arbuscular mycorrhizal fungi (AMF) and ectomycorrhizal fungi (EMF) can be beneficial for plants, their benefits depend on the plant species and the environmental conditions. Moreover, each type of mycorrhiza has its own importance in the environment.

DOI: 10.1201/9781003354109-4

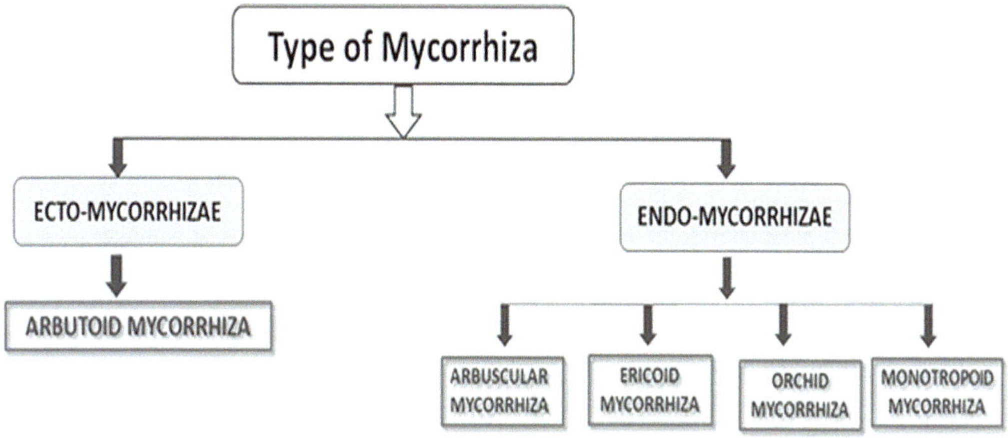

FIGURE 4.1 Different types of mycorrhiza.

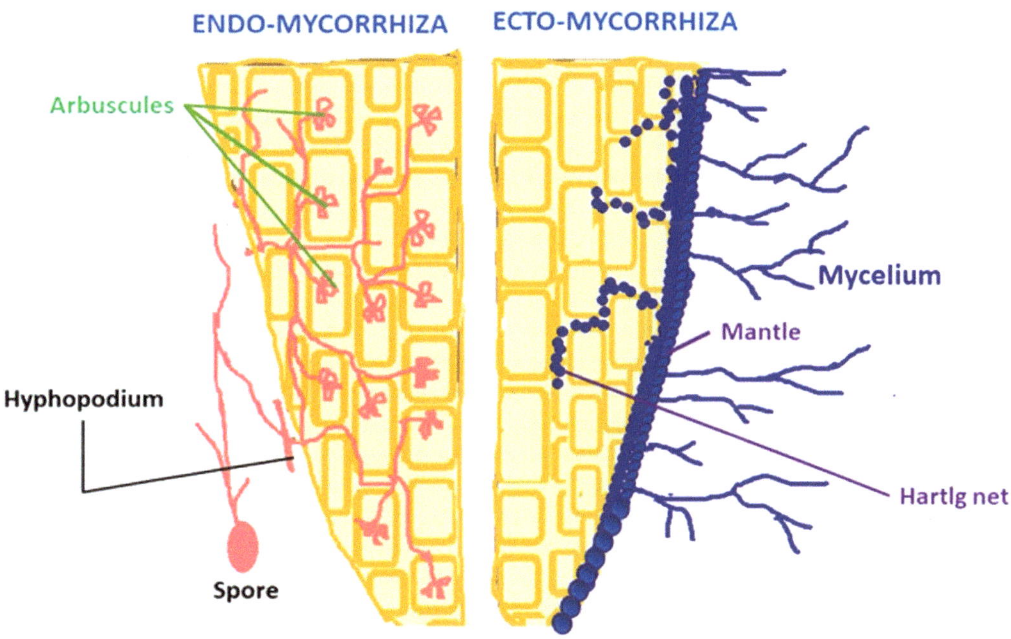

FIGURE 4.2 Root infection pattern by different types of mycorrhiza.

The AMF develops a dense web of hyphae that can enter root cells and improve a plant's absorption of water and minerals, particularly phosphorus (P), nitrogen (N), and potassium (K) (Fall et al., 2022). AMF also produces glomalin, a protein that improves soil structure and stabilizes soil aggregates, both of which are favorable for the plant root system. AMF has been demonstrated to boost plant growth and yield in many agricultural crops, and they are especially helpful for plants growing in nutrient-deficient soils. On the other hand, EMF and other types of the mycorrhiza have their own limitation because they have a very specific symbiotic relationship with the plant, but they also play an important role in nitrogen uptake and water uptake.

Mycorrhizal fungi significantly improve the plant tolerance to different stresses by lowering the uptake of toxic ions like sodium and encouraging the uptake of vital nutrients, namely potassium, phosphorus, and nitrogen. This may help plants develop more quickly and be more resilient to other stresses such as heat and drought. Additionally, they can promote the synthesis of stress-related proteins and antioxidants, which defend plants from harm brought on by dehydration. Mycorrhizal fungi also indirectly enhanced plant resistance against pathogens by stimulating the production of defense-related compounds and inducing systemic acquired resistance (SAR) in plants. SAR is a process by which plants develop a long-lasting resistance to a wide range of pathogens. Overall, mycorrhizal fungi can improve plant stress tolerance by enhancing plant nutrient uptake and use efficiency, regulating plant water balance, stimulating the production of stress-related compounds, and inducing systemic acquired resistance. By doing so, mycorrhizal fungi can help plants to cope with various stress factors and maintain their growth and productivity under adverse conditions. Hence, the application of mycorrhiza in agriculture for better plant production and maintaining soil health, is increasing progressively.

Agriculture-related mycorrhizal applications are a growing area of study and development. Although mycorrhizal fungi have been used in cultivation for a long time, there is now a greater interest in their use due to advances in our knowledge of how these fungi affect plant growth and development. Increased plant development and yield, better nutrient uptake, and greater resistance to environmental stresses like drought, salinity, and pathogens are just a few advantages of using mycorrhizal fungi in agriculture. Additionally, mycorrhizal fungi can aid in lowering the use of pesticides and chemical fertilizers, promoting more sustainable farming practices.

In this chapter, we focused on how mycorrhiza interacts with the plant system, how they respond to environmental stress, how they are involved in stress amelioration, and the overall impact of mycorrhiza in the agriculture sector.

4.2 MYCORRHIZA STIMULATES PLANT ROOT-HAIR GROWTH

The root hairs are specialized structures that extend from the root surface and significantly increase the surface area, available for nutrient and water absorption (Tominaga-Wada and Wada, 2014; Dolan, 2017). Usually, long, and dense root hairs are the fundamental trait in ensuring effective acquirement of both macro- and micro-nutrients in the early establishment of the plants within nutrient-deficient soils with low nutrient involvement cropping systems (Wang et al., 2016). Additionally, root hairs are considered as the preferred site for infection of free-living nitrogen-fixing bacterium (Hossain et al., 2015). Hence, root hairs have a significant role in nutrient exchanges between the rhizosphere and plants. Root hairs are considered as a determinant factor of the soil rhizosphere for root–microbe interaction (Robertson-Albertyn et al., 2017). For example, *Pseudomonas* sp., is a beneficial bacterium, could effectively colonize within barley root and *Enterobacter* sp. infected the root hairs of finger millet (*Eleusine coracana*) (Buddrus-Schiemann et al., 2010).

When mycorrhizae colonize within plant roots, they form a network of fungal hyphae that can extend much farther into the soil than the plant roots alone. Both root hairs and extraradical hyphae of mycorrhiza cooperatively exist in the root surface to increase surface areas of root systems for nutrient acquisition. These fungal hyphae increase the availability of soil nutrients and water for the plant. In response to this increased nutrient availability, the plant may produce more root hairs to further increase its absorptive surface area and enhance nutrient uptake (Hashem et al., 2018). Additionally, mycorrhizae can produce enzymes and other substances that help break down complex organic molecules in the soil, making them more accessible to the plant. This can further stimulate root-hair growth by providing the plant with additional nutrients to support its growth and development. The mycelia can also extend beyond the root system, forming a network that can connect multiple plants together, allowing for the exchange of nutrients and other compounds. As a result, plant growth improvement and subsequent yield enhancement is reported. Another way that mycorrhizae can stimulate root-hair growth is by releasing growth-promoting substances like hypathorine and indole-3-acetic acid (IAA), which are considered as the plant growth regulator or promotor (Liu et al., 2018).

Universally, root hairs are wider than mycorrhizal fungal hyphae (Φ 30 μm and Φ 10 μm, respectively) (Allen, 2009), accordingly, mycorrhizal hyphae can enter into smaller soil pores and take up more soil nutrients (Peterson and Farquhar, 1996). Generally, plants having abundant root hairs are much less dependent on mycorrhizal symbiosis for the nutrient uptake (Baylis, 1975; Zou et al., 2019). For example, plants, having wider roots with less root hairs, such as *Citrus* sp., are convincingly mycorrhiza dependent. In comparison, plants with thinner roots along with a huge number of root hairs, are comparatively less dependent on mycorrhizal symbioses (for example, *Echinops exaltatus*) (Novero et al., 2008). Brown et al. (2013) noticed that plants without root hairs may recruit AM fungi in an endeavor to improve both P acquisition and plant growth. Under drought stress, AM fungi mainly enhanced root P and water uptake, in contrast, root hairs ostensibly increased shoot P levels. These approaches strongly indicate the complex relationship between root hairs and mycorrhizas (Li et al., 2014).

Root hairs are typically a preferred adhesion site not only for AM but also for ectomycorrhizal hyphae, which developed between root epidermal cells and in the first layers of the root cortex during the first stage of ectomycorrhizal association (Novero et al., 2008).

Mycorrhizal hyphae can colonize root hairs during the AM growth of plants without the appressorium-like structure (Novero et al., 2008). Several plants are actively participating in symbiosis with AM fungi such as *Avena* sp., *Festuca rubra (Red Fescue), Lilium formologo, Clintonia borealis (yellow blue-bead lily), Panax quinquefolius*, *Pisum sativum*, *Taxus baccata*, and *Taxus canadensis*, while few plant (e.g. *Hordeum* sp., *Melilotus alba*, *Narcissus* sp., and *Triticum*) root hair occasionally acts as the entry point for AM mycorrhizal hyphae (Guinel and Hirsch, 2000). Nicolson (1959) exhibited that up to 92% of root hairs in *Festuca rubra* were colonized by AM fungi. In contrast, Baylis (1975) reported that mycorrhizal-dependent degree is negatively proportionate with root-hair number and the length of the root hair. For example, citrus plants having less root hairs with high mycorrhizal colonization (Wu et al., 2016). For colonization, mycorrhizal hyphae release some diffusible factors like hydrolases, a group of enzymes that catalyse bond cleavage, indole-3-acetic acid (IAA) and some small molecules, assists to loosen the bonds between plant cell walls (Mousa et al., 2016) and plant root hairs are highly sensitive to those diffusible factors (Novero et al., 2008). Mycorrhizal fungal exudate namely hypathorine inhibits the host plant root-hair growth and IAA can counteract the negative effects of it (Ditengou et al., 2000). In contrast, host plant root exudates namely rutin stimulate the mycorrhizal fungal growth and development (Lagrange et al., 2001).

4.3 IMPROVE SOIL HEALTH

Microorganisms are considered as an important soil component and soil performs as a biological reactor for these microorganisms, where various biochemical reactions occur to solubilize organic matter, biogeochemical cycles of soil micro and macroelements also materialize due to the interaction between various microorganisms and the respective environment (Gessner et al., 2010). The competition and parasitism interaction between microorganism determine the soil fertility (Topalovic and Vestergard, 2021). Arbuscular mycorrhizal fungi (AMFs) are recognized as being major soil microorganism, involved in biogeochemical cycles of soils such as P, N, and C cycle. AMFs contribute to improve the soil fertility by improving the chemical, physical, and biological properties of the soil.

AMFs produce organic acids and glomalin, which participates to protect the soil from erosion. AMFs chelate the heavy metals, improve carbon sequestration, and ultimately stabilize soil macroaggregation (Fall et al., 2022). It also can stabilize inorganic phosphate into acceptable soluble form by producing H^+ and metabolites (Relwani et al., 2008; Behera et al., 2014). Liu et al. (2013)

demonstrated that the AMF generates alkaline phosphatases metabolites, which cleave the soil substrates and make the phosphate accessible for plant. Additionally, AMFs helps to release P from fertilizer (Billah et al., 2019). As like phosphorus, AMFs also help to convert the inaccessible form of nitrogen, present in soil (nitrites, nitrates, and ammonium ions) to accessible form for the plant and increase the mobilization of inorganic form of nitrogen (Casieri et al., 2013). The availability of the nitrogen is required to activate the transporters of AMF hyphae. Hence, AMF plays a significant position in decomposition of plant organic matter and mobilize the nutrient, especially nitrogen for host plant (Lambers et al., 2008).

Physically, mycorrhizal hyphae acts as a tunneling tool, applying considerable pressure on soil particles and bringing together organic matter and clay particles leading to the formation of microaggregate. These burrows will provide the adequate aeration within the soil particles and make easier for water penetration and movement. Biologically AMFs have the positive, neutral, or negative relationship with other soil microorganism. The synergistic relationship of AMFs with other microorganism, promotes the root growth of the host plants, involve the nutrient acquisition, biological control of the soil-borne root pathogens, enhancement of plant tolerance level against abiotic stress, and subsequently soil fertility (Gryndler, 2000; Barea et al., 2002, Alimi et al., 2021). Symbiotic association of mycorrhiza affects the root exudates quantitatively and qualitatively, which influence the composition of soil microflora (Baltrus, 2017). However, the relationship between AMFs and other bacteria and/or fungus depend on various factors such as the availability of phosphorus, nitrogen in the soil and the type of host plant (Wang et al., 2011; Larimer et al., 2014; Xu et al., 2018). Mycorrhizal species not only specific with another microorganism but also, they are highly host specific. Huang et al., (2018) revealed that the association of mycorrhiza with their host orchid plants are highly host specific. Host specificity was also reported in ectomycorrhizal association (Bruns et al., 2002). However, there is a lack of information about the exact mechanism of symbiosis the AMFs with other microorganism, the signaling molecules, which determine the symbiotic relationship, the role of AMF in the trophic chain, the cause of host dependency. Therefore, further research on mycorrhiza with agricultural crop rotation, the growth of mycorrhiza at various soil type may help to understand the AMFs in a better way.

4.4 ARBUSCULAR MYCORRHIZA STRESS AMELIORATION

Abiotic stresses are considered as a foremost hurdle for the sustainable growth in agriculture for fulfilling the increasing food demand worldwide (Sah et al., 2016). Stress can be described as any adverse condition affecting the growth and development of plant, quality, and yield of crops (Chen et al., 2022). Environmental stresses are categorized into two major groups: biotic and abiotic. Biotic stresses include biological factors such as weeds, microbes, pathogens, and predators. Sources of abiotic stress include drought, salinity, light intensity, extreme temperatures, pollutants, and mechanical damages. When these parameters exceed a critical limit, plants are considered to be under stress (Bartwal et al., 2012). Due to their sessile nature, plants have to confront the various stresses and develop potential adaptive strategies to either avoid or tolerate the adverse effects for survival. Drastic fluctuations in these abiotic conditions may lead to a series of biochemical, physiological, and morphological changes in plants (Zhu, 2016). Adverse abiotic factors further affect plant reproduction and yield thus, responsible for substantial losses in agricultural and economical aspects (Hasanuzzaman et al., 2013; De Storme and Geelen, 2014).

4.4.1 Mitigation of Salinity Stress in Plants by Arbuscular Mycorrhizal Symbiosis

Salinity is considered as one of the major factors affecting crop production throughout the world. Salinity either in water or soil represents one of the major abiotic stresses specifically in arid and semi-arid regions, which can severely limit the production of agricultural crops (Nayem et al.,

2020). High concentration of salt leads to hyper osmotic stress and ionic imbalance in the cellular system, which consequently leads to oxidative damages. Such damages in plants are responsible for growth retardation, membrane disruption, molecular alterations, and even death (Rahneshan et al., 2018; Kumar et al., 2021). The continuous increase in salinity in arable land due to poor cultivation practices and climate change have devastating global effects, and it is estimated that about 50% of arable land will be lost by the middle of the 21st century (Islam et al., 2019). To date, about 1,125 million hectares of agricultural lands have already been seriously affected by salinity, thus it is considered a serious threat to agriculture (Islam et al., 2019; Sanower-Hossain, 2019).

Plants are well equipped with various morphological and physiological adaptations for combating the adversities of salinity stress. However, these adaptive strategies may become inefficient to cope with the intense and prolonged exposure to salinity condition (Chen et al., 2022; Evelin et al., 2019). Among various techniques being implemented for mitigating the adverse effects of saline condition, the application of arbuscular mycorrhizal fungus (AMF) is considered as an efficient method for bio-amelioration of salinity stress in plants (Evelin et al., 2012). These fungi reinforce salinity tolerance by deploying a spectrum of molecular and physiological mechanism such as better nutrient acquisition, enriched rhizospheric parameters, defence against soil-borne pathogens, maintenance of ionic homeostasis, enhanced water utilizing capacity and root hydraulic conductivity, better photosynthetic efficiency and a competent antioxidant system. Thus, this leads to better biomass accumulation in associated plants and improved resilience to saline stress (Ruiz-Lozano et al., 2012; Hidri et al., 2016; Hashem et al., 2018; Evelin et al., 2019).

The degree of saline adversity on plants is evaluated in terms of biomass accumulated. Plants when not in association with mycorrhiza exhibit significant loss of biomass and retarded growth as compared to those associated with mycorrhiza. As reported in various plants species, the presence of AMF in the rhizosphere results in enhanced biomass and fitness in *Trigonella foenum-graecum* (Evelin et al., 2012), *Medicago sativa* (Campanelli et al., 2013), *Oryza sativa* (Porcel et al., 2015), *Gossypium hirsutum* (Liu et al., 2016), and *Chrysanthemum morifolium* (Wang et al., 2018). Enhanced biomass of these plants has been partly attributed to mycorrhizically mediated enhanced nutrient acquisition subsequently leading to better manifestation of crop yield (Sharifi et al., 2007; Talaat and Shawky, 2011).

Previous studies have suggested that salinity stress leads to enhanced production of free radicals, which are well known for their destructive impact on major biomolecules like proteins, nucleic acids, and lipids (Han and Yang, 2021). AM are found to mitigate such free radicals mediated damages by enhancing both enzymatic and non-enzymatic antioxidants in plant systems (Pandey and Garg, 2017) (Figure 4.3). Increased salinity in the soil impairs the nutrient acquisition of important minerals like phosphorus, calcium, nitrogen, and magnesium thus increasing the dependency of plants on mycorrhiza (Evelin et al., 2019). Studies suggest that mycorrhizal association helps in maintaining better ionic ratio and increased nutrient mobilization. Better phosporous solubilization and uptake was observed in *Trifolium alexandrium* plants under saline conditions (Shokri and Maadi, 2009). The acquisition of N and its metabolism is disrupted under saline conditions. Inoculation of mycorrhiza has been found to be beneficial in better utilization and assimilation of N in associated plants. Higher uptake of N has been reported in *Sesbania grandiflora* and *Sesbania aegyptiaca* in the presence of mycorrhiza (Giri and Mukerji, 2004). Furthermore, higher N uptake under salinity stress was attributed to elevated expression of nitrate and ammonium transporters in mycorrhizal plant host (Felicia et al., 2017). An elevated level of Na^+ in the substratum interferes with the Na^+: K^+ ratio, which is essential for several cellular functions. Higher availability of Na^+ than K^+ increases the Na^+: K^+ ratio, which significantly disrupts the ionic balance thus affecting various metabolic pathways (Giri et al., 2007). Mycorrhiza colonized plants have exhibited a higher K^+: Na^+ ratio than their non-colonized counterparts under saline conditions. Likewise, low availability of Ca^{2+} and Mg^{2+} due to salinity stress in plants have been overcome under the influence of

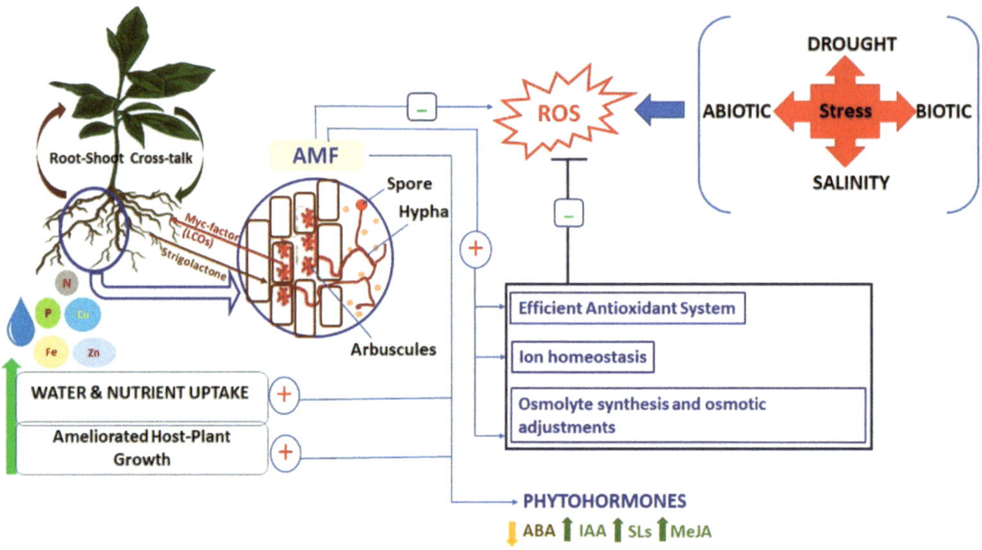

FIGURE 4.3 Stress alleviation by mycorrhiza.

mycorrhiza (Evelin et al., 2012; Garg and Bhandari, 2016) thus regulating active Ca^{2+} signaling and better photosynthetic rate due to increased uptake of Ca^{2+} and Mg^{2+}, respectively, in host plants.

4.4.2 Mycorrhizal Fungi to Alleviate Drought Stress

Drought stress is another major abiotic stress that considerably affects the agricultural aspects and jeopardizes global food security. According to the reports of the United Nations World Water Development (2018), approximately 55 million livelihoods are affected worldwide and further around 700 million individuals are at risk of being displaced by 2030 due to drought (WWAP, 2018). Such consequences cause huge monetary loss, which leads to drastic socio-economic downfall of affected nations. Considering the Indian scenario, out of 159.7 million hectares of agricultural land an estimated 42% land is drought affected (Gogoi and Tripathi, 2019).

Plants significantly rely on rhizospheric microflora for mitigation of negative effects of drought stress (Saia et al., 2014). Previous studies support the beneficial consequences of symbiotic association of plants with AM on the plant drought tolerance mechanism. The AM hyphae provides access to the water resource unavailable for plant roots (Liu et al., 2018), thus helping to develop better plant water relations and increase plant productivity under water deficit conditions (Huang et al., 2017). Moreover, evidence suggests AM being responsible for enhanced nutrient absorption, photosynthetic efficiency, root structure, reinforced antioxidant defense system, osmotic adjustment, aquaporin expression, and optimal hormone balance for resisting water deficit in soil (Wu et al., 2019; He et al., 2020; Zou et al., 2020; Cheng et al., 2021). AM is also found to improve stomatal conductance as reported in plant species, such as *Poncirus trifoliata* and *Rosmarinus officinalis* when inoculated with AM during drought stress (Sanchez-Blanco, 2004; Wu et al., 2006; Ruiz-Sanchez et al., 2011).

In response to water scarcity, the AMF-mediated mechanisms include modifications in the physiological status of important plant hormones, such as strigolactones, jasmonic acid, and abscisic acid (Fernández-Lizarazo et al., 2016). For instance, in *Solanum lycopersicum*, AM was reported to regulate the 14-3-3 genes (TFT1-TFT12) in the ABA signaling pathway, thus improving the plant water relation and enhancing the plant drought tolerance (Xu et al., 2018). Furthermore, considerable alleviation of drought, cold and salt stress was studied in *Digitaria eriantha* by mycorrhizal association through upregualtion of jasmonate biosynthesis and antioxidant enzyme system (Pedranzani et al., 2016).

In addition, studies have shown that hyphal H^+ effluxes and spores' growth rate were stimulated during the presymbiotic development of *Gigaspora margarita*, which is related to phosphorus and sucrose deficiency (Ramos et al., 2008). The plasma membrane H^+-ATPase, major membrane proteins present in the plant cell. It regulates the H^+ flux, which creates a driving force for transport of small molecules, absorption of mineral nutrient, cellular pH regulation and growth (Gaxiola et al., 2007). H^+-ATPase plays a vital role in acidification of the cell wall by regulating H^+ flux, which is essential for root elongation according to acid growth hypothesis (Staal et al., 2011). Hence the proper regulation of plasma memebrane H^+ATPase activity becomes essential for better growth and development of roots during drought stress for optimal nutrient acquisition. Liu et al. (2020) reported that the tomato AM-specific H^+-ATPase, SIHA8, was evolutionarily conserved in maintaining arbuscule development. Cheng et al. (2021) proposed that AM alleviated drought stress by regulating plasma membrane H^+-ATPase activity thus promoting root development and nutrient acquisition in *Poncirus trifoliate*. Considering the adverse effect of drought stress on agricultural aspects and food security and the beneficial effects of AM association with plants, it becomes vital for application of AM for the paradigm shift towards sustainable agriculture and better stress management during drought stress.

4.5 MYCORRHIZAL-BACTERIAL AMELIORATION OF PLANT ABIOTIC AND BIOTIC STRESS

One of the essential factors of a healthy soil are the microbes that inhabit the rhizosphere. These rhizospheric microbiota facilitate growth and development of the associated plants by recycling and mobilizing the vital nutrients (Trivedi et al., 2020). In natural ecosystems, plants interact with several rhizospheric microbes and successfully establish mutual beneficial alliances among them. One such alliance being the mycorhizzal association between fungi and plant roots where plants supply the photosynthetic assimilates to the fungus, which in turn, benefits plants with enhanced nutrient uptake, enriched soil, and better stress management (Chen et al., 2017; Mitra et al., 2019). Thus, they are considered as suitable biofertilizers for sustainable agricultural production, and beneficial replacement of chemical fertilizers and pesticides (Molina-Romero et al., 2021). These fungi are almost ubiquitous in the agricultural soil associating symbiotically with nearly 80% of terrestrial plants.

Among the plethora of microbes surrounding the plants another beneficial bacterial component present is the plant growth-promoting bacteria (PGPB), which significantly influences plant growth (Orozco-Mosqueda et al., 2020). These bacteria are known to exhibit an interactive relationship with mycorrhiza in the rhizosphere (Trivedi et al., 2020). The bacterial and fungal diversity of the plant rhizosphere is dynamic and varies depending on several biotic and abiotic factors. In 1994, Garbaye designated this specific microbial category as mycorrhization-helper bacteria (MHB) because of their ability to enhance mycorrhizal colonization as well as development of the mycelium during the symbiotic association. The different types of association has been reported between bacteria and ectomycorrhiza, such as bacterial isolates may exist as planktonic cells in the mycorrhizosphere, sometimes form specialized biofilms around the mycorrhiza or occupy the internal space of the hyphae (Frey-Klett et al., 2011).

PGPB and AM fungi are well known for their beneficial impact when co-inoculated into various agricultural crops. Such consortium is found to optimize soil quality, enhance nutrient acquisition, stimulate plant growth, induce tolerance towards stress, and increase the production and nutritional value (Berta et al., 2014; Raklami et al., 2019; Massa et al., 2020; Santoyo et al., 2021). Several reports have claimed the beneficial effects of both the mycorrhiza and PGPB in agricultural crops under biotic and abiotic stress. The co-inoculation of AM fungi and endophytic *Bacillus* strain induced drought tolerance in *Lavandula dentata* it better growth and nutrition (Pereira et al., 2016). In another study, the AM fungi *Glomus mosseae* and *G. etunicatum* were applied with bacteria *Azospirillium lipofrum* and *Azotobacter chroococcum* in walnut plants under water scarce conditions (Behrooz et al., 2019). It was observed that the co-inoculation of these microbes were significantly effective than their individual inoculation in mitigating the adverse effects of drought stress in walnut plants. Moreover, co-inoculation resulted in elevated synthesis of metabolites and osmolytes essential for stress management, like phenolics, soluble sugar, starch, and proline content.

Likewise combined application of AM and PGPB exhibiting amelioration of toxic effects of salinity stress in plants are reported previously. The authors have postulated that application of AM fungi and bacteria serve as an effective approach for rescuing plants from salinity stress. The application of bacteria *Pseudomonas* and *Pantoea* with the AM *Rhizoglomus irregular*, resulted in considerable decline in Na+ ion rather increase in nutrient accumulation like nitrogen and magnesium in *Zea mays* under saline conditions (Moreira et al., 2020). Moreover, co-inoculative applications of other mycorrhizal species like *Glomus etunicatum* and bacteria such as *Methylobacterium oryzae*, *R. intraradices*, and *Massilia* sp. RK4 have resulted in induced tolerance mechanism in plants under salinity stress. These microbes have significantly increased the biomass and nutrient accumulation in inoculated crops, attributed to better nutrient acquisition by mycorrhizal root colonization (Lee et al., 2015; Krishnamoorthy et al., 2016).

Besides regulating the tolerance mechanism in plants under abiotic stresses, mycorrhizal fungi is also known for its beneficial impacts on plants against biotic stress. Mycorrhiza provides a protective shield over the roots, involved in competitive exclusion of pathogenic microbes from the rhizosphere and induces proliferation of antagonistic bacteria against pathogens (Edwards et al., 1998). Furthermore, reports suggest mycorrhzal fungi are responsible for induction of the biochemical defense mechanism by regulating the synthesis of phenolic compounds, pathogenesis-related proteins, and other potential antimicrobial compounds in a plant's body under pathogenic attack (Edwards et al., 1998; Shaul et al., 1999; Al-Askar and Rashad, 2010; Song et al., 2015; Santoyo et al., 2021). The tomato plants when co-inoculated with AM (*Rhizophagus irregularis*) and to PGPB (*Pseudomonas jessenii* R62 and *Pseudomonas synxantha* R81) show potential resistance towards root-knot nematode *Meloidogyne incognita*. The co-inoculative approach resulted in significant elevation in functioning of enzymatic antioxidants, such as peroxidise, superoxide dismutase, and polyphenoloxidase along with increased accumulation non-enzymatic antioxidants, i.e., phenolic compounds, thus strengthening the defence against the nematode (Sharma and Sharma, 2017). In another study reported by Nepomuceno et al. (2019) the application of mycorrhizal fungi and growth promoting bacteria provided tolerance to onion crops against soil-borne pathogen *Sclerotium rolfsii*. It was observed that when both the microbes were used the infection by the pathogen was reduced significantly as compared to their separate inoculation, which suggest synergistic effect of mycorrhiza and bacteria on pathogen control. Similarly, co-inoculation with *Gigaspora margarita* (AM) and *Paenibacillus rhizospherae* (PGPB) was reported to mitigate the occurrence of white root rot in various fruit crops namely, grape, fig, apple and Japanese apricot by the pathogen *Rosellinia necatrix* Bearl ex Prill (Freire Cruz and de Oliveira Soares, 2014).

The studies on the application of AM and PGPB in the field has also supported their influence on the plant crops for adaptation against biotic factors. The application of AM (*Glomus fasciculatum*) and PGPB (*Pseudomonas monteilii*) in a two year field study exhibited protective effect on a medicinal plant, *Coleus forskohlii* against fungal pathogens *Ralstonia solanacearum* and *Fusarium chlamydosporum*. The pathogens are known to cause root rot and wilt and severely affect the growth and forskolin content, which were found to increase significantly under the influence of AM and PGPB (Singh et al., 2013). Additionally, in recent studies, the synergistic effect of PGPB and AM in interaction with plants cultivated in contaminated agricultural substratum and phytoremediation of soil has been reported (Guarino et al., 2020). The study performed on two plant species, *Oloptum miliaceum* and *Pennisetum setaceum* grown in petroleum hydrocarbons contaminated soil showed successful phytoremediation under the influence of AM and PGPB. The results suggested that selection of bacteria with optimal production of lipolysaccharides, ACC deaminase and abililty to synthesize siderophores, would be an effective strategy to combat recalcitrant molecules like petroleum hydrocarbons (Liu et al., 2015).

Mycorrhizae are also capable of acting as biocontrol against pyhtopathogenic microbes by conferring resistance to host plants through direct and indirect mechanisms. Mycorrhiza is known to regulate the pathogen resistance in plants via direct as well as indirect mechanisms. Besides changes in morphological and physiological parameters as discussed earlier, direct mechanism includes mycorrhizal competitive interaction with pathogenic microbes (Kalamulla et al., 2022). As reported by Chauhan et al., (2022), *Meloidogyne incognita*, a major pathogenic nematode was competitively excluded by AM thus, preventing root colonization in *Prunus perscisa* and inhibited gall formation. On the other hand, the indirect mechanism involves the activation of induced systemic resistance (ISR) of plants by AM under pathogenic attack. Such activation of ISR by AM was found in rice plants under symbiotic interaction with *Rhizophagus intraradices*, which conferred resistance to pathogen infection by regulating the expression of host plant genes (Golpayegani and Tilebeni, 2011). Moreover, mycorrhiza is known to mediate warning signal to other neighboring plants through their established mycorrhizal network. As observed under aphid attack in plants, AM was found to induce changes in volatile compounds like methysalicylate in plants and spreading the alert signal to the neighbors (Chauhan et al., 2022).

4.6 IMPORTANCE OF MYCORRHIZA FOR SUSTAINABLE AGRICULTURE AND AGRI-BUSINESSES

The agricultural production has increased gradually since the mid of the last century mainly due to implementation of new technology, advanced machinery, fertilizer, and pesticides. The degradation of the land, which is frequently a result of unsustainable farming methods, is a major reason why agricultural production have lately plateaued. Through the course of this century, food production must dramatically increase while sustainably using less arable land to satisfy future demand.

The control of field stress and agricultural productivity depend heavily on arbuscular mycorrhizal fungi. Through rhizobacteria and mycorrhizae, the rhizosphere microbiome has the potential to significantly improve agricultural yields and reduce biotic and abiotic stress factors, including disease. There will be an increase in the use of contemporary inoculum-based biotechnology. Messa and Savioli (2021) improve sustainability, lowering agriculture's damaging effects on the environment while also enhancing the productivity of agricultural systems by using mycorrhizal-based technology. Mycorrhizae can be used to raise horticulture crop yield and quality improvement. For example, AMF can be applied as biological fertilizers to increase crop output, boosting farmer revenue and reducing the use of chemical fertilizers and pesticides.

Recently, mycorrhiza-based products, including fertilizers and inoculants that include fungi, are currently offered on the agri-market. These products can be utilized in horticulture, crop production, forestry, and among other agricultural activities. Even though mycorrhizal treatment in agriculture may have benefits, it is currently not extensively practiced. Some of the causes of this include farmers' lack of knowledge regarding the benefits and limitations of mycorrhizal application, as well as inconsistent production and application of mycorrhizal products. Furthermore, mycorrhizal application may not always be beneficial because the effectiveness can vary based on the kind of soil, the type of crop, and other environmental factors. Overall, even though mycorrhizal fungi are currently being used in agriculture, the number of effective products for all types of crops are still insufficient.

4.7 CONCLUSION

While several research reports over many years have described the multi-complex mechanisms leading to plant-mycorrhiza symbiosis in enhancing stress tolerance in plants, still, there are several aspects that need to be focused to investigate for better understanding. There is limited information about the molecular mechanism directing the process of stress amelioration by AMFs in plants. However, the genes, responsible for symbiotic association, and the genes are involved in production of various antioxidants and enzymes, governing the synthesis of various osmo-regulators to regulates the stress, need to be uncovered. Additionally, the transporter systems functioning in the symbiotic association need to be elucidated.

REFERENCES

Alimi, A., Adeleke, R., & Moteetee, A. (2021). Soil environmental factors shape the rhizosphere arbuscular mycorrhizal fungal communities in South African indigenous legumes (Fabaceae). *Biodiversity Journal of Biological Diversity, 22*, 2466–2476.

Al-Askar, A.A., & Rashad, Y.M. (2010). Arbuscular mycorrhizal fungi: A biocontrol agent against common bean *Fusarium* root rot disease. *Plant Pathology Journal, 9*, 31–38.

Allen, M.F. (2009). Water relations in the mycorrhizosphere. In U. Lüttge, W. Beyschlag, B. Büdel & D. Francis (Eds.) *Progress in Botany* (pp. 257–276). Springer.

Baltrus, D.A. (2017). Adaptation, specialization, and coevolution within phytobiomes. *Current Opinion in Plant Biology, 38*, 109–116.

Barea, J.M., Azcón, R., & Azcón-Aguilar, C. (2002). Mycorrhizosphere interactions to improve plant fitness and soil quality. *Antonie Van Leeuwenhoek, 81*, 343–351.

Bartwal, A., Mall, R., Lohani, P., Guru, S.K., & Arora, S. (2012). Role of secondary metabolites and brassinosteroids in plant defense against environmental stresses. *Journal of Plant Growth Regulation, 32*, 216–232.

Baylis, G.T.S. (1975). The magnolioid mycorrhiza and mycotrophy in root systems derived from it. In F.E. Sanders, B. Mosse & P.B. Tinker (Eds.) *Endomycorrhizas* (pp. 373–389). Academic Press.

Behera, B.C., Singdevsachan, S.K., Mishra, R.R., Dutta, S.K., & Thatoi, H.N. (2014). Diversity, mechanism, and biotechnology of phosphate solubilising microorganism in mangrove – A review. *Biocatalysis and Agricultural Biotechnology, 3*, 97–110.

Behrooz, A., Vahdati, K., Rejali, F., Lotfi, M., Sarikhani, S., & Leslie, C. (2019). Arbuscular mycorrhiza and plant growth-promoting bacteria alleviate drought stress in walnuts. *HortScience,* 54 , 1087–1092.

Berta, G., Copetta, A., Gamalero, E., Bona, E., Cesaro, P., Scarafoni, A., & D'Agostino, G. (2014). Maize development and grain quality are differentially affected by mycorrhizal fungi and a growth-promoting pseudomonad in the field. *Mycorrhiza, 24*, 161–170.

Billah, M., Khan, M., Bano, A., Hassan, T.U., Munir, A., & Gurmani, A.R. (2019). Phosphorus and phosphate solubilizing bacteria: Keys for sustainable agriculture. *Geomicrobiology Journal, 36*, 904–916.

Brown, L.K., George, T.S., Barrett, G.E., Hubbard, S.F., & White, P.J. (2013). Interactions between root hair length and arbuscular mycorrhizal colonization in phosphorus deficient barley (*Hordeum vulgare*). *Plant and Soil, 372*, 195–205.

Bruns, T.D., Bidartondo, M.I., & Taylor, D.L. (2002). Host specificity in ectomycorrhizal communities: What do the exceptions tell us? *Integrative and Comparative Biology, 42*, 352–359.

Buddrus-Schiemann, K., Schmid, M. Schreiner, K. Welzl, G., & Hartmann, A. (2010). Root colonization by *Pseudomonas* sp DSMZ 13134 and impact on the indigenous rhizosphere bacterial community of barley. *Microbial Ecology, 60*, 381–393.

Campanelli, A., Ruta, C., De Mastro, G., & Morone-Fortunato, I. (2013). The role of arbuscular mycorrhizal fungi in alleviating salt stress in *Medicago sativa* L. var. icon. *Symbiosis, 59*, 65–76.

Casieri, L., Lahmidi, N.A., Doidy, J., Veneault-Fourrey, C., Migeon, A., Bonneau, L., Courty, P.E., Garcia, K., Charbonnier, M., Delteil, A., Brun, A., Zimmermann, S., Plassard, C., & Wipf, D. (2013). Biotrophic transportome in mutualistic plant-fungal interactions. *Mycorrhiza, 23*, 597–625.

Chauhan, P.K., Upadhyay S.K., Tripathi M., Singh, R., Krishna, D., Singh, S.K., & Dwivedi, P. (2022). Understanding the salinity stress on plant and developing sustainable management strategies mediated salt-tolerant plant growth-promoting rhizobacteria and CRISPR/Cas9. *Biotechnology and Genetic Engineering Reviews, 17*, 1–37.

Chen, D., Mubeen, B., Hasnain, A., Rizwan, M., Adrees, M., Naqvi, S.A.H., Iqbal, S., Kamran, M., El-Sabrout, A.M., Elansary, H.O., Mahmoud, E.A., Alaklabi, A., Sathish, M., & Din, G.M.U. (2022). Role of promising secondary metabolites to confer resistance against environmental stresses in crop plants: Current scenario and future perspectives. *Frontiers in Plant Science, 13*, 881032.

Chen, S., Zhao, H., Zou, C., Li, Y., Chen, Y., Wang, Z., Jiang, Y., Liu, A., Zhao, P., Wang, M., & Ahammed, G.J. (2017). Combined inoculation with multiple arbuscular mycorrhizal fungi improves growth, nutrient uptake and photosynthesis in cucumber seedlings. *Frontiers in Microbiology, 8*, 2516.

Cheng, H.Q., Zou, Y.N., Wu, Q.S., & Kuca, K. (2021). Arbuscular mycorrhizal fungi alleviate drought stress in trifoliate orange by regulating H^+-ATPase activity and gene expression. *Frontiers in Plant Science, 12*, 659694.

De Storme, N., & Geelen, D. (2014). The impact of environmental stress on male reproductive development in plants: Biological processes and molecular mechanisms. *Plant, Cell & Environment, 37*, 1–18.

Dhiman, M., Sharma, L., Kaushik, P., Singh, A., & Sharma, M.M. (2022). Mycorrhiza: An ecofriendly bio-tool for better survival of plants in nature. *Sustainability, 14*, 10220.

Ditengou, F.A., Beguiristain, T., & Lapeyrie, F. (2000). Root hair elongation is inhibited by hypaphorine, the indole alkaloid from the ectomycorrhizal fungus *Pisolithus tinctorius*, and restored by indole-3-acetic acid. *Planta, 211*, 722–728.

Dolan, L. (2017). Root hair development in grasses and cereals (Poaceae). *Current Opinion in Genetics & Development, 45*, 76–81.

Edwards, S.G., Young, J.P.W., & Fitter, A.H. (1998). Interactions between *Pseudomonas fluorescens* biocontrol agents and Glomus mosseae, an arbuscular mycorrhizal fungus, within the rhizosphere. *FEMS Microbiology Letters, 166*, 297–303.

Evelin, H., Devi, T.S., Gupta, S., & Kapoor, R. (2019). Mitigation of salinity stress in plants by arbuscular mycorrhizal symbiosis: Current understanding and new challenges. *Frontiers in Plant Science, 10*, 470.

Evelin, H., Giri, B., & Kapoor, R. (2012). Contribution of *Glomus intraradices* inoculation to nutrient acquisition and mitigation of ionic imbalance in NaCl stressed *Trigonella foenum-graecum. Mycorrhiza, 22*, 203–217.

Fall, A.F., Nakabonge, G., Ssekandi, J., Founoune-Mboup, H., Apori, S.O., Ndiaye, A., Badji, A., & Ngom, K. (2022). Roles of arbuscular mycorrhizal fungi on soil fertility: Contribution in the improvement of physical, chemical, and biological properties of the soil. *Frontiers in Fungal Biology, 3*, 723892.

Felicia, V., Russia, P., Ingraffia, R., Giambalvo, D., Frenda, A.S., & Martinelli, F. (2017). Arbuscular mycorrhizal symbiosis mitigates the negative effects of salinity on durum wheat. *PLoS One, 12*, e0184158.

Fernández-Lizarazo, J.C., & Moreno-Fonseca, L.P. (2016). Mechanisms for tolerance to water-deficit stress in plants inoculated with arbuscular mycorrhizal fungi. A review. *Agronomía Colombiana, 34*, 179–189.

Frank, A.B. (1885). On the nutrient providing root-symbiosis between underground fungi and certain trees. *Berichte der Deutschen botanischen Gesellschaft,* 5, 395–409.

Freire Cruz, A., & de Oliveira Soares, W.R. (2014). Impact of the arbuscular mycorrhizal fungi and bacteria on biocontrol of white root rot in fruit seedlings. *Journal of Plant Physiology & Pathology, 2*, 1–5.

Frey-Klett, P., Burlinson, P., Deveau, A., Barret, M., Tarkka, M., & Sarniguet, A. (2011). Bacterial-fungal interactions: Hyphens between agricultural, clinical, environmental, and food microbiologists. *Microbiology and Molecular Biology Reviews, 75*, 583–609.

Garbaye, J. (1994). Tansley review No. 76 Helper bacteria: A new dimension to the mycorrhizal symbiosis. *New Phytologist, 128*, 197–210.

Garg, N., & Bhandari, P. (2016). Silicon nutrition and mycorrhizal inoculations improve growth, nutrient status, K^+/Na^+ ratio and yield of *Cicer arietinum* L. genotypes under salinity stress. *Plant Growth Regulation, 78*, 371–387.

Gaxiola, R.A., Palmgren, M.G., & Schumacher, K. (2007). Plant proton pumps. *FEBS Letters, 581*, 2204–2214.

Gessner, M.O., Swan, C.M., Dang, C.K., McKie, B.G., Bardgett, R.D., Wall, D.H., & Hattenschwiler, S. (2010). Diversity meets decomposition. *Trends in Ecology and Evolution, 25*, 372–380.

Giri, B., & Mukerji K.G. (2004). Mycorrhizal inoculant alleviates salt stress in *Sesbania aegyptiaca* and *Sesbania grandiflora* under field conditions: Evidence for reduced sodium and improved magnesium uptake. *Mycorrhiza, 14*, 307–312.

Giri, B., Kapoor, R., & Mukerji, K.G. (2007). Improved tolerance of *Acacia nilotica* to salt stress by arbuscular mycorrhiza, *Glomus fasciculatum*, may be partly related to elevated K^+/Na^+ ratios in root and shoot tissues. *Microbial Ecology, 54*, 753–760.

Gogoi, A., & Tripathi, B. (2019). 42% India's Land Area Under Drought, Worsening Farm Distress In Election Year. www.indiaspend.com/42-indias-land-area-underdrought-worsening-farm-distress-in-election-year/.

Golpayegani, A., & Tilebeni, H. (2011). Effect of biological fertilizers on biochemical and physiological parameters of basil (*Ociumum basilicm* L.) medicine plant. *American-Eurasian Journal of Agricultural & Environmental Sciences, 11*, 411–416.

Gryndler, M. (2000). Interactions of arbuscular mycorrhizal fungi with other soil organisms, in arbuscular mycorrhizas. In Y. Kapulnik & D.D. Douds Dordrecht (Eds.) *Physiology and function* (pp. 239–262). Springer.

Guarino, C., Marziano, M., Tartaglia, M., Prigioniero, A., Postiglione, A., Scarano, P., & Sciarillo, R. (2020). Poaceae with PGPR bacteria and arbuscular mycorrhizae partnerships as a model system for plant microbiome manipulation for phytoremediation of petroleum hydrocarbons contaminated agricultural soils. *Agronomy, 10*, 1–17.

Guinel, F.C., & Hirsch, A.M. (2000). The involvement of root hairs in mycorrhizal associations. In R.W. Ridge & A.M.C. Emons (Eds.) *Root Hairs* (pp. 285–310). Springer.

Han, X., & Yang, Y. (2021). Phospholipids in salt stress response. *Plants, 10*, 2204.

Hasanuzzaman, M., Nahar, K., Alam, M.M., Roychowdhury, R., & Fujita, M. (2013). Physiological, biochemical, and molecular mechanisms of heat stress tolerance in plants. *International Journal of Molecular Sciences, 14*, 9643–9684.

Hashem, A., Alqarawi, A.A., Radhakrishnan, R., Al-Arjani, A.B.F., Aldehaish, H.A., Egamberdieva, D., & Abd Allah, E.F. (2018). Arbuscular mycorrhizal fungi regulate the oxidative system, hormones and ionic equilibrium to trigger salt stress tolerance in *Cucumis sativus* L. *Saudi Journal of Biological Sciences, 25*, 1102–1114.

He, J.D., Zou, Y.N., Wu, Q.S., & Kuca, K. (2020). Mycorrhizas enhance drought tolerance of trifoliate orange by enhancing activities and gene expression of antioxidant enzymes. *Scientia Horticulturae, 262*, 108745.

Hidri, R., Barea, J.M., Mahmoud, O.M.B., Abdelly, C., & Azcón, R. (2016). Impact of microbial inoculation on biomass accumulation by *Sulla carnosa* provenances, and in regulating nutrition, physiological and antioxidant activities of this species under non-saline and saline conditions. *Journal of Plant Physiology, 201*, 28–41.

Hossain, M.S., T. Joshi, & G. Stacey. (2015). System approaches to study root hairs as a single cell plant model: Current status and future perspectives. *Frontiers in Plant Science, 6*, 363.

Huang, H., Zi, XM., Lin, H., & Gao, J.Y. (2018). Host-specificity of symbiotic mycorrhizal fungi for enhancing seed germination, protocorm formation and seedling development of over-collected medicinal orchid, *Dendrobium devonianum*. *Journal of Microbiology, 56*, 42–48.

Huang, Y.M., Zou, Y.N., & Wu, Q.S. (2017). Alleviation of drought stress by mycorrhizas is related to increased root H_2O_2 efflux in trifoliate orange. *Scientific Reports, 7*, 42335.

Islam, F., Wang, J., Farooq, M.A., Yang, C., Jan, M., Mwamba, T.M., Hannan, F., Xu, L., & Zhou, W. (2019). Rice responses and tolerance to salt stress. In M. Hasanuzzaman, M. Fujita, K. Nahar & J. Biswas (Eds.) *Advances in Rice Research for Abiotic Stress Tolerance* (pp. 791–819). Woodhead Publishing.

Kalamulla, R., Karunarathna, S.C., Tibpromma, S., Galappaththi, M.C.A., Suwannarach, N., Stephenson, S.L., Asad, S., Salem, Z.S., & Yapa, N. (2022). Arbuscular mycorrhizal fungi in sustainable agriculture. *Sustainability, 14*, 12250.

Krishnamoorthy, R., Kim, K., Subramanian, P., Senthilkumar, M., Anandham, R., & Sa, T. (2016). Arbuscular mycorrhizal fungi and associated bacteria isolated from salt-affected soil enhances the tolerance of maize to salinity in coastal reclamation soil. *Agriculture Ecosystems & Environment, 231*, 233–239.

Kumar, S., Li, G., Yang, J., Huang, X., Ji, Q., Liu, Z., Ke, W., & Hou, H. (2021). Effect of salt stress on growth, physiological parameters, and ionic concentration of water dropwort (*Oenanthe javanica*) cultivars. *Frontiers in Plant Science, 12*, 660409.

Lagrange, H., Jay-Allgmand, C., & Lapeyrie, F. (2001). Rutin, the phenolglycoside from Eucalyptus root exudates, stimulates *Pisolithus* hyphal growth al picomolar concentration. *New Phytologist, 149*, 349–355.

Lambers, H., Raven, J.A., Shaver, G.R., & Smith, S.E. (2008). Plant nutrient acquisition strategies change with soil age. *Trends in Ecology & Evolution, 23*, 95–103.

Larimer, A.L., Clay, K., & Bever, J.D. (2014). Synergism and context dependency of interactions between arbuscular mycorrhizal fungi and rhizobia with a prairie legume. *Ecology, 95*, 1045–1054.

Lee, Y., Krishnamoorthy, R., Selvakumar, G., Kim, K., & Sa, T. (2015). Alleviation of salt stress in maize plant by co-inoculation of arbuscular mycorrhizal fungi and *Methylobacterium oryzae* CBMB20. *Journal of the Korean Society for Applied Biological Chemistry, 58*, 533–540.

Li, T., Lin, G. Zhang, X Chen, Y.L. Zhang, S.B., & Chen, B.D. (2014). Relative importance of an arbuscular mycorrhizal fungus (*Rhizophagus intraradices*) and root hairs in plant drought tolerance. *Mycorrhiza, 24*, 595–602.

Liu, C.Y., Zhang, F., Zhang, D.J., Srivastava, A.K., Wu, Q.S., & Zou, Y.N. (2018). Mycorrhiza stimulates root-hair growth and IAA synthesis and transport in trifoliate orange under drought stress. *Scientific Reports, 8*, 1978.

Liu, J., Chen, J., Xie, K., Tian, Y., Yan, A., Liu, J., Huang, Y., Wang, S., Zhu, Y., Chen, A., & Xu, G. (2020). A mycorrhiza-specific H^+-ATPase is essential for arbuscule development and symbiotic phosphate and nitrogen uptake. *Plant Cell & Environment, 43*, 1069–1083.

Liu, J.L., Xie, B.M., Shi, X.H., Ma, J.M., & Guo, C.H. (2015). Effects of two plant growth-promoting rhizobacteria containing 1-aminocyclopropane1-carboxylate deaminase on oat growth in petroleum-contaminated soil. *International Journal Environmental Science and Technology, 12*, 3887–3894.

Liu, L., Zhang, T., Gilliam, F.S., Gundersen, P., Zhang,W, Chen, H., & Mo, J. (2013). Interactive effects of nitrogen and phosphorus on soil microbial communities in a tropical forest. *PLoS ONE, 8*, e61188.

Liu, S., Guo, X., Feng, G., Maimaitiaili, B., Fan, J., & He, X. (2016). Indigenous arbuscular mycorrhizal fungi can alleviate salt stress and promote growth of cotton and maize in saline fields. *Plant Soil, 398*, 195–206.

Massa, N., Cesaro, P., Todeschini, V., Capraro, J., Scarafoni, A., Cantamessa, S., Copetta, A., Anastasia, F., Gamalero, E., Lingua, G., Berta, G., & Bona., E. (2020). Selected autochthonous rhizobia, applied in combination with AM fungi, improve seed quality of common bean cultivated in reduced fertilization condition. *Applied Soil Ecology, 148*, 103507.

Messa, V.R., & Savioli, M.R. (2021). Improving sustainable agriculture with arbuscular mycorrhizae. *Rhizosphere, 19*, 100412.

Mitra, D., Uniyal, N., Panneerselvam, P., Senapati, A., & Ganeshamurthy, A.N. (2019). Role of mycorrhiza and its associated bacteria on plant growth promotion and nutrient management in sustainable agriculture. *International Journal of Life Sciences and Applied Sciences, 1*, 1–10.

Molina-Romero, D., Juárez-Sánchez, S., Venegas, B., Ortíz-González, C.S., Baez, A., Morales-García, Y.E., & Muñoz-Rojas, J. (2021). A bacterial consortium interacts with different varieties of maize, promotes the plant growth, and reduces the application of chemical fertilizer under field conditions. *Frontiers in Sustainable Food Systems, 4*, 293.

Moreira, H., Pereira, S.I.A., Vega, A., Castro, P.M.L., & Marques, A.P.G.C. (2020). Synergistic effects of arbuscular mycorrhizal fungi and plant growth-promoting bacteria benefit maize growth under increasing soil salinity. *Journal of Environmental Management, 257*, 109982.

Mousa, W.K., Shearer, C., Limay-Rios, V., Ettinger, C.L., Eisen, J.A., & Raizada, M.N. (2016). Root-hair endophyte stacking in finger millet creates a physicochemical barrier to trap the fungal pathogen *Fusarium graminearum. Nature Microbiology, 1*, 16167.

Nayem, S.A., Salahuddin, M., Chowdhury, M., Sultana, N., Masum, G.Z.H., Rahman, M.S., & Jamal, M.A.H.M. (2020). Combined effect of salt stress and *Xanthomonas axonopodis* pv citri on citrus (*Citrus aurantifolia*). *Heliyon, 6*, e03403.

Nepomuceno, R.A., Brown, C.M.B., Mojica, P.N., & Brown, M.B. (2019). Biological control potential of Vesicular Arbuscular Mycorrhizal Root Inoculant (VAMRI) and associated phosphate solubilizing bacteria, *Pseudochrobactrum asaccharolyticum* against soilborne phytopathogens of Onion (*Allium cepa* L. var. Red Creole). *Archives of Phytopathoogy and Plant Protection, 52*, 714–732.

Nicolson, T.H. (1959). Mycorrhiza in the Gramineae. I. Vesicular arbuscular endophytes, with special reference to the external phase. *Transactions of the British Mycological Society, 42*, 421–438.

Novero, M., Genre, A. Szczyglowski, K., & Bonfante, P. (2008). Root hair colonization by mycorrhizal fungi. In A.M.C. Emons, & T. Ketelaar (Eds.) *Root Hairs* (pp. 315–338). Springer.

Orozco-Mosqueda, M.D.C., Glick, B.R., & Santoyo, G. (2020). ACC deaminase in plant growth-promoting bacteria (PGPB): An efficient mechanism to counter salt stress in crops. *Microbiological Research, 235*, 126439.

Pandey, R., & Garg, N. (2017). High effectiveness of *Rhizophagus irregularis* is linked to superior modulation of antioxidant defence mechanisms in *Cajanus cajan* (L.) Millsp. genotypes grown under salinity stress. *Mycorrhiza, 27*, 669–682.

Pedranzani, H., Rodríguezrivera, M., Gutiérrez, M., Porcel, R., Hause, B., & Ruizlozano, J.M. (2016). Arbuscular mycorrhizal symbiosis regulates physiology and performance of *Digitaria eriantha* plants subjected to abiotic stresses by modulating antioxidant and jasmonate levels. *Mycorrhiza, 26*, 141–152.

Pereira, S.I.A., Monteiro, C., Vega, A.L., & Castro, P.M.L. (2016). Endophytic culturable bacteria colonizing *Lavandula dentata* L. plants: Isolation, characterization and evaluation of their plant growth-promoting activities. *Ecological Engineering, 87*, 91–97.

Peterson, R.L., & Farquhar, M.L. (1996). Root hairs: Specialized tubular cells extending root surfaces. *The Botanical Review, 62*, 1–40.

Porcel, R., Redondo-Gómez, S., Mateos-Naranjo, E., Aroca, R., Garcia, R., & Ruiz-Lozano, J.M. (2015). Arbuscular mycorrhizal symbiosis ameliorates the optimum quantum yield of photosystem II and reduces non-photochemical quenching in rice plants subjected to salt stress. *Journal of Plant Physiology, 185*, 75–83.

Rahneshan, Z., Nasibi, F., & Moghadam, A.A. (2018). Effects of salinity stress on some growth, physiological, biochemical parameters and nutrients in two pistachio (*Pistacia* vera L.) rootstocks. *Journal of Plant Interactions, 13*, 73–82.

Raklami, A., Bechtaoui, N., Tahiri, A.I., Anli, M., Meddich, A., & Oufdou, K. (2019). Use of rhizobacteria and mycorrhizae consortium in the open field as a strategy for improving crop nutrition, productivity and soil fertility. *Frontiers in Microbiology, 10*, 1106.

Ramos, A.C., Façanha, A.R., & José, F. (2008). Proton (H^+) flux signature for the presymbiotic development of the arbuscular mycorrhizal fungi. *New Phytologist, 178*, 177–188.

Relwani, L., Krishna, P., & Reddy, M.S. (2008). Effect of carbon and nitrogen sources on phosphate solubilization by a wild-type strain and UV-induced mutants of *Aspergillus tubingensis*. *Current Microbiology, 57*, 401–406.

Robertson-Albertyn, S., Alegria Terrazas, R., Balbirnie, K., Blank, M., Janiak, A., Szarejko, I., Chmielewska, B., Karcz, J., Morris, J., Hedley, P.E., George, T.S., & Bulgarelli, D. (2017). Root hair mutations displace the barley rhizosphere microbiota. *Frontiers in Plant Science, 8*, 1094.

Ruiz-Lozano, J.M., Porcel, R., Azcón, C., & Aroca, R. (2012). Regulation by arbuscular mycorrhizae of the integrated physiological response to salinity in plants: New challenges in physiological and molecular studies. *Journal of Experimental Botany, 63*, 695–709.

Ruiz-Sanchez, M., Armada, E., Munoz, Y., Garcia de Salamone, I.E., Aroca, R., Ruiz-Lozano, J.M., & Azcon, R. (2011). *Azospirillum* and arbuscular mycorrhizal colonization enhance rice growth and physiological traits under well-watered and drought conditions. *Journal of Plant Physiology, 168*, 1031–1037.

Sah, S.K., Reddy, K.R., & Li, J. (2016). Abscisic acid and abiotic stress tolerance in crop plants. *Frontiers in Plant Science, 7*, 571.

Saia, S., Amato, G., Frenda, A.S., Giambalvo, D., & Ruisi, P. (2014). Influence of arbuscular mycorrhizae on biomass production and nitrogen fixation of berseem clover plants subjected to water stress. *PLoS ONE, 9*, e90738.

Sanchez-Blanco, M.J., Ferrandez, T., Morales, M.A., Morte, A., & Alarcon, J.J. (2004). Variations in water status, gas exchange, and growth in *Rosmarinus officinalis* plants infected with *Glomus deserticola* under drought conditions. *Journal of Plant Physiology, 161*, 675–682.

Sanower-Hossain, M. (2019). Present scenario of global salt affected soils, its management and importance of salinity research. *International Research Journal of Biological Sciences, 1*, 1–3.

Santoyo, G., Guzmán-Guzmán, P., Parra-Cota, F.I., Santos-Villalobos, S., de los Orozco-Mosqueda, M.d.C., & Glick, B.R. (2021). Plant growth stimulation by microbial consortia. *Agronomy, 11*, 219.

Sharifi, M., Ghorbanli, M., & Ebrahimzadeh, H. (2007). Improved growth of salinity-stressed soybean after inoculation with pre-treated mycorrhizal fungi. *Journal of Plant Physiology, 164*, 1144–1151.

Sharma, I.P., & Sharma, A.K. (2017). Physiological and biochemical changes in tomato cultivar PT-3 with dual inoculation of mycorrhiza and PGPR against root-knot nematode. *Symbiosis, 71*, 175–183.

Shaul, O., Galili, S., Volpin, H., Ginzberg, I., Elad, Y., Chet, I., & Kapulnik, Y. (1999). Mycorrhiza-induced changes in disease severity and PR protein expression in tobacco leaves. *Molecular Plant-Microbe Interactions, 12*, 1000–1007.

Shokri, S., & Maadi, B. (2009). Effects of arbuscular mycorrhizal fungus on the mineral nutrition and yield of *Trifolium alexandrium* plants under salinity stress. *Journal of Agronomy, 8*, 79–83.

Singh, R., Soni, S.K., & Kalra, A. (2013). Synergy between *Glomus fasciculatum* and a beneficial *Pseudomonas* in reducing root diseases and improving yield and forskolin content in *Coleus forskohlii* Briq. under organic field conditions. *Mycorrhiza, 23*, 35–44.

Song, Y., Chen, D., Lu, K., Sun, Z., & Zeng, R. (2015). Enhanced tomato disease resistance primed by arbuscular mycorrhizal fungus. *Frontiers in Plant Science, 6*, 786.

Staal, M., De-Cnodder, T., Simon, D., Vandenbussche, F., Vander Straeten, D., Verbelen, J., Elzenga, T., & Vissenberg, K. (2011). Apoplastic alkalinization is instrumental for the inhibition of cell elongation in the *Arabidopsis* root by the ethylene precursor 1-aminocyclopropane-1-carboxylic acid. *Plant Physiology, 155*, 2049–2055.

Talaat, N.B., & Shawky, B.T. (2011). Influence of arbuscular mycorrhizae on yield, nutrients, organic solutes, and antioxidant enzymes of two wheat cultivars under salt stress. *Journal of Plant Nutrition and Soil Science, 174*, 283–291.

Tominaga-Wada, R., & Wada, T. (2014). Regulation of root hair cell differentiation by R3 MYB transcription factors in tomato and *Arabidopsis*. *Frontiers in Plant Science, 5*, 91.

Topalovic, O., & Vestergard, M. (2021). Can microorganisms assist the survival and parasitism of plant-parasitic nematodes? *Trends in Parasitology, 37*, 947–958.

Trivedi, P., Leach, J.E., Tringe, S.G., Sa, T., & Singh, B.K. (2020). Plant– microbiome interactions: From community assembly to plant health. *Nature Reviews Microbiology, 18*, 607–621.

van der Heijden, M.G.A., Martin, F.M., Selosse, M.A., & Sanders, I.R. (2015). Mycorrhizal ecology and evolution: The past, the present, and the future. *New Phytologist, 205*, 1406–1423.

Wang, X., Pan, Q., Chen, F., Yan, X., & Liao, H. (2011). Effects of co-inoculation with arbuscular mycorrhizal fungi and rhizobia on soybean growth as related to root architecture and availability of N and P. *Mycorrhiza, 21*, 173–181.

Wang, Y., Thorup-Kristensen, K., Jensen, L.S., & Magid, J. (2016). Vigorous root growth is a better indicator of early nutrient uptake than root hair traits in spring wheat grown under low fertility. *Frontiers in Plant Science, 7*, 865.

Wang, Y., Wang, M., Li, Y., Wu, A., & Huang, J. (2018). Effects of arbuscular mycorrhizal fungi on growth and nitrogen uptake of *Chrysanthemum morifolium* under salt stress. *PLoS One, 13*, e0196408.

Wu, Q.S., He, J.D., Srivastava, A.K., Zou, Y.N., & Kuca, K. (2019). Mycorrhizas enhance drought tolerance of citrus by altering root fatty acid compositions and their saturation levels. *Tree Physiology, 39*, 1149–1158.

Wu, Q.S., Liu, C.Y., Zhang, D.J., Zou, Y.N. He, X.H., & Wu, Q.H. (2016). Mycorrhiza alters the profile of root hairs in trifoliate orange. *Mycorrhiza, 26*, 237–247.

Wu, Q.S., Xia, R.X., & Zou, Y.N. (2006). Reactive oxygen metabolism in mycorrhizal and non-mycorrhizal citrus (*Poncirus trifoliata*) seedlings subjected to water stress. *Journal of Plant Physiology, 163*, 1101–1110.

WWAP (United Nations World Water Assessment Programme). 2018. *The United Nations World Water Development Report 2018: Nature-Based Solutions*. UNESCO, Paris.

Xu, L., Li, T., Wu, Z., Feng, H., Yu, M., Zhang, X., & Chen, B. (2018). Arbuscular mycorrhiza enhances drought tolerance of tomato plants by regulating the 14-3-3 genes in the ABA signaling pathway. *Applied Soil Ecology, 125*, 213–221.

Zhu, J.K. (2016). Abiotic stress signaling and responses in plants. *Cell, 167*, 313–324.

Zou, Y.N., Wu, Q.S., & Kuca, K. (2020). Unravelling the role of arbuscular mycorrhizal fungi in mitigating the oxidative burst of plants under drought stress. *Plant Biology, 1*, 50–57.

Zou, Y.N., Zhang, D.J., Liu, C.Y., & Wu, Q.S. (2019). Relationships between mycorrhizae and root hairs. *Pakistan Journal of Botany, 51*, 727–733.

5 An Overview of the Role of Endophytic Bacteria as Plant Growth Promoters and Biocontrol Agents Against Pathogens of Different Crop Plants

Papan Chowhan and Arka Pratim Chakraborty

5.1 INTRODUCTION

Plants can form relationships with other members of the ecosystem to thrive in their natural environment. Microorganisms are one of the most important organisms that form beneficial associations with plants (Santoyo et al., 2016, Adeleke and Babalola, 2020). Microorganisms are abundant in nature and are found in almost every ecosystem. Plants also have a variety of microorganisms. Plant beneficial bacteria are a type of bacteria that provide various benefits to host plants, such as helping the plant survive various biotic and abiotic pressures that interfere with plant growth (Miliute et al., 2015). These bacteria can be found in the host plant both externally and internally. Bacteria that live outside of host plants are epiphytic, i.e., living in the leafy or rhizospheric area, i.e., living in the roots of plants within the soil (Compant et al., 2010). While bacteria live and thrive inside the host plant they are called endophytic bacteria (Hardoim et al., 2008). Many factors that promote plant growth are shared by all these bacterial categories (Compant et al., 2010). Bacteria can thus promote host plant growth, and plants can also "choose" their microbiome to infect colonias with beneficial bacteria, including those that live within plant tissues (Marasco et al., 2012; Hardoim et al., 2008; Rashid et al., 2012). Endophytes are plant-related microorganisms that reside in the tissues of the host plant without damaging the host plant (Ryan et al., 2008; Schulz and Boyle, 2006). Some researchers have proposed a definitive definition of endophytic activity, claiming that any bacterium is an endophyte that can be separated from plant tissue by water or extracted from inside the plant and does not affect the plant visually (Hallman et al., 1997). Endophytic bacteria are therefore bacteria that infect the inner tissues of plants without causing infection or injury to their host (Holliday, 1989; Schulz and Boyle, 2006), and almost all plant species in the world hold one or more endophyte (Strobel et al., 2004) . These endophytes can be found in almost every part of the host plant (roots, stem, leaf, flower, fruit, and seeds). Many bacteria are considered endophyte roots from the rhizosphere (Aloo et al., 2019). Some rhizosphere bacteria enter the internal tissue directly from the outer root zone to become endophytes (Nwachukwu et al., 2021). Endophytic bacteria that have the effect of promoting plant growth on hosted plants are known as endophytic bacteria that promote plant growth (PGPEB). They can also help plants thrive in drought-prone areas and limited

DOI: 10.1201/9781003354109-5

nutrient-rich soils (Banik et al., 2019). Endophytic bacteria *Bacillus cereus*, *Pseudomonas* sp., and *Pseudomonas otitidis* have been found to increase soy release under drought pressure by Dubey et al. (2021). *Pseudomonas brenneri, Ewingella Americana, Pantoea agglomerans, Bacillus cereus*, and *Pseudomonas otitidis* are some examples of endophytic bacteria genera (Babalola et al., 2021; Dubey et al., 2021; Rana et al., 2021).

5.2 DIVERSITY FOUND IN ENDOPHYTIC BACTERIA

Several factors influence the type of endophytic diversity found in a plant. Apart from bacterial ability to colonize plants as endophytes, the host plant and environmental conditions can have a significant impact on a plant's endophytic diversity. The sort of endophytic bacteria found in a host plant depends on its age, genotype, geographic region, and even the tissue being examined (Hallmann and Berg, 2006). Furthermore, host plant growth stages can influence a plant's endophytic diversity, with plant stages with more nutrient availability having more bacterial variety (Shi et al., 2014). Not only that, but climatic factors might have an impact on a plant's endophytic invaders (Penuelas et al., 2012). The nature of a plant's endophytic community is highly determined by its host species (Ding and Melcher, 2016). Endophytic diversity varies significantly among plant species growing in the same soil. In fact, different cultivars of a plant species grown in the same soil might differ in endophytic diversity as demonstrated by Granér et al. (2003) for four different cultivars of *Brassica napus* with various endophytic bacterial inhabitants (Table 5.1). Thus, the host plant species strongly influence the type of endophytic bacteria colonizing it. More interestingly, the type of soil in which a plant is grown might influence its endophytic community. Rashid et al. (2012) used one tomato cultivar in 15 different agricultural soils to isolate different types of endophytic bacteria. These results suggest that the presence of several endophytes is related to the heterogeneous nature of soil samples (Table 5.2).

TABLE 5.1
Some Common Endophytic Bacterial Genera Isolated From Agronomic Plants Reported in the Literature

Plant	Endophytic Bacterial Genera
Wheat (*Triticum aestivum*)	*Bacillus, Burkholderia, Flavobacterium, Klebsiella, Microbispora, Micrococcus, Micromonospora, Mycobacterium, Nacardiodes, Rathayibacter, Streptomyces*
Banana (*Musa* sp.)	*Azospirillum, Burkholderia, Citrobacter, Herbaspirillum, Klebsiella*
Radish (*Raphanus sativus*)	*Proteobacteria, Salmonella*
Tomato (*Solanum lycopersicum*)	*Brevibacillus, Escherichia, Pseudomonas, Salmonella*
Pineapple (*Ananas comosus*)	*Azospirillum, Burkholderia*
Maize (*Zea mays*)	*Achromobacter, Agrobacterium, Arthrobacter, Bacillus, Burkholderia, Corynebacterium, Curtobacterium, Enterobacter, Erwinia, Herbaspirillum, MicrobacteriumMicrococcus, Paenibacillus, Phyllobacterium, Pseudomonas, Rhizobium, Serratia*
Sugar cane (*Saccharum officinarum*)	*Acetobacter, Gluconacetobacter, Herbaspirillum, Klebsiella*
Cotton (*Gossypium herbaceum*)	*Bacillus, Burkholderia, Clavibacter, Erwinia, Phyllobacterium, Pseudomonas*
Soybean (*Glycine max*)	*Erwinia, Agrobacterium, Pseudomonas, Klebsiella, Enterobacter, Pantoea, Bacillus*
Alfalfa (*Medicago sativa*)	*Bacillus, Erwinia, Microbacterium, Pseudomonas, Salmonella*

(*continued*)

TABLE 5.1 (Continued)
Some Common Endophytic Bacterial Genera Isolated From Agronomic Plants Reported in the Literature

Plant	Endophytic Bacterial Genera
Clover (*Trifolium* sp.)	*Agrobacterium, Bacillus, Methylobacterium, Pseudomonas, Rhizobium*
Canola (*Brassica napus*)	*Acidovorax, Agrobacterium, Aureobacterium, Bacillus, Chryseobacterium, Cytophaga, Flavobacterium, Micrococcus, Pseudomonas, Rathayibacter,*
Potato (*Solanum tuberosum*)	*Acidovorax, Acinetobacter, Actinomyces, Agrobacterium, Alcaligenes, Arthrobacter, Bacillus, Capnocytophaga, Chryseobacterium, Comamonas, Corynebacterium, Curtobacterium, Enterobacter, Erwinia, Klebsiella, Leuconostoc, Methylobacterium, Micrococcus, Paenibacillus, Pantoea, Pseudomonas, Psychrobacter, Serratia, Shewanella, Sphinogomonas, Stenotrophomonas, Streptomyces, Vibrio, Xanthomonas*
Grapevine (*Vitis vinifera*)	*Comamonas, Enterobacter, Klebsiella, MoraxellaPantoea, Pseudomonas, Rahnella, Rhodococcus, Staphylococcus, Xanthomonas*
Black pepper (*Piper nigrum*)	*Arthrobacter, Bacillus, Curtobacterium, Micrococcus, Pseudomonas, Serratia*
Sugar beet (*Beta vulgaris*)	*Bacillus sp., Erwinia sp., Pseudomonas sp., Corynebacterium sp., Lactobacillus sp., Xanthomonas sp.*
Walnut (*Juglans regia*)	*Bacillus subtilis HB1310*
Lebanon oak (*Quercus libani*)	*B. firmus, Pseudomonas protegens, Stenotrophomonas maltophilia*
Rice (*Oryza sativa*)	*Agrobacterium, Azoarcus, Azorhizobium, Azospirillum, Bacillus, Bradyrhizobium, Burkholderia, Chromobacterium, Enterobacter, Herbaspirillum, Ideonella, Klebsiella, Micrococcus, Pantoea, Pseudomonas, Rhizobium, Serratia, Stenotrophomonas*
Cucumber (*Cucumis sativus*)	*Agrobacterium, Bacillus, Burkholderia, Chryseobacterium, Clavibacter, Curtobacterium, Enterobacter, Micrococcus, Paenibacillus, Phyllobacterium,Pseudomonas, Serratia, Stenotrophomonas*
Red clover (*Trifolium pratense*)	*Acidovorax, Agrobacterium, Arthobacter, Bacillus, Bordetella, Cellulomonas, Comamonas, Curtobacterium, Escherichia, Klebsiella, Methylobacterium, Micrococcus, Pantoea, Pasteurella, Phyllobacterium, Pseudomonas, Psychrobacter, Rhizobium, Serratia, Sphingomonas, Variovorax, Xanthomonas*
Carrot (*Daucus carota*)	*Agrobacterium, Bacillus, Klebsiella, Pseudomonas, Rhizobium, Salmonella, Staphylococcus*
Common bean (*Phaseolus vulgaris*)	*Acinetobacter radioresistens, Acinetobacter sp., Agromyces mediolanus, Agromyces* sp., *B. amyloliquefaciens, B. bataviensis, B. muralis, B. subtilis, B. thuringiensis, B. niacini, Bacillus sp., Brevibacillus agri, Brevundimonas vesicularis, Delftia tsuruhatensis, Dietzia cinnamea, Enterobacter asburiae, E. hormaechei, Frigoribacterium faeni, Kocuria palustris, Lysinibacillus sphaericus, Microbacterium foliorum, M. phyllosphaerae, M. testaceum, Microbacterium* sp., *Methylobacterium populi, Micrococcus luteus, Paenibacillus cineris, P. lautus, Paenibacillus* sp., *Pseudomonas aeruginosa, Rhizobium larrymoorei, Rhodococcus erythropolis, Staphylococcus caprae, S. epidermidis, S. kloosii, S. sanguinis, S. warneri, S. saprophyticus, Staphylococcus* sp., *Sphingobacterium multivorum, Sphingomonas dokdonensis, Sporosarcina aquimarina, Sporosarcina* sp., *Stenotrophomonas maltophilia, Stenotrophomonas* sp.

Source: Hallmann et al., 1997; Rosenblueth and Martínez-Romero, 2006; Miliute et al., 2015, Afzal et al., 2019; Yadav and Yadav, 2019.

TABLE 5.2
Diversity of Endophytic Bacteria Isolated from Some Wild Plants

Plant	Endophytic Bacteria	References
Cannabis sativa	*Acinetobacter gyllenbergii*	(Afzal et al., 2015)
	Acinetobacter nosocomialis	
	Acinetobacter parvus	
	Acinetobacter pittii	
	Bacillus anthracis	
	Chryseobacterium sp.	
	Enterobacter asburiae	
	Enterococcus casseliflavus	
	Nocardioides albus	
	Nocardioides kongjuensis	
	Pantoea vagans	
	Planomicrobium chinense	
	Pseudomonas taiwanensis	
	Rhizobium radiobacter	
	Streptomyces eurocidicus	
	Xanthomonas gardneri	
Halimione portulacoides	*Altererythrobacter*	(Fidalgo et al., 2016)
	Hoeflea	
	Labrenzia	
	Marinilactibacillus	
	Microbacterium	
	Salinicola	
	Vibrio	
Sedum alfredii Hance	*Burkholderia*	(Zhang et al., 2013)
	Sphingomonas	
	Variovorax	
Calystegia soldanella	*Acinetobacter*	(Park et al., 2005)
	Arthrobacter	
	Chryseobacterium	
	Curtobacterium	
	Enterobacter	
	Microbacterium	
	Pantoea	
	Pedobacter	
	Pseudomonas	
	Stenotrophomonas	
Commelina communis	*Arthrobacter*	(Sun et al., 2010)
	Arthrobacter	
	Bacillus	
	Bacillus pumilus	
	Herbaspirillum	
	Microbacterium	
	Sphingomonas	
Noccaea caerulescens	*Agreia*	(Visioli et al., 2014)
	Arthrobater	
	Bacillus	
	Kocuria	
	Microbacterium	
	Sthenotrophomonas	
	Variovorax	

(*continued*)

TABLE 5.2 (Continued)
Diversity of Endophytic Bacteria Isolated from Some Wild Plants

Plant	Endophytic Bacteria	References
Cressa cretica, Salicornia brachiate, Suadea nudiflora, Sphaeranthus indicus	*Acinetobacter* *Arthrobacter* *Bacillus* *Kocuria* *Oceanobacillus* *Paenibacillus* *Pseudomononas* *Virgibacilus*	(Arora et al., 2014)
Pinus contorta	*Bacillus* *Brevibacillus* *Brevundimonas* *Cellulomonas* *Kocuria* *Paenibacillus* *Pseudomonas*	(Bal et al., 2012)
Alyssum bertolonii	*Arthrobater* *Bacillus* *Curtobacterium* *Leifsonia* *Microbacterium* *Paenibacillus* *Pseudomonas* *Staphylococcus*	(Barzanti et al., 2007)
Polygonum pubescens	*Rahnella* sp. *JN6*	(He et al., 2013)
Thuja plicata	*Arthrobacter* *Bacillus* *Paenibacillus* *Pseudomonas* *Streptoverticillium*	(Bal et al., 2012)
Alnus firma	*Bacillus* sp.	(Shin et al., 2012)
Elymus mollis	*Acinetobacter* *Arthrobacter* *Chryseobacterium* *Enterobacter* *Exiguobacterium* *Flavobacterium* *Klebsiella* *Pedobacter* *Pseudomonas* *Stenotrophomonas*	(Park et al., 2005)
Pinus sylvestris	*Bacillus thuringiensis*	(Babu et al., 2013)

TABLE 5.2 (Continued)
Diversity of Endophytic Bacteria Isolated from Some Wild Plants

Plant	Endophytic Bacteria	References
Elsholtzia Splendens	*Xanthomonas translucens* *Acinetobacter calcoaceticus* *Acinetobacter junii* *Bacillus* *Bacillus firmus* *Bacillus megaterium* *Burkholderia* *Exiguobacterium aurantiacum* *Micrococcus luteus* *Moraxella* *Paracoccus* *Serratia marcescens*	(Sun et al., 2010)

Source: Afzal et al., 2019.

5.3 PLANT–ENDOPHYTE INTERACTIONS AND PROMOTION OF PLANT GROWTH

Endophytes benefit plants in two ways: directly (through phytohormone synthesis) and indirectly (by phytoremediation). However, the specific methods by which endophytes promote growth are unclear (Hardoim et al., 2008). Nitrogen fixation, modulation of plant hormone levels (auxin, cytokinin, ethylene, and gibberellin), phosphate, iron, and potassium solubilization, synthesis of secondary metabolites, antibiotic activities against various plant pathogens, and enhancing plant responses to abiotic stresses are all ways that endophytic bacteria help plant growth (Rajini et al., 2020). Phosphate-solubilizing bacteria such as *Pseudomonas*, *Burkholderia*, *Paraburkholderia*, *Novosphingobium*, and *Ochrobactrum* have been shown to increase the biomass output of Chinese fir seedlings due to their useful properties (Chen et al., 2021a). PGPEB uses direct mechanisms such as nitrogen fixation and the synthesis of phytohormones like auxin, cytokinin, gibberellin, and abscisic acid (Maheshwari et al., 2019). They can also reduce ethylene levels by using the enzyme 1-aminocyclopropane-1-carboxylate (ACC) deaminase, as well as boost mineral solubilization (zinc, iron, phosphorus, sulphur, and potassium) and increased survivability under stressful conditions including drought and soil salinity (Dubey et al., 2021). Endophytic bacteria can promote plant growth indirectly by activating plant responses or creating secondary metabolites that protect plants from different phytopathogens (Santoyo et al., 2012). Induced systemic resistance (ISR) is one such indirect method, which can be accomplished through certain plant response pathways, such as the jasmonic acid (JA) pathway (van Loon and Glick, 2004; Asghari et al., 2020). Montejano-Ramírez et al. (2020) reported that endophytic bacteria can stimulate plant defensive responses via different signaling pathways.

5.3.1 Different Mechanisms of Host Plant Growth Promotion

Endophytic bacteria have been found to directly benefit host plants by helping them in obtaining nutrients and improving plant growth by regulating growth-related hormones, which can help plants grow better under normal and stressful situations (Ma et al., 2016).

5.3.2 Nutrient Acquisition

Soils usually lack a sufficient amount of one or more of the macro and micronutrient compounds that are necessary for plant growth. Endophytic bacteria can assist their host plants in obtaining higher levels of limiting plant nutrients such as nitrogen, iron, and phosphorus, etc. (Glick, 2012).

5.3.2.1 Nitrogen Availability

Endophytic bacteria can boost the availability of nitrogen for their host plants. By exhibiting nitrogenase activity, these bacteria can supply fixed atmospheric nitrogen to their host plants (Montanez et al., 2012). According to Gupta et al. (2013) endophytic nitrogen-fixing bacteria may improve the rate of nitrogen fixation and its accumulation in plants growing in nitrogen-limited soils. *Paenibacillus* strain P22, a nitrogen-fixing endophyte identified in poplar trees, was reported to contribute to the host plant's total nitrogen pool (Scherling et al., 2009). Madhaiyan et al. (2013) reported a nitrogen-fixing endophyte, *Enterobacter* sp. R4-368, that colonized root and stem tissues and enhanced early plant growth and seed productivity in both sterilized and non-sterilized soils.

5.3.2.2 Phosphate Solubilization

One of the most typical characteristics of endophytic bacteria is phosphorus solubilization. Endophytic bacteria can improve phosphorus availability for plants by solubilizing precipitated phosphates through methods such as acidification, chelation, ion exchange, and organic acid synthesis (Nautiyal et al., 2000). They can also boost soil phosphorus availability by secreting acid phosphatase, which can mineralize organic phosphorus (Van Der Heijden et al., 2008). As for example, *Pantoea* sp., an endophytic bacteria from the *Enterobacteriaceae* family, has a phosphate-solubilizing property (Sulbaran et al., 2009).

5.3.2.3 Siderophores' Production

Endophytic bacteria create iron chelating compounds called siderophores that can bind insoluble ferric ions. Plants can obtain iron from these bound siderophores via root-based chelate breakdown or ligand exchange (Ma et al., 2016; Rajkumar et al., 2009). Endophytes with a siderophore-producing characteristic are frequent because free iron ions are scarce inside plant tissues (Sessitsch et al., 2004). Bacterial siderophores also improve the bioavailability of metals other than iron, allowing the plant to develop faster (Rajkumar et al., 2010).

5.3.3 Phytohormone Production

Endophytic bacteria can help host plants acquire nutrients and metabolism by releasing growth-regulating phytohormones. Endophytes increase legume crop output by creating indole acetic acid (Khan et al., 2014; Patel and Patel, 2014), ethylene (Kang et al., 2012; Long et al., 2010; Straub et al., 2013), gibberellic acid (Khan et al., 2014), and auxins (Dutta et al., 2014).

5.3.3.1 IAA Production

According to Tsavkelova et al. (2007) endophytic bacteria isolated from terrestrial orchids produced indole acetic acid (IAA). They discovered that the bacteria's culture supernatant boosted root production in kidney beans by considerably increasing root length and the quantity of developing roots, implying that bacterial IAA may have a function in plant root growth.

5.3.3.2 Production of Cytokinins and Gibberellins

Several studies reported that many plant beneficial endophytic bacteria can produce cytokinins and gibberellinshave. Bhore et al. (2010) identified cytokinin-like compounds in the broth-extracts

TABLE 5.3
Some Bacterial Endophytes and Their Plant Growth-Promoting Traits

Endophyte Species	Host Plant	Plant Growth-Promoting Traits	Reference
Stenotrophomonas maltophilia R551-3	Poplar	IAA synthesis, ACC deaminase	Taghavi et al., (2009)
Gluconacetobacter diazotrophicus Pal5	Sugarcane, rice, coffe, tea	Nitrogen fixation, auxin synthesis	Bertalan et al., (2009)
Serratia proteamaculans 568	Soybean	IAA synthesis, ACC deaminase, acetoin and 2,3-butanediol synthesis	Taghavi et al., (2009)
Klebsiella pneumoniae 342	Maize, wheat	Nitrogen fixation	Fouts et al., (2008)
Enterobacter sp. *638*	Poplar	Siderophore, IAA, acetoin and 2,3-butanediol synthesis	Taghavi et al., (2009)
Pseudomonas stuzeri A1501	Rice	Nitrogen fixation	Yan et al., (2008)
Burkholderia spp. *KJ006*	Rice	ACC deaminase, nif gene cluster, antifungal action (indirect PGP)	Kwak et al., (2012)
Azoarcus sp. BH72	Rice	Nitrogen fixation	Krause et al., (2006)
Pseudomonas putida W619	Poplar	IAA synthesis, ACC deaminase	Taghavi et al., (2009)
Burkholderia phytofirmans PsJN	Potato, tomato, maize, barley, onion, canola, grapevine	IAA synthesis, ACC deaminase	Weilharter et al., (2011)
Azospirillum lipoferum 4B	Rice, maize, wheat	Nitrogen fixation, phytohormone secretion	Wisniewski-Dyé et al., (2011)

Source: Santoyo et al., 2016.

of two endophytic bacteria identified as *Psuedomonas resinovorans* and *Paenibacillus polymaxa,* isolated from *Gynura procumbens*, using cucumber cotyledon greening bioassay.

5.3.3.3 Control of Ethylene

Endophytic bacteria can regulate ethylene levels of host plant by generating aminocyclopropane-1-carboxylic acid (ACC) deaminase, an enzyme that hydrolyzes ACC, a precursor to the plant hormone ethylene. Various plant growth-promoting endophytic bacteria have been reported to have ACC deaminase activity (Nikolic et al., 2011; Rashid et al., 2012; Zhang et al., 2011). A list of some bacterial endophytes and their plant-growth promoting traits have been represented in Table 5.3.

5.4 OTHER BENEFICIAL ACTIVITIES OF ENDOPHYTIC BACTERIA

5.4.1 Bioremediation

Bioremediation is the biological removal or breakdown of contaminants in the environment. Several non-neutralizable harmful metabolites are secreted by plants. Because such metabolites can affect the environment's ecology, they must be bioremediated using "associative bacteria." Endophytes assist plant bioremediation in a variety of ways. Endophytes in plant tissues reduce heavy metal stress (Zhang et al., 2012) and decompose harmful chemicals and metabolites (Han et al., 2011).

Entophytes remove greenhouse gases from the air (Stępniewska and Kuźniar, 2013) and prevent pest growth outside the plant tissues (Azevedo et al., 2000).

5.4.2 Phytoremediation

Plant-based remediation of environmental and soil contaminants is known as phytoremediation. Phytoremediation is a relatively new concept in agriculture, and it appears to be less expensive than existing engineering alternatives. This is a more "green" and pragmatic method that is attracting widespread attention from the scientific community throughout the world. Endophytes equip plants with the necessary breakdown pathways for enhanced biodegradation and phytotoxicity reduction (Weyens et al., 2009). By fixing nitrogen, solubilizing minerals, creating phytohormones, generating siderophores, converting nutrients and ACC, endophytic bacteria can promote phytoremediation and help the host plants (Germaine et al., 2009; Germaine et al., 2006; Rajkumar et al., 2009; Stępniewska and Kuźniar, 2013) (Figure 5.1). As compared with non-symbiotic ones. Endophytic bacteria are able to reduce metal toxicity by altering metal transport and accumulation in plants. Endophytic bacterium *Serratia nematodiphila* LRE07 reduced the Cd-induced changes by increasing biomass and higher photosynthetic pigment content in leaves of *Solanum nigrum* L. (Wan et al., 2012). Similar results were found for an endophyte *Bacillus* sp. SLS18 on sweet sorghum (Luo et al., 2012). Heavy metal photoextraction is reported to be assisted by many endobacteria (Rajkumar et al., 2009). The usage of genetically altered endophyte strains in scientific studies is expanding due to the great potentials in phytoremediation. As for example bioengineered *Pseudomonas putida VM1441* (pNAH7), was reported to protect the host plant from naphthalene's phytotoxic effects (Germaine et al., 2009). In another case, genetically modified *Burkholderia cepacia* L.S.2.4 was able to decompose toluene in plant tissues (Barac et al., 2004).

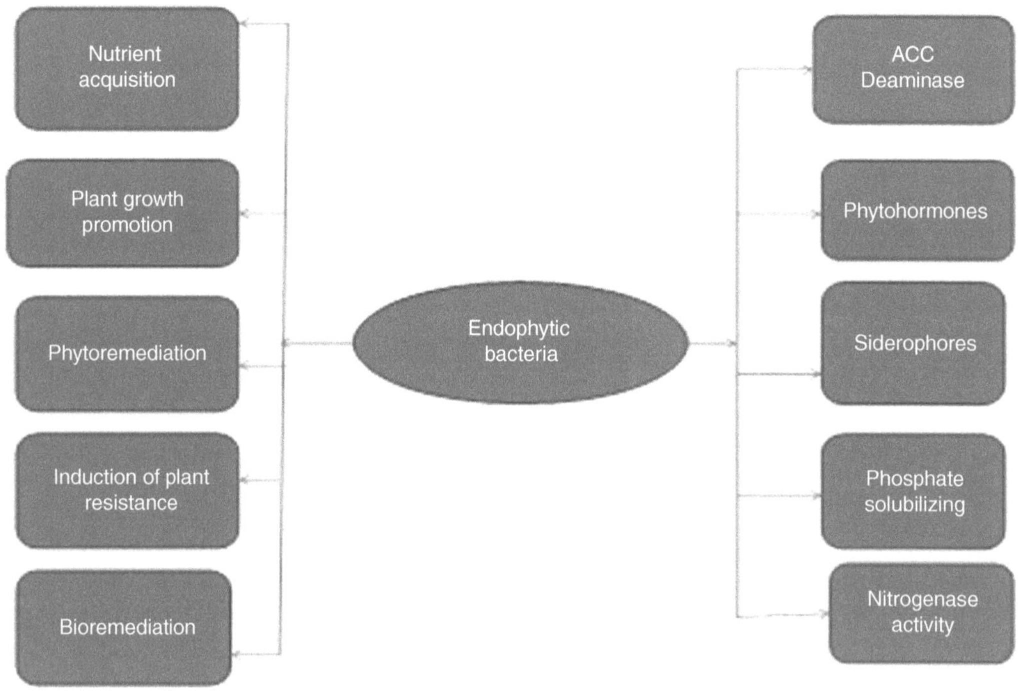

FIGURE 5.1 Direct and indirect mechanisms used by endophyte to enhance plant growth.

5.5 BIOLOGICAL CONTROL AND SUPPRESSION OF PLANT DISEASES BY ENDOPHYTIC BACTERIA

Endophytic bacteria produce a variety of metabolites that improve host tolerance to various stressors, making them beneficial to plants and able to act as promising biological agents in controlling many plant diseases. Many researchers have reported the biocontrol role of endophytes against various diseases. The following are the main mechanisms by which the endophyte suppresses certain diseases (Xia et al., 2022):

1. Compete with viruses for a niche and nutrients
2. To produce different lytic enzymes
3. Producing antimicrobial products, as well
4. To promote system resistance in hosted plants

Instead of simple structures toxic, plant endophyte metabolites incorporate various bioactive substances that strengthen the host's immunity to pathogens. Therefore, using one or more natural active ingredients as a lead ingredient has a promising mechanism for future green pesticides (Xia et al., 2022). However, there are some challenges you must overcome, such as the following:

(i) Many endophytes have never been enlarged and are unknown.
(ii) The effects of endophyte biocontrol are not always consistent with field testing.
(iii) The mechanisms of plant–endophyte interactions are not well understood.
(iv) Endophyte and its metabolites do not have available databases.
(v) Fermentation produces low amounts of metabolites.

5.5.1 Biocontrol Effects of Endophytic Bacteria

Biocontrol is a natural way of protecting our plants from various phytopathogens (Rybakova et al., 2015). Endophyte affects plant viruses directly or indirectly by altering the internal ecology (Gao et al., 2010). Endophytes, like rhizosphere bacteria, reduce phytopathogen resistance by competing with nutrients in the same natural environment and releasing antimicrobial chemicals. Some endophytes release antibiotics into the endosphere, limiting phytopathogens in the rhizosphere (Bara et al., 2013; Castillo et al., 2003; Franco et al., 2007). Various antibiotics such as coronamycin, ecomycins, kakadumycins, munumbicins, pseudomycins, and xiamycins are produced by a variety of endophytes (Castillo et al., 2003; Christina et al., 2013; Ezra et al., 2004). Different fungal pathogens are resistant to a variety of endophytes. Endophytes have been shown to reduce *Fusarium* wilt in bananas (Chen et al., 2011b), tomato (Benhamou et al., 1998), and capsicum (Sundaramoorthy et al., 2012) plants. *Pseudomonas fluorescens* PICF7, a root endophyte, was found to inhibit pathogenic *Verticillium dahlia* colonization in olive tissues (Prieto et al., 2009). Similarly the disease caused by *Phytophthora infestans* was reduced by the endophyte *Pseudomonas putida* P9, which was isolated from the potato plant (Andreote et al., 2009). Endophytes have also been shown to be potent in the biocontrol of the *Xanthomonas citri* subsp. *citri* (Xcc), a pathogen that causes citrus canker (Brunings and Gabriel, 2003) (Table 5.4). In many citrus and tropical citrus regions around the world, the disease affects all types of commercial oranges (Sharma and Sharma, 2009).

5.5.2 Suppression of Phytopathogens and Induction of Plant Resistance by Endophytic Bacteria

Endophytic bacteria indirectly promote host plant growth by inhibiting the growth of phytopathogens and pests. They have the ability to create chemicals that antagonize different phytopathogens.

TABLE 5.4
Biocontrol of Endophytic Bacteria Against Different Fungal Pathogens

Endophytic Bacterial Isolates	Host Plant	Pathogenic Fungi	References
Bacillus spp.	Cucumber	*Collectotrichum orbuculariae*	Raupach and Kloepper (1998)
B. pumilus and Pseudomonas spp.	Oak	*Fusarium* spp.	Brooks et al., (1994)
Bacillus spp.	Tomato	*F. oxysporum f.*sp. *lycopersici*	Benhamou et al., (1998)
P. aeruginosa 7 NSK2	Tomato	*Botrytis cinerea*	Audenaert et al., (2002a, 2002b)
P. fluorescens GRP3	Rice	*Rhizoctonia* sp.	Pathak et al., (2004)
Burkhloderia phytofirmans Ps JN	Grapevine	*Botrytis cinerea*	Compant et al., (2008)
P. fluorescens EP1	Sugarcane	*Colletotrichum falcatum*	Senthil et al., (2003)
Bacillus and *Pseudomonas*	Wheat	*F. graminearum*	Nourozian et al., (2006)
B. pumilus SE34	Pea	*F. oxysporum f.*sp. *pisi*	Chaudhary et al., (2009)
B. subtilis	Wheat	*Gaemanomyces graminis tritici*	Liu et al., (2009)
P. fluorescens CHA0	Tomato	*Pythium ultimum* and *F. oxysporum f.* sp. *pisi*	Ardebili et al., (2011)
Bacillus spp., *Pseudomonas* spp.	Peanut	*Sclerotinia sclerotiorum, S. minor, S. rolfsii* and *Fusarium solani*	Tonelli et al., (2010)
Bacillus spp., *Pseudomonas* spp.	Soyabean	*R. solani, F. oxysporum and S. rolfsii, C. truncatum, A. alternata, Macrophomina phaseolina*	Dalal and Kulkarni (2013)
Pseudomonas and *Burkholderia*	Banana	*F. oxysporum f.*sp. *cubense*	Fishal et al., (2010)
Pseudomonas spp.	Cucumber	*F. oxysporum f.*sp. *cucumerinum*	Ozaktan et al., (2015)

Source: Muthukumar et al., 2017.

Endophytic bacteria can target both bacterial and fungal infections (Lodewyckx et al., 2002). The most typically reported genera for antibacterial activity against phytopathogens are *Actinobacteria*, *Bacillus*, *Enterobacteor*, *Paenibacillus*, *Pseudomonas*, and *Serratia* (Aktuganov et al., 2008; Liu et al., 2010; Lodewyckx et al., 2002). Endophytic bacteria have been demonstrated to efficiently suppress fungal disease in plants such as black pepper, potato, and wheat (Aravind et al., 2009; Coombs et al., 2004; Sessitsch et al., 2004).The antimicrobial activities against fungi result from the synthesis of different fungal cell-wall targeting enzymes like chitinase, proteases, and glucanases (Zarei et al., 2011; Zhang et al., 2012). Plant beneficial bacteria can also defend their host plants from phytopathogens using a method known as induced systemic resistance (ISR) (Table 5.5). Endophytic bacteria-induced ISR can protect the host from fungal, bacterial, and viral pathogens

TABLE 5.5
Inducing Plant Resistance by Some Bacterial Endophytes

Endophytic Strain	Host Plant	Pathogen	References
Burkholderia gladioli E39CS3	*Crocus sativus Linn.*	*Fusarium oxysporum.*	Ahmad et al., (2021)
Bacillus sp. 2P2	Tomato (*Solanum lycopersicum L.*)	*Sclerotium rolfsii*	Sahu et al., (2019)
Paecilomyces Variotii SJ1	Tobacco (*Nicotiana tabacum)*	virus	Peng et al., (2020)
Klebsiella pneumoniae HR1	*Vigna mungo L.*	*Macrophomina phaseolina*	Dey et al., (2019)
Pseudomonas aeruginosa H40, Stenotrophomonas maltophila H8, Bacillus subtilis H18	*Pisum. sativum, Brassica oleracea, Capsicum annuum*	Rhizoctonia solani	Selim et al., (2017)
Streptomyces albidoflavus OsiLf-2	Rice (*Oryza sativa*)	*Magnaporthe oryzae,*	Gao et al., (2020)
Pseudomonas viridiflava	Canola (*Brassica napus*)	*Xanthomonas campestris, Sclerotinia sclerotiorum, Leptosphaeria maculans*	Romero et al., (2019)

Source: Xia et al., 2022.

(Alvin et al., 2014; Audenaert et al., 2002a,b). Salicylic acid (SA), jasmonic acid (JA), and ethylene (ET) mediated pathways, which are usually a network of interconnected signaling pathways are involved in ISR induction, and can be used by endophytic bacteria to induce ISR (Pieterse et al., 2012).

5.6 CONCLUSION

Many bacterial endophytes have gained a lot of attention because of their ability to support plant development either directly or indirectly through biocontrol agents. Despite the fact that the processes of plant growth promotion in rhizospheric and endophyte bacteria may be extremely similar, much of the study has been done in rhizosphere bacteria, thinking that the mechanism is identical in endophytes. The rhizosphere, on the other hand, is not the same as the internal plant tissues (Santoyo et al., 2016). Our understanding of what transforms a rhizospheric bacteria into a plant endophyte is still incomplete (Santoyo et al., 2016). Numerous endophyte genes have been identified, providing some insight into the bacteria's endophytic existence. Because of their unique microenvironment in the endosphere, our current understanding of endophyte function is limited. Endophyte-specific research is uncommon, and their metabolic ideas are not well understood (Ali et al., 2014). Plant-associated beneficial endophytic bacteria have a lot of potential as biofertilizers and biopesticides. Although many of these bacteria have been found, and they may infect a wide range of hosts, but they rarely produce consistent results in the field. One explanation for this is our lack of knowledge of the complex dynamics that govern plant–endophyte interactions. Finally, because endophytic diversity has been understudied, there is a good chance of discovering novel and interesting bacteria in unknown wild plants. Wild plants have a tendency for surviving severe environments and combating biotic and abiotic challenges, which may be aided by the unique endophytes they harbor (Afzal et al., 2019). It is significant to find these rare and promising bacterial endophytes with broad plant-beneficial properties. By better understand the mechanisms of function of endophytes, scientists may be able to modify bacterial endophytes to assist their potential to boost plant growth and development in the future.

REFERENCES

Adeleke BS, Babalola OO. (2020). The endosphere microbial communities, a great promise in agriculture. *Int. Microbiol.* 24: 1–17.

Afzal I, Shinwari ZK, Sikandar S, Shahzad S. (2019). Plant beneficial endophytic bacteria: mechanisms, diversity, host range and genetic determinants. *Microbiol. Res.* 221: 36–49.

Afzal I, Zabta KS, Irum I. (2015). Selective isolation and characterization of agriculturally beneficial endophytic bacteria from wild hemp using canola. *Pak. J. Bot.* 47: 1999–2008.

Ahmad T, Bashir A, Farooq S, Riyaz-Ul-Hassan S. (2021). *Burkholderia gladioli* E39CS3, an endophyte of *Crocus sativus* Linn., induces host resistance against corm-rot caused by *Fusarium oxysporum. J. Appl. Microbiol.* 132: 495–508.

Aktuganov G, Melentjev A, Galimzianova N, Khalikova E, Korpela T, Susi P. (2008). Wide-range antifungal antagonism of *Paenibacillus ehimensis* IB-Xb and its dependence on chitinase and β-1, 3-glucanase production. *Can. J. Microbiol.* 54: 577–587.

Ali S, Charles TC, Glick BR. (2014). Amelioration of high salinity stress damage by plant growth-promoting bacterial endophytes that contain ACC deaminase. *Plant. Physiol. Biochem.* 80: 160–167.

Aloo BN, Makumba BA, Mbega ER. (2019). The potential of Bacilli rhizobacteria for sustainable crop production and environmental sustainability. *Microbiol. Res.* 219: 26–39.

Alvin A, Miller KI, Neilan BA. (2014). Exploring the potential of endophytes from medicinal plants as sources of antimycobacterial compounds. *Microbiol. Res.* 169: 483–495.

Andreote FD, de Araujo WL, de Azevedo JL, van Elsas JD, da Rocha UN, van Overbeek LS. (2009). Endophytic colonization of potato (*Solanum tuberosum* L.) by a novel competent bacterial endophyte, *Pseudomonas putida* strain P9, and its effect on associated bacterial communities. *Appl. Environ. Microbiol.* 75: 3396–3406.

Aravind R, Kumar A, Eapen S, Ramana K. (2009). Endophytic bacterial flora in root and stem tissues of black pepper (*Piper nigrum* L.) genotype: isolation, identification and evaluation against *Phytophthora capsici. Lett. Appl. Microbiol.* 48: 58–64

Ardebili ZO, Ardebili NO, Mahdi Hamdi SM. (2011). Physiological effects of *Pseudomonas fluorescens* CHA0 on tomato (*'Lycopersicon esculentum* 'Mill.) plants and its possible impact on *Fusarium oxysporum* f. sp.'*Lycopersici*'. *Aust. J. Crop. Sci.* 5: 1631–1638.

Arora S, Patel PN, Vanza MJ, Rao GG. (2014). Isolation and characterization of endophytic bacteria colonizing halophyte and other salt tolerant plant species from coastal Gujarat. *Afr. J. Microbiol. Res.* 8: 1779–1788.

Asghari S, Harighi B, Ashengroph M, Clement C, Aziz A, Esmaeel Q, Ait Barka E. (2020). Induction of systemic resistance to *Agrobacterium tumefaciens* by endophytic bacteria in grapevine. *Plant Pathol.* 69: 827–837.

Audenaert K, De Meyer GB. Höfte MM. (2002a). Abscisic acid determines basal susceptibility of tomato to *Botrytis cinerea* and suppresses salicylic acid-dependent signaling mechanisms. *Plant Physiol.* 128: 491–501.

Audenaert K, Pattery T, Cornelis T, Höfte M. (2002b). Induction of systemic resistance to *Botrytis cinerea* in tomato by *Pseudomonas aeruginosa* 7NSK2: role of salicylic acid, pyochelin, and pyocyanin. *Mol. Plant-Microbe. Interact.* 15(11): 1147–1156.

Azevedo JL, Maccheroni W Jr, Pereira JO, de Araújo WL. (2000). Endophytic microorganisms: a review on insect control and recent advances on tropical plants. *Electron. J. Biotechnol.* 3: 15–16.

Babalola OO, Adeleke BS, Ayangbenro AS. (2021). Whole genome sequencing of sunflower root-associated *Bacillus cereus. Evol. Bioinf. Online* 17: 1–6.

Babu AG, Kim JD, Oh BT. (2013). Enhancement of heavy metal phytoremediation by *Alnus firma* with endophytic *Bacillus thuringiensis* GDB-1. *J. Hazard. Mater.* 250: 477–483.

Bal A, Anand R, Berge O, Chanway CP. (2012). Isolation and identification of diazotrophic bacteria from internal tissues of *Pinus contorta* and *Thuja plicata. Can. J. For. Res.* 42: 807–813.

Banik A, Dash GK, Swain P, Kumar U, Mukhopadhyay SK, Dangar TK. (2019). Application of rice (*Oryza sativa* L.) root endophytic diazotrophic *Azotobacter* sp. strain Avi2 (MCC 3432) can increase rice yield under green house and field condition. *Microbiol. Res.* 219: 56–65.

Bara R, Aly AH, Pretsch A, Wray V, Wang B, Proksch P, Debbab A. (2013). Antibiotically active metabolites from *Talaromyces wortmannii*, an endophyte of Aloe vera. *J. Antibiot.* 66: 491–493.

Barac T, Taghavi S, Borremans B, Provoost A, Oeyen L, Colpaert JV, Vangronsveld J, van der Lelie D. (2004). Engineered endophytic bacteria improve phytoremediation of water-soluble, volatile, organic pollutants. *Nat. Biotechnol.* 22: 583–588.

Barzanti R, Ozino F, Bazzicalupo M, Gabbrielli R, Galardi F, Gonnelli C, Mengoni A. (2007). Isolation and characterization of endophytic bacteria from the nickel hyperaccumulator plant Alyssum bertolonii. *Microb. Ecol.* 53: 306–316.

Benhamou N, Kloepper JW, Tuzun S. (1998). Induction of resistance against Fusarium wilt of tomato by combination of chitosan with an endophytic bacterial strain: ultrastructure and cytochemistry of the host response. *Planta* 204: 153–168.

Bertalan M, Albano R., de Pádua V, Rouws L, Rojas C, Hemerly A, Teixeira K, Schwab S, Araujo J, Oliveira A, Franc¸ a L, Magalhães V, Alquéres S, Cardoso A, Almeida W, Loureiro MM, Nogueira E, Cidade D, Oliveira D, Simão T, Macedo J, Valadão A, Dreschsel M, Freitas F, Vidal M, Guedes H, Rodrigues E, Meneses C, Brioso P, Pozzer L, Figueiredo D, Montano H, Junior J, de Souza Filho G, Martin Quintana Flores V, Ferreira B, Branco A, Gonzalez P, Guillobel H, Lemos M., Seibel L, Macedo J, Alves-Ferreira, M., Sachetto-Martins G, Coelho A, Santos E, Amaral G, Neves A, Pacheco AB, Carvalho D, Lery L, Bisch P, Rössle SC, Urményi T, Rael Pereira A, Silva R, Rondinelli E, von Krüger W, Martins O, Baldani JI, Ferreira PC. (2009). Complete genome sequence of the sugarcane nitrogen-fixing endophyte *Gluconacetobacter diazotrophicus* Pal5. *BMC Genomics* 10: 450.

Bhore SJ, Ravichantar N, Loh CY. (2010). Screening of endophytic bacteria isolated from leaves of *Sambung Nyawa* [*Gynura procumbens* (Lour.) Merr.] for cytokinin-like compounds. *Bioinformation* 5: 191.

Brooks DS, Gonzalez CF, Appel DN, Filer TH. (1994). Evaluation of endophytic bacteria as potential biological control agent for oak wilt. *Biologic Cont.* 4: 373–381.

Brunings AM, Gabriel DW. (2003). *Xanthomonas citri*: breaking the surface. *Mol. Plant Pathol.* 4: 141–157.

Castillo U, Harper JK, Strobel GA, Sears J, Alesi K, Ford E, Lin J, Hunter M, Maranta M, Ge H, Yaver D, Jensen JB, Porter H, Robison R, Millar D, Hess WM, Condron M, Teplow D. (2003). Plant growth-promoting endophytic bacteria and their potential to improve Kakadumycins, novel antibiotics from *Streptomyces* sp. NRRL 30566, an endophyte of Grevillea pteridifolia. *FEMS Microbiol. Lett.* 224: 183–190.

Chaudhary RG, Dhar V, Singh RK. (2009). Association of fungi with wilt complex of lentil at different crop growth stages and moisture regimes. *Arch. Phytopathol. Plant Protect.* 42: 340–343.

Chen B, Zhu J, Sun QG, Zheng YH, Huang HQ, Bao SX. (2011b). A bacterial endophyte from banana: its isolation, identification, activity to Fusarium wilt and PGPR effect to banana seedlings. *Microbiology/ Weishengwuxue Tongbao* 38: 199–205.

Chen J, Zhao G, Wei Y, Dong Y, Hou L, Jiao R. (2021a). Isolation and screening of multifunctional phosphate solubilizing bacteria and its growth-promoting effect on Chinese fir seedlings. *Sci. Rep.* 11: 9081.

Christina A, Christopher V, Bhore SJ. (2013). Endophytic bacteria as a source of novel antibiotics: an overview. *Pharmacogn. Rev.* 7: 11–16.

Compant S, Clément C, Sessitsch A. (2010). Plant growth-promoting bacteria in the rhizo-and endosphere of plants: their role, colonization, mechanisms involved and prospects for utilization. *Soil Biol. Biochem.* 42: 669–678.

Compant S, Nowak J, Coenye T, Clément C, Ait Barka E. (2008). Diversity and occurrence of *Burkholderia* spp. in the natural environment. *FEMS Microbiol. Rev.* 32: 607–626.

Coombs JT, Michelsen PP, Franco CM. (2004). Evaluation of endophytic actinobacteria as antagonists of Gaeumannomyces graminis var. tritici in wheat. *Biol. Control* 29: 359–366.

Dalal J, Kulkarni N. (2013). Antagonistic and plant growth promoting potentials of indigenous endophytic bacteria of soybean (*Glycine max* (L) Merril). *Curr. Res. Microbiol. Biotechnol.* 1: 62–69.

Dey S, Dutta P, Majumdar S. (2019). Biological control of *Macrophomina phaseolina* in *Vigna mungo* L. by Endophytic *Klebsiella pneumoniae* HR1. *Jordan J Biol Sci.* 12: 219–227.

Ding T, Melcher U. (2016). Influences of plant species, season and location on leaf endophytic bacterial communities of non-cultivated plants. *PLoS One* 11: e0150895.

Dubey A, Saiyam D, Kumar A, Hashem A, Abd_Allah EF, Khan, ML. (2021). Bacterial root endophytes: characterization of their competence and plant growth promotion in soybean (Glycine max (L.) Merr.) under drought stress. *Int. J. Environ. Res. Publ. Health* 18: 931.

Dutta D, Puzari KC, Gogoi R, Dutta P. (2014). Endophytes: exploitation as a tool in plant protection. *Braz. Arch. Biol. Technol.* 57: 621–629.

Ezra D, Castillo UF, Strobel GA, Hess WM, Porter H, Jensen JB, Condron MA, Teplow DB, Sears J, Maranta M. (2004). Coronamycins, peptide antibiotics produced by a verticillate *Streptomyces* sp.(MSU-2110) endophytic on Monstera sp. *Microbiology* 150: 785–793.
Fidalgo C, Henriques I, Rocha J, Tacão M, Alves A. (2016). Culturable endophytic bacteria from the salt marsh plant *Halimione portulacoides*. *Environ. Sci. Pollut. Res.* 23: 10200–10214.
Fishal EM, Meon S, Yun M. (2010). Induction of tolerance to fusarium wilt and defense-related mechanisms in the plantlets of susceptible Berangan Banana pre-inoculated with *Pseudomonas* sp. (UPMP3) and *Burkholderia* sp. (UPMB3). *Agric. Sci. China* 9: 1140–1149.
Fouts DE, Tyler HL, DeBoy RT, Daugherty S, Ren Q, Badger JH, Durkin AS, Huot H, Shrivastava S, Kothari S, Dodson RJ. (2008). Complete genome sequence of the N2-fixing broad host range endophyte *Klebsiella pneumoniae* 342 and virulence predictions verified in mice. *PLoS Genet.* 25(4): e1000141.
Franco C, Michelsen P, Percy N, Conn V, Listiana E, Moll S, Loria R, Coombs J. (2007). Actinobacterial endophytes for improved crop performance. *Australas Plant Pathol.* 36: 524–531.
Gao F, Dai C, Liu X. (2010). Mechanisms of fungal endophytes in plant protection against pathogens. *Afr. J. Microbiol. Res.* 4: 1346–1351.
Gao Y, Zeng XD, Ren B, Zeng JR, Xu T, Yang YZ, Zhu YH. (2020). Antagonistic activity against rice blast disease and elicitation of host defence response capability of an endophytic *Streptomyces albidoflavus* OsiLf2. *Plant Pathol.* 69: 259–271.
Germaine KJ, Keogh E, Ryan D, Dowling DN. (2009). Bacterial endophyte-mediated naphthalene phytoprotection and phytoremediation. *FEMS Microbiol. Lett.* 296: 226–234.
Germaine KJ, Liu X, Cabellos GG, Hogan JP, Ryan D, Dowling DN. (2006). Bacterial endophyte-enhanced phytoremediation of the organochlorine herbicide 2,4-dichlorophenoxyacetic acid. *FEMS Microbiol. Ecol.* 57: 302–310.
Glick BR. (2012). Plant growth-promoting bacteria: mechanisms and applications. *Scientifica* 2012: 963401.
Granér G, Persson P, Meijer J, Alström S. (2003). A study on microbial diversity in different cultivars of Brassica napus in relation to its wilt pathogen, *Verticillium longisporum. FEMS Microbiol. Lett.* 224: 269–276.
Gupta G, Panwar J, Jha PN. (2013). Natural occurrence of Pseudomonas aeruginosa, a dominant cultivable diazotrophic endophytic bacterium colonizing *Pennisetum glaucum* (L.) R. Br. *Appl. Soil Ecol.* 64: 252–261.
Hallmann J, Berg G. (2006). *Spectrum and population dynamics of bacterial root endophytes*. In Schulz, BJE, Boyle, CJC, Sieber, TN, (eds) *Microbial Root Endophytes* (pp. 15–31). Springer.
Hallmann, J, Quadt-Hallmann A, Mahaffee WF, Kloepper JW. (1997). Bacterial endophytes in agricultural crops. *Can. J. Microbiol.* 43: 895–914.
Han JI, Choi HK, Lee SW, Orwin PM, Kim J, Laroe SL, Kim TG, O'Neil J, Leadbetter JR, Lee SY, Hur CG, Spain JC, Ovchinnikova G, Goodwin L, Han C. (2011). Complete genome sequence of the metabolically versatile plant growth-promoting endophyte *Variovorax paradoxus* S110. *J. Bacteriol.* 193: 1183–1190.
Hardoim PR, van Overbeek LS, Elsas JD. (2008). Properties of bacterial endophytes and their proposed role in plant growth. *Trends Microbiol.* 16: 463–471.
He H, Ye Z, Yang D, Yan J, Xiao L, Zhong T, Yuan M, Cai X, Fang Z, Jing Y. (2013). Characterization of endophytic Rahnella sp. JN6 from *Polygonum pubescens* and its potential in promoting growth and Cd, Pb, Zn uptake by Brassica napus. *Chemosphere* 90: 1960–1965.
Holliday P. (1989). *A Dictionary of Plant Pathology*. Cambridge University Press, Cambridge.
Kang JW, Khan Z, Doty SL. (2012). Biodegradation of trichloroethylene by an endophyte of hybrid poplar. *Appl. Environ. Microbiol.* 78: 3504–3507.
Khan AL, Waqas M, Kang SM, Al-Harrasi A, Hussain J, Al-Rawahi A, Al-Khiziri S, Ullah I, Ali L, Jung HY, Lee IJ. (2014). Bacterial endophyte *Sphingomonas* sp. LK11 produces gibberellins and IAA and promotes tomato plant growth. *J. Microbiol.* 52: 689–695.
Krause A, Ramakumar A, Bartels D, Battistoni F, Bekel T, Boch J, Böhm, M.,Friedrich F, Hurek T, Krause L, Linke B, McHardy AC, Sarkar A, Schneiker S, Syed AA, Thauer R, Vorhölter FJ, Weidner S, Pühler A, Reinhold-Hurek, B, Kaiser O, Goesmann A. (2006). Complete genome of the mutualistic,N2-fixing grass endophyte *Azoarcus* sp. strain BH72. *Nat. Biotechnol.* 24: 1385–1391.
Kwak MJ, Song JY, Kim SY, Jeong H, Kang SG, Kim BK, Kwon SK, Lee CH, Yu DS, Park SH, Kim JF. (2012). Complete genome sequence of the endophytic bacterium *Burkholderia* sp. strain KJ006. *J. Bacteriol.* 194: 4432–4433.

Liu B, Qiao H, Huang L, Buchenauer H, Han Q, Kang Z, Gong Y. (2009). Biological control of take-all in wheat by endophytic *Bacillus subtilis* E1R-j and potential mode of action. *Biol. Control.* 49: 277–285.

Liu X, Jia J, Atkinson S, Cámara M, Gao K, Li H, Cao J. (2010). Biocontrol potential of an endophytic *Serratia* sp. G3 and its mode of action. *World J. Microbiol. Biotechnol.* 26: 1465–1471.

Lodewyckx C, Vangronsveld J, Porteous F, Moore ER, Taghavi S, Mezgeay M, der Lelie Dv. (2002). Endophytic bacteria and their potential applications. *Crit. Rev. Plant Sci.* 21: 583–606.

Long HH, Sonntag DG, Schmidt DD, Baldwin IT. (2010). The structure of the culturable root bacterial endophyte community of *Nicotiana attenuata* is organized by soil composition and host plant ethylene production and perception. *New Phytol.* 185: 554–567.

Luo S, Xu T, Chen L, Chen J, Rao C, Xiao X, Wan Y, Zeng G, Long F, Liu C, Liu Y. (2012). Endophyte-assisted promotion of biomass production and metal-uptake of energy crop sweet sorghum by plant-growth-promoting endophyte Bacillus sp. *SLS18. Appl. Microbiol. Biotechnol.* 93: 1745–1753.

Ma Y, Rajkumar M, Zhang C, Freitas H. (2016). Beneficial role of bacterial endophytes in heavy metal phytoremediation. *J. Environ. Manage.* 174: 14–25.

Madhaiyan M, Jin TY, Roy JJ, Kim SJ, Weon HY, Kwon SW. (2013). *Pleomorphomonas diazotrophica* sp. nov., an endophytic N-fixing bacterium isolated from root tissue of Jatropha curcas *L. Int. J. Syst. Evol. Microbiol.* 63: 2477–2483.

Maheshwari R, Bhutani N, Bhardwaj A, Suneja P. (2019). Functional diversity of cultivable endophytes from Cicer arietinum and Pisum sativum: bioprospecting their plant growth potential. *Biocatal. Agric. Biotechnol.* 20: 101229.

Marasco R, Rolli E, Ettoumi B, Vigani G, Mapelli F, Borin S, Abou-Hadid AF, El-Behairy UA, Sorlini C, Cherif A, Zocchi G, Daffonchio D. (2012). A drought resistance-promoting microbiome is selected by root system under desert farming. *PLoS One* 7: e48479.

Miliute I, Buzaite O, Baniulis D, Stanys V. (2015). Bacterial endophytes in agricultural crops and their role in stress tolerance: a review. *Zemdirbyste-Agric.* 102: 465–478.

Montanez A, Blanco AR., Barlocco C, Beracochea M, Sicardi M. (2012). Characterization of cultivable putative endophytic plant growth promoting bacteria associated with maize cultivars (Zea mays L.) and their inoculation effects in vitro. *Appl. Soil Ecol.* 58: 21–28.

Montejano-Ramírez V, García-Pineda E, Valencia-Cantero E. (2020). Bacterial compound N, N-dimethylhexadecylamine modulates expression of iron deficiency and defense response genes in *Medicago truncatula* independently of the jasmonic acid pathway. *Plants* 9: 624.

Muthukumar A, Udhayakumar R, Naveenkumar R. (2017). Role of bacterial endophytes in plant disease control. In Maheshwari, D, Annapurna, K, (eds) *Endophytes: Crop productivity and protection* (pp. 133–161). Springer.

Nautiyal CS, Bhadauria S, Kumar P, Lal H, Mondal R, Verma D. (2000). Stress induced phosphate solubilization in bacteria isolated from alkaline soils. *FEMS Microbiol. Lett.* 182: 291–296.

Nikolic B, Schwab H, Sessitsch A. (2011). Metagenomic analysis of the 1-aminocyclopropane- 1-carboxylate deaminase gene (acdS) operon of an uncultured bacterial endophyte colonizing *Solanum tuberosum* L. *Arch. Microbiol.* 193: 665–676.

Nourozian J, Etebarian HR, Khodakaramian G. (2006). Biological control of *Fusarium graminearum* on wheat by antagonistic bacteria. *Songklanakarin J. Sci. Technol.* 28(Suppl 1): 29–38.

Nwachukwu BC, Ayangbenro AS, Babalola OO. (2021). Elucidating the rhizosphere associated bacteria for environmental sustainability. *Agric. For.* 11: 75.

Ozaktan H, Çakır B, Gül A, Yolageldi L, Akköprü A, Fakhraei D, Akbaba M. (2015). *Isolation and evaluation of endophytic bacteria against Fusarium Oxysporum f. sp. cucumerinum* Infecting cucumber Plants. *Austin. J. Plant. Biol.* 1(1): 1003.

Park MS, Jung SR, Lee MS, Kim KO, Do JO, Lee KH, Kim SB, Bae KS. (2005). Isolation and characterization of bacteria associated with two sand dune plant species, *Calystegia soldanella* and *Elymus mollis. J. Microbiol.-Seoul.* 43: 219.

Patel MV, Patel RK. (2014). Indole-3-acetic acid (IAA) production by endophytic bacteria isolated from saline dessert, the Little Rann of Kutch. *CIBTech J. Microbiol.* 3:17–28.

Pathak A, Sharma A, Johri BN, Sharma AK. (2004). *Pseudomonas strain GRP3 induces systemic resistance to sheath blight in rice.* International Rice Research Notes.

Peng C, Zhang A, Wang Q, Song Y, Zhang M, Ding X, Zhu C. (2020). Ultrahigh-activity immune inducer from Endophytic Fungi induces tobacco resistance to virus by SA pathway and RNA silencing. *BMC Plant Biol.* 20: 1–14.

Penuelas J, Rico L, Ogaya R, Jump A, Terradas J. (2012). Summer season and long term drought increase the richness of bacteria and fungi in the foliar phyllosphere of *Quercus ilex* in a mixed Mediterranean forest. *Plant Biol.* 14: 565–575.

Pieterse CM, Van der Does D, Zamioudis C, Leon-Reyes A, Van Wees SC. (2012). Hormonal modulation of plant immunity. *Annu. Rev. Cell Dev. Biol.* 28 : 189–521.

Prieto P, Navarro-Raya C, Valverde-Corredor A, Amyotte SG, Dobinson KF, Mercado-Blanco J. (2009). Colonization process of olive tissues by *Verticillium dahliae* and its in planta interaction with the biocontrol root endophyte *Pseudomonas fluorescens* PICF7. *Microb. Biotechnol.* 2: 499–511.

Rajini SB, Nandhini M, Udayashankar AC, Niranjana SR, Lund OS, Prakash HS. (2020). Diversity, plant growth-promoting traits, and biocontrol potential of fungal endophytes of Sorghum bicolor. *Plant Pathol.* 69: 642–654.

Rajkumar M, Ae N, Freitas H. (2009). Endophytic bacteria and their potential to enhance heavy metal phytoextraction. *Chemosphere* 77: 153–160.

Rajkumar M, Ae N, Prasad MNV, Freitas H. (2010). Potential of siderophore-producing bacteria for improving heavy metal phytoextraction. *Trends Biotechnol.* 28: 142–149.

Rana KL, Kour D, Kaur T, Devi R, Yadav A, Yadav AN. (2021). Bioprospecting of endophytic bacteria from the Indian Himalayas and their role in plant growth promotion of maize (Zea mays L.). *J. Appl. Biol. Biotechnol.* 9: 41–50.

Rashid S, Charles TC, Glick BR. (2012). Isolation and characterization of new plant growth-promoting bacterial endophytes. *Appl. Soil Ecol.* 61: 217–224.

Raupach GS, Kloepper JW. (1998). Mixtures of plant growth promoting rhizobacteria enhance biological control of multiple cucumber pathogens. *Phytopathology* 88: 1158–1164.

Romero FM, Rossi FR, Gárriz A, Carrasco P, Ruíz OA. (2019). A bacterial endophyte from apoplast fluids protects canola plants from different phytopathogens via antibiosis and induction of host resistance. *Phytopathology* 109: 375–383.

Rosenblueth M, Martínez-Romero E. (2006). Bacterial endophytes and their interactions with hosts. *Mol. Plant Microbe. Interact.* 19: 827–837.

Ryan RP, Germaine K, Franks A, Ryan DJ, Dowling DN. (2008). Bacterial endophytes: recent developments and applications. *FEMS Microbiol. Lett.* 278: 1–9.

Rybakova D, Cernava T, Köberl M, Liebminger S, Etemadi M, Berg G. (2015). Endophytes-assisted biocontrol: novel insights in ecology and the mode of action of Paenibacillus. *Plant Soil* 405: 1–16.

Sahu PK, Singh S, Gupta A, Singh UB, Brahmaprakash GP, Saxena AK. (2019). Antagonistic potential of bacterial endophytes and induction of systemic resistance against collar rot pathogen Sclerotium rolfsii in tomato. *Biol. Control* 137: 104014.

Santoyo G, Moreno-Hagelsieb G, del Carmen Orozco-Mosqueda M, Glick BR. (2016). Plant growth-promoting bacterial endophytes. *Microbiol. Res.* 183: 92–99.

Santoyo G, Orozco-Mosqueda MdC, Govindappa M. (2012). Mechanisms of biocontrol and plant growth-promoting activity in soil bacterial species of Bacillus and Pseudomonas: a review. *Biocontrol Sci. Technol.* 22: 855–872.

Scherling C, Ulrich K, Ewald D, Weckwerth W. (2009). A metabolic signature of the beneficial interaction of the endophyte *Paenibacillus* sp. isolate and in vitro–grown poplar plants revealed by metabolomics. *Mol. Plant Microbe. Interact.* 22: 1032–1037.

Schulz B, Boyle C. (2006). *Microbial Root Endophytes*. Springer, Berlin.

Selim HM, Gomaa NM, Essa AM. (2017). Application of endophytic bacteria for the biocontrol of Rhizoctonia solani (Cantharellales: ceratobasidiaceae) damping-off disease in cotton seedlings. *Biocontrol Sci. Technol.* 27: 81–95.

Senthil N, Raguchander T, Viswanathan R, Samiyappan R. (2003). Talc formulated fluorescent pseudomonads for sugarcane red rot suppression and enhanced yield under field conditions. *Sugar Tech.* 5: 37–43.

Sessitsch A, Reiter B, Berg G. (2004). Endophytic bacterial communities of field-grown potato plants and their plant-growth-promoting and antagonistic abilities. *Can. J. Microbiol.* 50: 239–249.

Sharma SK, Sharma R. (2009). Citrus canker approaching century: a review. *TFSB* 2: 54–65.

Shi Y, Yang H, Zhang T, Sun J, Lou K. (2014). Illumina-based analysis of endophytic bacterial diversity and space-time dynamics in sugar beet on the north slope of Tianshan mountain. *Appl. Microbiol. Biotechnol.* 98: 6375–6385.

Shin MN, Shim J, You Y, Myung H, Bang KS, Cho M, Kamala-Kannan S, Oh BT. (2012). Characterization of lead resistant endophytic Bacillus sp. MN3-4 and its potential for promoting lead accumulation in metal hyperaccumulator Alnus firma. *J. Hazard. Mater.* 199: 314–320.

Stępniewska Z, Kuźniar A. (2013). Endophytic microorganisms—promising applications in bioremediation of greenhouse gases. *Appl. Microbiol. Biotechnol.* 97: 9589–9596.

Straub D, Yang H, Liu Y, Tsap T, Ludewig U. (2013). Root ethylene signalling is involved in *Miscanthus sinensis* growth promotion by the bacterial endophyte *Herbaspirillum frisingense* GSF30(T). *J. Exp. Bot.* 64: 4603–4615.

Strobel G, Daisy B, Castillo U, Harper J. (2004). Natural products from endophytic microorganisms. *J. Nat. Prod.* 67: 257–268.

Sulbaran M, Perez E, Ball MM, Bahsas A, Yarzabal LA. (2009). Characterization of the mineral phosphate-solubilizing activity of *Pantoea agglomerans* MMB051 isolated from an iron-rich soil in southeastern Venezuela (Bolivar State). *Curr. Microbiol.* 58: 378–383.

Sun LN, Zhang YF, He LY, Chen ZJ, Wang QY, Qian M, Sheng XF. (2010). Genetic diversity and characterization of heavy metal-resistant-endophytic bacteria from two copper-tolerant plant species on copper mine wasteland. *Bioresour. Technol.* 101: 501–509

Sundaramoorthy S, Raguchander T, Ragupathi N, Samiyappan R. (2012). Combinatorial effect of endophytic and plant growth promoting rhizobacteria against wilt disease of *Capsicum annum* L. caused by *Fusarium solani. Biol. Control* 60: 59–67.

Taghavi S, Garafola C, Monchy S, Newman L, Hoffman A, Weyens N, Barac T, Vangronsveld J, vander Lelie D. (2009). Genome survey and characterization of endophytic bacteria exhibiting a beneficial effect on growth and development of poplar trees. *Appl. Environ. Microbiol.* 75: 748–757.

Tonelli ML, Taurian T, Ibáñez F, Angelini J, Fabra A. (2010). Selection and in vitro characterization of biocontrol agents with potential to protect peanut plants against fungal pathogens. *J. Plant Pathol.* 92(1): 73–82.

Tsavkelova EA, Cherdyntseva TA, Botina SG, Netrusov AI. (2007). Bacteria associated with orchid roots and microbial production of auxin. *Microbiol. Res.* 162: 69–76.

Van Der Heijden MG, Bardgett RD, Van Straalen NM. (2008). The unseen majority: soil microbes as drivers of plant diversity and productivity in terrestrial ecosystems. *Ecol. Lett.* 11: 296–310.

van Loon LC, Glick BR. (2004). *Increased Plant Fitness by Rhizobacteria, Molecular Ecotoxicology of Plants.* Springer-Verlag, Berlin, pp. 177–205.

Visioli G, D'Egidio S, Vamerali T, Mattarozzi M, Sanangelantoni AM. (2014). Culturable endophytic bacteria enhance Ni translocation in the hyperaccumulator *Noccaea caerulescens. Chemosphere* 117: 538–544.

Wan Y, Luo S, Chen J, Xiao X, Chen L, Zeng G, Liu C, He Y. (2012). Effect of endophyte-infection on growth parameters and Cd-induced phytotoxicity of Cd-hyperaccumulator *Solanum nigrum* L. *Chemosphere* 89: 743–750.

Weilharter A, Mitter B, Shin MV, Chain PS, Nowak, J, Sessitsch A. (2011). Complete genome sequence of the plant growth-promoting endophyte *Burkholderia phytofirmans* strain PsJN. *J. Bacteriol.* 193: 3383–3384.

Weyens N, van der Lelie D, Taghavi S, Vangronsveld J. (2009). Phytoremediation: plant–endophyte partnerships take the challenge. *Curr. Opin. Biotechnol.* 20: 248–254.

Wisniewski-Dyé F, Borziak K, Khalsa-Moyers G, Alexandre G, Sukharnikov LO, Wuichet K, Hurst GB, McDonald WH, Robertson JS, Barbe V, Calteau A, Rouy Z, Mangenot S, Prigent-Combaret C, Normand P, Boyer M, Siguier P, Dessaux Y, Elmerich C, Condemine G, Krishnen G, Kennedy I, Paterson AH, González V, Mavingui P, Zhulin IB. (2011). Azospirillum genomes reveal transition of bacteria from aquatic to terrestrial environments. *PLoS Genet.* 7: e1002430.

Xia Y, Liu J, Chen C, Mo X, Tan Q, He Y, Wang Z, Yin J, Zhou G. (2022). The multifunctions and future prospects of edophytes and their metabolites in plant disease management. *Microorganisms* 10: 1072.

Yadav A, Yadav K. (2019). Plant Growth-Promoting Endophytic Bacteria and Their Potential to Improve Agricultural Crop Yields. In Singh, D, Prabha, R, (eds) *Microbial Interventions in Agriculture and Environment* (pp. 143–169). Springer, Singapore.

Yan Y, Yang J, Dou Y, Chen M, Ping S, Peng J, Lu W, Zhang W, Yao Z, Li H, Liu W, He S, Geng L, Zhang X, Yang F, Yu H, Zhan Y, Li D, Lin Z, Wang Y, Elmerich C, Lin M, Jin Q. (2008). Nitrogen fixation island and rhizosphere competence traits in the genome of root-associated Pseudomonas stutzeriA1501. *Proc. Nat. Acad. Sci.* 105: 7564–7569.

Zarei M, Aminzadeh S, Zolgharnein H, Safahieh A, Daliri M, Noghabi KA, Ghoroghi A, Motallebi A. (2011). Characterization of a chitinase with antifungal activity from a native *Serratia marcescens* B4A. *Braz. J. Microbiol.* 42: 1017–1029.

Zhang D, Spadaro D, Valente S, Garibaldi A, Gullino ML. (2012). Cloning, characterization, expression and antifungal activity of an alkaline serine protease of Aureobasidium pullulans PL5 involved in the biological control of postharvest pathogens. *Int. J. Food Microbiol.* 153: 453–464.

Zhang X, Li C, Nan Z. (2012). Effects of cadmium stress on seed germination and seedling growth of Elymus dahuricus infected with the Neotyphodium endophyte. *Sci China Life Sci.* 55: 793–799.

Zhang X, Lin L, Zhu Z, Yang X, Wang Y, An Q. (2013). Colonization and modulation of host growth and metal uptake by endophytic bacteria of Sedum alfredii. *Int. J. Phytoremed.* 15: 51–64.

Zhang YF, He LY , Chen ZJ, Wang QY, Qian M, Sheng XF. (2011). Characterization of ACC deaminase-producing endophytic bacteria isolated from copper-tolerant plants and their potential in promoting the growth and copper accumulation of *Brassica napus*. *Chemosphere* 83: 57–62.

6 Strategies of Rhizospheric Microbes in Heavy Metal Management and Remediation

Debamalla Pathak, Prabha Toppo and Piyush Mathur

6.1 INTRODUCTION

The advent of industrialization and various anthropogenic activities has increased the number of hazardous contaminants causing pollution in almost every stratum of our environment (Verma & Kuila, 2019). HMs are prominent among these contaminants due to their usage in numerous industrial, agricultural, pharmaceutical, and domestic activities. Elements of nature with atomic weight and 4000 kg/m^3 density or five times greater than water are considered HMs (Tchounwou et al., 2014; Kulshreshtha et al., 2014). Some examples of HMs include Hg, Pb, As, Co, Ni, and Cr. HMs can be introduced into the environment from either natural sources (volcanic eruptions, sea salts, soil particles blown by wind, forest fire) or anthropogenic sources (mining, irrigation of crop fields with untreated industrial wastewater, usage of pesticides, herbicides, and fertilizers) (Zaynab et al., 2022). In the environment, HMs can be divided into two broad categories: essential and non-essential HMs (Aslam et al., 2023). Many HMs are classified as essential HMs, such as Cu, Fe, Mn, Co, Zn, and Ni. These HMs are essential for the primary metabolism of plants such as growth, metabolism, and development of various plant organs, and also act as cofactors for various enzymes and proteins (El-Mageed et al., 2023). Essential elements are also referred to as micronutrients because they are often required in trace amounts to plants (about 10–15 ppm) (Barman et al., 2022). In contrast, non-essential HMs are those elements that do not trigger biological metabolic processes in plants, even in trace amounts. Examples of some of the non-essential HMs are Co, Si, Se, Cd, Pb, Hg, Cr, and Al (Nunes da Silva et al., 2022; Tang et al., 2023). These non-essential HMs are also known to be toxic and without any necessary or positive role in organisms, making their accumulation a severe concern (Singh et al., 2011; Islam et al., 2021). Toxicity induced by HMs is a severe problem for the environment and plants as HMs, unlike pollutants of organic nature, do not turn into more superficial structures instead, they change into a form with less toxicity like Hg^{2+} converts into Hg^0 (Priyadarshanee & Das, 2021). HMs cause a reduction in the production of the crop and lessen the quality of food due to the extensive use of fertilizers, herbicides, and pesticides. Due to their undegradable nature, HMs persist, accumulate, and remain in nature for an extended amount of time (Medfu Tarekegn et al., 2020). In this situation, the necessity of HM removal is undeniable.

The rhizosphere, the soil zone around plant roots, acts as a sink of nutrients and shows a high level of bioactivity. Many heavy-metal-resistant and tolerant microbial populations can be found in the rhizosphere of plants that have grown in heavy-metal-contaminated soil, and such microbial communities help to remedy metal toxicity with the help of siderophores and phytochelators (Kamaludeen & Ramasamy, 2008). The strategy for alleviating heavy metal stress involving rhizobacteria and mycorrhizae is termed rhizoremediation (Ganesan, 2012). By altering their synthetic characteristics,

DOI: 10.1201/9781003354109-6

Rhizospheric microbes influence metals' dissolvability and elevate the biological accessibility of HMs by secreting various organic substances and boosting the intake of non-essential HMs along with nutrients (Kumar et al., 2017). Symbiotic associations with rhizospheric microbes help the plant in alleviating the heavy metal stress and the potential removal of heavy metal contaminants (Deng & Cao, 2017). Several rhizospheric microbes have been observed to play an important role in bioremediation of HMs.

There are several kinds of rhizospheric microbes found in nature. These organisms can be bacteria, fungi, oomycetes, etc. (Mendes et al., 2013). One of the prominent examples of rhizospheric microbe is plant growth-promoting rhizobacteria (PGPRs) that can stimulate defence mechanisms and plant development. PGPR activity is a significant factor in assessing the fertility of the rhizospheric soil as well as plant growth (Riseh et al., 2023). PGPR can be both extracellular or ePGPR (*Agrobacterium, Azotobactor*) and intracellular or iPGPR (*Allorhizobium*, *Azorhizobium* of the *Rhizobiaceae* family) (Giannelli et al., 2023). Arbuscular mycorrhizal fungi or AM fungi is another notable example of a rhizospheric microbe that provides the plant HM and salinity tolerance, increases nutrient uptake, and defends from pathogens (Wang et al., 2023). Mycoparasitic fungi also can be mentioned as these opportunistic type of fungi (from the genera *Trichoderma* and *Clonostachys*) are used as biocontrol agents on plant pathogenic fungi (Nygren et al., 2018).

An important class of such microbes are endophytes, which are found to have symbiotic associations with their host plant tissues and have gained considerable attention in ameliorating plant stress in recent decades (Chouhan et al., 2022). As described, endophytes (primarily fungi and bacteria) offer a bio-based sustainable strategy for removing HM contamination from soil and enhancing plant tolerance for improved adaptability and survival (Kumawat et al., 2022; Tiwari & Bae, 2023).

While there has been significant progress in using plant-associated endophytes in HM tolerance and mitigation with translational success, endophytes; efficacy in phytoremediation can be improved further by genetic manipulation and the establishment of high-value endophyte strains (Tiwari & Bae, 2023). Furthermore, genomics biology (including metagenomics, metaproteomics, metatranscriptomics, and metabolomics) has significantly contributed to the characterization of plant microbial communities, boosting our knowledge of plant–microbe dynamics and functional processes (Diwan et al., 2022).

Various chemical and physical techniques have been available to remediate the HMs pollution but the process is costly and leaves behind some byproducts that are harmful to the environment. Currently, endophytes have gained considerable attention due to their unique ability to help plants eliminate HMs from cells or reduce the negative effects of HMs on them (Zheng et al., 2023; Tovar-Sánchez et al., 2023).

6.2 HEAVY METALS AND THEIR EFFECTS ON THE ENVIRONMENT

Due to their recalcitrance regarding biodegradation, the removal of HMs from the environment is a pressing concern as most of these HMs are toxic when exposed to organisms (Ojuederie & Babalola, 2017). The progression of industrial and urban activities such as mining, ore refining, metal plating, manufacturing of electrical appliances, and disposal of sludges has increased the chances of such exposure and contamination of HMs mostly in aquatic ecosystems (Medfu Tarekegn et al., 2020; Choudhary et al., 2017). In these situations, the propensity to accumulate in biological organisms is the foremost concern as HMs are known to be highly mobile and toxic in aquatic environments (Gautam et al., 2016). Excessive levels of HMs can halt plant physiological processes such as photosynthesis, respiration, transpiration, nitrogen metabolism, and mineral uptake, impair cell growth, reduce biomass, and even cause plant death (Raza et al., 2021; Kordrostami & Rabiei, 2019). Metals ions such as Pb^{2+}, As^{3+}, Cd^{2+}, and Hg^{2+} tend to react with various biological compounds causing the formation of toxic substances that are very stable and toxicity such as these snowballs depending upon the increased concentration of HMs due to biomagnification (Carolin et al., 2017; Kulshreshtha et al., 2014). When levels of HMs cross a certain limit, interruption of cell metabolisms like enzyme inhibition and cell structure damage is a possibility due to high concentrations of HMs as well as the exchange of essential nutrients at certain sites in plants (Carolin et al., 2017, Asati et al., 2016). High concentrations of HMs also endanger the microbial diversity of beneficial kinds, lessening the breakdown of organic matter, which is also responsible for the increase in soil infertility (Asati et al., 2016). The toxicity induced by HMs affects protein and DNA production, disrupts the regulations of ions and functions of fatal enzymes, helps in the formation of reactive oxygen species (ROS), and changes the properties of microbes biochemically and physiologically (Medfu Tarekegn et al., 2020). HMs are inherently carcinogenic to the environment and negatively affect living organisms by generating free radicals in DNA, proteins, and lipids that cause serious health problems and drastic environmental impacts (Priyadarshanee et al., 2022; Mukherjee et al., 2022).

Although every heavy metal shows different kinds of disruptive and toxic effects on the environment, sometimes the same HMs but of different forms or isotopes can also show varying levels of toxicity. As an example, Cr (III) and Cr (VI) are two different forms of chromium in nature where Cr (VI) show greater levels of toxicity compared to Cr (III) (Ghosh et al., 2016; Carolin et al., 2017). Despite being essential nutrients, elements like Cu and Zn can show detrimental effects in elevated concentrations. Copper can cause leaf chlorosis, induce oxidative stress, ROS and cytotoxicity, reduce both plant and root growth, and negatively affect various mechanisms in photosynthesis in the leaves of many tree species (Asati et al., 2016). High levels of Zn are known to inhibit the metabolic activity of plants, cause a reduction in chlorophyll, carotenoid, sugar and other growth parameters and induce phytotoxicity and chlorosis in young leaves (Asati et al., 2016). According to Karaca et al. (2010), due to the varying chemical affinity of enzymes, every metal affects the activities of various enzymes in soil systems, in different ways. The toxicity of Cd is greater than Pb, owing to the superior mobility and low soil colloid affinity of Cd in comparison to Pb (Singh & Kalamdhad, 2011). Cd disrupts various mechanisms regarding the growth and development of plants, which results in arrested plant growth, chlorosis, PEP carboxylase inhibition, and rolling of leaves. Increased Cd levels also trigger senescence and sometimes even death (Kiran et al., 2021).

HMs at higher concentrations negatively impact soil microbial population and some important activities resulting in decreased soil fertility (Ahmed et al., 2005). In elevated levels, HMs show a significant amount of toxicity in microbes due to their high affinity, which, by inactivating several essential biological macromolecules (like enzyme activity centers), causes different problems and death (Chu, 2018). For example, Cu and Cd are known to bind with residues of amino acids present in the enzymes and disrupt the normal functions of enzymes (Singh & Kalamdhad, 2011). Bioaccumulation of HMs by soil microbes also shows disruption in the fixation and assimilation of nitrogen along with interference in organic residue breakdown (Nwuche & Ugoji, 2008). In soils containing high levels of HMs, carbon utilization and biomass growth of microorganisms were found to be less whereas energy consumption and release of CO_2 were notably higher in comparison to soils containing low levels of HMs (Abdu et al., 2017). In a study, the composition and activity of soil microbial communities were found to be affected by HMs and these parameters have an intimate relationship with the fertility of the soil (Wang et al., 2007). Another study by Wang et al. (2010) revealed that exposure to HMs diminished the number of fungi and bacteria, whereas the decline in the number of bacteria was notably sharp. This unequal decline between fungi and bacteria also proves the possible altercation of the fungi–bacteria ratio in soil due to heavy metal stress. Heavy metal stress negatively impacts microbial activity, metabolism, and respiration, which in turn influences soil respiration (Jin et al., 2018).

6.3 RHIZOSPHERIC MICROBIOME AND HM ALLEVIATION

Rhizospheric microbiome is composed of all the microorganisms that affect plant growth and development both positively and negatively. The consortia of all microorganisms that enhance plant growth and alleviate them from stress conditions together constitute beneficial microorganisms. A large number of benefical microbes help in HM elimation from rhizospheric region and augment bioremediation process. The term bioremediation entails removing hazardous organic or inorganic substances present in the terrestrial and aquatic environment with the help of biological organisms and bringing them to safe levels (Sharma et al., 2021). This approach of involving microbes in the remediation of HMs ensures an eco-friendly solution that alleviates the concerns regarding HMs without producing secondary pollutants like sludges (Verma & Sharma, 2017).

6.3.1 Endophytes

6.3.1.1 Endophytic Bacteria

A number of studies have shown that the endophytic bacteria isolated from the plants assisted in tolerating different HM stresses. In a study by Chen et al. (2020), different endophytic bacteria namely *Stenotrophomonas* sp. S20, *Pseudomonas* sp. P21, and *Sphingobium* sp. S42 isolated from the roots of *Tamarix chinensis.* All these isolates when co-inoculated with the same host plant in the presence of HMs, the plants inoculated with isolate *Sphingobium* sp. S42 revealed the highest tolerant ability, i.e., up to 200 mg l^{-1}. In a different study, the efficacy of HM accumulation of endophytic bacterial strains *Enterobacter ludwigii* SAK5 and *Exiguobacterium indicum* SA22 isolated from the roots of *Artemisia princeps*, *Chenopodium ficifolium*, *Oenothera biennis*, and *Echinochloa crus-galli* were evaluated through inoculation in rice plants. Results showed that the treated inoculated plants have enhanced biomass including root and shoot length, fresh and dry weight, and increased chlorophyll content (Figure 6.1). Moreover, inoculation with endophytic bacterial strains upregulated the stress-responsive genes such as *OsGST*, *OsMTP1*, and *OsPCS1* (Jan et al., 2019). It has been observed that endophytic bacteria *Bacillus* sp. and *Lysinibacillus* sp., isolated from the medicinal plant *Silybum marianum* grown in HM-polluted sites successfully translocated and degraded Cr from *Brassica napus* plants grown under Cr-treated soil (Anwar et al., 2023). Ullah et al. (2022) stated that rice plants inoculated with endophytic bacteria *Serratia* sp. AI001 and *Klebsiella* sp. AI002 isolated from the *Solanum nigrum* induced Cd tolerance up to 25 mg l^{-1}. Concomitantly, there was increased IAA production, as well as the other phytohormones that improved plant growth dynamics (i.e., plant length, biomass, chlorophyll contents), relieved electrolyte leakage, and reduced glutathione significantly. Additionally, inoculation of AI001 and AI002 significantly induced the expression of HM ATPase genes, i.e., "*HMA2*, *HMA3*, and *HMA4*" and led to increased Cd translocation compared to uninoculated plants. A similar study was carried out by Rolón-Cárdenas et al. (2021) in which some Cd-tolerant endophytic bacteria *Pseudomonas rhodesiae* strains (GRC065, GRC066, GRC093, and GRC140) were isolated from the roots of *Typha latifolia* and their HM tolerance was studied by inoculating them in *Arabidopsis thaliana* Col-0 seedlings. A representation of different endophytic bacteria in regulation of gene expression under HM stress have been shown in Table 6.1.

Previous studies have shown that endophytes have the ability to manipulate hormone-regulatory pathways to benefit their host plants under metal stress conditions (Ameen et al., 2024; Cui et al., 2024). One of the studies has discussed the mechanisms where root endophytic bacteria *Lysinibacillus fusiformis* Cr33 have the ability to control Cd stress through increased acetic acid, and Fe bioavailability in the roots. It has been observed that tomato plants inoculated with *L. fusiformis* Cr33 provoked the JA signaling pathway to inhibit root NO burst, thus reducing the root Fe deficiency responses imposed by Cd stress. These combined effects made it more difficult for Cd to enter root cells by increasing Fe availability, which made Fe more competitive with Cd for metal transporters like IRT1. IRT1, which is a broad-spectrum transporter of metal, belongs to the ZIP family and is known to be able to transport Fe, Mn, and Zn. IRT1 also transports Cd and this gene has been reported to have hypersensitivity towards Cd in Fe-deficient plants (Spielmann et al., 2022). Furthermore, under these conditions, suppression of NO signaling downregulated genes related to Fe uptake, which in turn prevented IRT1 from supporting Cd uptake (Zhu et al., 2021).

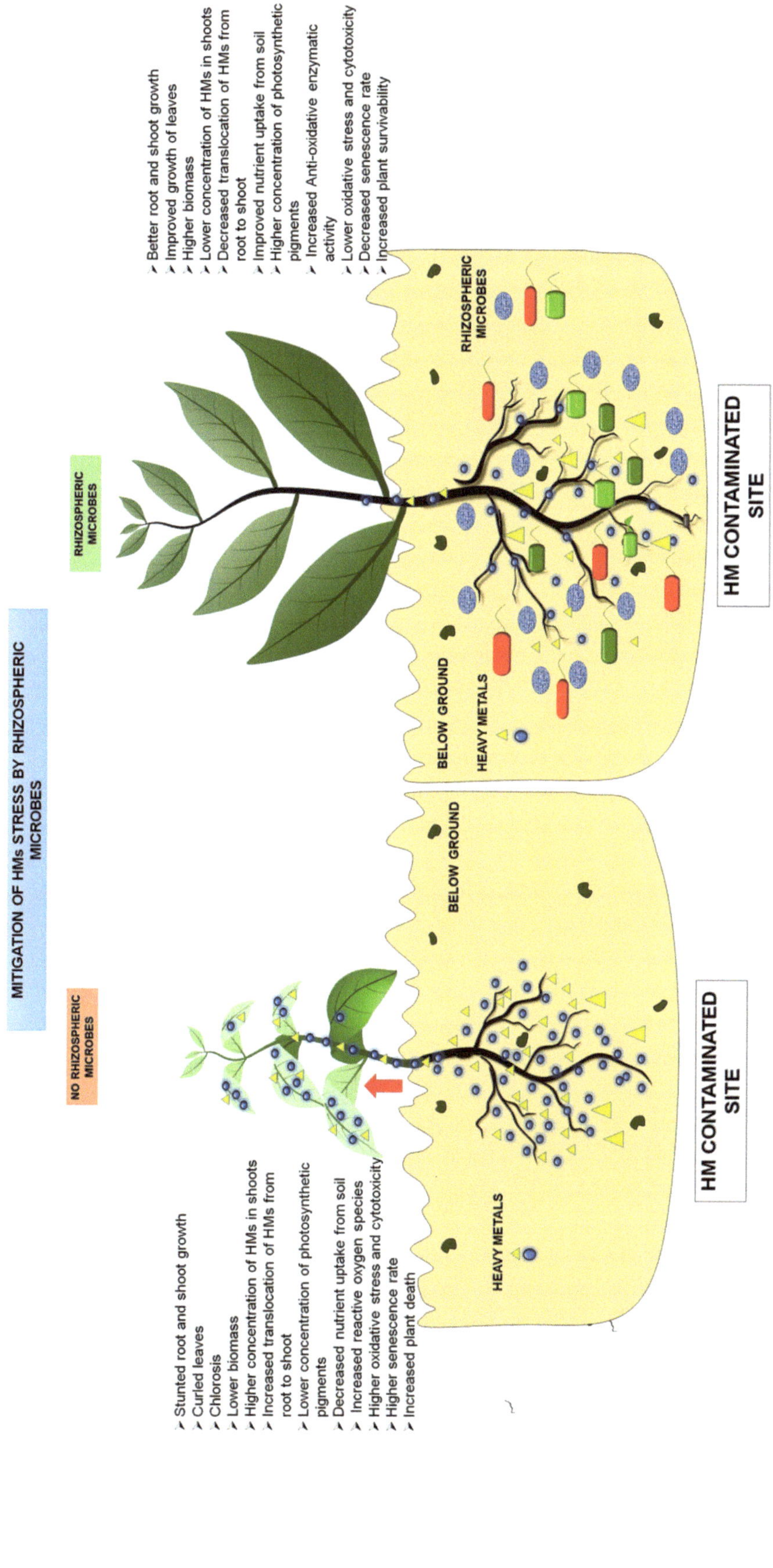

FIGURE 6.1 Rhizospheric microbes mediate the allevation of HM stress in plants.

TABLE 6.1
List of Studies Showing the Endophyte-Mediated HM Stress Mitigation Mechanism at the Cellular Level *via* Endophytic Fungi

Endophytes	HMs	Application Methods	Techniques Use	Key Genes, Proteins and Metabolites, and Their Functions	References
Enterobacter ludwigii SAK5 and *Exiguobacterium indicum* SA22	Cadmium (Cd), Nickel (Ni)	*Oryza Sativa* cv. Ilmi	Hydroponics	• Increased activity of *OsGST, OsMTP1,* and *OsPCS1*	Jan et al., 2019
Serratia sp. AI001 and *Klebsiella* sp. AI002.	Cd	*Oryza Sativa*	-	• Increased expression of ATPase genes ie., "*HMA2, HMA3*, and *HMA4*"	Ullah et al., 2022
Pseudomonas aeruginosa and *Burkholderia gladioli*	Cd	*Lycopersicon esculentum* seeds var. Pusa Ruby	Seed soaking	• Increased levels of protein and non-protein-bound thiols	Khanna et al., 2019
Bacillus megaterium BM18-2, and *B. megaterium* BM18	Cd	Seeds of Hybrid *Pennisetum*	Soaking of roots	• Expression of BM18-2 are involved	Wu et al., 2019
Chaetomium cupreum	Cu, Zn, Pb, As, and Cd	*Eucalyptus globulus* Roots	Pot inoculation	• Increased expression of genes of metal transporters, transcription factors, stress and defence response proteins • Increased expression of genes involved in auxin biosynthesis and metabolism	Ortiz et al., 2019
Penicillium funiculosum LHL06	Ni, Cu, Pb, Cr, Al	*Glycine max*	Spore suspension	• Downregulation of HM ATPase genes *GmHMA13, GmHMA14, GmHMA19*) and *GmMATE1* • Upregulation of stress-related protein expression	Bilal et al., 2019
Pseudomonas qingdaonensis ZCR6 s	Cd, Zn, and Cu	*Zea mays*	Soil inoculation	• Differential expression of genes	Chlebek et al., 2021

In a similar study, Rolón-Cárdenas et al. (2021) showed that endophytic bacteria *Pseudomonas rhodesiae* isolated from the roots of *Typha latifolia* growing on a Cd-contaminated area when inoculated with *Arabidopsis thaliana* Col-0 in the presence of Cd, altered the root architecture of Arabidopsis plant thereby preventing damage to the plant. Isolation of some arsenic (As)-resistant endophytic bacteria was carried out from the roots of alfalfa (*Medicago sativa* var. Hamedani) and chickpea (*Cicer arietinum* var. Kabuli ILC482) plants. Among the isolated endophyte strains, *Pseudomonas* sp. QNC1, showed the highest tolerance up to a Cd concentration of 350 mM and lower NPR3 gene expression in chickpea and alfalfa roots (Tashan et al., 2021). In a similar study, three endophytic bacteria of the genus *Pseudomonas* were isolated from the roots of *Pityrogramma calomelanos*, which tolerated Pb concentrations of up to 1850 mg l^{-1}. To observe their efficacy roots of *Acacia mangium* and *Eucalyptus camaldulensis* were inoculated with the above strain in a hydroponic experiment with Pb. The results delineated that the *A. mangium* plant inoculated with endophytic bacteria showed no effect on the plant, but in *E. camaldulensis*, inoculation showed a decrease in Pb content (3763 ± 592 mg/kg) compared to non-inoculated (4233 ± 264 mg/kg) plant (Yongpisanphop et al., 2019). In a study, the researchers isolated 82 endophytic bacteria from the root nodules of *Robinia pseudoacacia* out of which two symbiotic isolates *Mesorhizobium loti* HZ76 and *Agrobacterium radiobacter* HZ6 showed the highest potential for HM tolerance. In a successive experiment, these isolates were inoculated in *R. pseudoacacia* seedlings grown in pots with different concentrations of Cd, Zn, Pb, or Cu, where HZ76 and it was seen that plant growth in terms of shoot biomass was significantly enhanced compared to non-inoculated control plant seedlings (Fan et al., 2018).

6.3.1.2 Endophytic Fungi

Saad et al. (2022) showed in their study that endophytic fungi *Fusarium falciforme* (MW898136) and *Fusarium* sp. (MW898137) isolated from the roots of *Convolvulus arvensis* could tolerate high metal concentrations of cadmium (Cd). 276.5 and 101.6 μg g^{-1} with bioremediation percentages of 82.4% and 76.5%. These isolates when inoculated to *Triticum aestivum* and *Phaseolus vulgaris* showed higher amounts of Cd uptake by roots of respective plants and immobilized cadmium from roots to shoots. Another study documented that the endophytic fungi *Aspergillus welwitschiae* reduced Cr and As uptake in the *Glycine max* plant and promoted higher root shoot length, fresh/dry weight under stress and simultaneously increased enzymatic antioxidants like catalases, peroxidases as well non-enzymatic antioxidants including ascorbic acid oxidase and 1,1-diphenyl-2-picrylhydrazyl (DPPH) (Husna et al., 2021). It is now well known that endophytic fungi associated with hyperaccumulators eliminate HMs through the presence of some key genes, proteins and metabolites. Gu et al. (2024) revealed in their study that endophytic fungi *Fusarium* sp. Fo10 present symbiotically with rice plants results in high Cd uptake due to the action of OsCAL1 protein that helps in vacuolar sequestration, efflux to the environment, and formation of Fe plaques in the rhizosphere. A study demonstrated that soybean plants inoculated with endophytic fungi *Penicillium funiculosum* LHL06 showed modulation of physio-biochemical, molecular, and protein level responses to combined HMs (Ni, Cu, Pb, Cr, and Al) toxicity. The symbiotic association between fungus and plant resulted in downregulation of HM transporters ATPase genes *GmHMA13*, *GmHMA14*, *GmHMA19*, and *GmMATE1* and increased stress and defence-related proteins expression (glutathione S-transferase L3, isoflavone reductase-such as chalcone isomerase A, NAD(P)H dehydrogenase, FQR1-like 1 isoform X2, and Peroxidase 3) (Bilal et al., 2019). In a different work, endophytic fungi isolated from roots of *Hordeum vulgare* L belong to both dark septate (DSE) *Alternaria alternata*, *Microdochium bolleyi*, *Bipolaris zeicola* (Tw26), *Alternaria* sp. (TBR5), and *Pleosporales* sp., and non-DSE including *Fusarium redolens*, *F. tricinctum*, *F. monliforme*, *Clonostachys rosea*, and *Epicoccum nigrum*. Among the isolated endophytic fungi, *Alternaria* sp. (TBR5) and *Bipolaris zeicola* (Tw26) showed improved efficiency of phytoremediation of Cd with a MIC value of 2000 mg l^{-1} and 1800 mg l^{-1}, respectively (Shadmani et al., 2021). Musa et al. (2023) the HMs alleviation capability of the endophytic fungus, *Paecilomyces lilacinus* (MRF), isolated from the roots of *Justicia adathoda.* It was observed that the endophytic fungal strain MRF substantiated the growth of *Solanum lycopersicum* grown in the presence of Pb and Co, there was increased root and shoot growth, and plant biomass (Figure 6.1). Additionally, there was an increased concentration of the sugar, protein, plant growth promoter hormone IAA, secondary metabolites like phenols and flavonoids, higher proline content, and increased relative water content in the fungus-inoculated plant compared to the control plants.

Furthermore, Yamaji et al. (2016) isolated endophytic fungi *Phialocephala fortinii, Rhizodermea veluwensis*, and *Rhizoscyphus* from the roots of *Clethra barbinervis* that is a naturally grown plant found at several mine sites that have high concentrations of HMs such as Cu, Zn, and Pb in rhizospheric soils. In further work, the isolated endophytic fungi were investigated for alleviating the growth and nutrient uptake of *C. barbinervis* seedlings growing in sterilized mine soil with root fungal endophytes. The results showed that the endophytes promoted seedling growth, increased K uptake in the shoots and reduced HM concentrations in the roots. In contrast, the control plant *C. barbinervis* could hardly grow and showed chlorosis and toxicity. The dark septate endophytes *Exophiala pisciphila*, first isolated from the surface-sterilized root of *Arundinella bengalensis* grew naturally as an indigenous Cd hyperaccumulator in an abandoned mining area in China. These endophytes were found to promote maize plant growth, protect the plant from Cd damage by significantly affecting the expression of genes involved in signal transduction and polar transport of phytohormones, and then affect the content of abscisic acid (ABA) as well as indole-3-acetic acid (IAA) in maize roots. In addition, endophytes also increase lignin content by 27% by regulating the expression of genes and also activate glutathione metabolism through the upregulation of genes related to glutathione S-transferase (Wang et al., 2023). Similarly, *Prosopis laevigata* is a hyperaccumulating legume that bioaccumulates Pb, Cu, and Zn and is cultivated in mine tailings located in Mexico. Further investigations revealed that the most potent endophytic fungus *Aspergillus luchuensis* C7 has a high tolerance to Cu, Zn, and Pb. The C7 strain is a growth-promoting agent for the *P. laevigata* plant that increases the Cu translocation of metals (Tovar-Sánchez et al., 2023). Nevertheless, the endophytic fungus *Talaromyces pinophilus* MR1 has shown significantly improved Cd tolerance in pot studies. Isolates were obtained from the roots of *Miscanthus floridulus* and identified by 18S rRNA sequence analysis with a similarity rate of 99.83%. Strain MR1 was tested on *Arabidopsis thaliana* for growth and Cd uptake at low and high concentrations. The experimental results showed that MR1 successfully increased the dry weight of the tissue by 25.95–107.21%. At low Cd concentration, the underground tissue was reduced by 39.28%, while at high Cd concentration it increased by 28.28%. MR1 inoculations increased the total amount of Cd eliminated (17,080 μg) by 129.77% at high concentrations (Xiao et al., 2023). A list of studies showing application of endophytic fungi in HM mitigation in different plants have been listed in Table 6.2.

TABLE 6.2
List of Studies Summarizing Bioremediation of Metal Pollutants Using Endophytic Fungi and Plant Growth Promotion

Host Plant	Endophytes	Name of Plant HM Mitigation was Done	Pollutants	Mechanisms of Action	Reference
Glycine max L.	*Penicillium funiculosum* LHL06	*Glycine max* L.	Ni, Cu, Pb, Cr, and Al	• Increased gibberellins and indole-3-acetic acid (IAA) production. • Decreased abscisic acid and jasmonic acid levels • Downregulate heavy metal ATPase genes (*GmHMA13, GmHMA14, GmHMA19)* and *GmMATE1*	Bilal et al., 2023

TABLE 6.2 (Continued)
List of Studies Summarizing Bioremediation of Metal Pollutants Using Endophytic Fungi and Plant Growth Promotion

Host Plant	Endophytes	Name of Plant HM Mitigation was Done	Pollutants	Mechanisms of Action	Reference
Justicia adathoda	*Paecilomyces lilacinus*	*S. lycoprsicum*	Pb and Co	• Increased root and shoot growth, fresh and dry weight • Increased IAA, phenols, flavonoids, sugar, proline, relative water content • Increased protein contents	Musa et al., 2023
Prosopis laevigata	*Aspergillus luchuensis*	*Prosopis laevigata*	Pb, Cu and Zn	• Increased metal translocation from roots to leaves	Tovar-Sánchez et al., 2023
Convolvulus arvensis	*Fusarium falciforme* (MW898136) and *Fusarium* sp. (MW898137)	*Triticum aestivum, Phaseolus vulgaris*	Cd	• Higher biomass • Increased uptake of Cd • Decreased translocation of Cd from roots to shoots	Saad et al., 2022
Eupatorium triplinerve	Eighteen fungal isolates, dominant fungi are *Colletotrichum* Sp. and *Fusarium* Sp.	*Oryza sativa*	Cd	• Increased IAA and GA production, • Better phosphate solubilization, increased, siderophore production under Cd stress	Mukherjee et al., 2022
Chlorophytum comosum L.	*Aspergillus welwitschiae* (Bk)	*Glycine max* L.	Cr-VI and As-V in the form of $K_2Cr_2O_7$ and Na_3AsO_4	• Higher root/shoot length and fresh/dry weight • Increased enzymatic antioxidants, catalases (CAT), ascorbic acid oxidase (AAO), peroxidase activity (POD) • Higher ROS scavenging	Hussain et al., 2021
-	*Paecilomyces formosus* LHL10 and *Penicillium funiculosum* LHL06	*Glycine max* L.	Al, Cr, Cu, Ni, Cd, and Al	• Increased GA and IAA • Improved nutritional status	Bilal et al., 2021
Hordeum vulgare cv. Arna and *Glycine max* cv. Almaty	Two hundred fifty-three endophytic fungal strains	-	Cd	-	Ignatova et al., 2021
Agrostis stolonifera	*Penicillium lanosum, Tolypocladium cylindrosporum, Phialocephala* sp., *P. Canescens, P. Sanguifluum, Acremonium* sp., *Sarocladium kiliense*	*Festuca arundinacea*	Pb	• Higher germination and increased biomass.	Soldi et al., 2021

(*continued*)

TABLE 6.2 (Continued)
List of Studies Summarizing Bioremediation of Metal Pollutants Using Endophytic Fungi and Plant Growth Promotion

Host Plant	Endophytes	Name of Plant HM Mitigation was Done	Pollutants	Mechanisms of Action	Reference
Parthenium sp.	*Mucor* sp. MHR-7	NA	Cr, Mn, Co, Cu and Zn	• Increased metal biotransformation • Increased IAA, ACC deaminase • Better phosphate solubilization	Zahoor et al., 2017
Roots of tomato, cucumber, and frankincense plant leaves	*Penicillium janthinellum*, *Paecilomyces formosus*, *Exophiala* sp., *Preussia* sp.	*Glycine max* L.	Ni	• Enhanced translocation of HMs from root to the shoot. • Reduced lipid peroxidation • Increased accumulation of linolenic acid, • Increased enzymatic antioxidants glutathione, • peroxidase, polyphenol oxidase, catalase, and superoxide dismutase.	Bilal et al., 2017
Hedysarum pallidum Desf.	*Aspergillus tubingensis* MH189391	-	Sb	• Increased growth • Increased antioxidant enzymatic activity CAT, superoxide dismutase (SOD) • Higher proline content	Meghnous et al., 2019
Polygonum acuminatum and Aeschynomene fluminensis	32 endophytic fungi isolated	*Aeschynomene fluminensis*	Hg	• Improved growth	Pietro-Souza et al., 2017
Brassica napus	*Fusarium* sp. CBRF44, *Penicillium* sp. CBRF65, *Alternaria* sp. CBSF68	-	Cd and Pb	• Increased IAA production • Better phosphate solubilization	Shi et al., 2017
Astragalus adsurgens Pall.	*Gaeumannomyces cylindrosporus*	*Zea mays*	Pb, and Zn	• Increased growth. • Improved photosynthetic efficiency • Changed translocation of Pb	Yihui et al., 2017

6.3.2 Arbuscular Mycorrhizzal Fungi

A number of previous reports stated that AMF plays a major role in mitigation of HM stress in plants (Basu et al., 2018; Dhalaria et al., 2020; Khaliq et al., 2022). Hnini et al. (2024) comprehensively reviewed the role of AMF in maintenance of soil health and their role in lowering phytotoxic effects of HMs. In a study carried out by Adeyemi et al. (2021), inoculation with *Rhizophagus intraradices* improved the growth of soybean plants grown in presence of HMs like Pb, Cu, and Zn. Inoculated plants higher bioacculmulation of HMs in roots and lesser translocation into other above ground parts (Figure 6.1). A similar type of work carried out by Luxmi et al. (2023) demonstrated a significant improvement in growth parameters of *Monarda citriodora* plants inoculated with *Glomus fasciculatum* in the presence of Pb, As, Ni as compared to non-inoculated plants. The mitigating effect of AMF was also depicted in a work of Rasouli-Sadaghiani et al. (2019) in which growth of plant *Acroptilon repens* inoculated with *Glomus* spp. showed better chlorophyll content, higher shoot biomass, higher relative water content in existence of Cd-treated soil. AMF mediated HM stress amoleriation was also observed in a study by Zhan et al. (2019) where growth of grass *Cynodon dactlyon* was significantly mitigated by AMF in the presence of Pb-Zn mine wasteland. There was decreased translocation of HMs in shoots, increased bioavailability of P and S in soil comparison to HMs. Positive effects of AMF inoculation in mitigation of HM stress was observed in aromatic plant lavender in which AMF *Funnelliformis mosseae* improved the growth of plants in the presence of Ni and Pb as well there was augmentation of essential oil quality and contents (Pirsarandib et al., 2022). A detailed review by Tiwari et al. (2022) on AMF mediated HM stress tolerance in plants presented a different mechanism through which AMF lowers the HM concentration. The paper stated that AMF secretes various organic acids, polyphosphate granules, as well as modulates the physio-chemical properties of the rhizospheric soil region that alter the pH and bring about redox changes in the surrounding region (Tiwari et al., 2022). At the same time, AMF also secretes low molecular weight compounds that have metal binding properties, which are reffered to as glomalin and these glomalin binds to HMs in soil and forms a complex that prevents translocation of HMs into the roots of plants (Riaz et al., 2021). These all above further assist in AMF-mediated phtyoextraction, phytostabilization, and phytotransformation of HMs, respectively. Additionally, it has been observed that various transporters are located in fungal membrane that help in HM sequesteration in fungal cells (Ferrol et al., 2016).

6.3.3 Plant Growth-Promoting Microorganisms

A number of PGPR strains have the ability for phytoremediation and have been employed for plant growth promotion of plants under HM stress (Gupta et al., 2024; Nazli et al., 2020; Kumar et al., 2019; Hansda et al., 2014). Manoj et al. (2020) comprehensively reviewed the mechanism of PGPR-mediated HM removal and demonstrated that PGPR secretes a number of different compounds, such as organic acid, phytohormones, chelators, etc., in the rhizospheric region that help in phytoextraction, phytostabilization as well as phytostabilization. Similar reports were made by Wang et al. (2022) and Qin et al. (2024) in their detailed review where they presented the PGPR-stimulated HM detoxification mechanism and provided application of these PGPR strains for enhancing the growth of plants under HM stressed conditions. In a study, HM-tolerant PGPR bacteria *Acinetobacter beijerinckii* and *Raoultella planticola* substantiated growth of pea seedlings grown in the presence of As and Cr *via* modulation of different physiochemical parameters of plant such as sugars, proteins, stress compunds like proline etc. and augment growth of seedlings. Additionally these strains were found to stimulate production of salicylic acid, IAA and gibberellic acid, etc. (Husna et al., 2023). Another study showed that two salt-tolerant PGPR stains, namely *Pseudomonas moraviensis* and *Bacillus cereus* when singly or co-inoculated to wheat seedlings in the presence of HMs Cd Co, Cr, Cu, Mn, Ni, Pb alleviate seedlings from the damage caused by these HMs and augment the growth of seedlings. The study showed that there was much lower accumulation of these HMs in both roots and leaves of wheat seedlings when co-inoculated as compared to inoculated singly (Hassan et al., 2017). The potential ability of the PGPR strain in enhancing HM detoxification of an hyperacculumator plant *Spartina maritima* was observed by Mesa-Marín et al. (2018), and it was seen that the PGPR strain *Bacillus methylotrophicus*, *Bacillus aryabhattai*, SMT48, *Bacillus aryabhattai*, SMT50, and *Bacillus licheniformis* isolated previously from the rhizosphere of *S. maritima* led to increased accumulation of HMs, Cu, Cd, and Pb, in the roots rather than the shoots of *S. maritima* seedlings grown in the presence of the above HMs. Additionally there was increased growth of seedlings, higher antioxidant activities as well as an increased root respiration rate with these PGPR inoculations. One of the study strains isolated from the rhizosphere of *Stipa tenuissima*, *Sulla spinosissima*, and *Acacia cyanophylla*, well adapted in the mining site, showed plant growth-promoting traits and inoculation of these strains to seedlings of *S. spinosissima* in the presence of As, Cu, Zn, Cd, and Pb, escalating their growth by increasing the root and shoot lengths, which increased chlorophyll and carotenoid content as well as higher catalase, peroxidase and ascorbate peroxidase enzyme thereby indicationg the roles of these strains in phytoremediation (Oubohssaine et al., 2022) (Figure 6.1). In a study by Liu et al. (2015), the PGPR *Phyllobacterium myrsinacearum* strain RC6b augment the phytoextraction efficiency of two hyperaccumulator plants *Sedum alfredii* and *Medicago sativa* grown in the presence of Pb, Zn, Cu, and Cd. It has been observed that inoculation with this PGPR strain led to increased growth, biomass as well as increased content of Zn, Cd, Pb in shoots of both the plants in comparison to non-inoculated plants (Liu et al., 2015). Similar reports were made by Sorour et al. (2022) in which different bacteria were isolated from rhizosphere as well as from soil and marine water samples contaminated with HMs like Zn, Ni, Cd, and Cu and it was observed that two of the bacterial genera were found to be most dominant, i.e., *Pseudomonas* and *Bacillus* and augment growth and phytoextraction efficiency of *Helianthus* plants grown in the presence of HMs Zn and Cd. Tirry et al. (2021) demonstrated PGPR isolate *Pseudomonas* assisted in overcoming Cr effects from *Medicago sativa* plants and led to increased growth.

6.4 MECHANISM OF BIOREMEDIATION FROM HM STRESS

The different mechanisms through which rhizospheric microbes carries out bioremediation have been discussed in this section.

6.4.1 Biosorption

Due to their small size and exceptional adaptability to different environments, bacteria are considered among the most efficient biosorbents for HMs (Ghosh et al. 2016). Because of their small size, bacteria have a high surface area-to-volume ratio for biosorption. Bacterial biomass has many redeeming qualities like a high revival rate and low cost and being environment-friendly (Priyadarshanee & Das, 2021). In a study, Cd-resistant isolates *Burkholderia* sp. 1–22 and *Bacillus* sp. 6–6, which were collected from rice rhizosphere showed prominent plant growth promoting activity in the absence of Cd stress. In the presence of Cd, these strains alleviated Cd toxicity by biosorption and pH regulation. On the cell walls of these bacteria, carboxyl and amino groups were present, which are known to facilitate Cd adsorption. Further testing on these strains also showed the presence of the MICP mechanism by which these strains can form $CdCO_3$ (Yan et al., 2024). In a study by Hasylmah et al. (2021), a novel bacteria *Aureispira* sp. CCB-QB1 was found capable of adsorbing almost 90% of Cu^{2+} and Fe^{3+} below 50 mg l^{-1} concentration after an incubation period of 10 minutes.

6.4.2 Bioaccumulation

Dependency on energy is an important feature of bioaccumulation because in this process microbes accumulate HMs inside the cell and turn them into inactive complexes with other ligands of high affinity and this whole process requires active participation of the cell that costs energy (Pratush et al., 2018). Bioaccumulation has two main steps; the absorption of metal ions in the cell wall (which does not require energy or any metabolic activity) and the transfer of metal ions inside the cell through an energy-dependent pathway (Nnaji et al., 2023). Sometimes due to the reduction reaction inside the cell, the toxicity of these heavy metal ions reduces to less toxic forms (Durve et al., 2013). Akhter et al. (2017) isolated rhizospheric bacteria *B. cereus* BDBC01, *B. cereus* AVP12, and *B. cereus* NC7401 from the rhizosphere of *Tagetes minuta*. These three strains, *B. cereus* BDBC01, *B. cereus* AVP12, and *B. cereus* NC7401 removed 118.2 mg g^{-1}, 121.87 mg g^{-1} and 90 mg g^{-1} Cr (VI), 135 mg g^{-1}, 127.5 mg g^{-1} and 146.25 mg g^{-1} Ni (II) and 120 mg g^{-1}, 129.37 mg g^{-1}, 135 mg g^{-1} Cd(II) via bioaccumulation.

6.4.3 Solubilization of HMs or Bioleaching

Solubilization of metal sulphides and oxides that are sourced from deposits of ore and various wastes through bacteria and fungi is considered bioleaching (Tayang & Songachan, 2021). In bacteria, this process directly or indirectly involves bacterial metabolism and due to the nature of the process, this process can be considered a bio-hydrometallurgical method. Fungi such as *Aspergillus niger* produces a variety of acids that helps in the solubilization of metals and can reportedly solubilize almost 70% of copper when tests were done (Mulligan et al., 2004). Bioleaching can be used to retrieve both precious and harmful metals alike, such as Cu, Ar, Pb, Au, Sb, V, Co, and Bi (Sreedevi et al., 2022). Chemolithotropic bacteria like *Thiobacillus thiooxidants*, *T. ferooxidants*, and *Acidthiobacillus ferooxidants*, are known to use iron and sulfur, as an energy source in acidic environments and are good examples of bioleaching bacteria (Roy & Roy, 2015, Sreedevi et al., 2022).

6.4.4 Bioprecipitation and Biocrystallization

The process of separating HMs from solutions using precipitation and crystallization through microbial means is termed bioprecipitation and biocrystallization, respectively. This phenomenon is not dependent upon cellular metabolism as the basis of this process is the interaction between metal ions and cell walls (Sreedevi et al., 2022). This process changes the metals and metalloids from soluble to insoluble precipitates under the catalysation of reductive and oxidative processes, which results in the precipitation of HMs like Fe and Cr (Sahinkaya et al., 2017). Microbially induced carbonate precipitation (MICP) is an example of this process that sustainably immobilizes over 60 types of minerals and HMs. Fang et al. (2021) studied the MICP process of Cd^{2+} where *Sporosarcina pasteurii* was able to cause bioremediation of Cd^{2+} in both low and high levels (0.1–50 mM).

6.4.5 Biotransformation

Biotransformation of heavy metals causes the conversion of metals from toxic to less toxic or readily retrievable forms. Biotransformation via microbes can occur via two mechanisms: conversion of inorganic forms via redox reactions and conversion between organic and inorganic forms (can happen in both ways). These kinds of biotransformations play a significant role in the biogeochemical cycles of metals and can be used in bioremediate soil polluted with metals (Juwarkar & Yadav, 2010).

6.5 CONCLUSION

The rhizospheric region is a dynamic region rich in nutrients and offers an interface between soil and plant through roots. A unique diversity of microbes has been found associated with this interface that play a key role in sustaining plant growth and development as well as maintenance of soil quality and fertility. Heavy metals (HMs) have hampered the growth and productivity of crop plants to a very high level. Due to extreme usage of fertilizers, herbicides and pesticides, bioaccumulation of HMs in plant and its products pose a serious threat to mankind. Presently, to prevent losses incurred due to HM utilization of rhizospheric microbes for the mitigation of HM stress in plants have gained considerable attention. A number of studies have shown the diversity of microbes, i.e., PGPR, endophytes, AMF, etc., alleviate plants from HM stress and their further application in agriculture will help in improving crop productivity in the present scenario of severe HM contamination. These microbes detoxify HMs through different mechanisms and also sequester them. However, most of these studies have been carried out under in vitro conditions and very few of them have reported their utilization in field-based experiments. Concomitantly, the complete cellular pathway for detoxification of HMs by these rhizospheric microbes shall be elucidated so as to develop microbial consortia that can degrade these HMs at a faster rate and is eco-friendly.

REFERENCES

Abdu, N., Abdullahi, A.A., & Abdulkadir, A. (2017). Heavy metals and soil microbes. *Environmental Chemistry Letters, 15*, 65–84. https://doi.org/10.1007/s10311-016-0587-x

Adeyemi, N.O., Atayese, M.O., Sakariyawo, O.S., Azeez, J.O., Sobowale, S.P.A., Olubode, A., Mudathir, R., Adebayo, R., & Adeoye, S. (2021). Alleviation of heavy metal stress by arbuscular mycorrhizal symbiosis in *Glycine max* (L.) grown in copper, lead and zinc contaminated soils. *Rhizosphere, 18*, 100325.

Akhter, K., Ghous, T., Andleeb, S., Ejaz, S., Khan, B.A., & Ahmed, M.N. (2017). Bioaccumulation of heavy metals by metal-resistant bacteria isolated from *Tagetes minuta* rhizosphere, growing in soil adjoining automobile workshops. *Pakistan Journal of Zoology, 49*(5), 1841–1846.

Ameen, M., Mahmood, A., Sahkoor, A., Zia, M.A., & Ullah, M.S. (2024). The role of endophytes to combat abiotic stress in plants. *Plant Stress*, 100435.

Anwar, Y., Khan, S.U., Ullah, I., Hemeg, H.A., Ashamrani, R., Al-sulami, N., Ghazi Alniami, E., Hashem Alqethami, M. & Ullah, A. (2023). Molecular Identification of Endophytic Bacteria from *Silybum marianum* and Their Effect on Brassica napus Growth under Heavy Metal Stress. *Sustainability,* 15, 3126.

Asati, A., Pichhode, M., & Nikhil, K. (2016). Effect of heavy metals on plants: an overview. *International Journal of Application or Innovation in Engineering & Management, 5*(3), 56–66.

Aslam, A., Naz, A., Shah, S.S.H., Rasheed, F., Naz, R., Kalsom, A., Mukhtar, N., Niaz, A., Aftab, M., Rasheed, I., & Haq, M.A.U. (2023). Heavy metals contamination in vegetables irrigated with wastewater: a case study of underdeveloping regions of Pakistan. *Environmental Geochemistry and Health, 45*, 8911–8927. https://doi.org/10.1007/s10653-023-01662-0

Barman, P., Kadam, R, & Kumar, A. (2022). Lignin-based adsorbent for effective removal of toxic heavy metals from wastewater. *Emergent Materials, 5*, 923–943. https://doi.org/10.1007/s42247-021-00311-5

Basu, S., Rabara, R.C., & Negi, S. (2018). AMF: the future prospect for sustainable agriculture. *Physiological and Molecular Plant Pathology, 102*, 36–45.

Bilal, S., Shahzad, R., Asaf, S., Imran, M., Al-Harrasi, A., & Lee, I. J. (2023). Efficacy of endophytic SB10 and glycine betaine duo in alleviating phytotoxic impact of combined heat and salinity in *Glycine max* L. via regulation of redox homeostasis and physiological and molecular responses. *Environmental Pollution, 316*, 120658.

Bilal, S., Khan, A. L., Shahzad, R., Asaf, S., Kang, S. M., & Lee, I. J. (2017). Endophytic Paecilomyces formosus LHL10 augments Glycine max L. adaptation to Ni-contamination through affecting endogenous phytohormones and oxidative stress. *Frontiers in Plant Science*, *8*, 870, 1–17.

Bilal, S., Shahzad, R., Khan, A.L., Al-Harrasi, A., Kim, C.K., & Lee, I.J. (2019). Phytohormones enabled endophytic *Penicillium funiculosum* LHL06 protects *Glycine max* L. from synergistic toxicity of HMs by hormonal and stress-responsive proteins modulation. *Journal of Hazardous Materials, 379*, 120824.

Bilal, S., Shahzad, R., & Lee, I. J. (2021). Synergistic interaction of fungal endophytes, *Paecilomyces formosus* LHL10 and *Penicillium funiculosum* LHL06, in alleviating multi-metal toxicity stress in Glycine max L. *Environmental Science and Pollution Research, 28*(47), 67429–67444.

Carolin, C.F., Kumar, P.S., Saravanan, A., Joshiba, G.J., & Naushad, M. (2017). Efficient techniques for the removal of toxic heavy metals from aquatic environment: a review. *Journal of Environmental Chemical Engineering, 5*(3), 2782–2799.

Chen, J., Li, N., Han, S., Sun, Y., Wang, L., Qu, Z., Dai, M. & Zhao, G. (2020). Characterization and bioremediation potential of nickel-resistant endophytic bacteria isolated from the wetland plant Tamarix chinensis. *FEMS Microbiology Letters, 367*, p.fnaa098.

Chlebek, D., Płociniczak, T., Gobetti, S., Kumor, A., Hupert-Kocurek, K., & Pacwa-Płociniczak, M. (2021). Analysis of the genome of the heavy metal resistant and hydrocarbon-degrading rhizospheric *Pseudomonas qingdaonensis* ZCR6 strain and assessment of its plant-growth-promoting traits. *International Journal of Molecular Sciences, 23*(1): 214, 1–28.

Choudhary, M., Kumar, R., Datta, A., Nehra, V., & Garg, N. (2017). Bioremediation of Heavy Metals by Microbes. In: Arora, S., Singh, A., Singh, Y. (eds) *Bioremediation of Salt Affected Soils: An Indian Perspective*. Springer, Cham. https://doi.org/10.1007/978-3-319-48257-6_12

Chouhan, S., Agrawal, L., & Prakash, A. (2022). Amelioration in traditional farming system by exploring the different plant growth-promoting attributes of endophytes for sustainable agriculture. *Archives of Microbiology, 204*, 151. https://doi.org/10.1007/s00203-021-02637-4

Chu, D. (2018). Effects of heavy metals on soil microbial community. *In IOP Conference Series: Earth and environmental science* (Vol. 113, p. 012009). IOP Publishing.

Cui, J., Nie, F., Zhao, Y., Zhang, D., Zhou, D., Wu, J., Qu, L., Xiao, L., & Liu, L. (2024). A review on plant endophytes in response to abiotic stress. *Environmental Pollutants and Bioavailability, 36*, 2323123.

Deng, Z., & Cao, L. (2017). Fungal endophytes and their interactions with plants in phytoremediation: a review. *Chemosphere, 168*, 1100–1106. https://doi.org/10.1016/j.chemosphere.2016.10.097

Dhalaria, R., Kumar, D., Kumar, H., Nepovimova, E., Kuča, K., Torequl Islam, M., & Verma, R. (2020). Arbuscular mycorrhizal fungi as potential agents in ameliorating heavy metal stress in plants. *Agronomy, 10*, 815.

Diwan, D., Rashid, M.M., & Vaishnav, A. (2022). Current understanding of plant-microbe interaction through the lenses of multi-omics approaches and their benefits in sustainable agriculture. *Microbiological Research, 265*, 127180.

Durve, A., Naphade, S., Bhot, M., Varghese, J., & Chandra, N. (2013). Quantitative evaluation of heavy metal bioaccumulation by microbes. *Journal of Microbiology and Biotechnology Research, 3*(6), 21–32.

El-Mageed, T.A., Semida, W.M., Abdou, N.M., & El-Mageed, S.A.A. (2023). Coupling effects of potassium fertilization rate and application time on growth and grain yield of wheat (*Triticum aestivum* L.) plants grown under Cd-contaminated saline soil. *Journal of Soil Sciences and Plant Nutrition, 23, 1070–1084.* https://doi.org/10.1007/s42729-022-01104-3

Fan, M., Liu, Z., Nan, L., Wang, E., Chen, W., Lin, Y., & Wei, G. (2018). Isolation, characterization, and selection of heavy metal-resistant and plant growth-promoting endophytic bacteria from root nodules of *Robinia pseudoacacia* in a Pb/Zn mining area. *Microbiological Research, 217*, 51–59.

Fang, L., Qijian, N., Liang, C., Jianxiong, J., Yang-Yang, Yu., Jian, C., Varenyam, A., & Tianyan, Y. (2021). Ca-mediated alleviation of Cd2+ induced toxicity and improved Cd2+ biomineralization by *Sporosarcina pasteurii. Science of The Total Environment, 787*, 147627.

Ferrol, N., Tamayo, E., & Vargas, P. (2016). The heavy metal paradox in arbuscular mycorrhizas: from mechanisms to biotechnological applications. *Journal of Experimental Botany, 67*(22), 6253–6265.

Ganesan, V. (2012). Rhizoremediation: A Pragmatic Approach for Remediation of Heavy Metal-Contaminated Soil. In: Zaidi, A., Wani, P., Khan, M. (eds) *Toxicity of Heavy Metals to Legumes and Bioremediation.* Springer, Vienna. https://doi.org/10.1007/978-3-7091-0730-0_9

Gautam, P.K., Gautam, R.K., Banerjee, S., Chattopadhyaya, M.C., & Pandey, J.D. (2016). Heavy metals in the environment: fate, transport, toxicity and remediation technologies. *Nova Sci Publishers, 60*, 101–130.

Ghosh, A., Ghosh Dastidar, M., & Sreekrishnan, T.R. (2016). Recent advances in bioremediation of heavy metals and metal complex dyes. *Journal of Environmental Engineering, 142*(9), C4015003.

Giannelli, G., Potestio, S., & Visioli, G. (2023). The contribution of PGPR in salt stress tolerance in crops: unravelling the molecular mechanisms of cross-talk between plant and bacteria. *Plants, 12*, 2197. https://doi.org/10.3390/plants12112197

Gu, T., Qi, Z., Wang, Y., Chen, S., Yan, J., Qiu, H., Yu, Y., Fang, Z., Wang, J., & Gong, J. (2024). An endophytic fungus interacts with the defensin-like protein OsCAL1 to regulate cadmium allocation in rice. *Molecular Plant, 17*, 312–324.

Gupta, R., Khan, F., Alqahtani, F.M., Hashem, M., & Ahmad, F. (2024). Plant growth–promoting Rhizobacteria (PGPR) assisted bioremediation of Heavy Metal Toxicity. *Applied Biochemistry and Biotechnology, 196*, 2928–2956.

Hansda, A., Kumar, V., Anshumali, A., & Usmani, Z. (2014). Phytoremediation of heavy metals contaminated soil using plant growth promoting rhizobacteria (PGPR): a current perspective. *Recent Research in Science and Technology*, *6*(1), 131–134.

Hassan, T.U., Bano, A., & Naz, I. (2017). Alleviation of heavy metals toxicity by the application of plant growth promoting rhizobacteria and effects on wheat grown in saline sodic field. *International Journal of Phytoremediation, 19*, 522–529. https://doi.org/10.1080/15226514.2016.1267696

Hasyimah, N.A.R., Furusawa, G., & Amirul, A.A. (2021). Biosorption of a dye and heavy metals using dead cells of filamentous bacterium, *Aureispira* sp. *CCB-QB1. International Journal of Environmental Science and Technology, 18*, 1627–1636. https://doi.org/10.1007/s13762-020-02918-3

Hnini, M., Rabeh, K., & Oubohssaine, M. (2024). Interactions between beneficial soil microorganisms (PGPR and AMF) and host plants for environmental restoration: a systematic review. *Plant Stress*, 11, 100391.

Husna, Hussain, A., Shah, M., Hamayun, M., Qadir, M., & Iqbal, A. (2021). HM tolerant endophytic fungi *Aspergillus welwitschiae* improves growth, ceasing metal uptake and strengthening antioxidant system in Glycine max L. *Environmental Science and Pollution Research, 1*, 15.

Husna, Hussain, A., Shah, M., Hamayun, M., Iqbal, A., Qadir, M., Alataway, A., Dewidar, A.Z., Elansary, H.O. & Lee, I.J. (2023). Phytohormones producing rhizobacteria alleviate heavy metals stress in soybean through multilayered response. *Microbiological Research, 266,* 127237.

Ignatova, L., Kistaubayeva, A., Brazhnikova, Y., Omirbekova, A., Mukasheva, T., Savitskaya, I., Karpenyuk, T., Goncharova, A., Egamberdieva, D. & Sokolov, A. (2021). Characterization of cadmium-tolerant endophytic fungi isolated from soybean (*Glycine max*) and barley (*Hordeum vulgare*). *Heliyon*, *7*(11), e08240, 1–9.

Islam, M.S., Idris, A.M., Islam, A.R.M.T., Ali, M.M., & Rakib, M.R.J. (2021). Hydrological distribution of physicochemical parameters and heavy metals in surface water and their ecotoxicological implications in the Bay of Bengal coast of Bangladesh. *Environmental Science and Pollution Research, 1*, 15.

Jan, R., Khan, M.A., Asaf, S., Lubna, Lee, I.J., & Kim, K.M. (2019). Metal resistant endophytic bacteria reduces cadmium, nickel toxicity, and enhances expression of metal stress related genes with improved growth of *Oryza sativa*, via regulating its antioxidant machinery and endogenous hormones. *Plants, 8*, 363.

Jin, Y., Luan, Y., Ning, Y., & Wang, L. (2018). Effects and mechanisms of microbial remediation of heavy metals in soil: a critical review. *Applied Sciences, 8*, 1336. https://doi.org/10.3390/app8081336

Juwarkar, A.A., & Yadav, S.K. (2010). Bioaccumulation and Biotransformation of Heavy Metals. In: Fulekar, M.H. (eds) *Bioremediation Technology*. Springer, Dordrecht. https://doi.org/10.1007/978-90-481-3678-0_9

Kamaludeen, S.P.B., & Ramasamy, K. (2008). Rhizoremediation of metals: harnessing microbial communities. *Indian Journal of Microbiology, 48*, 80–88. https://doi.org/10.1007/s12088-008-0008-3

Karaca, A., Cetin, S.C., Turgay, O.C., & Kizilkaya, R. (2010). Effects of Heavy Metals on Soil Enzyme Activities. In: Sherameti, I., Varma, A. (eds) *Soil Heavy Metals. Soil Biology*, vol 19. Springer, Berlin, Heidelberg. https://doi.org/10.1007/978-3-642-02436-8_11

Khaliq, A., Perveen, S., Alamer, K.H., Zia Ul Haq, M., Rafique, Z., Alsudays, I.M., Althobaiti, A.T., Saleh, M.A., Hussain, S., & Attia, H. (2022). Arbuscular mycorrhizal fungi symbiosis to enhance plant–soil interaction. *Sustainability, 14*, 7840.

Khanna, K., Jamwal, V. L., Gandhi, S. G., Ohri, P., & Bhardwaj, R. (2019). Metal resistant PGPR lowered Cd uptake and expression of metal transporter genes with improved growth and photosynthetic pigments in *Lycopersicon esculentum* under metal toxicity. *Scientific Reports*, *9*(1), 5855, 1–14.

Kiran, A., Wakeel, A., Ishaq, R., Mubaraka, R., Ishfaq, M., & Mahmood, A. (2021). Zinc priming of maize seed enhances root to shoot Zn translocation but not of analogous heavy metals. *JAPS: Journal of Animal & Plant Sciences, 31*(4).

Kordrostami, M., & Rabiei, B. (2019). Salinity stress tolerance in plants: physiological, molecular, and biotechnological approaches. In: Hasanuzzaman M., Hakeem K., Nahar K., Alharby H. (eds) *Plant Abiotic Stress Tolerance,* 101–127. Springer, Berlin, Germany.

Kulshreshtha, S., Mathur, N., & Bhatnagar, P. (2014). Mushrooms as a product and their role in mycoremediation. *AMB Express, 4,* 29. https://doi.org/10.1186/s13568-014-0029-8

Kumar, A., Patel, J.S., Meena, V.S., & Srivastava, R. (2019). Recent advances of PGPR based approaches for stress tolerance in plants for sustainable agriculture. *Biocatalysis and Agricultural Biotechnology, 20,* 101271.

Kumar, S.S., Kadier, A., Malyan, S.K., Ahmad, A., & Bishnoi, N.R. (2017). Phytoremediation and Rhizoremediation: Uptake, Mobilization and Sequestration of Heavy Metals by Plants. In: Singh, D., Singh, H., Prabha, R. (eds) *Plant-Microbe Interactions in Agro-Ecological Perspectives.* Springer, Singapore. https://doi.org/10.1007/978-981-10-6593-4_15

Kumawat, K.C., Singh, I., Nagpal, S., Sharma, P., Gupta, R.K., & Sirari, A. (2022). Co-inoculation of indigenous *Pseudomonas oryzihabitans* and *Bradyrhizobium* sp. modulates the growth, symbiotic efficacy, nutrient acquisition, and grain yield of soybean. *Pedosphere,* 32, 438–451.

Liu, Z.F., Ge, H.G., Li, C., Zhao, Z.P., Song, F.M. & Hu, S.B. (2015). Enhanced phytoextraction of heavy metals from contaminated soil by plant co-cropping associated with PGPR. *Water, Air, & Soil Pollution, 226,* 1–10.

Luxmi, S., Singh, R., Ahmed, S., Gandhi, S.G., & Bhanwaria, R. (2023). *Glomus fasciculatum* an arbuscular mycorrhizal fungus alleviate the adverse effects of lead, arsenic, nickel, and improves growth parameters of *Monarda citriodora* Cerv. ex Lag (lemon beebalm). *Rhizosphere, 27,* 100753.

Manoj, S.R., Karthik, C., Kadirvelu, K., Arulselvi, P.I., Shanmugasundaram, T., Bruno, B., & Rajkumar, M. (2020). Understanding the molecular mechanisms for the enhanced phytoremediation of heavy metals through plant growth promoting rhizobacteria: a review. *Journal of Environmental Management, 254,* 109779.

Medfu Tarekegn, M., Zewdu Salilih, F., & Ishetu, A.I. (2020). Microbes used as a tool for bioremediation of heavy metal from the environment. *Cogent Food & Agriculture, 6,* 1783174.

Meghnous, O., Dehimat, L., Doumas, P., Kassa-Laouar, M., Mosbah, F., & Rached, O. (2019). Oxidative and antioxidative responses to antimony stress by endophytic fungus Aspergillus tubingensis isolated from antimony accumulator Hedysarum pallidum Desf. *Biologia, 74,* 1711–1720.

Mendes, R., Garbeva, P., & Raaijmakers, JM. (2013). The rhizosphere microbiome: significance of plant beneficial, plant pathogenic, and human pathogenic microorganisms. *FEMS Microbiology Reviews, 37,* 634–663, https://doi.org/10.1111/1574-6976.12028

Mesa-Marín, J., Del-Saz, N.F., Rodríguez-Llorente, I.D., Redondo-Gómez, S., Pajuelo, E., Ribas-Carbó, M., & Mateos-Naranjo, E. (2018). PGPR reduce root respiration and oxidative stress enhancing *Spartina maritima* root growth and heavy metal rhizoaccumulation. *Frontiers in Plant Science, 9,* 1500.

Mondal, S., Pramanik, K., Ghosh, S.K., Pal, P., Mondal, T., Soren, T. & Maiti, T.K. (2021). Unraveling the role of plant growth-promoting rhizobacteria in the alleviation of arsenic phytotoxicity: a review. *Microbiological Research, 250,* 126809.

Mukherjee, A., Pal, S., Das, P., & Mukhopadhyay, S.K. (2022). Heavy metal exposure to a migratory waterfowl, Northern Pintail (*Anas acuta*), in two peri-urban wetlands. *Science of The Total Environment, 851,* 158238.

Mulligan, C.N., Kamali, M., & Gibbs, B.F. (2004). Bioleaching of heavy metals from a low-grade mining ore using *Aspergillus niger. Journal of Hazardous Materials, 110*(1–3), 77–84.

Musa, M., Jan, FG., Hamayun, M., Jan, G., Khan, SA., Rehman, G., Ali, S., & Lee, IJ. (2023). An endophytic fungal isolate *Paecilomyces lilacinus* produces bioactive secondary metabolites and promotes growth of *Solanum lycopersicum* under HM stress. *Agronomy, 16,* 883.

Nazli, F., Mustafa, A., Ahmad, M., Hussain, A., Jamil, M., Wang, X., Shakeel, Q., Imtiaz, M., & El-Esawi, M.A. (2020). A review on practical application and potentials of phytohormone-producing plant growth-promoting rhizobacteria for inducing heavy metal tolerance in crops. *Sustainability, 12,* 9056.

Nnaji, N.D., Onyeaka, H., Miri, T., & Ugwa, C. (2023). Bioaccumulation for heavy metal removal: a review. *SN Applied Sciences, 5,* 125. https://doi.org/10.1007/s42452-023-05351-6

Nunes da Silva, M., Machado, J., Osorio, J., Duarte, R., & Santos, CS. (2022). Non-essential elements and their role in sustainable agriculture. *Agronomy, 12*, 888. https://doi.org/10.3390/agronomy12040888

Nwuche, C.O., & Ugoji, E.O. (2008). Effects of heavy metal pollution on the soil microbial activity. *International Journal of Environmental Science & Technology, 5*, 409–414.

Nygren, K., Dubey, M., Zapparata, A., Iqbal, M., Tzelepis, G.D., Durling, M.B., Jensen, D.F., & Karlsson, M. (2018). The mycoparasitic fungus *Clonostachys rosea* responds with both common and specific gene expression during interspecific interactions with fungal prey. *Evolutionary Applications, 11*(6), 931–949.

Ojuederie, O.B., & Babalola, O.O. (2017). Microbial and Plant-Assisted Bioremediation of Heavy Metal Polluted Environments: a Review. *International Journal of Environmental Research and Public Health, 14*(12), 1504. https://doi.org/10.3390/ijerph14121504

Ortiz, J., Soto, J., Almonacid, L., Fuentes, A., Campos-Vargas, R., & Arriagada, C. (2019). Alleviation of metal stress by *Pseudomonas orientalis* and *Chaetomium cupreum* strains and their effects on *Eucalyptus globulus* growth promotion. *Plant and Soil, 436*, 449–461.

Oubohssaine, M., Sbabou, L., & Aurag, J. (2022). Native heavy metal-tolerant plant growth promoting rhizobacteria improves *Sulla spinosissima* (L.) growth in post-mining contaminated soils. *Microorganisms,* 10, 838.

Pietro-Souza, W., Mello, I. S., Vendruscullo, S. J., Silva, G. F. D., Cunha, C. N. D., White, J. F., & Soares, M. A. (2017). Endophytic fungal communities of *Polygonum acuminatum* and *Aeschynomene fluminensis* are influenced by soil mercury contamination. *PloS One, 12*(7), e0182017, 1–24.

Pratush, A., Kumar, A., & Hu, Z. (2018). Adverse effect of heavy metals (As, Pb, Hg, and Cr) on health and their bioremediation strategies: a review. *International Microbiology, 21*, 97–106. https://doi.org/10.1007/s10123-018-0012-3

Priyadarshanee, M., & Das, S. (2021). Biosorption and removal of toxic heavy metals by metal tolerating bacteria for bioremediation of metal contamination: a comprehensive review. *Journal of Environmental Chemical Engineering, 9*(1), 104686. https://doi.org/10.1016/j.jece.2020.104686

Priyadarshanee, M., Chatterjee, S., Rath, S., Dash, H.R., & Das, S. (2022). Cellular and genetic mechanism of bacterial mercury resistance and their role in biogeochemistry and bioremediation. *Journal of Hazardous Materials, 423*, 126985.

Qin, H., Wang, Z., Sha, W., Song, S., Qin, F., & Zhang, W. (2024). Role of Plant-Growth-Promoting Rhizobacteria in Plant Machinery for Soil Heavy Metal Detoxification. *Microorganisms, 12*, 700.

Pirsarandib, Y., Hassanpouraghdam, M.B., Rasouli, F., Aazami, M.A., Puglisi, I., & Baglieri, A. (2022). Phytoremediation of soil contaminated with heavy metals via arbuscular mycorrhiza (*Funneliformis mosseae*) inoculation ameliorates the growth responses and essential oil content in lavender (*Lavandula angustifolia* L.). *Agronomy, 12*, 1221.

Rasouli-Sadaghiani, M.H., Barin, M., Khodaverdiloo, H., Siavash Moghaddam, S., Damalas, C.A., & Kazemalilou, S. (2019). Arbuscular mycorrhizal fungi and rhizobacteria promote growth of Russian knapweed (*Acroptilon repens* L.) in a Cd-contaminated soil. *Journal of Plant Growth Regulation, 38,* 113–121.

Raza, A., Charagh, S., Najafi-Kakavand, S., & Siddiqui, M.H. (2021). The Crucial Role of Jasmonates in Enhancing Heavy Metals Tolerance in Plants. In: Aftab, T., Yusuf, M. (eds) *Jasmonates and Salicylates Signaling in Plants. Signaling and Communication in Plants.* Springer, Cham. https://doi.org/10.1007/978-3-030-75805-9_8

Riaz, M., Kamran, M., Fang, Y., Wang, Q., Cao, H., Yang, G., Deng, L., Wang, Y., Zhou, Y., Anastopoulos, I., & Wang, X. (2021). Arbuscular mycorrhizal fungi-induced mitigation of heavy metal phytotoxicity in metal contaminated soils: a critical review. *Journal of Hazardous Materials, 402*, 123919.

Riseh, R.S., Vazvani, M.G., Hajabdollahi, N., & Thakur, V.K. (2023). Bioremediation of heavy metals by Rhizobacteria. *Applied Biochemistry and Biotechnology, 195*(8), 4689–4711. https://doi.org/10.1016/j.rhisph.2022.100565

Rolón-Cárdenas, G.A., Arvizu-Gómez, J.L., Pacheco-Aguilar, J.R., Vázquez-Martínez, J., & Hernández-Morales, A. (2021). Cadmium-tolerant endophytic *Pseudomonas rhodesiae* strains isolated from *Typha latifolia* modify the root architecture of *Arabidopsis thaliana* Col-0 in presence and absence of Cd. *Brazilian Journal of Microbiology, 52*, 349–361.

Roy, S., & Roy, M. (2015). Bioleaching of heavy metals by sulfur oxidizing bacteria: a review. *International Research Journal of Environmental Sciences, 4*(9), 75–79.

Saad, M.M., Saad, M.A., Saad, B.S., Zakaria, F.A., Husain, A.R.A., & Abdelgaleil, S.A. (2022). Bioremediation and microbial-assisted phytoremediation of HMs by endophytic Fusarium species isolated from *Convolvulus arvensis*. *Bioremediation Journal, 28*(2), 202–212.

Sahinkaya, E., Uçar, D., & Kaksonen, A.H. (2017). Bioprecipitation of Metals and Metalloids. In: Rene, E., Sahinkaya, E., Lewis, A., Lens, P. (eds) *Sustainable Heavy Metal Remediation. Environmental Chemistry for a Sustainable World*, vol 8. Springer, Cham. https://doi.org/10.1007/978-3-319-58622-9_7

Shadmani, L., Jamali, S. & Fatemi, A. (2021). Isolation, identification, and characterization of cadmium-tolerant endophytic fungi isolated from barley (*Hordeum vulgare* L.) roots and their role in enhancing phytoremediation. *Brazilian Journal of Microbiology*, *52,* 1097–1106.

Sharma, J., Goutam, J., Dhuriya, Y.K., & Sharma, D. (2021). Bioremediation of Industrial Pollutants. In: Panpatte, D.G., Jhala, Y.K. (eds) *Microbial Rejuvenation of Polluted Environment. Microorganisms for Sustainability*, vol 26. Springer, Singapore. https://doi.org/10.1007/978-981-15-7455-9_1

Shi, Y., Xie, H., Cao, L., Zhang, R., Xu, Z., Wang, Z., & Deng, Z. (2017). Effects of Cd-and Pb-resistant endophytic fungi on growth and phytoextraction of *Brassica napus* in metal-contaminated soils. *Environmental Science and Pollution Research*, *24*, 417–426.

Singh, A., & Prasad, S.M. (2011). Reduction of heavy metal load in food chain: technology assessment. *Reviews in Environmental Science and Bio/Technology, 10*, 199–214. https://doi.org/10.1007/s11157-011-9241-z

Singh, J., & Kalamdhad, A.S. (2011). Effects of heavy metals on soil, plants, human health and aquatic life. *International Journal of Research in Chemistry and Environment, 1*(2), 15–21.

Sorour, A.A., Khairy, H., Zaghloul, E.H. & Zaghloul, H.A. (2022). Microbe-plant interaction as a sustainable tool for mopping up heavy metal contaminated sites. *BMC Microbiology, 22*, 174.

Spielmann, J., Cointry, V., Devime, F., Ravanel, S., Neveu, J., & Vert, G. (2022). Differential metal sensing and metal--dependent degradation of the broad spectrum root metal transporter IRT1. *The Plant Journal, 112*, 1252–1265.

Sreedevi, P.R., Suresh, K., & Jiang, G. (2022). Bacterial bioremediation of heavy metals in wastewater: a review of processes and applications. *Journal of Water Process Engineering, 48*, 102884.

Tang, Z., Liu, X., Niu, X., Yin, H., Liu, M., Zhang, D., & Guo, H., (2023). Ecological risk assessment of aquatic organisms induced by heavy metals in the estuarine waters of the Pearl River. *Scientific Reports, 13*(1), 9145. https://doi.org/10.1038/s41598-023-35798-x

Tashan, H., Harighi, B., Rostamzadeh, J., & Azizi, A. (2021). Characterization of arsenic-resistant endophytic bacteria from alfalfa and chickpea plants. *Frontiers in Plant Science, 12,* 696750.

Tayang, A., & Songachan, L.S. (2021). Microbial bioremediation of heavy metals. *Current Science*, *120*, 1013–1025.

Tchounwou, P.B., Yedjou, C.G., Patlolla, A.K., & Sutton, D.J. (2014). *Heavy metals toxicity and the environment.* Berlin: National Institute of Health, USA.

Tirry, N., Kouchou, A., El Omari, B., Ferioun, M. & El Ghachtouli, N. (2021). Improved chromium tolerance of *Medicago sativa* by plant growth-promoting rhizobacteria (PGPR). *Journal of Genetic Engineering and Biotechnology, 19*, 1–14.

Tiwari, P., & Bae, H. (2023). Trends in harnessing plant endophytic microbiome for heavy metal mitigation in plants: a perspective. *Plants*, *12*(7), 1515. https://doi.org/10.3390/plants12071515

Tiwari, J., Ma, Y., & Bauddh, K. (2022). Arbuscular mycorrhizal fungi: an ecological accelerator of phytoremediation of metal contaminated soils. *Archives of Agronomy and Soil Science, 68*, 283–296.

Tovar-Sánchez, E., Concepción-Acosta, C.M., Sánchez-Reyes, A., Sánchez-Cruz, R., Folch-Mallol, J.L., & Mussali-Galante, P. (2023). Aspergillus luchuensis, an endophyte fungus from the metal hyperaccumulator plant Prosopis laevigata, promotes its growth and increases metal translocation. *Plants, 12*, 1338. https://doi.org/10.3390/plants12061338

Ullah, I., Mateen, A., Ahmad, M.A., Munir, I., Iqbal, A., Alghamdi, K.M., Al-Solami, H.M. & Siddiqui, M.F. (2022). Heavy metal ATPase genes (HMAs) expression induced by endophytic bacteria,"AI001, and AI002" mediate cadmium translocation and phytoremediation. *Environmental Pollution, 293*, 118508.

Verma, N., & Sharma, R. (2017). Bioremediation of toxic heavy metals: a patent review. *Recent Patents on Biotechnology, 11*, 171–187.

Verma, S., & Kuila, A. (2019). Bioremediation of heavy metals by microbial process. *Environmental Technology & Innovation, 14*, 100369. https://doi.org/10.1016/j.eti.2019.100369

Wang, F., Zhang, L., Zhou, J., Rengel, Z., George, T.S., & Feng, G. (2022). Exploring the secrets of hyphosphere of arbuscular mycorrhizal fungi: processes and ecological functions. *Plant and Soil, 481*(1), 1–22. https://doi.org/10.1016/j.scienta.2022.111591

Wang, L., Li, Z., Zhang, G., Liang, X., Hu, L., Li, Y., He, Y. and Zhan, F. (2023). Dark septate endophyte *Exophiala pisciphila* promotes maize growth and alleviates cadmium toxicity. *Frontiers in Microbiology, 14*, 1165131.

Wang, X., He, M., Xie, J., Xi, J., & Lu, X. (2010). Heavy metal pollution of the world largest antimony mine-affected agricultural soils in Hunan province (China*). Journal of Soils and Sediments, 10*, 827–837.

Wang, Y.P., Shi, J.Y., Qi, L.I.N., Chen, X.C., & Chen, Y.X. (2007). Heavy metal availability and impact on activity of soil microorganisms along a Cu/Zn contamination gradient. *Journal of Environmental Sciences, 19*(7), 848–853.

Wang, Y., Narayanan, M., Shi, X., Chen, X., Li, Z., Natarajan, D., & Ma, Y. (2022). Plant growth-promoting bacteria in metal-contaminated soil: current perspectives on remediation mechanisms. *Frontiers in Microbiology, 13*, 966226.

Wu, J., Kamal, N., Hao, H., Qian, C., Liu, Z., Shao, Y., Zhong, X. & Xu, B. (2019). Endophytic *Bacillus megaterium* BM18-2 mutated for cadmium accumulation and improving plant growth in Hybrid Pennisetum. *Biotechnology Reports*, 24, p.e00374, 1–9.

Xiao, Y., Chen, R., Chen, L., Yang, B., Jiang, L., & Fang, J. (2023). Endophytic fungus Talaromyces sp. MR1 promotes the growth and cadmium uptake of *Arabidopsis thaliana* L. under cadmium stress. *Current Microbiology, 80*, 346.

Yamaji, K., Watanabe, Y., Masuya, H., Shigeto, A., Yui, H., & Haruma, T. (2016). Root fungal endophytes enhance heavy-metal stress tolerance of *Clethra barbinervis* growing naturally at mining sites via growth enhancement, promotion of nutrient uptake and decrease of heavy-metal concentration. *PloS One, 11*, e0169089.

Yan, Z.X., Li, Y., Peng, S.Y., Wei, L., Zhang, B., Deng, X.Y., Zhong, M., & Cheng, X. (2024). Cadmium biosorption and mechanism investigation using two cadmium-tolerant microorganisms isolated from rhizosphere soil of rice. *Journal of Hazardous Materials, 470*, 134134.

Yihui, B. A. N., Zhouying, X. U., Yurong, Y. A. N. G., Zhang, H., Hui, C. H. E. N., & Ming, T. A. N. G. (2017). Effect of dark septate endophytic fungus Gaeumannomyces cylindrosporus on plant growth, photosynthesis and Pb tolerance of maize (Zea mays L.). *Pedosphere, 27*(2), 283–292.

Yongpisanphop, J., Babel, S., Kurisu, F., Kruatrachue, M., & Pokethitiyook, P. (2019). Isolation and characterization of Pb-resistant plant growth promoting endophytic bacteria and their role in Pb accumulation by fast-growing trees. *Environmental Technology, 41*(27), 3598–3606.

Zahoor, M., Irshad, M., Rahman, H., Qasim, M., Afridi, S. G., Qadir, M., & Hussain, A. (2017). Alleviation of heavy metal toxicity and phytostimulation of *Brassica campestris* L. by endophytic *Mucor* sp. MHR-7. *Ecotoxicology and Environmental Safety*, *142*, 139–149.

Zaynab, M., Al-Yahyai, R., Ameen, A., Sharif, Y., Ali, L., Fatima, M., Khan, K.A. & Li, S. (2022). Health and environmental effects of heavy metals. *Journal of King Saud University-Science, 34*, 101653. https://doi.org/10.1016/j.jksus.2021.101653

Zhan, F., Li, B., Jiang, M., Li, T., He, Y., Li, Y., & Wang, Y. (2019). Effects of arbuscular mycorrhizal fungi on the growth and heavy metal accumulation of bermudagrass [*Cynodon dactylon* (L.) Pers.] grown in a lead-zinc mine wasteland. *International Journal of Phytoremediation, 21*, 849–856.

Zheng, J., Xie, X., Li, C., Wang, H., Yu, Y., & Huang, B. (2023). Regulation mechanism of plant response to heavy metal stress mediated by endophytic fungi. *International Journal of Phytoremediation, 25*(12), 1596–1613.

Zhu, L., Guo, J., Sun, Y., Wang, S., & Zhou, C. (2021). Acetic acid-producing endophyte *Lysinibacillus fusiformis* orchestrates jasmonic acid signaling and contributes to repression of cadmium uptake in tomato plants. *Frontiers in Plant Science, 12*, 670216.

7 Role of Phytohormone-Secreting Microorganisms in the Mitigation of Abiotic Stress in Plants

Raja Ghosh and Swarnendu Roy

7.1 INTRODUCTION

Plants cannot avoid unfavorable environmental conditions like other organisms due to their sessile nature. Instead, they have developed resistance mechanisms for their survivability against adverse environmental conditions (Gimenez et al. 2018). With many other limiting factors in the environment, abiotic stress is the one that majorly interferes with plant growth and development. The major abiotic stresses like drought, salinity, heavy metals, and temperature have a vast impact on crop production. Among all stresses, drought causes the highest destruction worldwide; salinity comes second, followed by others (Akram et al. 2019). The human population is estimated to reach 9.7 billion by 2050, raising food demand by 70% (Randive et al. 2021). Therefore, the reduction in crop production and uncontrolled explosion in the human population have raised a vital question regarding food security (Sharma et al. 2020). Thus, sustainable agricultural practices demand a lot of attention nowadays for maintaining the food supply.

Plants have developed various resistance mechanisms to cope with abiotic factors, involving phytohormones as an integral part of their resistance response. Major phytohormones like auxin (IAA) help in root growth, leading to better nutrients, water uptake, and mycorrhizal colonization, assisting the plant to achieve better stress tolerance (Grover et al. 2021). Foliar application of gibberellic acid (GA3) has been known to enhance the overall growth parameters like branch count, chlorophyll content, relative water content, and dry weight in salt-stressed *Solanum lycopersicum* and *Brassica napus* (Jayasinghe et al. 2019; Elahi et al. 2022). Overexpression of genes like *Isopentenyl Transferase* (IPT) helps in the biosynthesis of cytokinin (CK), which in turn activates stress-related genes like Arabidopsis AHK1 (*Arabidopsis histidine kinase 1*) and helps plants under drought stress (Peleg and Blumwald 2011; Wang et al. 2015). It also helps plants by stopping the premature senescence of leaves, enhancing grain filling and cellular proliferation (Egamberdieva et al. 2017). Abscisic acid (ABA) is another stress-related hormone that controls many signaling pathways, helping to achieve better stress tolerance. ABA-mediated stomatal closure and accumulation of osmoprotectants help in drought and heat stress tolerance (Sah et al. 2016). Nutrient depletion is another abiotic stress in which the role of strigolactones (SL) can be beneficial for plants. SL was found to improve the biomass and fresh weight of *Solanum lycopersicum* in phosphate-deficient soil by inducing lateral and primary root production (Santoro et al. 2020). Another study reported on *Arabidopsis* sp. that SL signaling mutant plants are susceptible towards salinity stress compared to wild-type plants suggesting the role of SL in salt stress also (Bhoi et al. 2021). Ethylene, besides its fruit ripening activity, is also known for its pivotal role in the alleviation of abiotic stress in plants. In this connection, a study was performed on *Marchantia polymorpha*, where they knocked out

DOI: 10.1201/9781003354109-7

two genes involved in ethylene signaling, Mpein3 (*Marchantia polymorpha* ethylene insensitive 3) and Mpctr1 (*Marchantia polymorpha* constitutive ethylene response 1), responsible for conferring ethylene insensitivity and overexpression, respectively. Mutants were then exposed to several stresses like salinity, nutrients, heat and constant far-red light, and it was found that Mpctr1 mutants showed stress tolerance while Mpein3 mutants, due to lack of ethylene, showed more sensitivity towards the mentioned stress (Bharadwaj et al. 2022).

Most microorganisms associated with plants in a symbiotic relationship also produce phytohormones that impart various plant growth and defense-related functions (Kashyap et al. 2019). Bacteria like *Azospirillum* sp., *Bacillus* sp., and *Pseudomonas* sp. are found to tackle drought stress in plants, accelerating the production of phytohormones such as auxin, abscisic acid (ABA), cytokinin (CK), and others (Koza et al. 2022). Talaat (2019) reported applying a combination of microorganisms (*Rhodopseudomonas* sp., *Lactobacillus* sp., *Aspergillus* sp., *Streptomyces griseus, S. albus*) to salt-stressed *Phaseolus vulgaris* plants has significantly reduced the adverse effects of salt stress compared to the uninoculated plants. The plants inoculated with these beneficial microorganisms have been exhibited to retrieve the photosynthetic efficiency by enhancing photosystem-I and II activity, chlorophyll fluorescence, and improved gas exchange. In this context, new strains of *Rhizobacteria* sp. (SH-9 and SH-19) produce phytohormones like IAA, SA, and ABA, which then impart oxidative stress tolerance in plants (Shaffique et al. 2023). Phytohormone-producing microorganisms (PPMs), therefore, can be very useful in supporting plant survival during the period of vulnerability due to abiotic stresses. They also play a pivotal role in the betterment of plant growth and production quality. In this context, phytohormone-producing PPMs have the potential to be used as biofertilizers. This chapter thus focuses on the details of the stress amelioration and growth-promoting abilities of various PPMs.

7.2 PHYTOHORMONES AS GROWTH-ENHANCING AND STRESS-AMELIORATING AGENTS

7.2.1 Growth Enhancement

Phytohormones play a multidimensional role in plants, touching almost every aspect of plant life, starting from germination and growth to protect against various stresses (Figure 7.1). In an experiment, Werner et al. (2001) intentionally lowered the cytokinin (CK) level in tobacco plants through genetic modification, which resulted in reduced growth of shoots but increased growth of roots, suggesting the role of CK in governing the meristematic activity. In a different study, overexpression of PIN (ZmPIN1α) protein showed more significant accumulation of auxin in maize roots, forming a bulky root system resulting in better nutrient uptake and water absorption, following overall improvement in yield (Li et al. 2018). The ratio of cytokinin/auxin ratio in a plant plays a pivotal role in the determination of root or shoot growth; when the ratio is >1, the growth of shoot apical meristem is noted, and if the ratio is <1, branching and growth of the root is observed (Bielach et al. 2017). Similarly, gibberellic acid (GA) seems to enhance the overall biomass of plants subjected to different light irradiance. In the case of low light intensity, GA promotes light absorption and accelerates internode elongation (Falcioni et al. 2018). Another experiment shows the long-lasting effect of seed priming with GA_3 on the growth and development of the *Leymus chinensis* plant. The GA_3-primed seeds enhanced biomass accumulation in the seedlings compared to the non-primed seeds (Ma et al. 2018). A lesser-known phytohormone, strigolactone, is known to help plants shape their root and shoot structure and accelerate the symbiotic association of endomycorrhizal fungi (Liu et al. 2018). A study has also shown that synthetic strigolactone GR24 helps primary root elongation by suppressing adventitious root growth. Acceleration in hyphal growth in arbuscular mycorrhizal fungi associated with plant roots was noticed upon SL application, resulting in better plant growth (Alvi et al. 2022). Another essential phytohormone, brassinosteroid (BR), improves stress

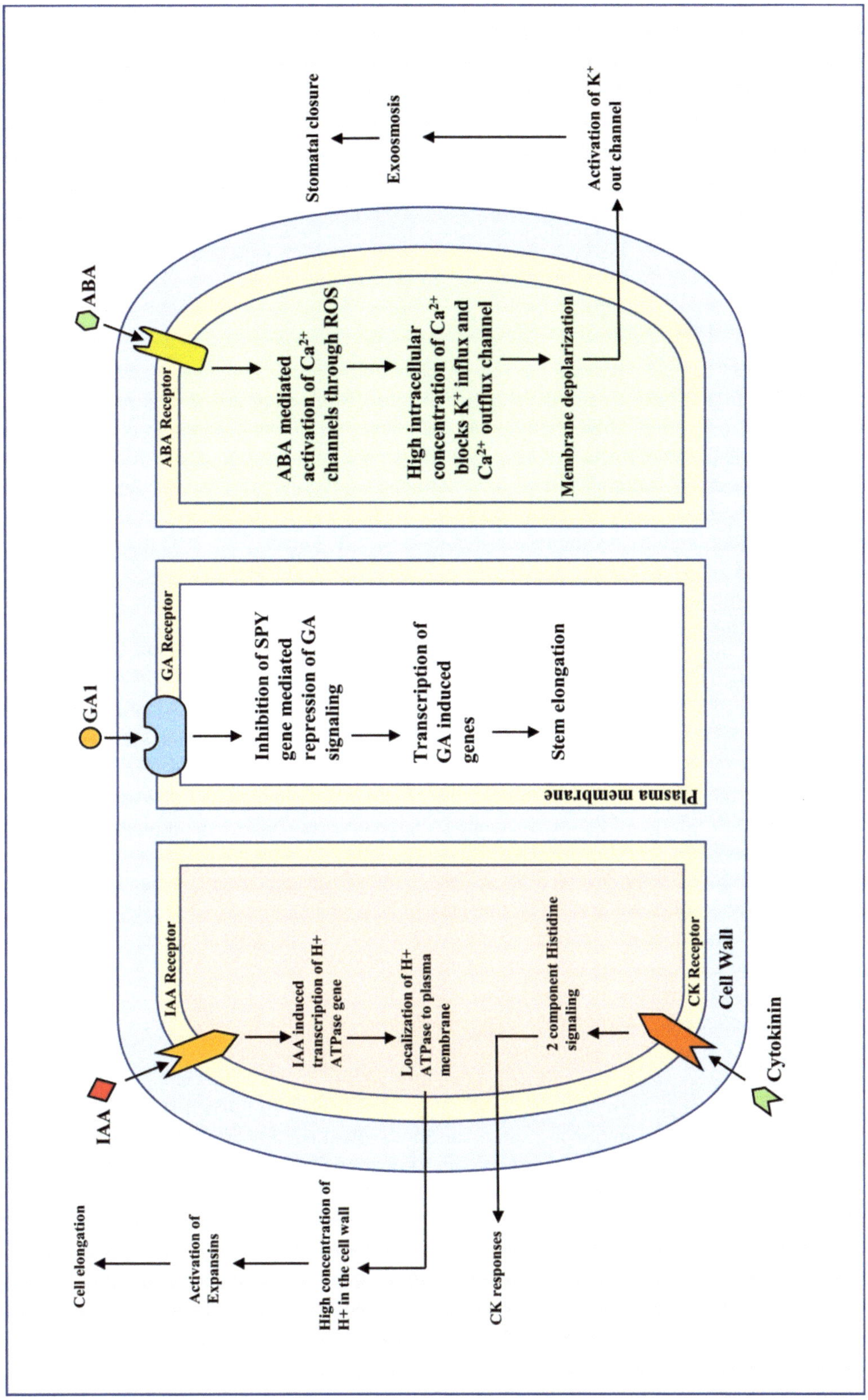

FIGURE 7.1 Signaling pathways of phytohormones (IAA, CK, GA, ABA) involved in plant growth.

response, chlorophyll content in strawberry plants and fruit quality, and shelf life in cherry plants (Hussain et al. 2020). At the molecular level, Chen et al. (2017a) found that *BRASSINOSTEROID INSENSITIVE1-EMS-SUPPRESSOR 1* (BES_1)/ *BRASSINAZOLE-RESISTANT 1* (BZR_1) transcription factors are regulated by BR signaling, which is involved with growth in *Arabidopsis*. In this context, the involvement of BR is found in the regulation of plant height and spikelet length in rice. BR is also responsible for pollination in maize, as lowering the concentration of BR brings male sterility (Manghwar et al. 2022). Ethylene, the gaseous hormone, is important for the transition of plant developmental phases, fruit ripening, and leaf senescence (Iqbal et al. 2017). Ethylene response factors (ERF) are found to have a relation with the growth of *Arabidopsis* plant, as knocking out of overexpression of the same resulted in changes in growth attributes (Dubois et al. 2018).

7.2.2 Stress Amelioration

Several researchers have emphasized the role of phytohormones in inducing greater stress tolerance in addition to their growth and developmental roles (Bhatt et al. 2020). The phytohormones have been shown to respond differentially under the effect of different stresses. In this connection, Cao et al. (2017) analyzed the effect of salinity stress on barley roots through LC-MS analysis and exhibited the differential accumulation of phytohormones. According to this study, ABA increased in all the cultivars. In contrast, the enhanced level of root SA was found only in the case of the tolerant cultivars, as exhibited by the increase in chlorophyll content. Similarly, the application of Salicylic acid was shown to enhance the level of GA and lower the ABA content of the sunflower seedlings along with the regulation of Zn^{2+} transporter genes, thereby restricting the uptake of Zn^{+2} ions, which help in alleviation of zinc stress (Huang et al. 2021). Brassinosteroids are an important class of hormone that are at the center stage of regulating the biosynthesis of other hormones in drought-stressed barley cultivars (Gruszka et al. 2018). It has been shown that drought stress induces the accumulation of plant steroid hormones like castasterone and 24-epi-brassinolide simultaneously regulates the biosynthesis of other phytohormones like GA, SA, JA, and ABA (Iqbal et al. 2022). Elevated CO_2 is known to improve plants' photosynthetic efficiency and is also involved in improving thermotolerance. In this context, the phytohormone-dependent pathways have been shown to regulate CO_2-induced thermotolerance (Ahammed et al. 2021). Auxin plays a major role in drought stress as GH3 (The Gretchen Hagen 3) gene family enzymes are responsive to auxin concentration and are found to be involved in drought stress tolerance. Late embryogenesis abundant (LEA) protein expression is accelerated by the OsGH3.13 gene belonging to the GH3 family, which in turn enhances drought stress tolerance (Pandey et al. 2019). The exogenous application of GA also helps mitigate lead stress and accelerates plant growth by enhancing the total amount of chlorophyll and biomass of carrot plants (Ghani et al. 2021). The restoration and alleviation of the effects of several stresses on plants through the exogenous application of phytohormones are also well-known (Kosakivska et al. 2020). Another study shows GA3-mediated seed priming alleviate fluoride toxicity in rice cultivars (Banerjee and Roychoudhury 2020). Besides increasing the endogenous accumulation of the phytohormone, the application of GA also stimulated the biosynthesis of indole-3-acetic acid, thereby improving shoot and root growth and lowering fluoride-induced oxidative stress. Seed priming with phytohormones also imparts tolerance against heavy metal toxicity. For instance, SA priming has been shown to subside the effect of cadmium stress in wheat plants (Gul et al. 2020). Plants subjected to heavy metal stress are found to have reduced primary and secondary root growth as well as leaf area, which makes the plant more sensitive towards the stress. In this case, auxin homeostasis genes (cytochrome p450 monooxygenase, transport inhibitor response 1) were found to be regulated differently in *Arabidopsis* shoots and roots conferring heavy metal (Pb^{2+}, Zn^{2+}) stress tolerance (Mathur et al. 2022). Moreover, seed priming with GA3 significantly negated the adverse impact of salinity stress in *Pisum sativum* (Ahmad et al. 2021). In this case, GA-mediated stress tolerance was manifested by enhanced activity of antioxidant enzymes, total phenol and flavonoid content, and upregulation of *SOS1* and *NHX1* genes for ion homeostasis. Moreover, seed pre-treatment with lesser-known phytohormones like strigolactones and karrikins has also been shown to positively affect stress tolerance in plants subjected to salinity stress. In this connection, seed treatment with a synthetic strigolactone – GR24 – has been shown to improve the growth, enzymatic antioxidants (superoxide dismutase, peroxidase, catalase), total phenolics, and compatible solutes in *Trachyspermum ammi* plants grown under the effect of salinity stress (Mujahid et al. 2023). Similarly, foliar application of jasmonic acid (JA) has been shown to impart enhanced tolerance in soybean seedlings exposed to salinity stress (Sheteiwy et al. 2021). In this study, the foliar application of JA has been found to enhance the endogenous levels ABA, GA, and JA, along with the upregulation of antioxidant genes (SOD, POX, CAT, APX) that co-ordinately helped in alleviating the impact of salinity stress. Exogenous application of GA improved parameters like stomatal conductance, biomass, photosynthesis in *Corchorus capsularis* plant, which was collected from a copper mining area containing a very high amount of Cu (Saleem et al. 2020).

7.3 PHYTOHORMONE-SECRETING MICROORGANISMS FACILITATE PLANT GROWTH AND DEVELOPMENT

Plant growth and development are directly related to its surrounding environment, and microorganisms play a pivotal role in that by secreting important phytohormones. Certain bacterial species like *Agrobacterium*, *Pseudomonas*, *Enterobacter*, and *Rhizobium* are found in the root zone of plants that produce a high amount of auxin that helps in plant growth and development (Pathania et al. 2020). Khan et al. (2020), in their experiment, inoculated soybean plants with the *B. cereus* SA1 strain, which produces hormones like GA and auxin, was shown to improve growth aspects like chlorophyll content, overall biomass, and amino acid content. In this perspective, they proposed the utility of this strain for the development of biofertilizers. It has been reported from Brazil that the maize plants show better nitrogen fixation along with enhanced root growth and water absorption due to the presence of *Azospirillum brasilense*, which secrete several hormones like IAA, GA, and cytokinin-like substances (Tien et al. 1979; Zeffa et al. 2019). *Azotobacter* or yeast, when applied in combination with compost or sludge, has been shown to promote the overall yield of maize plants by enhancing the number of grains per spike, spike length and weight of grains. This upliftment of growth parameters is mainly due to some chemical compounds secreted by *Azotobacter* sp. that, in turn, induce phytohormones like IAA, GA, and CK (Mohamed et al. 2019). Phytohormones like IAA, ABA, and GA production in high amounts have been noticed when endophytic strains like *Acinetobacter johnsonii*, *Bacillus pumilus*, and *Chryseobacterium indologene* are inoculated to *Beta* sp. These phytohormones improved the plant's root growth, shoot growth, and overall biomass (Shi et al. 2009). Bacterial isolates like *Azotobacter* sp. and *Pantoea* sp. from nodules of the wheat plant were also reported to produce IAA, which maximizes lateral root formation and colonization (Narula et al. 2006).

Also, *Bacillus* sp. is known to produce a high amount of gibberellins, enhancing nutrient uptake, overall biomass, and pigment synthesis in rice plants (Kang et al. 2019). *Brassica oleracea* plant showed expansion of root area and better shoot growth when inoculated with *Pantoea* sp. and *B. megaterium*. This acceleration of plant growth is achieved due to the ability of the strain to produce a high amount of hormones like IAA, SA, and GA (Ekinci et al. 2014). In their study, Turan et al. (2014) also found that *Pantoea* sp. inoculation in cabbage plants increased the accumulation of phytohormones like IAA, GA, and SA, which accelerated chlorophyll content and biomass of the plant. Strains of *Trichoderma* sp. are also found to accelerate plant growth and defense by accumulating different plant-related hormones like ethylene and auxin in the rhizosphere of melon plants (Medina et al. 2014). Similarly, Patel and Saraf (2017) reported the isolation of 70 rhizobacterial species in which they found *Pseudomonas stutzeri* MTP40 strain to produce the highest amount of essential phytohormones like GA, IAA, CK that improve plant health.

Priyadharsini and Muthukumar (2017) reported that *Curvularia* sp. – an endophytic fungus isolated from *Parthenium hysterophorus* – was responsible for the improvement of the growth attributes of pea plants when inoculated. This fungus was found to secrete IAA and was also able to solubilize phosphate, which helped the plants to grow better than the non-inoculated plants. Apart from these, two strains of *Serratia* sp. seem to accelerate chickpea growth under nutrient stress by the production of IAA (Zaheer et al. 2016). According to Asaf et al. (2017), three species of bacterial endophytes – *B. subtilis*, *Sphingomonas* sp., and *Serratia* sp. – were found to produce IAA in culture broth. These endophytes when inoculated to soybean plants effectively modulated the endogenous phytohormone levels of ABA, GA, and JA accumulation, as well as IAA production, resulting in better plant growth than the uninoculated plants (Asaf et al. 2017).

7.4 PHYTOHORMONE-SECRETING MICROORGANISMS IN THE MITIGATION OF ABIOTIC STRESSES

It has been very well realized that abiotic stresses impart a multitude of adverse effects on crop plants, hampering their growth and productivity. The use of chemical fertilizers and modern technology is insufficient to provide a permanent solution to this. Therefore, the exploration of microorganisms becomes crucial as there are several of them present in the rhizosphere and also inside the plant tissues, naturally providing the plant with the much-needed strength to survive under adverse conditions. One of the vital ways in this regard is their ability to regulate the host metabolic pathways through the secretion of phytohormones. Microorganisms can play an influential role in modulating the biosynthesis of phytohormones and work towards mitigating the negative impacts of several stresses in plants (Figure 7.2). In this connection, manipulating stress-tolerant microorganisms can be effectively inculcated in stress-management practices. The following sections will, therefore, discuss the ability of several microorganisms to impart tolerance to different abiotic stresses. In this context, Table 7.1 also enlists several microorganisms that impart enhanced stress tolerance in plants by secreting hormones in the plant rhizosphere.

7.4.1 Salinity Stress

Salinity is one of the major abiotic stresses that brings a serious problem with it. The rate of worldwide destruction of crops due to salinity is so high that it will soon become a barrier between humans and their daily food requirements if it remains unsolved. According to a recent report by FAO, more than 833 million hectares of cultivated land contain a concentration of salt that is unsuitable for cultivation (FAO 2021). However, using salt-tolerant microorganisms seems promising for mitigating salinity stress (Nadeem et al. 2013). *Brachybacterium paraconglomeratum* – an ACC deaminase-producing salt-tolerant endophyte that was isolated from the roots of *Chlorophytum borivilianum* – was used to induce salinity tolerance in the salt-stressed plants (Barnawal et al. 2016). In this case, the inoculation of the bacterium imparted the deamination of ACC in *Chlorophytum* roots, thereby lowering the accumulation of ethylene (ET) that reduced the extent of chlorosis and senescence to improve the yield of plants. The application of PGPR strains like *Arthrobacter protophormiae*, *Dietzia natronolimnaea*, and *Bacillus subtilis* has also been shown to impart drought and salt stress tolerance in wheat (Barnawal et al. 2017). This study has emphasized that all three PGPR strains induced stress tolerance by enhancing the biosynthesis of IAA and reducing the accumulation of ABA and ethylene, modulating the ethylene signaling pathway and DREB2 transcription factor (Barnawal et al. 2017). In a similar study, three strains of *Pseudomonas* sp. were found to accelerate root growth of wheat seedlings under salinity stress by producing IAA (Egamberdieva 2009). Apart from this, the microorganism was found to enhance the expression levels of several genes encoding antioxidant enzymes (*GR1* and *APX1*), sodium ion transportation (*NHX1*, *NHX2*, *NHX3*), and other salt-responsive genes (*WRKY58*, *DREB2A*), thereby, assisting the phytohormone-mediated growth

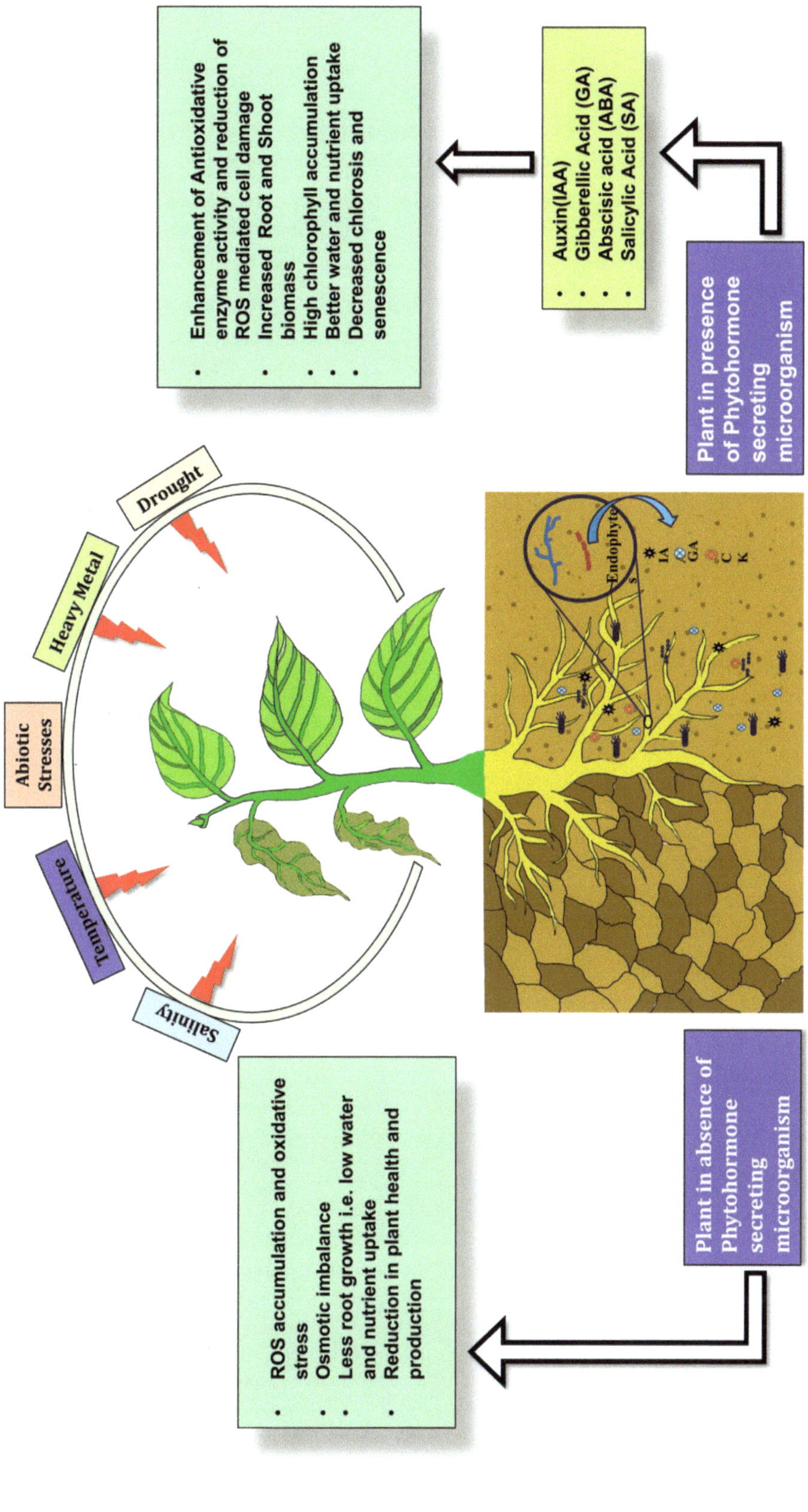

FIGURE 7.2 Overall depiction of mechanisms imparting abiotic stress tolerance in a plant under the effect and presence of phytohormone-secreting microorganisms.

TABLE 7.1
List of Phytohormone-Secreting Microorganisms that Assist in the Alleviation of Abiotic Stress in Plants

Name of Microorganism	Related Abiotic Stress	Isolated From	Hormone in Concern	Mode of Action	References
Bacillus velezensis FMH2	Salinity	*Solanum lycopersicum*	IAA	Root structure, leaf area, fresh and dry weight got increased	Masmoudi et al. 2021
Curtobacterium **sp.** *SAK1*	Salinity	*Glycine max*	IAA and ABA	Number of root tips increased, and as a result better nutrient uptake was noticed	Khan et al. 2019
Azospirilum **sp.**	Drought	*Zea mays*	ABA and GA	Leaf area and shoot length enhanced in drought-stressed plants	Cohen et al. 2009
Phoma **sp.**	Drought	*Pinus tabulaeformis*	ABA	Enhanced antioxidative enzyme activity and proline level, also enhance ABA mediated seed protective mechanism	Zhou et al. 2021
Kocuria rhizophila	Heavy metal	*Zea mays*	IAA	Modulated endogenous IAA and ABA levels, thereby improving nutrient uptake	Li et al. 2020
Leclercia adecarboxylata	Heavy metal	*Cucumis sativus*	IAA	Secreted siderophores that chelate zinc and silicates and reduce endogenous H_2O_2 accumulation	Kang et al. 2021
Paecilomyces Formosus	Heavy metal	*Zea mays*	IAA and GA	Reduce ABA and JA levels in rice	Bilal et al. 2018
Providencia rettgeri TCR22	Heavy metal	*Sorghum bicolor*	IAA	Produce siderophores and reduce Cr^{6+} to Cr^{3+}	Bruno et al. 2020
Bacillus aryabhattai SRB02	Temperature	*Glycine max*	ABA	ABA-mediated stomatal closure, and IAA-mediated root growth	Park et al. 2017
Bacillus cereus SA1	Temperature	*Glycine max*	GA and	Reduction of ABA, Increased SA and Antioxidative enzymes	Khan et al. 2020
Myroides odoratimimus TCR22	Temperature	*Sorghum bicolor*	IAA	Produce IAA, upregulate SOD, APX1, CAT and reduces oxidative stress	Bruno et al. 2020
Pseudomonas putida	Temperature	*Triticum aestivum*	IAA and GA	Increased antioxidative enzymes activity, lowers plasma membrane injury	Ali et al. 2011

and development in the plants under salinity stress (Li et al. 2020). In another experiment, Hidri et al. (2016) isolated *Bacillus* sp. from saline environment and noticed production of a high amount of auxin with the increase in salt concentration. So, they inoculated the *Sulla carnosa* plant with these auxin-producing bacteria and observed reduction in adverse effects of salinity stress due to enhance root and shoot growth in treated plants compared to the control. Another experiment by Hamayun et al. (2017) shows an enhancement of growth activities of the soybean plant under salinity stress. According to the study, *Porostereum spadiceum* produces high amounts of active GA, which is mainly responsible for stress tolerance. Some other examples of salinity stress tolerance achieved by phytohormone-producing bacteria are *Bacillus velezensis* FMH2, which produces IAA and promotes root growth of the tomato plant; *Curtobacterium* sp. producing ABA and IAA; and *Nocardiodies* sp. producing IAA that help the plant to withstand saline conditions (Khan et al. 2019; Masmoudi et al. 2021; Meena et al. 2020). Similarly, inoculation of salt-stressed plants with *B. aryabhattai* ALT29 and *Planococcus rifietoensis* RT4 reduces salt stress by helping the potassium uptake and enhancing EPS production, respectively. They also promoted synthesis of ACC deaminase, which promote salt stress tolerance by lowering the ethylene concentration (Krishnamoorthy et al. 2022). Likewise, the inoculation of *Kocuria rhizophila* to maize roots subjected to NaCl stress has been reported to modulate the levels of IAA and ABA, thereby improving the acquisition of nutrients (Li et al. 2020). Similarly, two endophytes, *Phoma glomerata* and *Penicillium* sp., are found to bring GA and IAA-mediated drought and salinity stress tolerance in cucumber plants. Tolerance is mainly achieved by accumulating several nutrients, upregulation of SA, and downregulation of ABA (Ali et al. 2017).

7.4.2 Drought Stress

Scarcity of water is a global problem with no borders, and thus, the impact of drought stress hampers plant growth, biomass, photosynthesis, and productivity worldwide. This stress can occur for many reasons, such as low rainfall, excessive heat, and changes in the intensity of light (Seleiman et al. 2021). According to Ghosh et al. (2019), the phytohormone (IAA, GA, CK, ABA) secreting strain of *Pseudomonas putida* GAP-P45 can alter the endogenous accumulation of the phytohormones in plants under drought stress along with the biosynthesis of proline and polyamines, thereby maintaining the osmotic potential of the plant cells. Similarly, IAA and SA-secreting microbial strains like *Enterobacter* sp. and *Bacillus* sp. have been shown to positively regulate root growth and development of plants under drought stress (Jochum et al. 2019). In a different study, the *Zea mays* plant was exposed to chemicals that inhibit ABA and GA synthesis, resulting in drought stress-like conditions. However, adverse effects were reversed when the plants were inoculated with *Azospirilum* sp., as the bacteria enhanced the production of ABA and GA (Cohen et al. 2009). *Platycladus orientalis* plant, when inoculated with cytokinin-producing *Bacillus subtilis*, exhibited higher relative water content, total dry weight, root growth, and stomatal conductance compared to non-inoculated plants under drought stress (Liu et al. 2013). Apart from this, Zhang et al. (2019) performed a pot experiment where they found better growth of orange plants under drought stress due to the acceleration of root growth. They also found higher concentrations of phytohormones like IAA, ABA, GA, BRs, and methyl jasmonate in the roots due to the inoculation of the mycorrhizal fungus *Funneliformis* sp. In this context, they concluded that the adaptability of drought stress in orange plants is mainly achieved by mycorrhizal inoculation. Apart from these, Ashry et al. (2022) isolated 20 bacterial strains from less water-containing soil, checked their drought tolerance ability and selected two strains. They found high production of IAA and GA by those two strains (*Bacillus cerus* and *B. albus*), which showed better plant growth under drought stress upon inoculation. When tomato plants were inoculated with endophytic fungi *Talaromyces omanensis* isolated from the desert area, it was found to uplift GA concentration, which in turn increased root growth, biomass, and photosynthetic capacity of the plants exposed to drought stress. They found it alleviated GA3 concentration in the treated plant, suggesting GA3 as one of the significant factors for drought tolerance (Halo et al. 2020).

7.4.3 Heavy Metal Stress

Heavy metal-contaminated areas are not only problematic for crop production as they hamper plant growth in many ways, but food safety parameters are also threatened due to the contamination of heavy metals that are bad for human health (Edelstein and Ben-Hur 2018). Phytohormone-secreting microorganisms, in this case, play a pivotal role in giving plants enough strength to overcome this adverse situation. In this connection, the application of microbial consortia of the fungus – *Paecilomyces formosus* and the bacteria – *Sphingomonas* sp. was shown to improve the heavy metal stress tolerance in *Glycine max* significantly by secretion of hormones like GA and IAA (Bilal et al. 2018). In this study, the inoculation of the microorganisms enhanced the endogenous accumulation of GAs and lowered the JA and ABA contents, supporting plant growth and development. In a different study, Kang et al. (2021) have shown that an IAA-producing Gram-negative bacterium *Leclercia adecarboxylata,* negated the toxic effects of zinc stress in cucumber plants due to their ability to secrete siderophores. The production of siderophores enabled the bacterium to utilize zinc and silicates. It simultaneously reduced the endogenous accumulation of H_2O_2 and phytohormones like ABA and SA to alleviate the impact of stress on plants. Three potential strains *Myroides odoratimimus, Bacillus cerus*, and *Providencia rettgeri*, which were found to be IAA secreting, when inoculated to *Sorghum bicolor* plants, promoted plant growth by strengthening the root zone and helped in the reduction of chromium to a less toxic form (Bruno et al. 2020). Nickle in high amounts imparts stress and has been shown to affect different growth parameters of the *Vinca rosea* plant. Auxin and ACC deaminase producing *Bacillus megaterium* inoculated with *Vinca rosea* helped to reverse the adverse effects largely by improved auxin and reduced ethylene accumulation in the plants (Khan et al. 2017). Also, *Rhizoglomus* sp. – an arbuscular mycorrhizal fungus – acted exceptionally well to save the inoculated plant under arsenic stress. Enhancement in levels of IAA and ABA, as well as diminution in levels of GA observed in this experiment, resulted in better stress tolerance and growth of the inoculated plants of *Robinia pseudoacacia* (Zhang et al. 2020).

7.4.4 Temperature Stress

Worldwide growth and productivity of plants are influenced by several abiotic stresses, and temperature stress is one of those. Through various mechanisms, phytohormone-secreting microbes can potentially mitigate it (Shekhawat et al. 2022). Park et al. (2017) reported better growth of soybean plants under heat stress when inoculated with *Bacillus aryabhattai* SRB02. The strain produces a high amount of ABA, which regulates stomatal closure to minimize water loss, and IAA accelerates the growth and development of roots for water absorption under heat stress. Phytohormones like JA in high concentrations can activate the defense system in plants but halt the growth by delaying the breakdown of DELLA protein. This protein is a family of transcription regulator that hamper plant growth by inhibiting the cell proliferation and cell expansion. In this connection, the endophyte *Paecilomyces formosus* LWL1 has been reported to improve plant growth under heat stress by sustaining the balance of JA in the plant in such a way so that DELLA proteins remain silent and plants can achieve stress tolerance (Waqas et al. 2015). *Paecilomyces formosus* was found to decrease the endogenous levels of ABA and JA in rice and produced IAA and GA that significantly improved the plant growth parameters like plant height and fresh and dry weight under the effect of heat stress (Waqas et al. 2015). Similarly, a higher ethylene level initiates the heat stress response in plants, but in the long term, it can negatively affect plant growth. However, bacteria like *Bacillus* sp. producing ACC deaminase that cleaves ACC – an immediate precursor of ethylene – into α-ketobutyrate and ammonia and maintains the optimum level of ethylene (Notununu et al. 2022). Another study has shown that bacteria in the rhizosphere secrete exopolysaccharides containing 97% of water, that improve the soil structure and enhance the water availability to plants, which in turn help the plant to tolerate heat stress (Shaffique et al. 2022). Heat stress-mediated ROS generation and cell damage are also known to be minimized by the alternation of antioxidant signaling by the IAA and GA producing bacteria *Bacillus cereus* SA1 in the soybean plant (Khan et al. 2020). Apart from this, heat-tolerant bacteria *B. tequilensis*, when inoculated to the cabbage plant, improved photosynthetic activity, biomass, and overall growth of the plant undergoing heat stress. This kind of stress tolerance in plants was primarily achieved by the actions performed by phytohormones like GA, IAA, JA, and SA, which were secreted by *B. tequilensis* (Kang et al. 2019).

7.5 STATUS OF BIOFERTILIZERS BASED ON PHYTOHORMONE-SECRETING MICROORGANISMS AND FUTURE PERSPECTIVES

The ever-expanding human population demands more and more productivity of food crops, leading to the usage of ample amounts of inorganic fertilizers everywhere, which has been causing severe consequences like depletion in soil fertility and a rise in environmental pollution (Asoegwu et al. 2020). Biofertilizers, in this case, can be a problem solver; certain microorganisms can make essential nutrients available to plants already present in the soil in an inaccessible form (Nosheen et al. 2021). According to Putriani et al. (2019), an endophytic bacterial strain EAP3, having 99% similarity with *Enterobacter* sp. has the potential ability to be used as a biofertilizer because it produces IAA and accelerates various growth parameters of *Oryza sativa* as well as save the plants from pathogen attack. Masniawaty et al. (2019) isolated bacterial strains from the chilli pepper's rhizosphere, showing the production of phosphate and GA, thus making it a probable biofertilizer. *Azospirillum* sp. produces a wide range of essential plant hormones (IAA, GA, ET), helping plants to grow better under adverse conditions (Suhameena et al. 2020). *Azospirillum*-based biofertilizers are commercially available and marketed as 'Azospirillum' by Anushika Agri Products and 'Azospirillum' by IFFCO India (IFFCO (2017)). Ismail et al. (2021) isolated different auxin-producing microorganisms from the root zone of bean plants and checked for better growth of plants using an external hormone source. Interestingly, they found better growth of plants under the influence of endophytic bacteria and suggested *Brevibacillus agri* as a potential biofertilizer.

A sustainable biofertilizer must have both the qualities of plant improvement and pesticide resilience. In this context, Shen et al. (2019) recorded five endophytic strains with plant growth-promoting capability in which they also found *B. aryabhattai* to produce the highest amount of auxin and fungicide resistance; thus, they considered this strain of bacteria as a probable biofertilizer. Sukiman et al. (2019) showed that combining multiple microorganisms as biofertilizers promotes plant growth. They have taken endophytes like *Rhizobium* as an auxin producer, *Bradyrhyzobium* sp. as a nitrogen fixer, and *Pseudomonas* sp. as a phosphate solubilizer. This combination of biofertilizers helped the soybean plant to grow better. Fungal endophytes *Fusarium* sp. and *Aspergillus* sp. that can produce phytohormones like auxin and gibberellin have been shown to help rice plants grow under adverse conditions; thus, they can also be used as biofertilizers (Bilal et al. 2018).

Although very few phytohormone-producing microorganism-related biofertilizers are available commercially, there is great potential hidden in these PPMs. Many initial levels of studies are present that have claimed the potential of these microorganisms' ability to confer stress tolerance, but it needs the immediate attention of big players in the fertilizer industry and the government for major funding so that ground-breaking research in this area can be conducted. Another aspect is the combined application of different microorganisms, which may be beneficial for the fight against various abiotic stresses. Similarly, transporting these microorganisms is another gray area where a lot of work needs to be done to make these strains environmentally strong enough to survive.

7.6 CONCLUSION

Sustainable agricultural practices with less harmful environmental consequences are the need of the hour. Globally used chemical fertilizers are slowly degrading the quality of soil, which will impact crop production and food security in the near future. Biofertilizers, on the other hand, can be a good alternative for the long-term sustainability of agriculture as they are found promising in accelerating plant growth without causing pollution. Many phytohormone-producing microorganisms have already been isolated and marketized and are performing well. However, limiting factors like transportation, isolation, and viability need to be eliminated for better commercial and global applications. There is an immense capability of phytohormone-producing microorganisms and their abiotic stress-related activities. Therefore, with a few more efforts, we can develop phytohormone-secreting microorganism fertilizers and reduce chemical fertilizer usage.

REFERENCES

Ahammed, G. J., Guang, Y., Yang, Y., & Chen, J. (2021). Mechanisms of elevated CO2-induced thermotolerance in plants: the role of phytohormones. *Plant Cell Reports*, *40*(12), 2273–2286.

Ahmad, F., Kamal, A., Singh, A., Ashfaque, F., Alamri, S., Siddiqui, M. H., & Khan, M. I. R. (2021). Seed priming with gibberellic acid induces high salinity tolerance in Pisum sativum through antioxidants, secondary metabolites and up-regulation of antiporter genes. *Plant Biology, 23*, 113–121.

Akram, R., Fahad, S., Masood, N., Rasool, A., Ijaz, M., Ihsan, M. Z., Maqbool, M.M., Ahmad, S., Hussain, S., Ahmed, M. and Kaleem, S. (2019). Plant growth and morphological changes in rice under abiotic stress. In *Advances in Rice Research for Abiotic Stress Tolerance*, ed. M. Hasanuzzaman, M. Fujita, K. Nahar, J. K. Biswas (pp. 69-85). Woodhead Publishing.

Ali, S. Z., Sandhya, V., Grover, M., Linga, V. R., & Bandi, V. (2011). Effect of inoculation with a thermotolerant plant growth promoting Pseudomonas putida strain AKMP7 on growth of wheat (Triticum spp.) under heat stress. *Journal of Plant Interactions*, *6*(4), 239–246.

Ali, S., Charles, T. C., & Glick, B. R. (2017). Endophytic phytohormones and their role in plant growth promotion. *In* S.L. Doty (Ed.) *Functional importance of the plant microbiome: implications for agriculture, forestry and bioenergy,* Springer International Publishing, 89–105.

Alvi, A. F., Sehar, Z., Fatma, M., Masood, A., & Khan, N. A. (2022). Strigolactone: an emerging growth regulator for developing resilience in plants. *Plants*, *11*(19), 2604.

Asaf, S., Khan, M. A., Khan, A. L., Waqas, M., Shahzad, R., Kim, A. Y., ... & Lee, I. J. (2017). Bacterial endophytes from arid land plants regulate endogenous hormone content and promote growth in crop plants: an example of Sphingomonas sp. and Serratia marcescens. *Journal of Plant Interactions, 12*(1), 31–38.

Ashry, N. M., Alaidaroos, B. A., Mohamed, S. A., Badr, O. A., El-Saadony, M. T., & Esmael, A. (2022). Utilization of drought-tolerant bacterial strains isolated from harsh soils as a plant growth-promoting rhizobacteria (PGPR). *Saudi Journal of Biological Sciences*, *29*(3), 1760–1769.

Asoegwu, C. R., Awuchi, C. G., Nelson, K. C. T., Orji, C. G., Nwosu, O. U., Egbufor, U. C., & Awuchi, C. G. (2020). A review on the role of biofertilizers in reducing soil pollution and increasing soil nutrients. *Himalayan Journal of Agriculture*, *1*(1), 34–38.

Banerjee, A., & Roychoudhury, A. (2020). Gibberellic acid-priming promotes fluoride tolerance in a susceptible indica rice cultivar by regulating the antioxidant and phytohormone homeostasis. *Journal of Plant Growth Regulation, 39*, 1476–1487.

Barnawal, D., Bharti, N., Pandey, S. S., Pandey, A., Chanotiya, C. S., & Kalra, A. (2017). Plant growth--promoting rhizobacteria enhance wheat salt and drought stress tolerance by altering endogenous phytohormone levels and TaCTR1/TaDREB2 expression. *Physiologia Plantarum*, *161*(4), 502–514.

Barnawal, D., Bharti, N., Tripathi, A., Pandey, S. S., Chanotiya, C. S., & Kalra, A. (2016). ACC-deaminase-producing endophyte Brachybacterium paraconglomeratum strain SMR20 ameliorates Chlorophytum salinity stress via altering phytohormone generation. *Journal of Plant Growth Regulation, 35*, 553–564.

Bharadwaj, P. S., Sanchez, L., Li, D., Enyi, D., Van de Poel, B., & Chang, C. (2022). The plant hormone ethylene promotes abiotic stress tolerance in the liverwort Marchantia polymorpha. *Frontiers in Plant Science, 13*, 998267.

Bhatt, D., Nath, M., Sharma, M., Bhatt, M. D., Bisht, D. S., & Butani, N. V. (2020). Role of growth regulators and phytohormones in overcoming environmental stress. *In* A. Roychoudhury, D. K. Tripathi (Eds.) *Protective chemical agents in the amelioration of plant abiotic stress: biochemical and molecular perspectives,* John Wiley & Sons Ltd. 254–279.

Bhoi, A., Yadu, B., Chandra, J., & Keshavkant, S. (2021). Contribution of strigolactone in plant physiology, hormonal interaction and abiotic stresses. *Planta, 254*(2), 28.

Bielach, A., Hrtyan, M., & Tognetti, V. B. (2017). Plants under stress: involvement of auxin and cytokinin. *International Journal of Molecular Sciences, 18*(7), 1427.

Bilal, S., Shahzad, R., Khan, A. L., Kang, S. M., Al-Harrasi, A., & Lee, I. J. (2018). Endophytic microbial consortia of phytohormones-producing fungus Paecilomyces formosus LHL10 and bacteria Sphingomonas sp. LK11 to Glycine max L. regulates physio-hormonal changes to attenuate aluminum and zinc stresses. *Frontiers in Plant Science, 9*, 395863.

Bruno, L. B., Karthik, C., Ma, Y., Kadirvelu, K., Freitas, H., & Rajkumar, M. (2020). Amelioration of chromium and heat stresses in Sorghum bicolor by Cr6+ reducing-thermotolerant plant growth promoting bacteria. *Chemosphere, 244*, 125521.

Cao, D., Lutz, A., Hill, C. B., Callahan, D. L., & Roessner, U. (2017). A quantitative profiling method of phytohormones and other metabolites applied to barley roots subjected to salinity stress. *Frontiers in Plant Science, 7*, 233111.

Chen, J., Nolan, T. M., Ye, H., Zhang, M., Tong, H., Xin, P., ... & Yin, Y. (2017). Arabidopsis WRKY46, WRKY54, and WRKY70 transcription factors are involved in brassinosteroid-regulated plant growth and drought responses. *The Plant Cell, 29*(6), 1425–1439.

Cohen, A. C., Travaglia, C. N., Bottini, R., & Piccoli, P. N. (2009). Participation of abscisic acid and gibberellins produced by endophytic Azospirillum in the alleviation of drought effects in maize. *Botany, 87*(5), 455–462.

Dubois, M., Van den Broeck, L., & Inzé, D. (2018). The pivotal role of ethylene in plant growth. *Trends in Plant Science, 23*(4), 311–323.

Edelstein, M., & Ben-Hur, M. (2018). Heavy metals and metalloids: sources, risks and strategies to reduce their accumulation in horticultural crops. *Scientia Horticulturae, 234*, 431–444.

Egamberdieva, D. (2009). Alleviation of salt stress by plant growth regulators and IAA producing bacteria in wheat. *Acta Physiologiae Plantarum, 31*(4), 861–864.

Egamberdieva, D., Wirth, S. J., Alqarawi, A. A., Abd_Allah, E. F., & Hashem, A. (2017). Phytohormones and beneficial microbes: essential components for plants to balance stress and fitness. *Frontiers in microbiology, 8*, 278255.

Ekinci, M., Turan, M., Yildirim, E., Gunes, A., Kotan, R., & Dursun, A. (2014). Effect of plant growth promoting rhizobacteria on growth, nutrient, organic acid, amino acid and hormone content of cauliflower (Brassica oleracea L. var. botrytis) transplants. *Acta Scientiarum Polonorum. Hortorum Cultus, 13*(6), 71–85.

Elahi, N. N., Raza, S., Rizwan, M. S., Albalawi, B. F. A., Ishaq, M. Z., Ahmed, H. M., ... & Ditta, A. (2022). Foliar application of gibberellin alleviates adverse impacts of drought stress and improves growth, physiological and biochemical attributes of canola (Brassica napus L.). *Sustainability, 15*(1), 78.

Falcioni, R., Moriwaki, T., Benedito, E., Bonato, C. M., de Souza, L. A., & Antunes, W. C. (2018). Increased gibberellin levels enhance light capture efficiency in tobacco plants and promote dry matter accumulation. *Theoretical and Experimental Plant Physiology, 30*, 235–250.

Food and Agriculture Organization of the United Nations. (2021, 20th October) *World map of salt-affected soils launched at virtual conference.* www.fao.org/newsroom/detail/salt-affected-soils-map-symposium/en

Ghani, M. A., Abbas, M. M., Ali, B., Aziz, R., Qadri, R. W. K., Noor, A., ... & Ali, S. (2021). Alleviating role of gibberellic acid in enhancing plant growth and stimulating phenolic compounds in carrot (Daucus carota L.) under lead stress. *Sustainability, 13*(21), 12329.

Ghosh, D., Gupta, A., & Mohapatra, S. (2019). Dynamics of endogenous hormone regulation in plants by phytohormone secreting rhizobacteria under water-stress. *Symbiosis, 77*, 265–278.

Gimenez, E., Salinas, M., & Manzano-Agugliaro, F. (2018). Worldwide research on plant defense against biotic stresses as improvement for sustainable agriculture. *Sustainability, 10*(2), 391.

Grover, M., Bodhankar, S., Sharma, A., Sharma, P., Singh, J., & Nain, L. (2021). PGPR mediated alterations in root traits: way toward sustainable crop production. *Frontiers in Sustainable Food Systems, 4*, 618230.

Gruszka, D. (2018). Crosstalk of the brassinosteroid signalosome with phytohormonal and stress signaling components maintains a balance between the processes of growth and stress tolerance. *International Journal of Molecular Sciences*, *19*(9), 2675.
Gul, F., Arfan, M., Shahbaz, M., & Basra, S. M. (2020). Salicylic acid seed priming modulates morphology, nutrient relations and photosynthetic attributes of wheat grown under cadmium stress. *International Journal of Agriculture & Biology*, 197–204.
Halo, B. A., Al-Yahyai, R. A., & Al-Sadi, A. M. (2020). An endophytic Talaromyces omanensis enhances reproductive, physiological and anatomical characteristics of drought-stressed tomato. *Journal of Plant Physiology, 249*, 153163.
Hamayun, M., Hussain, A., Khan, S. A., Kim, H. Y., Khan, A. L., Waqas, M., ... & Lee, I. J. (2017). Gibberellins producing endophytic fungus Porostereum spadiceum AGH786 rescues growth of salt affected soybean. *Frontiers in Microbiology, 8*, 246589.
Hidri, R., Barea, J. M., Mahmoud, O. M. B., Abdelly, C., & Azcón, R. (2016). Impact of microbial inoculation on biomass accumulation by Sulla carnosa provenances, and in regulating nutrition, physiological and antioxidant activities of this species under non-saline and saline conditions. *Journal of Plant Physiology, 201*, 28–41.
Huang, Y. T., Cai, S. Y., Ruan, X. L., Chen, S. Y., Mei, G. F., Ruan, G. H., & Cao, D. D. (2021). Salicylic acid enhances sunflower seed germination under Zn2+ stress via involvement in Zn2+ metabolic balance and phytohormone interactions. *Scientia Horticulturae, 275*, 109702.
Hussain, M. A., Fahad, S., Sharif, R., Jan, M. F., Mujtaba, M., Ali, Q., ... & Hou, J. (2020). Multifunctional role of brassinosteroid and its analogues in plants. *Plant Growth Regulation, 92*, 141–156.
Indian Farmers Fertilizer Cooperative Limited. (2017). Iffco India website, Government Of India. Accessed 10th December 2022. https://www.iffco.in
Iqbal, N., Khan, N. A., Ferrante, A., Trivellini, A., Francini, A., & Khan, M. I. R. (2017). Ethylene role in plant growth, development and senescence: interaction with other phytohormones. *Frontiers in plant science, 8*, 235913.
Iqbal, S., Wang, X., Mubeen, I., Kamran, M., Díaz, G. A., Abbas, A., & Fahad, S. (2022). Phytohormones trigger drought tolerance in crop plants: outlook and future perspectives. *Frontiers in Plant Science, 12*, 799318.
Ismail, M. A., Amin, M. A., Eid, A. M., Hassan, S. E. D., Mahgoub, H. A., Lashin, I., ... & Fouda, A. (2021). Comparative Study between exogenously applied plant growth hormones versus metabolites of microbial endophytes as plant growth-promoting for Phaseolus vulgaris L. *Cells, 10*(5), 1059.
Jayasinghe, T., Perera, P., & Wimalasekera, R. (2019, December). Effect of Foliar application of Gibberellin in mitigating salt stress in tomato (Solanum lycopersicum),'Thilina'Variety. In *Proceedings of the 6th International Conference on Multidisciplinary Approaches (iCMA)*.
Jochum, M. D., McWilliams, K. L., Borrego, E. J., Kolomiets, M. V., Niu, G., Pierson, E. A., & Jo, Y. K. (2019). Bioprospecting plant growth-promoting rhizobacteria that mitigate drought stress in grasses. *Frontiers in Microbiology, 10*, 466447.
Kang, S. M., Hamayun, M., Khan, M. A., Iqbal, A., & Lee, I. J. (2019). Bacillus subtilis JW1 enhances plant growth and nutrient uptake of Chinese cabbage through gibberellins secretion. *Journal of Applied Botany & Food Quality, 92,* 172–178.
Kang, S. M., Khan, A. L., Waqas, M., Asaf, S., Lee, K. E., Park, Y. G., ... & Lee, I. J. (2019). Integrated phytohormone production by the plant growth-promoting rhizobacterium Bacillus tequilensis SSB07 induced thermotolerance in soybean. *Journal of Plant Interactions, 14*(1), 416–423.
Kang, S. M., Shahzad, R., Khan, M. A., Hasnain, Z., Lee, K. E., Park, H. S., ... & Lee, I. J. (2021). Ameliorative effect of indole-3-acetic acid-and siderophore-producing Leclercia adecarboxylata MO1 on cucumber plants under zinc stress. *Journal of Plant Interactions, 16*(1), 30–41.
Kashyap, B. K., Solanki, M. K., Pandey, A. K., Prabha, S., Kumar, P., & Kumari, B. (2019). Bacillus as plant growth promoting rhizobacteria (PGPR): a promising green agriculture technology. In R. A. Ansari, I. Mahamood (Eds.) *Plant Health Under Biotic Stress: Volume 2:* Microbial Interactions, Springer Nature Singapore, 219–236.
Khan, M. A., Asaf, S., Khan, A. L., Jan, R., Kang, S. M., Kim, K. M., & Lee, I. J. (2020). Thermotolerance effect of plant growth-promoting Bacillus cereus SA1 on soybean during heat stress. *BMC Microbiology, 20*, 1–14.

Khan, M. A., Asaf, S., Khan, A. L., Ullah, I., Ali, S., Kang, S. M., & Lee, I. J. (2019). Alleviation of salt stress response in soybean plants with the endophytic bacterial isolate Curtobacterium sp. *SAK1. Annals of Microbiology, 69*, 797–808.

Khan, W. U., Ahmad, S. R., Yasin, N. A., Ali, A., Ahmad, A., & Akram, W. (2017). Application of Bacillus megaterium MCR-8 improved phytoextraction and stress alleviation of nickel in Vinca rosea. *International Journal of Phytoremediation, 19*(9), 813–824.

Kosakivska, I. V., Vasyuk, V. A., Voytenko, L. V., Shcherbatiuk, M. M., Romanenko, K. O., & Babenko, L. M. (2020). Endogenous phytohormones of fern Polystichum aculeatum (L.) Roth gametophytes at different stages of morphogenesis in vitro culture. *Cytology and Genetics, 54*, 23–30.

Koza, N. A., Adedayo, A. A., Babalola, O. O., & Kappo, A. P. (2022). Microorganisms in plant growth and development: roles in abiotic stress tolerance and secondary metabolites secretion. *Microorganisms, 10*(8), 1528.

Krishnamoorthy, R., Roy Choudhury, A., Walitang, D. I., Anandham, R., Senthilkumar, M., & Sa, T. (2022). Salt stress tolerance-promoting proteins and metabolites under plant-bacteria-salt stress tripartite interactions. *Applied Sciences, 12*(6), 3126.

Li, X., Sun, P., Zhang, Y., Jin, C., & Guan, C. (2020). A novel PGPR strain Kocuria rhizophila Y1 enhances salt stress tolerance in maize by regulating phytohormone levels, nutrient acquisition, redox potential, ion homeostasis, photosynthetic capacity and stress-responsive genes expression. *Environmental and Experimental Botany, 174*, 104023.

Li, Z., Zhang, X., Zhao, Y., Li, Y., Zhang, G., Peng, Z., & Zhang, J. (2018). Enhancing auxin accumulation in maize root tips improves root growth and dwarfs plant height. *Plant Biotechnology Journal, 16*(1), 86–99.

Liu, F., Xing, S., Ma, H., Du, Z., & Ma, B. (2013). Cytokinin-producing, plant growth-promoting rhizobacteria that confer resistance to drought stress in Platycladus orientalis container seedlings. *Applied Microbiology and Biotechnology, 97*, 9155–9164.

Liu, G., Pfeifer, J., de Brito Francisco, R., Emonet, A., Stirnemann, M., Gübeli, C., ... & Borghi, L. (2018). Changes in the allocation of endogenous strigolactone improve plant biomass production on phosphate-poor soils. *New Phytologist, 217*(2), 784–798.

Ma, H. Y., Zhao, D. D., Ning, Q. R., Wei, J. P., Li, Y., Wang, M. M., ... & Liang, Z. W. (2018). A multi-year beneficial effect of seed priming with gibberellic acid-3 (GA3) on plant growth and production in a perennial grass, Leymus chinensis. *Scientific Reports, 8*(1), 13214.

Manghwar, H., Hussain, A., Ali, Q., & Liu, F. (2022). Brassinosteroids (BRs) role in plant development and coping with different stresses. *International Journal of Molecular Sciences, 23*(3), 1012.

Martínez-Medina, A., Del Mar Alguacil, M., Pascual, J. A. and Van Wees, S. C., (2014). Phytohormone profiles induced by Trichoderma isolates correspond with their biocontrol and plant growth-promoting activity on melon plants. *Journal of Chemical Ecology, 40*, 804–815.

Masmoudi, F., Tounsi, S., Dunlap, C. A., & Trigui, M. (2021). Endophytic halotolerant Bacillus velezensis FMH2 alleviates salt stress on tomato plants by improving plant growth and altering physiological and antioxidant responses. *Plant Physiology and Biochemistry, 165*, 217–227.

Masniawaty, Mustari, K., Astuti, Gusmiaty, Larekeng, H., Yani, A., & Rahim, I. (2019, October). Exploration of bacteria associated with chili peppers' rhizosphere and their capacity to absorb and produce gibberellin hormone. In *IOP Conference Series: Earth and Environmental Science* (Vol. 343, No. 1, p. 012059). IOP Publishing.

Mathur, P., Tripathi, D. K., Baluška, F., & Mukherjee, S. (2022). Auxin-mediated molecular mechanisms of heavy metal and metalloid stress regulation in plants. *Environmental and Experimental Botany, 196*, 104796.

Meena, K. K., Bitla, U. M., Sorty, A. M., Singh, D. P., Gupta, V. K., Wakchaure, G. C., & Kumar, S. (2020). Mitigation of salinity stress in wheat seedlings due to the application of phytohormone-rich culture filtrate extract of methylotrophic actinobacterium Nocardioides sp. NIMMe6. *Frontiers in Microbiology, 11*, 2091.

Mohamed, M. F., Thalooth, A. T., Elewa, T. A., & Ahmed, A. G. (2019). Yield and nutrient status of wheat plants (Triticum aestivum) as affected by sludge, compost, and biofertilizers under newly reclaimed soil. *Bulletin of the National Research Centre, 43*, 1–6.

Mujahid, N., Shahbaz, M., Kiran, A., & Wahid, M. A. (2023). Modulations Induced by seed priming of strigolactone (GR24) in morpho-physiological and biochemical attributes of ajwain (Trachyspermum ammi L.) under salt stress. *Journal of Plant Growth Regulation*, *42*(5), 2893–2906.

Nadeem, S. M., Zahir, Z. A., Naveed, M., & Nawaz, S. (2013). Mitigation of salinity-induced negative impact on the growth and yield of wheat by plant growth-promoting rhizobacteria in naturally saline conditions. *Annals of Microbiology, 63*, 225–232.

Narula, N., Deubel, A., Gans, W., Behl, R. K., & Merbach, W. P. S. E. (2006). Paranodules and colonization of wheat roots by phytohormone producing bacteria in soil. *Plant Soil and Environment*, *52*(3), 119.

Nosheen, S., Ajmal, I., & Song, Y. (2021). Microbes as biofertilizers, a potential approach for sustainable crop production. *Sustainability*, *13*(4), 1868.

Notununu, I., Moleleki, L., Roopnarain, A., & Adeleke, R. (2022). Effects of plant growth-promoting rhizobacteria on the molecular responses of maize under drought and heat stresses: a review. *Pedosphere*, *32*(1), 90–106.

Pandey, V., Bhatt, I. D., & Nandi, S. K. (2019). Role and regulation of auxin signaling in abiotic stress tolerance. In M. I. R. khan, P. S. Reddy, A. Ferrante, N. A. Khan (Eds.) *Plant signaling molecules,* 319–331. Woodhead Publishing.

Park, Y. G., Mun, B. G., Kang, S. M., Hussain, A., Shahzad, R., Seo, C. W., ... & Yun, B. W. (2017). Bacillus aryabhattai SRB02 tolerates oxidative and nitrosative stress and promotes the growth of soybean by modulating the production of phytohormones. *PloS One, 12*(3), e0173203.

Patel, T. and Saraf, M., (2017). Biosynthesis of phytohormones from novel rhizobacterial isolates and their in vitro plant growth-promoting efficacy. *Journal of Plant Interactions*, *12*(1), 480–487.

Pathania, P., Rajta, A., Singh, P. C., & Bhatia, R. (2020). Role of plant growth-promoting bacteria in sustainable agriculture. *Biocatalysis and Agricultural Biotechnology, 30*, 101842.

Peleg, Z., & Blumwald, E. (2011). Hormone balance and abiotic stress tolerance in crop plants. *Current Opinion in Plant Biology*, *14*(3), 290–295.

Priyadharsini, P., & Muthukumar, T. (2017). The root endophytic fungus Curvularia geniculata from Parthenium hysterophorus roots improves plant growth through phosphate solubilization and phytohormone production. *Fungal Ecology, 27*, 69–77.

Putriani, P., Fitri, L., & Ismail, Y. S. (2019). The Potential Endophytic Bacteria Isolated from Rice (Oryza sativa) as Biofertilizer. *Biosaintifika: Journal of Biology & Biology Education*, *11*(2), 178–185.

Randive, K., Raut, T., & Jawadand, S. (2021). An overview of the global fertilizer trends and India's position in 2020. In *Mineral Economics,* Springer Nature, 371–384.

Sah, S. K., Reddy, K. R., & Li, J. (2016). Abscisic acid and abiotic stress tolerance in crop plants. *Frontiers in Plant Science, 7*, 190245.

Saleem, M. H., Fahad, S., Adnan, M., Ali, M., Rana, M. S., Kamran, M., ... & Hussain, R. M. (2020). Foliar application of gibberellic acid endorsed phytoextraction of copper and alleviates oxidative stress in jute (Corchorus capsularis L.) plant grown in highly copper-contaminated soil of China. *Environmental Science and Pollution Research, 27*, 37121–37133.

Santoro, V., Schiavon, M., Gresta, F., Ertani, A., Cardinale, F., Sturrock, C. J., ... & Schubert, A. (2020). Strigolactones control root system architecture and tip anatomy in Solanum lycopersicum L. plants under P starvation. *Plants, 9*(5), 612.

Seleiman, M. F., Al-Suhaibani, N., Ali, N., Akmal, M., Alotaibi, M., Refay, Y., ... & Battaglia, M. L. (2021). Drought stress impacts on plants and different approaches to alleviate its adverse effects. *Plants, 10*(2), 259.

Shaffique, S., Alomrani, S. O., Injamum-Ul-Hoque, M., Peter, O., Imran, M., & Lee, I. J. (2023). Unlocking the potential of newly isolated phytohormone-producing bacterial strains for enhanced plant growth and stress tolerance. *Plant Stress, 10*, 100260.

Shaffique, S., Khan, M. A., Wani, S. H., Pande, A., Imran, M., Kang, S. M., ... & Lee, I. J. (2022). A review on the role of endophytes and plant growth promoting rhizobacteria in mitigating heat stress in plants. *Microorganisms, 10*(7), 1286.

Sharma, A., Kumar, V., Shahzad, B., Ramakrishnan, M., Singh Sidhu, G. P., Bali, A. S., ... & Zheng, B. (2020). Photosynthetic response of plants under different abiotic stresses: a review. *Journal of Plant Growth Regulation, 39*, 509–531.

Shekhawat, K., Almeida-Trapp, M., García-Ramírez, G. X., & Hirt, H. (2022). Beat the heat: plant-and microbe-mediated strategies for crop thermotolerance. *Trends in Plant Science*, *27*(8), 802–813.

Shen, F. T., Yen, J. H., Liao, C. S., Chen, W. C., & Chao, Y. T. (2019). Screening of rice endophytic biofertilizers with fungicide tolerance and plant growth-promoting characteristics. *Sustainability*, *11*(4), 1133.

Sheteiwy, M. S., Shao, H., Qi, W., Daly, P., Sharma, A., Shaghaleh, H., ... & Lu, H. (2021). Seed priming and foliar application with jasmonic acid enhance salinity stress tolerance of soybean (Glycine max L.) seedlings. *Journal of the Science of Food and Agriculture, 101*(5), 2027–2041.

Shi, Y., Lou, K., & Li, C. (2009). Promotion of plant growth by phytohormone-producing endophytic microbes of sugar beet. *Biology and Fertility of Soils, 45*, 645–653.

Suhameena, B., Devi, S., Gowri, R., & Kumar, S. D. (2020). Utilization of Azospirillum as a Biofertilizer–an overview. *International Journal of Pharmaceutical Sciences Review and Research*, *62*(2), 141–145.

Sukiman, H., Lekatompessy, S., Widowati, T., & Simarmata, R. (2019, August). Application of arbuscular mycorrhizal fungi in combination with nitrogen fixing bacteria and other potential soil microbes as biofertilizer for soybean plant. In *IOP Conference Series: Earth and Environmental Science* (Vol. 308, No. 1, p. 012059). IOP Publishing.

Talaat, N. B. (2019). Effective microorganisms: an innovative tool for inducing common bean (Phaseolus vulgaris L.) salt-tolerance by regulating photosynthetic rate and endogenous phytohormones production. *Scientia Horticulturae, 250*, 254–265.

Tien, T. M., Gaskins, M. H., & Hubbell, D. (1979). Plant growth substances produced by Azospirillum brasilense and their effect on the growth of pearl millet (Pennisetum americanum L.). *Applied and Environmental Microbiology*, *37*(5), 1016–1024.

Turan, M., Ekinci, M., Yildirim, E., Güneş, A., Karagöz, K., Kotan, R., & Dursun, A. (2014). Plant growth-promoting rhizobacteria improved growth, nutrient, and hormone content of cabbage (Brassica oleracea) seedlings. *Turkish Journal of Agriculture and Forestry*, *38*(3), 327–333.

Wang, Y., Shen, W., Chan, Z., & Wu, Y. (2015). Endogenous cytokinin overproduction modulates ROS homeostasis and decreases salt stress resistance in Arabidopsis thaliana. *Frontiers in Plant Science*, *6*, 1004.

Waqas, M., Khan, A. L., Shahzad, R., Ullah, I., Khan, A. R., & Lee, I. J. (2015). Mutualistic fungal endophytes produce phytohormones and organic acids that promote japonica rice plant growth under prolonged heat stress. *Journal of Zhejiang University. Science. B*, *16*(12), 1011.

Werner, T., Motyka, V., Strnad, M., & Schmülling, T. (2001). Regulation of plant growth by cytokinin. *Proceedings of the National Academy of Sciences*, *98*(18), 10487–10492.

Zaheer, A., Mirza, B. S., Mclean, J. E., Yasmin, S., Shah, T. M., Malik, K. A., & Mirza, M. S. (2016). Association of plant growth-promoting Serratia spp. with the root nodules of chickpea. *Research in Microbiology*, *167*(6), 510–520.

Zeffa, D. M., Perini, L. J., Silva, M. B., de Sousa, N. V., Scapim, C. A., Oliveira, A. L. M. D., Amaral Júnior, A. T. D. and Azeredo Gonçalves, L. S., (2019). Azospirillum brasilense promotes increases in growth and nitrogen use efficiency of maize genotypes. *Plos One*, *14*(4), p.e0215332.

Zhang, F., Wang, P., Zou, Y. N., Wu, Q. S., & Kuča, K. (2019). Effects of mycorrhizal fungi on root-hair growth and hormone levels of taproot and lateral roots in trifoliate orange under drought stress. *Archives of Agronomy and Soil Science*, *65*(9), 1316–1330.

Zhang, Q., & Gong, M. (2020). Rhizoglomus intraradices improves plant growth, root morphology and Phytohormone balance of Robinia pseudoacacia in arsenic-contaminated soils. *Frontiers in Microbiology, 11*, 547547.

Zhou, X. R., Dai, L., Xu, G. F., & Wang, H. S. (2021). A strain of Phoma species improves drought tolerance of Pinus tabulaeformis. *Scientific Reports*, *11*(1), 7637.

8 Redox Balance in Roots Induces Tolerance for Rice Plants Under Anaerobic Stress

Malay Kumar Adak, Nilakshi Chakraborty, Abir Das, Bipul Sarkar, Sayan Pal, Sudipta Kumar Sil, Trisha Ghosh and Santanu Gupta

8.1 INTRODUCTION

Soil flooding in rice fields is considered as wet-land, which is also accompanied by changes in the physical and chemical characteristics of soil. The typical changes under flooding are predominately characterized by lowering of the soil redox potential (E_h). This often leads to an enhancement of oxygen demand within biological and non-biological environments as well as by the production of specific soil chemical residues (Rani et al. 2020). These are the reaction byproducts of reductive, deaminative, decarboxylative pathways from soil-borne microorganism, micro flora, and other non-culturable bacterial consortium. In submerged rice soil, these are predominated by methanogenic bacteria, which accumulates a significant amount of methane gas production, causing a decrease in environmental pH. Technically, the rice rhizosphere is most predominant with bacteria and other microorganisms, where a redox shift from acidity to alkalinity is the most important. This is also aggravated by redial oxygen loss (ROL) from the submerged roots and thereby roots are more adopted with enzymatic responses of anaerobic respiration. Still, the chemical properties and the reducing nature of the soil environment, where the plants are faced with an oxygen deficit and other processes, developed alteration of plant responses (Kant et al. 2011). Therefore, an anaerobic stress coupled with ample chances of oxidative environment may draw attention to the importance of chemical reactions in the assessment of soil reduction. If this is properly done, it would evaluate the wet-land plant responses, more specifically at the roots following the whole plant level. Moreover, the oxygen deficit would partially account for the plant stress responses and therefore the importance of low soil En would be another criterion for the sustainability of rice roots in submerged conditions. Wet-land rice production in Asia ranks as a major contribution to world rice production, where flooded or saturated water logging is common both for direct seeded and transplanted conditions. Water productivity in rice, which clarifies the economic or grain yield production per unit of rain fall and irrigation as water-supplying agencies, may be of the utmost importance. The shortened situation in the rice-growing season would reduce soil fertility and depletion of essential elements, which are mostly recovered under submerged conditions (Zayed et al. 2011). Submergence is characterized by a gradual decrease but an increase finally in pH values of the soil solution. Under submerged soil, field capacity also observes similar trends for pH; however, the changes are much less realized. The redox potential in rice fields under highly saturated conditions is decreased and a negative value attained on prolonged submergence. Submergence, on the other hand, elevates the concentration of water-soluble iron and manganese, both are redox metals. The concentration of phosphorus initially becomes maximum but gradually decreased. The changes in nitrate gradually decrease and a continued flux of oxidation (NO_3 to NO_2) may characterize the

DOI: 10.1201/9781003354109-8

accumulation of more free electrons for easily available oxidation of the rhizosphere (Luo et al. 2015). The increase in ammonia-derived nitrogen is another index for submergence but its increase is less pronounced. Submergence can induce other free metals (Ca, Mg, and K), which can harbor the root membrane for better capability of redox reactions. So, redox reaction is apprehended as better pathways for plants nutrient assimilation and its conversion into macro molecules. During aerobic microbial respiration, the final electron acceptor is oxygen, which is commonly depleted under submerged soil. The aerobic microbial process is replaced by the anaerobic process where a number of reduced compounds are found. The anaerobic respiration can produce the low energy equivalence in the form of ATP than aerobic respiration and thereby organic matter decomposition is accelerated. Submergence following de-submergence can induce an oxygen turn over for the rice roots and thereby an oxidative redox is ensured (Liu et al. 2022). Collectively all these phenomena are targeted for analysis and compared with the oxygenated and de-oxygenated conditions of rice roots whereby a number of cellular and physiological activities are concerned. Thus, this chapter aims to highlight the major area of roots' responses to the redox potential with oxidative stress.

8.2 ROOTS' RESPONSE TO SUBMERGENCE THROUGH SOIL OXIDATION REDUCTION REACTION

When roots of rice under submerged conditions are exposed to de-submergence or anaerobic flooding stress, plants receive a reduced (low) redox potential (E_h) under such a condition (Bridgemohan et al. 2020). A number of reactions occur on the root's plasma membrane, covering physical, chemical, and biological processes, which imply significant activity on the survival of wet-land plants (Khan et al. 2020). Under anaerobic conditions, the soil is restricted for atmospheric gas diffusion and thereby results in the depletion of oxygen but a high concentration of carbon dioxide accumulation (Song et al. 2022). Meanwhile, on thewater level receding from flooding, a minimum supply of oxygen in the soil pore is immediately depleted by the capillary actions of the roots of plants, microorganisms, and soil reductants. These physical processes, on one hand, can deplete the oxygen and bring down the soil E_h (Ganot et al. 2021). This situation induces a number of soil chemical changes, which most predominantly covers de-nitrification and iron reduction ($Fe^{3+} \rightarrow Fe^{2+}$). Changes in the redox state for manganese (Mn^{2+}), sulfate (SO_4^-), and others, may collectively be the main causes for changes in the soil pH and E_h (Wang et al. 2021). The reduction of nitrate NO_3^- to N_2, Mn^{4+} to Mn^{2+}, SO_4^{2-} to S^{2-}, H_2S, SH^- would be other reactions to deplete the soil pH. Roots suffer with acidification mostly by acetic acids and butyric acid, where acidolysis of membrane proteins takes place and is coupled with the production of other organic acids from the microbial metabolism. Therefore, submerged soil represents a complete anaerobic condition where redox is more negative towards the development of oxidative stress. The redox potential of the submerged soil depends on the speed in which the sequence occurs based on organic matter as electron donors. The concentration of oxidants as well as temperature are the other factors to maintain the reduced E_h in submerged soil. After the electron acceptor as the higher redox potential is exhausted, the next redox potential acceptor is oxidized. Electron-accepting species like nitrates can limit sulfate reduction, which mostly happens when organic carbon is low in availability (Westermann et al. 1987). It has been shown that nitrate inhibition for reducing sulfur can be reversed with organic carbon inputs. Under anaerobic submerged conditions, few hydrolytic and fermentative bacteria can gain energy from oxygen in aerobic respiration, but are shifted to anaerobic metabolism when oxygen is low with decomposition of organic matters to simple sugars. The organic compounds cover organic acids, alcohol (which may have undertaken acetate), hydrogen, and carbon dioxide. Therefore, in submerged soil, the roots receive a decreased redox potential with denitrifying microbes using nitrate to reduce nitrous oxide and nitrogen. Finally, the redox process in rice roots under submergence becomes a methane production with a number of organic residues, such as acetate, methanol, and even CO_2 (Yue et al. 2007).

8.3 MEASUREMENT OF THE ROOT REDOX POTENTIAL AND ITS EFFECTS ON SUBMERGENCE TOLERANCE

Under submerged soil, oxidation reduction can highlight the intensities of the root-reducing power to adapt to anaerobic stress. Still, following flooding, reduction of inorganic salts can induce the capacity of root hydraulic conductivity to absorb water and dissolved minerals (Dong et al. 2023) (Figure 8.1). The threshold value of critical E_h, which otherwise signifies the intensity of reduction, can provide valuable information to the major metabolic pathways, including the respiration or any other catabolic process demanding oxygen in the root zone (Wei et al. 2022). In contrast, redox capacity is the reflection of the number of electrons received by O_2, NO_3, Mn^{2+}, Fe^{3+}, and SO_4^{2-} (the oxidants). These also provide support for the microorganisms with oxygen decomposing the organic matters (Yi et al. 2022).

Therefore, the rhizospheric zones are the most important domains of the overall plant body, particularly, in submerged rice roots where the relative high buffering capacity exists for redox potential (E_h). This otherwise explains the reason for wet-land soil, where the maximum number of bio-reducible agents like Fe^{3+} and SO_4^{2-} are most important to be downregulated following submergence (Kieush et al. 2022). This varies with the genotypic capacity to reduce the root zone that may compete for oxygen transportation or radial oxygen loss (ROL) from the roots into the rhizosphere. Under these circumstances, a high soil reduction capacity can be supported to reduce the rhizospheric oxygen demand through aerenchymatous tissues that potentially can increase the anaerobic stress tolerance both under anoxia of submergence as well as under the post-submergence

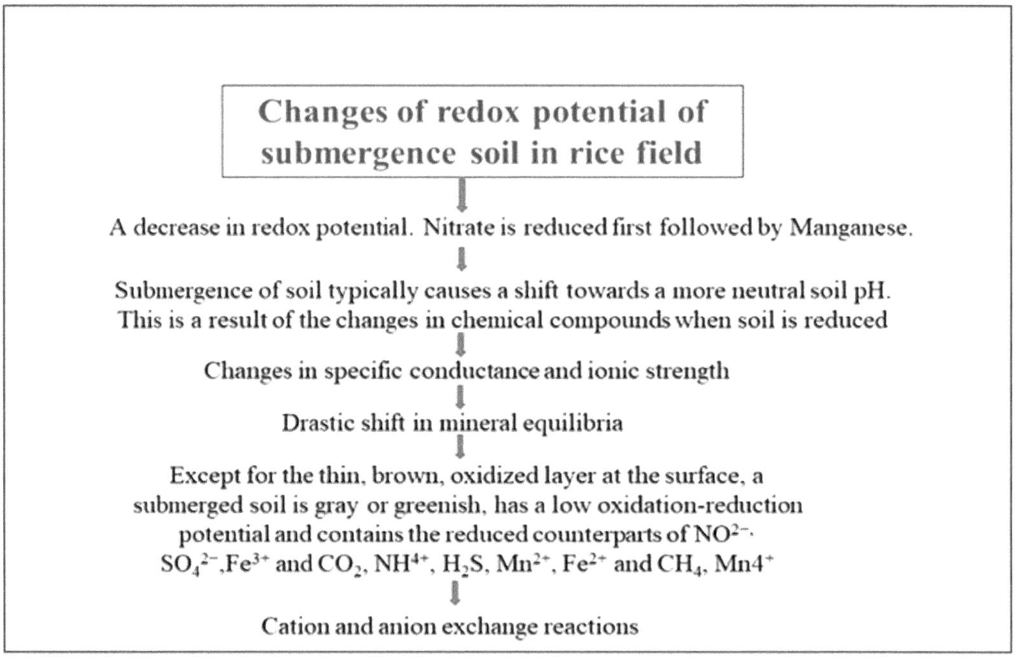

FIGURE 8.1 Changes of oxidation reduction potential for the availability of minerals by the roots of submerged rice plants. Root depleting in rice culture for oxygen can induce the availability of NO^{2-}, SO_4^{2-}, Fe^{3+}, CO_2, NH^{4+}, H_2S, Mn^{2+}, Fe^{2+}, CH_4, Mn^{4+} under anaerobic conditions by oxidation of respective counterparts. The availability of the metal in rice roots is also dependent on the membrane permeability as well as the root hydraulic conductivity of the soil. Finally, an increase cation exchange capacity is the feature of submerged rice roots.

period (Martínez-Arias et al. 2022). The direct correlation through an increase in reduction capacity, i.e., the capacity of the root system to receive electrons with a decrease in E_h , i.e., an increase in the reduction potential directly proposes the magnitude of plant stress (Nardi et al. 2021). Oxygen consumption by the root water interphase may be mathematically interpreted as follows:

$$E_h = E_0 + 2.3\ [RT/nF]\ \log^e\ [\text{ox/red}]$$

where ox and red are the ratio of oxidized and reduced forms of any substrate under a given condition.

E_0 = electrode potential of 50% oxidized substance on that specific system. n = electron number transferred, F = Faraday constant, R = gas constant, and T = absolute temperature (K) (Richardson et al. 2022).

8.4 FACTORS RELATED TO THE INTENSITY OF SOIL REDUCTION TO INFLUENCE ROOT SURVIVAL

In waterlogged soil, it is the redox reaction that is used to measure the intensity of the reduction. Roots exposed to a mixture of diverse inorganic ions are naturally perturbed with redox, including oxygen deficiency under submergence is described as "intensity" and "capacity," with these two modules. The intensity of the reduction is described as a critical E_h value at which the redox system became unstable to provide any electron accept or donate from the roots or rhizosphere (Ao et al. 2022). This typically happens under the condition of oxygen tension or its partial pressure in the rhizosphere when aquaporin or the water channels are sacrificed. In contrast, redox capacity is a description where a few elements, such as O_2, NO_3, Mn^+, Fe, SO_4^{3-} like oxidants are accepted as electrons, supporting the respiratory activities of microorganisms in the lysis of organic residues (Tong et al. 2023). The oxidants should be represented with a high redox buffering capacity, mostly from the small molecular weights of organic residues in the roots. These otherwise state that the causes for wet-land soil with large amounts of bio-degradable iron residues (siderophore, ferrochelatin) do not rapidly reduce in redox intensity under submergence (Kumar et al. 2023). Therefore, high soil reduction activity at the rhizosphere remains always in a competitive mood for oxygen. Transportation of oxygen through root aerenchymatous tissues, particularly, under submergence rice can increase E_h and thereby moderate the plant stress, which is due to a reduction of E_h (Zandi et al. 2023).

8.5 REDOX REGULATION FOR HYPOXIA TOLERANCE IN ROOTS

With reference to rice cultivars, specific quantitative trait loci (QTL) are the elements that have already been identified where an elevated oxidative render the tolerance to submergence stress. This QTL is known as submergence 1A (SUB 1A), located at the ninth chromosome of rice (Singh et al. 2020). The rice cultivars under submergence stress meet with a reduced cellular redox, however, under inundated condition it is regardless of genotypic specificity. Still, when plants are exposed to direct oxygen contact during the de-submergence period, the redox shunt takes on oxidative stress, which is compounded with high light intensities. Under this condition, SUB 1A is activated by the interaction of a few growth regulators like abscisic acid, ethylene, jasmonic acid, etc. (Chen et al. 2021). In fact, irrespective of the rice cultivar under submergence, cultivars develop a significant amount of ethylene from the biosynthetic pool where methionine is the precursor. Ethylene may act discriminately to induce two strategies: escape and quiescence strategy. In the case of escape, an elongation of shoot and leaf sheath is the characteristic feature where ethylene can trigger two genes: SNORKEL 1 and SNORKEL 2 (*SK1* and *SK2*) (Nagai et al. 2021). The co-operativity of two genes is expressed in such a way that root cells under anoxia can induce the GA metabolism by suppressing the ABA activity. The suppressor is Abscinazole-E3M, which inhibits the biosynthesis of ABA or its early catabolism into inactivated form (Castro-Camba et al. 2022). Therefore, *SUB1A* QTL, which is absent in the escape follower cultivars, is induced for cell elongation in the roots and internodes by the energy available from GA induce starch hydrolysis. In contrast, SUB 1A, which also inhibits ethylene in a feedback manner is synthesized in another path. Ethylene in SUB 1A QTL possesses cultivar and induces two SLENDER LIKE RICE 1 (*SLR1*) and SLENDER LIKE RICE LIKE1 (*SLRL1*), where inhibition of GA is achieved by a repressor protein. The DELLA protein which is the product of genes like *SLR1* and *SLRL1* can inhibit any of the biosynthetic steps of the gene. Moreover, *SLR1* and *SLRL1* become an inducer for ABA in these biosynthetic pathways, a gene 9-cis-epoxycarotenoid dioxygenase (*NCED*) (Das et al. 2024). On the other hand, two genes on anaerobic carbohydrate metabolism: ADH and PDC are also induced by *SLR1* and *SLRL1* whereby the root redox can accelerate the fermentation metabolism. Under anoxia the reduced redox potential is also involved in the hydrolysis of sugar in alcoholic fermentation pathways where the minimum amount of energy is exhausted under inadequate survival of rice root.

At the chemical level ABA is a terpenoid, basically 15C that is sesquiterpene in nature. ABA is produced via carotenoids (C^{14}) with oxidative cleavage. Biosynthesis of ABA and its overall metabolism is mediated by several modifications like oxidation and conjugation. On the root membrane, a number of proteins have been identified as well as reported from other crop species like rice, maize, sugarcane, etc., where PYR, PYL, RCAR are the common receptors that play a crucial function for ABA signaling (Miyakawa et al. 2013). A number of phosphorylating proteins also participated in the regulation of these receptors by phosphorylation and de-phosphorylation reactions. The most common of those, which inhibit the ABA perception by receptors is group A protein phosphatase 2C (PP2C). In fact, this protein behaves as the common receptors of ABA signaling (Rodriguez et al. 1998). ABA is basically highly soluble, acidic in nature ,and thereby freely diffused by influx and efflux carrier proteins in the root membrane. ABA is also evident in activating apoplastic NADP(H) oxidase, cytosolic xanthine oxidase, amine oxidase like proteins, which induces the ROS generation (Mandal et al. 2022). ABA is also found to be diffused through apoplastic as well as symplastic spaces and become a positive regulator of these enzymes to generate ROS. Moreover, the catabolism of a few organic residues, like polyamine into hydrogen peroxide, is also induced by ABA. Therefore, for both intracellular and extracellular pathways, ABA may independently be related to ROS biosynthesis or in connection with ROS that transmits the signaling of any abiotic responses.

Abscisic acid (ABA) is mostly involved in water balance and osmoregulation, particularly under abiotic as well as biotic stress. The site of ABA generation in the root also sensitizes the root membrane for the signaling of different stresses as well as translocation of the signal into the root epidermal cell following amplification through the stems. In the stem, the role of ABA is not restricted only for seed germination, seedling growth, stomatal closure, and others, facilitating water stress and salinity but also developing a cellular redox, which is important. In fact, ABA and ROS are the most possible endogenous secondary messengers for any signal to perceive its amplification and its manifestation into different gene expressions (Wang et al. 2020). Therefore, the role of ABA for its redox regulation would be important under water, salinity, and other stress, which ultimately is converted into oxidative stress. The production of ROS either in apoplastic or symplastic in nature may participate in ABA biosynthesis through different genes in the roots of plants. In rice, an anaerobic condition of the roots under submergence would be another signaling to generate ROS that may otherwise connect ABA metabolism and finally its activity in a plant's survival. ABA also connects other redox-dependent proteins like ferredoxin, thioredoxin, etc., where the disulfide bridges are reduced by NADP(H)/$FADH_2$-dependent pathways in chloroplast under ABA interference (Sevilla et al. 2023). Root responses to anoxia are also induced to a few proteins to generate the redox potential to encounter oxidative stress under post-submergence stress.

This observation may possibly reveal the insights of oxidative stress and its interaction with ABA. ABA, being a universal growth substance, is mostly concerned with osmotic stress and other related biological activities. Since ABA is biosynthesized in the roots, a depletion of oxygen under submergence stress must, therefore, be involved with oxidative phenomenon where ABA may be a governing residue (Adak et al. 2023). So, by studying ABA metabolism through its biosynthesis and catabolism, the selected pathways and concerned genes are to be deciphered for oxidative stress tolerance in plants. Therefore, a selection on the basis of the recruiting genes for ABA metabolism, rice cultivars possessing SUB 1 QTL may be possible. In another study where plant growth-promoting rhizobacteria (PGPR) would be exercised to interact with ethylene metabolism is expected to be beneficial for another aspect. This work would be focusing on isolation, characterization and identification of a few bacterial strains, which may colonize rice rhizosphere under anaerobic condition (Beneduzi et al. 2008). The isolated organisms are expected to serve to downregulate the ethylene metabolism whereby SUB1 QTL would be beneficial for its escape and quiescence strategy (Pradhan et al. 2015). So, by culturing these organisms, alternative strategies may be developed to regulate the oxidative stress as well as tolerance under anoxic or hypoxic conditions of submergence. This is mostly dealt with by an enhanced accumulation of ROS in roots, altering the redox state of cells and thereby through a common cascade triggering the activation of multiple defenses. ROS through redox signaling in submerged or non-submerged roots become an acclimation mechanism whereby pre- or post-stress induced pathways are regulated. Activation of ROS and redox responses from the direct exposure to stress (may be described as "local" tissues), perceptions of stress signals that switches on different systematic signaling through cellular or non-cellular paths (described as systematic tolerance) and finally activation of the acclimation for defense responses within them. This phenomenon may be operative before the roots are subjected to anaerobic stress. With regards to perception mechanisms of the roots, the systematic signaling under hypoxic or anoxic conditions may cover electric impulse, metal (Ca^+), redox status, ROS (H_2O_2), gaseous substance (NO, ethylene), volatile compounds (jasmonic acid), phenolics (salicylic acid), and a few others, which are mobilized in a cell-to-cell manner from root tip to endodermis via vascular conducts (Medrano-Macías et al. 2021). Therefore, there should be an additional highlight on recent studies for unrevealing the modalities of different systematic signaling through ROS and redox in plants' stress roots.

From the point of ROS, a number of variants like H_2O_2, O_2^-, 1/2 O_2, and OH^- are the byproducts of redox reactions on cell chloroplast, mitochondrial and cell membrane bound apoplastic spaces are important in this regard (Mandal et al. 2022). A number of studies reveal that a steady-state concentration of ROS both in cytosol and other organelles are of most concern for triggering the ROS accommodation. There are two kinds of ROS inflammation: normal electron transfer to oxygen and spillage of electrons from the electron transport chain to oxygen. Irrespective of the cases, both apoplastic and symplastic ROS are considered as metabolic signaling during abiotic stress, sensing the stress pathogenic infection through roots. A number of dedicated enzymes for ROS production is involved, of which the preliminary are "respiratory burst oxidase homologues" (RBOHs) and are equivalent to NADP(H) oxidase (NOX) (Du et al. 2023). The inductions of RBOHs in roots is the major active ROS producer during stress and together with ROS generated as a byproduct of normal plant metabolism that constitutes total redox balance in the root cell (Zhang et al. 2023). In a number of cases, a few redox regulators are reported during the acclimation and defense mechanism at the cellular level, while ROS can oxidize a number of proteins of the cell membrane as well as develop their corresponding carbonylated residues. Still, a few regulators are also important for protein oxidation in a protein-specific manner under submerged roots, including peroxi-redoxin (PRXs) receptors like kinase (cystine-rich receptor like kinase), H_2O_2 induced calcium1 and few thio-redoxin (TRXS) are impotent. In other cases, roots exposed to aluminum stress can also induce glutathione-reductase (GRXs) and other small molecular weights of non-thiol residues (glutathione) are also important (Cassier-Chauvat et al. 2023). These proteins expressed variably in root membranes synergistically with ROS accumulation also alter the function of their targeted proteins. The later includes different transcription factors, kinases, carrier and channel proteins, phosphatases and a few others (Myc 30 phytoalexin deficit 4) could activate acclimation and cell-to-cell transmission of different signals under stress (Figure 8.2).

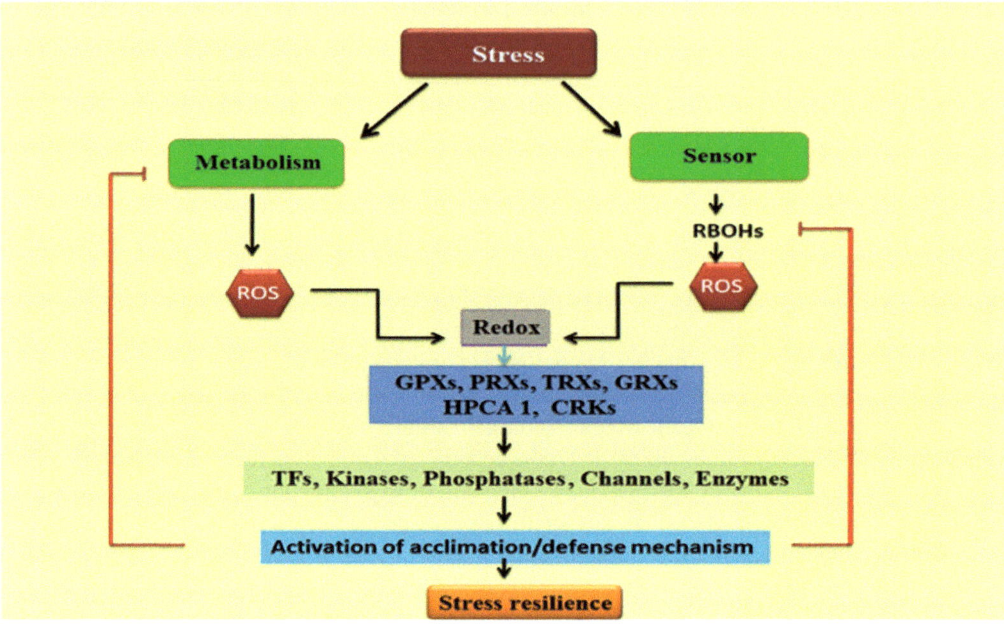

FIGURE 8.2 Anaerobic stress-induced metabolism in rice roots initially sensed by respiratory burst oxidase homologues (RBOH) in apoplastic spaces. Accumulated ROS can induce a change in redox that activates few proteins including guaiacol peroxidase, protein thioredoxins, thioredoxin, glutathione reductase. The changes in redox can also be induced for channel proteins, phosphorus metabolism, and transcription factors. These collectively induce activation or acclimation for stress resilience reactions in plants.

8.6 SIGNALING FROM ROOTS FOR RHIZOSPHERIC CHEMICAL COMMUNICATION

Submergence in rice rhizosphere is undoubtedly created by a complex consortium where root exudates play a major role for chemical elicitation in mediating plant–plant and plant rhizo–microorganism interaction. The interactions include regulation of biochemical or physiological aspects for plant microorganism association that may include and enhance host plant growth and adaptation to the environment. Root exudates containing much organic residues may act as the signal that also plays a negative allelopathy from neighboring plants and recruiting a diverse array of rhizo-microbiomes (Kong et al. 2018). In fact, the exudates' production by submerged roots is undoubtedly a redox-dependent dynamic physiochemical as well as physiological process where most of the metabolites are unknown. The biochemical pathways to produce those metabolites are also redox dependent where the regulation of a few key enzymes requires changes of chemical modification by accepting/rejecting electrons. In most of the cases, those enzymes are dependent on reversible phosphorylation- dephosphorylation and adenylation-deadenylation and sulfhydryl-thiol changes. The symbiotic relationship between plants and rhizo-microorganism species may be additive to grow by root exudates; a few are culturable and can be grown in vitro by the activity of redox-dependent pH. Therefore, redox signaling may add unknown insights of how and where root signaling may accelerate the microbe species. Under climate changing the isolation, purification and culture of those microbes may reveal the possibility of sustainable development of crop and other economic species (Olabi et al. 2023). Therefore, the root exudates under submergence rice-microbiome may be another domain where a microbial redox alteration influences the sustainability of bio-molecules. Consortium may be dependent on production according to redox changes, particularly anaerobic conditions. Root exudates for rice species may include the signal exchange for inter-organismal interaction including plant microbes in the rhizosphere within the same ecosystem. There are ample chances to exercise different omics techniques to decipher the chemical nature, concentration, or dynamics of the root exudates' production. Additionally, under changing redox, the beneficial microbes may facilitate the positive relationship that enhances plant development and specific elements of root exudates for growth of other phyto-microbion species for the same or different ecosystem (Lyu et al. 2021).

8.7 SUBMERGENCE AND MICROBIOME-MEDIATED SIGNAL TRANSDUCTION THROUGH ROOTS

By the colonizing bacteria where rhizosphere and phyllosphere are benefited in plant growth and development could perceive specific signals. The signal under submergence would be hypoxic or anoxic in nature, which communicates long distance (root xylem to shoot, both vertically and redially) in connection with commensally grown microbiome with the recent development of omics study. Specific plant-associated microbial taxa may contribute systematic plant responses that associates escalation plant growth even under biotic and abiotic stress. Although the nature of signaling residues, and their paths, the inductions are not properly understood under submergence, still, it is the hypoxia-induced anaerobic conditions of the root with development of lactate fermentation mechanisms that are released as signaling induction. Therefore, plant microbe interaction and thereby changes the rhizosphere alteration would be the basis of submergence tolerance through survival of the rice roots. Submerged soil are basically reducing in the redox potential, which also favors the soil plant interface colonizing bacteria, protozoa, and fungal species that exerts growth promotion and adaptive benefits. The relationship with plant microbiome may depend upon accumulation of some volatile, however, non-toxic organic compounds (Weis et al. 2004). These may include some non-housekeeping but inductive canonical phytohormones (derivation of auxin, cytokinin, aminopurines, benz-aminopurine) few microbial metabolites (mostly quorum sensing compounds like N-acyl-homoserine lactones (AHL), cyclodipeptides, etc). Alternatively submerged rice roots when infected with *Trichoderma* species of *Hypocreaceae* can release highly active volatile residue, including 6-pentyl-2H-pyran-2-one (6-PP), β-caryophyllene required for root organogenesis (Chopa et al. 2023). These residues are also responsible for root existing endophytes of submerged plants, but not restricted to rice and related cereals. The recent findings for the mechanism of beneficial microbes, fungi affecting root development and anaerobically grown may extend signaling events that mediates microbial establishment. For submerged roots like rice, there is an association of ethylene and auxin that interact in emission and establishment of volatile sensing residues (Figure 8.3).

8.8 ROOT RESPONSES TO VOLATILE SUBSTANCES UNDER ANOXIA

It has already been reported that the development of volatile organic compounds from fungi and other organic microorganisms may affect plant growth and development through precise signals. A number of recognized volatile substances have been identified and of those 6-pentyl-2H-pyran-2-one (6-PP) is the principal component mostly obtained from *Trichoderma* sp. (Torres-Ortega et al. 2022). Under submergence, this is mostly induced by auxin-mediated ethylene development, which in rice not only opens different transmembrane-gated proteins but also induces lateral roots. In a number of cases, submergence conditions that induces anaerobiosis or hypoxic can be influenced by auxin-induced ethylene production. In general, under terrestrial conditions, the formation of lateral roots is common for *Trichoderma*-mediated signaling, which is manifested into auxin-mediated ethylene production (Ji et al. 2021). However, under submergence, more precisely in rice instead of lateral root, the rice genotype can overexpress the fibrous root development. Auxin induces ethylene

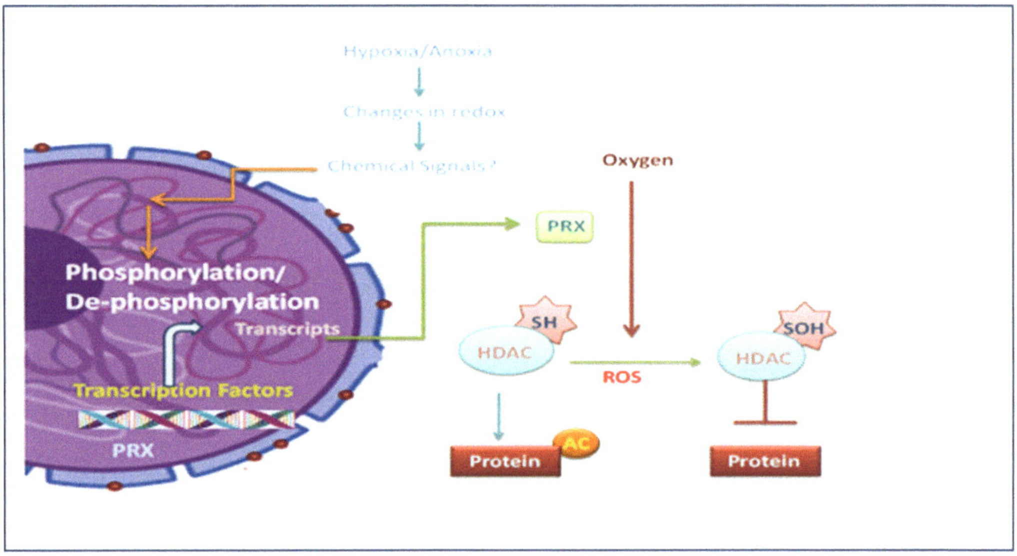

FIGURE 8.3 The signaling initiates some chemical elicitors' generation, which otherwise causes phosphorylation/de-phosphorylation, adenylation/de-adenylation of the transcription factors (PRX). The m-RNA is translated into specific proteins like HDAC. Redox changes of those proteins are brought by sulfation with the –SH group of the thiols bringing the conformational as well as redox changes and finally conversion of proteins-AC.

production, which is also a 6PP modulated expression of PIN, an auxin transporter that id expressed in a dose-dependent manner of anoxia or hypoxia. Therefore, this is a complex network where a number of auxin receptors may be induced by hypoxia or anoxia as well as by chemical residues like 6PP. There are a few auxin receptors under submergence rice roots, including TIR1, AFB2, AFB3, and other interacting transcription factors like auxin response factor 7 (ARF7), and ARF19 (Leftley et al. 2021). Ethylene has its other role in this chemical singling where EIN2 can modulate the synthesis and action of 6PP. Therefore, submergence along with its redox imbalance is related to both auxin and ethylene biosynthesis, their modulation with different transcription factors responding to redox where 6PP would be the most important.

8.9 NITRATE-INDUCED HYPOXIA/ANOXIA SENSITIVITY IN RICE SUBMERGENCE TOLERANCE

Oxygen is undoubtedly a crucial substance for sustainability of plants through biochemical reactions. Oxygen deprivation happens to be crucial for plant growth and development where its prolonged exposure can induce redox turnover. In the case of rice cultivation, submergence belongs to a compound stress or can induce hypoxia or anoxia in roots and can turn out many biochemical reactions where fermentation is most important. To avoid or to sustain plant life under hypoxia or anoxia, the normal oxidative mechanism of sugar is converted into fermentation pathways along with development of reactive oxygen species (ROS), reactive nitrogen species (RNS), and development of special tissues like aerenchyma and adventitious root formation are important (Shikov et al. 2020). The mineral nutrition for nitrogen in the form of nitrate (NO_3^-), nitrite (NO_2^-), and nitric oxide (NO) are shown to be beneficial to modulate the cellular mechanism in adjustment of submergence tolerance. Therefore, nitric oxide formation through reductive pathways may be beneficial to reduce the deleterious effect of oxidative stress with modulation of adaptation capacity under such conditions (Thirupathi et al. 2020). In contrast, nitric oxide over production through reductive pathways under hypoxic rice roots can lead to dysfunction and cell death. Thus, inhibition of scavenging of nitric oxide could be more important for plants' survival under submergence. Additionally, submerged rice fallows produce greenhouse gas, like nitrous oxide (N_2O), as a result of over accumulation of nitrate and nitrite in the soil. Even under such conditions, denitrifying microflora in the rhizosphere can enrich N_2O and support the tolerance against hypoxia and anoxia (Martínez-Arias et al. 2022). In roots, NO reduction through the phytoglobin NO cycle is important for NO irradiation and thereby nitro oxidative stress is reduced. This also enhances the root survival during oxygen deficits or limited conditions under submergence. Therefore, submerged or water logging conditions are important criteria where the modification of nitrogen metabolism becomes crucial. Hypoxia undoubtedly in rice roots can inhibit the energy releasing pathways' nitrogen assimilation, biological nitrogen metabolism, water/nutrients uptake, where a reduced concentration of ATP and ratio of NAD^+ to $NADH^+$ would demark the cell viability (Ghorbaniaghdam et al. 2014). This is accompanied by a substantial amount of ROS, and RNS in affected tissues where severely damaged cell membranes are the results. The situation of oxidative redox under nitrate stress is more compounded with cytoplasmic acidolysis (low pH) inducing cell death (Mondal et al. 2023). Therefore, the oxygen deficiency would be more detrimental for rice genotypes intolerant in nature and would be facing the nitrate deficiency and its corresponding metabolism. Oxygen stress can reduce germination and coleoptile growth in sensitive barley and oats species but in rice where anoxia is common, it favors the growth.

8.10 SIGNALING OF SUBMERGED RICE ROOT THROUGH LYSIGENOUS CAVITIES

In rice, ethylene is the most implicated inducer for the formation of lysigenous vacuoles. In roots particularly, under submergence the cavities made by aerenchyma are the constitutive expression under non-submerged, however, under well aerated conditions. Formation of a lysigenous cavity can further be extended by exogenous ethylene application even under water moisture deficits as also induced in rice roots (Pan et al. 2021). By application of biosynthetic inhibitors for ethylene like Ag^{2+} under stagnant deoxygenated water can limit the development of aerenchyma formation and leads the plants to not survive under submergence conditions. Cell lysis though lysigenous aerenchyma formation may be similar to apoptosis where DNA fragmentation, nuclear material condensation, and blebbing of plasma membrane might be the causes. The redox for lysigenous cavity formation can lead initially to the cell wall modification. Changes in esterification of pectin on the cell wall of the cortical tissue may lead to the cell lysis. A number of enzymes in combination with pectolytic and xylolytic are important along with cellulase. The activity of cellulase is increased with ethylene application and there is a continued flux of intercellular Ca^{2+} in the cytosol from vacuoles (Gómez-Maqueo et al. 2020). This is reversed by the application of some inhibitors of Ca^{2+} metabolism. An oxidative redox, particularly, during the de-submergence period can induce the gene encoding xyloglucan endotransglycosylase (XET) in rice roots and these inductions can be reversed by ethylene biosynthetic inhibitors amino-oxy acetate (AOA). Therefore, ethylene must be involved in lysigenous aerenchyma formation through cell wall modification. In the context of cell lysis in rice roots, when lysigenous aerenchyma formation takes place, it may follow the corresponding five steps:

1. A redox signaling perception of hypoxia that initiates ethylene biosynthesis.
2. The recognition of ethylene signaling by EIN, like receptors of the cells in the cortex.
3. Initiation of cell death with significant ion loss on the environment, plasma membrane invagination and development of vesicle-like structures.

4. Nuclear material condensation, induction of xyloglucan endotransglucosylase (XET), cellulose and finally lysis of the cell wall residues.
5. Absorption of water by the surrounding cells following compaction of other tissues.

8.11 EXPRESSION OF GENES FOR REDOX SENSING AND OXIDATIVE STRESS REGULATION

From the physiological perspective, redox regulated lysigenous aerenchyma formation was more complicated when a few genes were expressed and not properly identified in aerenchyma formation in the roots. The microarray techniques indicate it is the Ca^{2+} signaling-related genes encoding calcineurin beta-like protein and calmodulin that are upregulated under submergence with increased expression in cortical cells than stellar regions, and gradually they are induced for cell wall modifications (Liu et al. 2022). Therefore, calcium signaling and wall modification is under ethylene control, which may be reversed by 1-MCP. This result is undoubtedly supporting the ethylene activation and aerenchyma formation. ROS accumulation and thereby its changing redox may also be related to specific gene expression for lysigenous cavities' development. Specific respiratory burst oxidase homologue (RBOH), which is related to the homologous gene gp^{91phox} is responsible for almost the cases in development of ROS. These have the similarity with NADPH oxidase in apoplastic spaces of plant tissue, required for ROS generation (Zhang et al. 2022). Rice roots under submergence are more upregulated with the generation of ROS in the cortical cell that extends in the stellar regions and the outer cell layer of roots under submergence. Zinc encoding some metal binding protein like metallothionein, which has a similar impact in ROS scavenging activities, is inducible expressed mostly in cortical, stellar cells under anaerobic conditions (Ma et al. 2023). In contrast, submergence may induce metallothionein mostly in stellar cells extruding outer cortical layers. This is also suggestive of another way that H_2O_2 and metallothionein are inducted in the concerned tissues. Finally, the tissues are more activated towards subsequent pathways of programmed cell death by lysigenous aerenchyma formation. The root expressing RBOH also can downregulate the metallothionein biosynthesis in rice during anaerobic conditions. In a similar manner, ethylene promotes the suppression of the metallothionein gene and thereby indirectly enhances the accumulation of H_2O_2 under NADP(H) oxidase regulation. In rice cultivation with elongation strategy, this gene activity can induce epidermal cell lysis or aerenchyma formation in rice internodes. Therefore, the metallothionein gene and its downregulation may play an important role in tissue-specific program cell death in rice roots.

8.12 REDOX SENSING AND ADJUSTMENT OF SUGAR FLUX UNDER ANOXIC CONDITION

Sugar metabolism, in general, is strongly inhibited when oxygen is depleted or limited, particularly, hydrolysis of starch and its inflow in glycolysis in the form of simple sugars. Most of the sugar metabolizing enzymes is less active under low oxygen partial pressure for its reduced activities. The major reason for this inhibition happens to be the activation of energy conservation. Still, in some anoxia-tolerant cultivars, few rice genotypes can withstand hypoxia during germination including RAmy3D, sucrose phosphate synthase, aldolase, etc. Non-tolerant genotypes are much more sensitized for those genes under anoxia. The hydrolysis of pyruvate through the TCA cycle is also a limiting step under oxidative redox as well as anaerobic conditions (Ke et al. 2022). The TCA cycle is inhibited by depletion of oxygen and terminal electron acceptors. Few aquatic plants even semi-aquatic species like rice can activate the economy of energy and utilization through alternation of glycolysis into fermentative metabolism. The submerge coleoptiles can undergo the fermentation mechanism using alcohol, lactate, and alanine as the substrate to regenerate the NAD^+ required for glycolysis. Of those, three alcoholic fermentations are the most important and 92% of the pyruvate is converted to ethanol production in comparison to 7 and 1% of alanine pathway (Miro et al. 2013). Plants' response to such a condition would be more sensitive to oxidative stress when they are subjected to de-submergence. Alcoholic fermentation is activated at a high rate in rice during germination under oxygen deprivation with over expression of pyruvate decarboxylase (PDC) and alcohol dehydrogenase (ADH). Alanine metabolism is also important in rice with regards to synthesis of glutamine and glutamate. Hypoxic or anoxic conditions as created by submergence plus water logging may cause rapid gene expression, which have been demonstrated as variants of tissue specificity, spatial distribution, as well as diurnal changes also. Initially plants under submergence may downregulate or stall the protein synthesis following encoding of special polypeptides, regarded as transition peptides, being induced. Anaerobic proteins are the subsequent outcomes to be transcribed following translation in submergence rice (Tsuji et al. 2000). A few consecutive pathways are involved in metabolic activities in the absence of oxygen, including glycolytic reactions, oxidative decarboxylation of organic acids through the citric acid cycle, and oxidative phosphorylation. These pathways of anaerobic metabolism leading to alcohol and lactate fermentation with a smaller number of ATP can solidify the basis anaerobic adjustment when plants are subjected to submergence stress. Two genes are the most important, such as pyruvate decarboxylase and alcohol dehydrogenase, which are involved in the production of a few amounts of ATP, alcohol, or lactic acid that may directly support the quiescence strategy of rice root under submergence.

8.13 SPECIAL BIOCHEMICAL PATHWAYS OF SUBMERGED ROOTS UNDER REDOX CONTROL

ROS are generated constantly in submerged rice roots through symplastic or non-symplastic NAD(P)H oxidase coupled with spillage of electrons from chloroplast, mitochondrial leakage on electron transport chain. These electrons have a high potential to oxidize bio-molecules like lipid, proteins, nucleic acid as well as metabolic enzymes and transcription factors (Korac et al. 2021). As electrons are the byproducts of the aerobic respiration, ROS can also stimulate signaling molecules, controlling plant growth and development under abiotic or environmental fluctuations. ROS as signaling residues can support a number of cellular processes, covering enzyme conformation, cell wall elongation, floral bud development, seed germination, induction of other growth regulators in thr abscission or senescence process (Mahakham et al. 2017). Many other biochemical reactions where ROS levels are stringently and spatially maintained in tissues may meet protein carbonylation, nucleic acid acetylation, etc. For example, transcription factors (UPBEAT1) are upregulated in the root meristem to promote the cell division, proliferation, differentiation with distribution of superoxide and hydrogen peroxide. Another transcription factor (P-loop NTPs 1) and its expression can modify the ROS homeostasis to control the activity of stem cell division and differentiation in the root tips. With regards to growth factor 1 under regulation of ROS distribution, a few enzymes in glutathione metabolism like glutathione reductase 2 can modulate the glutathione redox for root meristem development. In rice also, salicylic acids can induce the synthesis of abnormal inflorescence meristem 1 mutation that may also modulate the ROS accumulation affecting root growth (Xu et al. 2023). Rhizosphere that modifies the ROS accumulation in the root cell can also reduce the actual redox required for survival of rice roots. In comparison to the other metabolites, hormones also play an important role in ROS signaling, regulation, and homeostasis in submerged roots. In addition to ABA, roots may develop other metabolites for asymmetric distribution of auxin that supports the gravitropism and hydrotropism in responses (Taniguchi et al. 2010). The interaction between nitric oxide, ROS, and auxin develops the complete root formation through auxin signaling pathways. Still, more insights regarding the ROS homeostasis and interaction of transcription factors' mediated gene expression are awaited for enriching the proper redox regulation for submerged rice roots. These are more valid when the homeo-box family transcription factors play a vital role in adventitious root immergence under waterlogging or anaerobic conditions (Jiang et al. 2023). Such transcription factors can regulate the expression of cytokinin and auxin signaling genes. Therefore, ROS-related genes are overcontrolled in redox signaling and its perception through plant organs, which may develop in the root system following its transmission to shoot.

8.14 CONCLUSION

Root redox, therefore, would be the key element for plant homeostasis that may communicate between submerged plant parts and above aerial shoots. In the case of rice where submergence is a common note, creating anoxia or hypoxia would be better in response to redox changes, particularly, when oxidative redox is concerned. Root signaling may affect interaction with associated phytohormones, stress compounds, secondary metabolites and other bio residues directly or indirectly influencing plant growth. Polyamines like compounds are more amenable to be overexpressed under submergence stress, which directly interacts with ROS and retrieves the actual redox for plant growth. Evaluating the rhizospheric coverage for redox sustenance, there is a note for two-way communication between plants and microbe's consortium, which are dependent on various compounds or bio-residues. Application of those bio-molecules prior to proper identification and their exercise for sustainable development of agricultural crops would be the targets. Likewise, the application of root exudates may serve as better bio-inoculants to trigger the various cascades of signaling and extract the beneficial aspects. Finally, root redox would be a proper choice to justify the cellular and bio-chemical reactions directly or indirectly supportable for submergence tolerance with reference to anoxic or hypoxic stress.

ACKNOWLEGEMENTS

MKA is acknowledging the support from DST PURSE II applicable to the University of Kalyani. NC, AD, SP are acknowledged for financial support from URS fellowship, UGC NET JRF and SVMCM fellowship, respectively.

REFERENCES

Adak MK, Das A, Kundu A, Chatterjee M, Hasanuzzaman M. 2023. "Molecular Mechanisms in Understanding Anoxia Tolerance in Rice Seeds Under Submergence and Their Implication in Rice Biotechnology." *Seeds* 2(3):246–58. https://doi.org/10.3390/seeds2030019

Ao M, Chen X, Deng T, Sun S, Tang Y, Morel JL, Qiu R and Wang S. 2022. "Chromium Biogeochemical Behaviour in Soil-Plant Systems and Remediation Strategies: A Critical Review." *Journal of Hazardous Materials* 15(424):127233. https://doi.org/10.1016/j.jhazmat.2021.127233

Beneduzi A, Peres D, Vargas LK, Bodanese-Zanettini MH, Passaglia LM. 2008. "Evaluation of Genetic Diversity and Plant Growth Promoting Activities of Nitrogen-Fixing Bacilli Isolated from Rice Fields in South Brazil." *Applied Soil* 39(3):311–20. https://doi.org/10.1016/j.apsoil.2008.01.006

Bridgemohan P, Mohammed M, Bridgemohan RS and Mohammed Z. 2020. "Ecophysiology of Anaerobiosis Stress Due to Flood and Waterlogging in Rice." *Journal of Horticulture and Postharvest Research* 1(3):113–28. https://doi.org/10.22077/jhpr.2020.2962.1110

Cassier-Chauvat C, Marceau F, Farci S, Ouchane S and Chauvat F. 2023. "The Glutathione System: A Journey from Cyanobacteria to Higher Eukaryotes." *Antioxidants* 12(6):1199. https://doi.org/10.3390/antiox12061199

Castro-Camba R, Sánchez C, Vidal N and Vielba JM. 2022. "Plant Development and Crop Yield: The Role of Gibberellins." *Plants* 11(19):2650. https://doi.org/10.3390/plants11192650

Chen H, Bullock Jr DA, Alonso JM and Stepanova AN. 2021. "To Fight or To Grow: The Balancing Role of Ethylene in Plant Abiotic Stress Responses." *Plants* 11(1):33. https://doi.org/10.3390/plants11010033

Chopra A, Mongad D, Satpute S, Mazumder PB and Rahi P. 2023. "Quorum Sensing Activities and Genomic Insights of Plant Growth-Promoting Rhizobacteria Isolated from Assam Tea." *World Journal of Microbiology and Biotechnology* 39(6):160. https://doi.org/10.1007/s11274-023-03608-1

Das A, Pal S, Chakraborty N, Hasanuzzaman M and Adak MK. 2024. "Regulation of Reactive Oxygen Species Metabolism and Oxidative Stress Signaling by Abscisic Acid Pretreatment in Rice (*Oryza sativa* L.) Seedlings through *sub1A* QTL Under Salinity." *Plant Stress* 11(2024):100422. https://doi.org/10.1016/j.stress.2024.100422

Dong H, Leung AK, Chen R, Lourenco S and Kamchoom V. 2023. "Changes in Pore-Size Distribution and Hydraulic Conductivity of Compacted Soils by Grass-Derived Hydrochar." *Journal of Geotechnical and Geoenvironmental Engineering* 149(9):04023079. https://doi.org/10.1061/JGGEFK.GTENG-11158

Du L, Jiang Z, Zhou Y, Shen L, He J, Xia X, Zhang L and Yang X. 2023. "Genome-Wide Identification and Expression Analysis of Respiratory Burst Oxidase Homolog (RBOH) Gene Family in Eggplant (*Solanum melongena* L.) Under Abiotic and Biotic Stress." *Genes* 14(9):1665. https://doi.org/10.3390/genes14091665

Ganot Y and Dahlke HE. 2021. "Natural and Forced Soil Aeration During Agricultural Managed Aquifer Recharge." *Vadose Zone Journal* 20(3):20128. https://doi.org/10.1002/vzj2.20128

Ghorbaniaghdam A, Chen J, Henry O, Jolicoeur M. 2014. "Analyzing Clonal Variation of Monoclonal Antibody-Producing CHO Cell Lines Using an In Silico Metabolomic Platform." *PloS One* 9(3):90832. https://doi.org/10.1371/journal.pone.0090832

Gómez-Maqueo A, Welti-Chanes J and Cano MP. 2020. "Release Mechanisms of Bioactive Compounds in Fruits Submitted to High Hydrostatic Pressure: A Dynamic Microstructural Analysis Based on Prickly Pear Cells." *Food Research International* 130:108909. https://doi.org/10.1016/j.foodres.2019.108909

Ji S, Liu Z and Wang Y. 2021. "Trichoderma-Induced Ethylene Responsive Factor Mserf105 Mediates Defense Responses in *Malus sieversii*." *Frontiers in Plant Science* 12:708010. https://doi.org/10.3389/fpls.2021.708010

Jiang Z, Xia X, Liu Y, He J and Yang X. 2023. "Integrated miRNA and mRNA Transcriptome Analysis Reveals Eggplant's (*Solanum melongena* L.) Responses to Waterlogging Stress." *Agronomy* 13(9):2215. https://doi.org/10.3390/agronomy13092215

Kant S, Bi YM, Rothstein SJ. 2011. "Understanding Plant Response to Nitrogen Limitation for the Improvement of Crop Nitrogen Use Efficiency." *Journal of Experimental Botany* 62(4):1499–509. https://doi.org/10.1093/jxb/erq297

Ke N, Kumka JE, Fang M, Weaver B, Burstyn JN and Bauer CE. 2022. "Redox Brake Regulator Redb and Fnrl Function as Yin-Yang Regulators of Anaerobic-Aerobic Metabolism in *Rhodobacter capsulatus*." *Microbiology Spectrum* 10(5):02354–22. https://doi.org/10.1128/spectrum.02354-22

Khan MI, Trivellini A, Chhillar H, Chopra P, Ferrante A, Khan NA and Ismail AM. 2020. "The Significance and Functions of Ethylene in Flooding Stress Tolerance in Plants." *Environmental and Experimental Botany* 179:104188. https://doi.org/10.1016/j.envexpbot.2020.104188

Kieush L, Rieger J, Schenk J, Brondi C, Rovelli D, Echterhof T, Cirilli F, Thaler C, Jaeger N, Snaet D and Peters K. 2022. "A Comprehensive Review of Secondary Carbon Bio-Carriers for Application in Metallurgical Processes: Utilization of Torrefied Biomass in Steel Production." *Metals* 12(12):2005. https://doi.org/10.3390/met12122005

Kong CH, Zhang SZ, Li YH, Xia ZC, Yang XF, Meiners SJ, Wang P. 2018. "Plant Neighbor Detection and Allelochemical Response are Driven by Root-Secreted Signaling Chemicals." *Nature Communications* 9(1):3867. https://doi.org/10.1038/s41467-018-06429-1

Korac B, Kalezic A, Pekovic-Vaughan V, Korac A and Jankovic A. 2021. "Redox Changes in Obesity, Metabolic Syndrome, and Diabetes." *Redox Biology* 42:101887. https://doi.org/10.1016/j.redox.2021.101887

Kumar A, Kumari N, Singh A, Kumar D, Yadav DK, Varshney A and Sharma N. 2023. "The Effect of Cadmium Tolerant Plant Growth Promoting Rhizobacteria on Plant Growth Promotion and Phytoremediation: A Review." *Current Microbiology* 80(5):153. https://doi.org/10.1007/s00284-023-03267-3

Leftley N, Banda J, Pandey B, Bennett M and Voß U. 2021."Uncovering How Auxin Optimizes Root Systems Architecture in Response to Environmental Stresses." *Cold Spring Harbor Perspectives in Biology* 13(11):040014. https://doi.org/10.1101/cshperspect.a040014

Liu S, Gao Y, Quan L, Yang M, Wang YZ and Hou C. 2022. "Improvement of Lignocellulolytic Enzyme Production Mediated by Calcium Signaling in *Bacillus subtilis* Z2 Under Graphene Oxide Stress." *Applied and Environmental Microbiology* 88(19):00960–22. https://doi.org/10.1128/aem.00960-22

Liu Z, Hartman S, van Veen H, Zhang H, Leeggangers HA, Martopawiro S, Bosman F, de Deugd F, Su P, Hummel M, Rankenberg T. 2022. "Ethylene Augments Root Hypoxia Tolerance Through Amelioration of Reactive Oxygen Species and Growth Cessation." *bioRxiv*:2022-01. https://doi.org/10.1101/2022.01.21.477196

Luo YH, Chen R, Wen LL, Meng F, Zhang Y, Lai CY, Rittmann BE, Zhao HP, Zheng P. 2015. "Complete Perchlorate Reduction Using Methane as the Sole Electron Donor and Carbon Source." *Environmental Science & Technology* 49(4):2341–9. https://doi.org/10.1021/es504990m

Lyu D, Zajonc J, Pagé A, Tanney CA, Shah A, Monjezi N, Msimbira LA, Antar M, Nazari M, Backer R and Smith DL. 2021. "Plant Holobiont Theory: The Phytomicrobiome Plays A Central Role in Evolution and Success." *Microorganisms* 9(4):675. https://doi.org/10.3390/microorganisms9040675

Ma Y, Xue M, Zhang X and Chen S. 2023. "Genome-Wide Analysis of the Metallothionein Gene Family in Cassava Reveals its Role in Response to Physiological Stress Through the Regulation of Reactive Oxygen Species." *BMC Plant Biology* 23(1):227. https://doi.org/10.1186/s12870-023-04174-2

Mahakham W, Sarmah AK, Maensiri S, Theerakulpisut P. 2017. "Nanopriming Technology for Enhancing Germination and Starch Metabolism of Aged Rice Seeds Using Phytosynthesized Silver Nanoparticles." *Scientific Reports* 7(1):8263. https://doi.org/10.1038/s41598-017-08669-5

Mandal M, Sarkar M, Khan A, Biswas M, Masi A, Rakwal R, Agrawal GK, Srivastava A and Sarkar A. 2022. "Reactive Oxygen Species (ROS) and Reactive Nitrogen Species (RNS) in Plants–Maintenance of Structural Individuality and Functional Blend." *Advances in Redox Research* 5:100039. https://doi.org/10.1016/j.arres.2022.100039

Martínez-Arias C, Witzell J, Solla A, Martin JA and Rodríguez-Calcerrada J. 2022. "Beneficial and pathogenic plant-microbe interactions during flooding stress." *Plant, Cell & Environment* 45(10):2875–97. https://doi.org/10.1111/pce.14403

Medrano-Macías J, Flores-Gallegos AC, Nava-Reyna E, Morales I, Tortella G, Solís-Gaona S and Benavides-Mendoza A. 2022. "Reactive Oxygen, Nitrogen, and Sulfur Species (RONSS) as a Metabolic Cluster for Signaling and Biostimulation of Plants: An Overview." *Plants* 11(23):3203. https://doi.org/10.3390/plants11233203

Miro B and Ismail AM. 2013. "Tolerance of Anaerobic Conditions Caused by Flooding During Germination and Early Growth in Rice (*Oryza sativa* L.)." *Frontiers in Plant Science* 4:269. https://doi.org/10.3389/fpls.2013.00269

Miyakawa T, Fujita Y, Yamaguchi-Shinozaki K and Tanokura M. 2013. "Structure and Function of Abscisic Acid Receptors." *Trends in Plant Science* 18(5):259–66. https://doi.org/10.1016/j.tplants.2012.11.002

Mondal S, Mukherjee SK and Hossain ST. 2023. "Exploration of Plant Growth Promoting Rhizobacteria (PGPRs) for Heavy Metal Bioremediation and Environmental Sustainability: Recent Advances and Future Prospects. Modern Approaches in Waste Bioremediation: Environmental Microbiology." *Plant, Cell & Environment* 45(10):2875–97. https://doi.org/10.1007/978-3-031-24086-7_3

Nagai K, Kurokawa Y, Mori Y, Minami A, Reuscher S, Wu J, Matsumoto T and Ashikari M. 2022. "SNORKEL Genes Relating to Flood Tolerance were Pseudogenized in Normal Cultivated Rice." *Plants* 11(3):376. https://doi.org/10.3390/plants11030376

Nardi S, Schiavon M and Francioso O. 2021. "Chemical Structure and Biological Activity of Humic Substances Define their Role as Plant Growth Promoters." *Molecules* 26(8):2256. https://doi.org/10.3390/molecules26082256

Olabi AG, Shehata N, Sayed ET, Rodriguez C, Anyanwu RC, Russell C and Abdelkareem MA. 2023. "Role of Microalgae in Achieving Sustainable Development Goals and Circular Economy." *Science of The Total Environment* 854:158689. https://doi.org/10.1016/j.scitotenv.2022.158689

Pan J, Sharif R, Xu X and Chen X. 2021. "Mechanisms of waterlogging tolerance in plants: Research progress and prospects." *Frontiers in Plant Science* 11:627331. https://doi.org/10.3389/fpls.2020.627331

Pradhan SK, Barik SR, Sahoo J, Pandit E, Nayak DK, Pani DR, Anandan A. 2015. "Comparison of Sub1 Markers and their Combinations for Submergence Tolerance and Analysis of Adaptation Strategies of Rice in Rainfed Lowland Ecology." *Comptes Rendus Biologies* 338(10):650–9. https://doi.org/10.1016/j.crvi.2015.06.010

Rani P, Kumar V, Singh PP, Matharu AS, Zhang W, Kim KH, Singh J, Rawat M. 2020. "Highly Stable Agnps Prepared via a Novel Green Approach for Catalytic and Photocatalytic Removal of Biological and Non-Biological Pollutants." *Environment International* 143:105924. https://doi.org/10.1016/j.envint.2020.105924

Richardson KH, Seif-Eddine M, Sills A and Roessler MM. 2022. "Controlling and Exploiting Intrinsic Unpaired Electrons in Metalloproteins." *In Methods in Enzymology*, edited by R. David Britt, vol 666: 233–296. Academic Press. https://doi.org/10.1016/bs.mie.2022.02.014

Rodriguez PL. 1998. "Protein Phosphatase 2C (PP2C) Function in Higher Plants." *Plant Molecular Biology* 38:919–27. https://doi.org/10.1023/A:1006054607850

Sevilla, MF, Martí C, De Brasi-Velasco S and Jiménez A. 2023. "Redox Regulation, Thioredoxins and Glutaredoxins in Retrograde Signalling and in Gene Transcription." *Journal of Experimental Botany* 74: erad270. https://doi.org/10.1093/jxb/erad270

Shikov AE, Chirkova TV and Yemelyanov VV. 2020. "Post-Anoxia in Plants: Reasons, Consequences, and Possible Mechanisms." *Russian Journal of Plant Physiology* 67:45–59. https://doi.org/10.1134/S1021443720010203

Singh A , Singh Y, Mahato AK, Jayaswal PK, Singh S, Singh R, Yadav N, Singh AK, Singh PK, Singh R and Kumar R. 2020. "Allelic Sequence Variation in the Sub1A, Sub1B and Sub1C Genes Among Diverse Rice Cultivars and its Association with Submergence Tolerance. "*Scientific Reports* 10(1):8621. https://doi.org/10.1038/s41598-020-65588-8

Song X, Wei H, Rees RM and Ju X. 2022. "Soil Oxygen Depletion and Corresponding Nitrous Oxide Production at Hot Moments in an Agricultural Soil." *Environmental Pollution* 292:118345. https://doi.org/10.1016/j.envpol.2021.118345

Taniguchi YY, Taniguchi M, Tsuge T, Oka A, Aoyama T. 2010. "Involvement of *Arabidopsis thaliana* Phospholipase Dζ2 in Root Hydrotropism through the Suppression of Root Gravitropism." *Planta* 231:491–7. https://doi.org/10.1007/s00425-009-1052-x

Thirupathi A, Pinho RA and Chang YZ. 2020. "Physical Exercise: An Inducer of Positive Oxidative Stress in Skeletal Muscle Aging." *Life Sciences* 252:117630. https://doi.org/10.1016/j.lfs.2020.117630

Tong D, Wang Y, Yu H, Shen H, Dahlgren RA and Xu J. 2023. "Viral Lysing can Alleviate Microbial Nutrient Limitations and Accumulate Recalcitrant Dissolved Organic Matter Components in Soil." *The ISME Journal* 29:1–0. https://doi.org/10.1038/s41396-023-01438-5

Torres-Ortega R, Guillén-Alonso H, Alcalde-Vázquez R, Ramírez-Chávez E , Molina-Torres J and Winkler R. 2022. "In Vivo Low-Temperature Plasma Ionization Mass Spectrometry (LTP-MS) Reveals Regulation of 6-pentyl-2H-pyran-2-one (6-PP) as a Physiological Variable During Plant-Fungal Interaction." *Metabolites* 12(12):1231. https://doi.org/10.3390/metabo12121231

Tsuji H, Nakazono M, Saisho D, Tsutsumi N, Hirai A. 2000. "Transcript Levels of the Nuclear-Encoded Respiratory Genes in Rice Decrease by Oxygen Deprivation: Evidence for Involvement of Calcium in Expression of the Alternative Oxidase 1a Gene." *FEBS Letters* 471(2–3):201–4. https://doi.org/10.1016/S0014-5793(00)01411-3

Wang L, Sadeghnezhad E and Nick P. 2020. "Upstream of Gene Expression: What is The Role of Microtubules in Cold Signalling?" *Journal of Experimental Botany* 71(1):36–48. https://doi.org/10.1093/jxb/erz419

Wang Y, Xu Y, Liang X, Wang L, Sun Y, Huang Q, Qin X and Zhao L. "Soil Application of Manganese Sulfate Could Reduce Wheat Cd Accumulation in Cd Contaminated Soil by the Modulation of the Key Tissues and Ionomic of Wheat." *Science of the Total Environment* 770:145328. https://doi.org/10.1016/j.scitotenv.2021.145328

Wei S, Li Z, Sun Y, Zhang J, Ge Y and Li Z. 2022. "A Comprehensive Review on Biomass Humification: Recent Advances in Pathways, Challenges, New Applications, and Perspectives." *Renewable and Sustainable Energy Reviews* 170:112984. https://doi.org/10.1016/j.rser.2022.112984

Weis JS, Weis P. 2004. "Metal Uptake, Transport and Release by Wetland Plants: Implications for Phytoremediation and Restoration." *Environment International* 30(5):685–700. https://doi.org/10.1016/j.envint.2003.11.002

Westermann P, Ahring BK. 1987. "Dynamics of Methane Production, Sulfate Reduction, and Denitrification in a Permanently Waterlogged Alder Swamp." *Applied and Environmental Microbiology* 53(10):2554–9. https://doi.org/10.1128/aem.53.10.2554-2559.1987

Xu Q, Wang Y, Chen Z, Yue Y, Huang H, Wu B, Liu Y, Zhou DX and Zhao Y. 2023. "ROS-stimulated Protein Lysine Acetylation is Required for Crown Root Development in Rice." *Journal of Advanced Research* 48:33–46. https://doi.org/10.1016/j.jare.2022.07.010

Yi L, Sheng R, Wei W, Zhu B and Zhang W. 2022. "Differential Contributions of Electron Donors to Denitrification Along the Proceeding of Flooding-Drying Process in a Paddy Soil." *Applied Soil Ecology 177*:104527. https://doi.org/10.1016/j.apsoil.2022.104527

Yue J, Shi Y, Zheng X, Huang G, Zhu J. 2007. "The Influence of Free-Air CO2 Enrichment on Microorganisms of a Paddy Soil in the Rice-Growing Season." *Applied Soil Ecology* 35(1):154–62. https://doi.org/10.1016/j.apsoil.2006.05.005

Zandi P, Yang J, Darma A, Bloem E, Xia X, Wang Y, Li Q and Schnug E. 2023. "Iron Plaque Formation, Characteristics, and Its Role as A Barrier and/or Facilitator to Heavy Metal Uptake in Hydrophyte Rice (*Oryza sativa* L.)." *Environmental Geochemistry and Health* 45(3):525–59. https://doi.org/10.1007/s10653-022-01246-4

Zayed BA, Salem AK, El Sharkawy HM. 2011. "Effect of Different Micronutrient Treatments on Rice (*Oriza sativa* L.) Growth and Yield Under Saline Soil Conditions." *World Journal of Agricultural Sciences* 7(2):179–84. ISSN 1817-3047.

Zhang H, Wang X, Yan A, Deng J, Xie Y, Liu S, Liu D, He L, Weng J and Xu J. 2023. "Evolutionary Analysis of Respiratory Burst Oxidase Homolog (RBOH) Genes in Plants and Characterization of ZmRBOHs." *International Journal of Molecular Sciences* 24(4):3858. https://doi.org/10.3390/ijms24043858

Zhang Y, Zhang Y, Luo L, Lu C, Kong W , Cheng L, Xu X and Liu J. 2022. "Genome Wide Identification of Respiratory Burst Oxidase Homolog (Rboh) Genes in *Citrus sinensis* and Functional Analysis of Csrbohd in Cold Tolerance." *International Journal of Molecular Sciences* 23(2):648. https://doi.org/10.3390/ijms23020648

9 Amalgamation of Hydroponics with the Biofilm to Enhance Plant Performance and Nutrient Management

Astha Gupta, Tripti Sharma, Dasputre Yash Sanjaykumar, Suthar Shubhamkumar Manojkumar and Dolly Wattal Dhar

9.1 INTRODUCTION

Hydroponics is the practice of growing plants where the roots absorb all essential nutrients from aerated aqueous nutrition formulations, as inorganic ions either under natural or controlled growth conditions. Plants may be grown either with substrates (may be organic-coconut coir, peat, bark or inorganic-perlite, vermiculite, hydrogel etc.) or without substrates. A hydroponics system is also a system with considerable potential and an alternative approach for farmers or concerned people, having limited agricultural land. In addition to that, this cultivation technique does not need any specific climatic conditions as required for a particular crop in an experimental field for growth and can be regulated manually in artificial conditions (water level, light intensity, temperature, and nutrients). The elements, which a plant takes through the soil, are replaced by a nutrient aqueous solution in this soil-less growth method. As a result, plants are able to absorb nutrients considerably more effectively, producing higher yields with less inputs. In order to absorb and digest all essential nutrients, the roots of hydroponically grown plants are kept in a pH-balanced nutrient solution, which needs the least amount of energy. The yield of flowers and fruits is increased due to the decreased energy requirements. Although an inert medium is typically used in hydroponics to support the roots, other techniques, like aeroponics, may do this without the need of any media at all.

Hydroponics, also known as hydroculture has been used to grow many crops such as lettuce (*Latuca sativa* L.), strawberries (*Fragaria ananassa*), tomato (*Solanum lycopersicum* L.), bell peppers (*Capsicum annuum* L.), spinach (*Spinacia oleracea* L.), microgreens amongst many others (Domingues et al., 2012; de Carvalho et al., 2015; Ramirez-Arias et al., 2017; Malik et al., 2018; Rosa-Rodríguez et al., 2020; Tong et al., 2021; Gao et al., 2020; Maneejantra et al., 2016). This technique is also useful to study the effects of heavy metal and abiotic stress responses under controlled conditions in many crop species. For example, mitigation of salt stress by application of potassium and silicon, in different sugarcane (*Saccharum officinarum* L.) genotypes varying in salt stress tolerance, has been studied in a hydroponics setup (Ashraf et al., 2010). In another study, hydroponic solution supplemented with different concentrations of iron, enabled screening of rice genotypes for their response to toxic levels of iron (Suma et al., 2022). Genotypes of sorghum have been screened for tolerance to salinity using a hydroponics system (Vigneshwaran et al., 2022). Further, crops may be grown hydroponically by adding nutrients to irrigation water, a process known as fertigation. In addition to supporting monoculture, i.e., growing a single crop in a season, soil-less agriculture is better than in-ground agriculture and can be employed on non-suitable or destroyed land (eroded,

DOI: 10.1201/9781003354109-9

arid, or even infertile land), and is also typically less labor intensive. Feasibility to practice monoculture encourages the grower to focus more on market demand rather than practising crop rotation. For handling soil-free cultures adequately, nevertheless, it requires considerable expertise, precise setup, and optimized nutrient composition.

9.2 SOIL: A NATURAL RESERVOIR OR REPOSITORY OF NUTRIENTS

"Soil is an entity – an object in nature which has characteristics that distinguish it from all other objects in nature." It is that integral element of biosphere that enables cultivation of food, fiber and maintenance of environmental quality at local, regional, as well as global level (Glanz, 1995). It is the foundation of agriculture and natural flora on the earth. Thus, it has a major role in governing the survival and extinction of several terrestrial life forms on earth (Doran, 1996). These attributes of soil are mainly governed by soil quality and health, both of which may be used synonymously.

9.2.1 Soil Health and Fertility

Soil health is demonstrated by its capacity to function within ecological limitations, to sustain biological production and maintain environmental quality (Karlen et al., 1997; Arshad & Martin, 2002). It is the ability of soil to promote growth and health of growing plants and inhabiting animals and microbes. A healthy soil would provide optimal water and nutrient retention along with their release, fostering and sustaining root growth, protecting the biotic habitat of the soil, responding to management, and resisting degradation. Further, soil health may be assessed via a variety of physical, chemical, and biological properties and processes. These properties may be expressed both quantitatively and qualitatively, as well as results dictated by the soil, including production, nutrient and water usage efficiency, and produce quality. Apart from properties governing soil health, several other factors like soil fertility, soil productivity are also essential for crop production.

Soil fertility is defined as the earth's potential or intrinsic capacity to give plant nutrients in the quantities, types, and proportions essential for crop growth and development (Hartemink, 2006; Roba, 2018). The amount of chemical components or compounds necessary for plant development is used to calculate fertility. Moreover, soil productivity is regarded as its ability to produce plants under a specific management scheme. Agricultural yield per unit area of land is a determinant of soil productivity. However, fertility is one of the determinants of soil productivity. Sometimes a soil may be rich in nutrients but not equally productive. A healthy, fertile and productive soil is the secret behind curing the problem of hunger, consequently achieving an important goal of sustainable development.

9.2.2 Ecological Balance of Soil: A Heterogeneous Entity

Soil is a natural reservoir of nutrients for both plants and residing microbial population. All three work synergistically for survival and maintenance of ecological balance. It is a heterogeneous entity existing in solid, liquid, and gaseous phases (Mengel et al., 2001). Although, all three phases individually have an impact on nutrient supply to roots, but solid phase is the prime nutrient source. The content of nutrients in this phase of soil majorly includes inorganic particles (comprising of potassium, sodium, calcium, magnesium, iron, manganese, zinc, and copper) and organic particles (mainly nitrogen, followed by sulfur and phosphorus) (Mengel et al., 2001). While the liquid phase is involved in transport of nutrients within the soil, the gaseous phase facilitates the exchange of gases between the biotic components of soil (plants and microbes) and the atmosphere. The exchange in the gaseous phase of soil is responsible for the supply of oxygen to soil biota and removal of carbon dioxide, the byproduct of respiration. Soil atmosphere also facilitates the transfer of nitrogen to the root nodules of plant species forming a symbiotic relationship with nitrogen-fixing bacteria.

Thus, all three phases of soil discussed above are essential components for soil to act as a nutrient reservoir. For plant growth and development, nitrogen, phosphorus, and potassium (NPK) are the main nutrients (Zhu et al., 2020). Zhu et al. (2020) have studied sorghum seedlings to understand the difference in growth, nutrients, and gene expression profiles under single and triple NPK deficient conditions. Further, microRNAs involved in the regulation of response in single and triple NPK deficiency have also been identified and compared in sorghum (grown hydroponically) (Zhu et al., 2021).

Soil may meet its nutrient gains from atmospheric depositions, biological stabilization, addition of manures, such as farm yard manures (FYM), synthetic fertilizers, and plant byproducts. Nutrients deposited on land or water from the atmosphere are referred to as atmospheric deposition. Additionally, inaccessible air nutrients are converted physiologically into plant-available nutrients through biological stabilization. The inability to meet the nutrient requirements naturally necessitates the need to supplement manures like FYM (byproduct of animal production and a promising source of nutrients for crop cultivation), green manures (green material that has not been decomposed) etc. and synthetic fertilizers. These help to boost crop yields and quality, when sprayed to agricultural soils. Crop leftovers or plant byproducts possess a large amount of nutrients that are returned to the soil.

Nutrient cycling is required to maintain the equilibrium of all essential organic and inorganic components of the soil nutrient profile. It is defined as nutrient flux within and in between the abiotic or biotic pool of nutrients in soil (Brady et al., 2008). Cycling of nutrients particularly impacted by the action of soil microorganisms, including bacteria, archaea, and fungi. Mineral nitrogen, possibly mineralizable nitrogen, soil nitrate, soil test phosphorus, potassium, sulfur, calcium, magnesium, boron, zinc, and copper, and soil salinity are all fertility indicators, and can be used to measure the nutrient cycle.

Interference with the soil environment due to anthropogenic interventions, excessive cultivation, and change in land use patterns affects the inhabiting microbial communities. Since microbial communities have a potential impact in the soil processes and properties, this may result in alteration in soil nutrient cycling (French et al., 2009). Plants absorb nutrients in the form of soluble, inorganic free ions from the soil solution and remove them from the soil in harvested materials that leave the field. Dissolved nutrients may be lost in water that flows through the soil to groundwater or out of the field via drain lines, by a process called leaching (Ng et al., 2022). It is aggravated by excess of rainfall and may lead to acidification of soil (Lehmann et al., 2003). Other mechanisms of nutrient loss may include fertilizer loss either from flowing water across the soil surface – as runoff or from crops due to wind or water movement – as erosion. Moreover, various N types may be lost as gases to the atmosphere due to volatilization and denitrification.

9.2.3 Nutrient Retention Capacity of Soil

An important soil attribute for crop production is the nutrient retention capacity. It refers to the capacity of soil to retain supplemented nutrients against losses occurring due to leaching. It is governed by several soil characteristics including cation exchange capacity of soil as well as clay fraction, base saturation, soil texture, and organic carbon content of soil.

Cation exchange capacity (CEC) of soil is the number of negatively charged sites on its surface that can electrostatically retain positively charged ions (cations) such as calcium (Ca^{2+}), magnesium (Mg^{2+}), and potassium (K^{+}), also known as base cations (Rayment & Higginson, 1992). In exchange for other cations, cations can join to and detach from certain sites. CEC is expressed as milliequivalents of charge per 100 grammes of soil (meq/100 g or cmol/kg) (Rengasamy & Churchman, 1999; van Reeuwijk, 2002; Orji & Obasi, 2012). Certain attributes of soil such as pH, structural stability, and nutrient availability are influenced by CEC (Hazelton & Murphy, 2016). Nutrient ions are transferred from soil solution to plant roots and replaced from soil solids and soil surface storage. Because electrostatically maintained cations are easily exchangeable with cations in soil solution, soil with a higher CEC has a greater capacity to maintain acceptable levels of Ca^{2+}, Mg^{2+}, and K^+ than soil with a low CEC. Because acid cations such as hydrogen (H^+) and aluminium (Al^{3+}) can occupy soil CEC, a soil with a higher CEC may not necessarily be more productive. When combined with other soil fertility measures, CEC is a great predictor of soil quality in terms of fertility and output. Cation exchange sites are typically found on the surfaces of clay minerals and organic materials (OM). Soil OM has a greater CEC at near-neutral pH than at acidic pH (pH-dependent CEC). As a result, adding organic material to a soil will most likely increase its CEC over time. The CEC of a soil, on the other hand, might decline with time owing to natural or fertilizer-induced acidity and/or OM breakdown.

The total number of exchangeable anions that a soil may adsorb is known as its anion exchange capacity (AEC). In general, the size of the CEC in soils is greater than the magnitude of the AEC. Soils that lack the ability to retain anions, nitrate, and sulfate leach readily and are no longer available to support plant growth. The weather has an effect on the exchangeable ions in soils. Fe^{3+}, Al^{3+} , complex aluminium hydroxyl ions, and H^+ are the most prevalent in humid environments. In low-rainfall soil, Ca^{2+}, Mg^{2+}, and Na^+ prevail. In a particular soil, the saturation percentage for a certain cation is the fraction of the cation exchange capacity satisfied by that cation. The exchange complex is considered to have a calcium saturation percentage of 50% if Ca^{2+} ions fulfil 50% of the CEC. This terminology is particularly useful in assessing the relative amounts of acidity and alkalinity sources in the soil solution. Also, CEC is a crucial determinant of exchangeable cations (K^+, Na^+, Mg^{2+}, and Ca^{2+}) available in soil. Moreover, CEC is considered as an important soil quality due to instant availability of exchangeable cations -K^+, Na^+, Mg^{2+}, and Ca^{2+} for uptake by plant. Additionally, the cations exchanged at the exchange sites have the ability to resist leaching into the soil.

9.3 HYDROPONICS: CULTURING PLANTS IN WATER

The term hydroponics was given by William Frederick Gericke in 1937, when he first grew tomato vines in a solution composed of minerals. Thereafter he was regarded as the "Father of Hydroponics." Earlier, "water culture" was used to describe the practice of culturing plants in water. This term was later substituted by a more sophisticated terminology, i.e., "Hydroponics," which is a combination of two Greek words – "Hydro" implying water and "Ponos" implying labor or working (Jayachandran et al., 2022). As mentioned earlier, hydroponics in the simplest terms is the growing of plants in aqueous solution containing essential minerals or nutrients, thus providing an appreciable control over the root's immediate nutritive environment. It became a useful tool to decipher plant nutritional requirements in a quantitative manner and could also be employed in physiological experiments to understand the status of known nutrients being provided to plants. Early studies on plants grown in water mainly enabled researchers to determine that the main physiological attribute of plant roots is to absorb and accumulate water and minerals from an aqueous form of soil.

Hydroponics is often referred to as soil-less cultivation of plants wherein immersion of roots in aerated nutrient suspension is a prerequisite (Maharana & Koul, 2011). Therefore, it is mandatory to consider providing all essential mineral components in the nutrient solution as available in the soil – the natural source for plant minerals. In simpler words, the composition of aqueous growth medium should mimic that of the soil for desired plant growth and development while other growth conditions like pH, temperature, humidity, oxygen levels, etc., are efficiently maintained (Maluin et al., 2021). However, for providing mechanical support to growing roots, several inert media such as peat moss, coconut fiber, gravel, vermiculite, etc., may be considered for use. Early studies were based on determining the composition of nutrient solution and its management. The first nutrient solution used to grow plants hydroponically was formulated by Knops (1865), which provided the template for improvement and modification. An ideal nutrient solution was desired to have a balanced composition of elements and osmotic potential, no precipitation occurred during use and the plants could grow without encountering any stress. Following modifications, the nutrient media composed by Arnon and Hoagland has found wide acceptance and utiltity (Arnon & Hoagland, 1940). Ample aeration of the nutrient medium and the need to renew it at appropriate intervals, were identified critical for its thorough management.

With the advent of technology, there arises considerable opportunities as well as drawbacks. However, several advantages of hydroponics are included below.

a. Hydroponics can be used to grow crops ubiquitously even in limited space available, wherever control over growing conditions can be exercised. This benefit has allowed it to be considered for supplying food to spaceship crews during space explorations.
b. Vertical arrangement for crop production may be explored depending upon the type and nature of crop for enhancing yield and productivity.
c. It assures the quality and quantity of the produce as the nutrients are administered in the growing media keeping in view the physiological requirements of the crop. This assures the quality as well as quantity of the crop yield.
d. It is a sustainable method of growing plants, as the environment for growth is controlled it protects the plants against attack by pests and pathogens. Thus, eliminating the need for the use of chemical pesticides, insecticides, etc.

Further, the drawbacks associated with hydroponics include high preliminary investment for setting up the infrastructure (Velazquez-Gonzalez et al., 2022). It requires trained labour or professionals to efficiently look after the system, which may vary with the crop under cultivation. The nutrient solution used for growing crops is rich in salts of nitrates and phosphorus, therefore requires proper management and disposal after use. Unplanned and improper disposal can lead to environment pollution by promoting growth of microorganisms in water bodies.

9.3.1 Advantages of a Hydroponics System

Hydroponics offer several advantages over conventional agriculture practices. With hydroponics, the farmer can keep an eye on, manage, and modify the plant's growth environment to ensure ideal pH levels, water supply, and nutritional balances in perfect synchronization. Additionally, there is no weed interference, and the plants benefit from improved pest and disease management. Therefore, hydroponics provides excellent potential for crop production in regions with limited or expensive water resources, simultaneously saving water, a precious natural resource. The solvent may be customized to the demands of the plant at a certain development period since the nutrients essential for plant growth are provided straight to the roots. Contrarily, with in-ground farming, farmers are limited in their ability to completely control the nutrient availability and distribution to the plants due to the mineralization activity and leaching of nutrients taking place in the soil, thus rendering hydroponics an environmentally sustainable practice. Additionally, some fertilizers could be dropped to runoff, which reduces effectiveness and raises environmental issues. Plants raised hydroponically often have fewer root systems and may focus greater attention on the development of their leaves, stems, and fruits because they can relatively easily access the nutrients they require than plants raised in soil. Also, the plant roots may be harvested as such for any further root based studies or analysis (Nguyen et al., 2016). As a result, compared to soil-medium culture, hydroponic culture can provide outputs that are 5 to 25% greater and with improved nutritional quality. Growing plants in hydroponics may prove beneficiary for both soil and water conservation (de Carvalho et al., 2015). In addition, growing plants hydroponically occupies about 20% less space as compared to traditional agriculture. This may be because in hydroponics the nutrients reach directly to the roots allowing them to grow compactly without the need to spread out to absorb nutrients unlike when grown in soil. As a result, investment in infrastructure is highly reduced while yields are comparatively higher. The edge hydroponics has over conventional agricultural practices has increased its applicability at domestic as well as commercial scale. Apart from these, growing crops in hydroponics allows for off-season availability of many fruits and vegetables. Additionally, hydroponic cultivation does away with the use of harmful chemicals and pesticide sprays as it does not encounter the pests and diseases like that in an open farm or soil-based agriculture. However, soil-less growing methods relying on aerated nutrient solution also include aquaponics and aeroponics. In aeroponics (generally regarded as a sub-group of hydroponics) the nutrients are provided as mist to the plant roots suspended in a dark enclosure at definite intervals and the plant grows without the requirement of a substrate at all. Contrary to this, aquaponics is the amalgamation of principle of aquaculture (i.e., practice of raising fish) and hydroponics together. Both the fish and plant complement each other's growth where the fish provides organic matter for the growth of the plant and the plant in return provides oxygen and filters the water for the fish. However, each of these have their own advantages and drawback associated with them but a common root lies in the fact that they are environment friendly and sustainable alternatives to conventional farming. These soil-less cultivation techniques facilitate vertical farming that increases yield per unit area (Marginson, 2010)

9.3.2 Types of Hydroponics Systems

Three primary categories of hydroponic systems exist: (1) First in which plants grow with their roots in large pipelines fed by drip irrigation. (2) Media bed hydroponics, which involves growing plants in a medium. (3) Deep water culture (DWC) or floating raft systems, where the plants are hung over a container of water that uses a floating raft. The plants are planted in planters with some medium and then in the NFT channel. There are typically two ways to manage water in hydroponic systems:

9.3.2.1 Non-Circulating System

To produce the best output, non-recovery methods are frequently used in corporate production. The plants are constantly provided with a fresh, balanced nutrition mixture. None of it is shared, for example, to prevent the transmission of viruses. Nevertheless, more nutrients are lost throughout this action, and therefore more water is used.

9.3.2.2 Recirculating System

A pump is used to gather the fertilizer mixture and deliver it back to the plants. In doing so, less resources are used. A recirculating system, although, is more difficult to manage and more difficult to handle with problems like disease or insect infestations. Regarding how the hydroponic element's design affects the general performance and water use of aquaponic systems, a literature review by Maucieri et al. (2018) discovered that NFT is less effective than hydroponic systems using media beds or DWC, yet the findings were not conclusive. In addition to being one of the major contributors to water loss via plant evapotranspiration, the hydroponic component has a clear impact on the water value, and that is crucial for fish raising. Hence, the hydroponic element's layout has an impact on the whole processes' sustainability, either directly through water use or indirectly through system administration expenses. An aquaponic system's hydroponic component selection will also affect how the system is designed as a whole. For instance, the base in media bed systems normally has adequate surface coverage for bacterial growth and purification, but in NFT channels, the surface area is minimal and bio-filters must be placed (Maucieri et al., 2018).

9.3.3 Media Bed Hydroponics

In this system, to assist the roots in bearing the pressure of the plant, a base or soil-less growth media is employed. Additionally acting as a biological and physical filter is the medium bed. Because media beds have a vast surface area wherein microorganisms containing nitrifying and other bacteria may grow, they have the most effective biological purification of all the hydroponic sub-systems. Additionally, the medium traps floating organic atoms and hard and suspended fish waste, albeit the efficiency of this physical filter will vary depending on the substrate's particle and grain size and the water flow velocity. The organic particles are gradually converted into simple atoms and ions that the plants may consume over time through biological and physical processes.

There are several methods for supplying media beds with nutrient-enriched water. It can merely drip via drippers connected to pipes that are evenly spaced across the medium. Alternatively, a technique known as "flood-and-drain" (or "ebb-and-flow") involves regularly flooding the media beds with water, which further drains again to a reservoir. By alternately flooding and draining, it is assured that the roots of the plants receive sufficient air movement, new nutrients, and oxygen. Additionally, it assures that there is always adequate moisture in the medium bed for the bacteria to grow under the best possible circumstances. Three distinct zones, that are distinguished by both water and oxygen contents, are produced by the characteristics of a flood-and-drain media bed (Somerville et al., 2014). The top 2 to 5 cm are the dry zone, that serves as a stopper to reduce evaporation and keep light from directly reaching the water, that might encourage the development of algae. Additionally, it reduces the likelihood that mosquito larvae will form and limits the growth of hazardous bacteria and fungi near the base of plant stems. Moisture and strong gas exchange are both present in the dry/wet zone. The medium bed sometimes floods and drains in this 10–20 cm zone. The whole zone will be the path across which the water travels through the medium if flood-and-drain procedures are not used. Inside this area the majority of bioactivity takes place.

The lowest 3–5 cm of the bed, which is always wet, is known as the "wet zone." The microbes that are more energetic in mineralization, such as heterotrophic bacteria and other microorganisms, are also present inside this boundary because slight particulate solid wastes accumulate there. These organisms split down the waste into narrower percentages and particles that might be sucked up by the plants via mineralization (Somerville et al., 2014).

9.3.4 Nutrient Film Technique (NFT)

In this system, the base of tiny tubes containing the root systems are continuously flooded with a fine nutritional solution layer between two and three millimeters' depth (Figure 9.1). The goal of NFT is to have some of the emerging root mat in the nutritional flow while keeping the other roots floating beyond it in the humid air so they may receive oxygen without being submerged. Pipes are

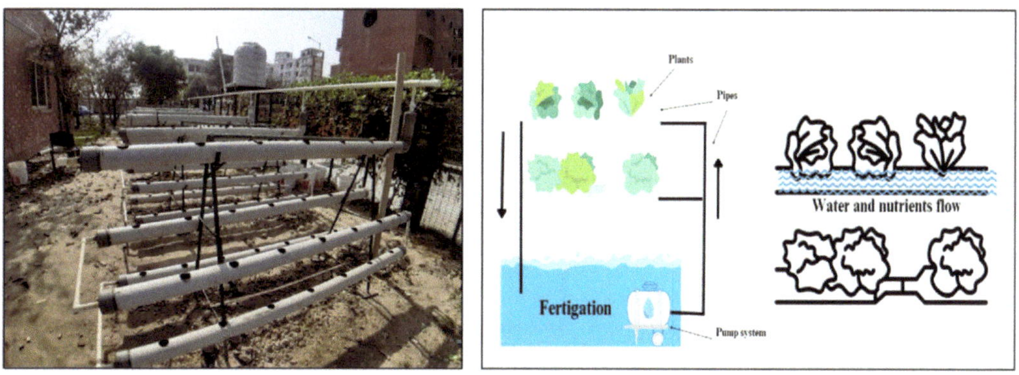

FIGURE 9.1 Representation of nutrient film technique established at Crop Cafeteria, Sharda University, Greater Noida.

a popular network shape. The finest pipes have a rectangular form and a breadth that is greater than the elevation because this causes more water to reach the roots, promoting greater nutrient uptake and plant development. The majority of crops produced with NFT systems are lettuce, herbs, green vegetables, fodder, and micro greens.

Wider and bigger pipes are needed for bigger fruiting crops and polycultures as opposed to smaller veggies and quickly developing leafy greens. Although the span of the pipe may differ, it is essential to keep in mind that because the initial plants have already depleted the elements, nutritional deficits may develop in plants at the end of particularly lengthy pipes. White pipes are recommended because they throw back the sun's radiation, keeping the interiors cool. In order for the fertilizer solution to flow at a good flow rate, which for most setups is approximately one liter per minute, the networks must be placed on a slope. Moreover, NFT along with coconut coir medium is the most appropriate forms of hydroponics for commercial crop growth with accuracy of 93% (Mamatha & Kavitha, 2023).

9.3.5 EBB AND FLOW SYSTEM

This system could be built up in a variety of methods, yet they all function according to the same principle. This hydroponic system slowly floods the plant plate with liquid fertilizer (Figure 9.2). Usually, a timer is used to regulate the water pump to flood the grow bed at certain distances. The mixture runs through overflowing into the reservoir again when hydroponic solution reaches the required level. The roots are then revealed to the required amount of oxygen as the mixture flushes

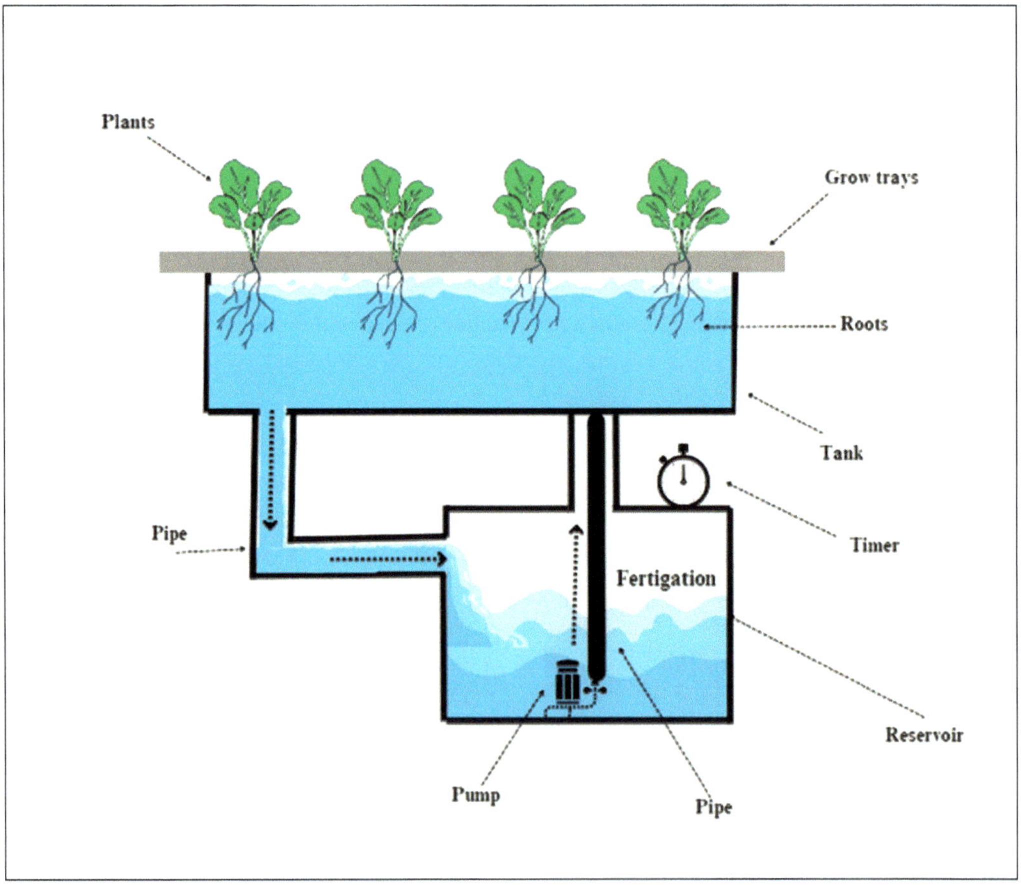

FIGURE 9.2 Representation of ebb and flow system.

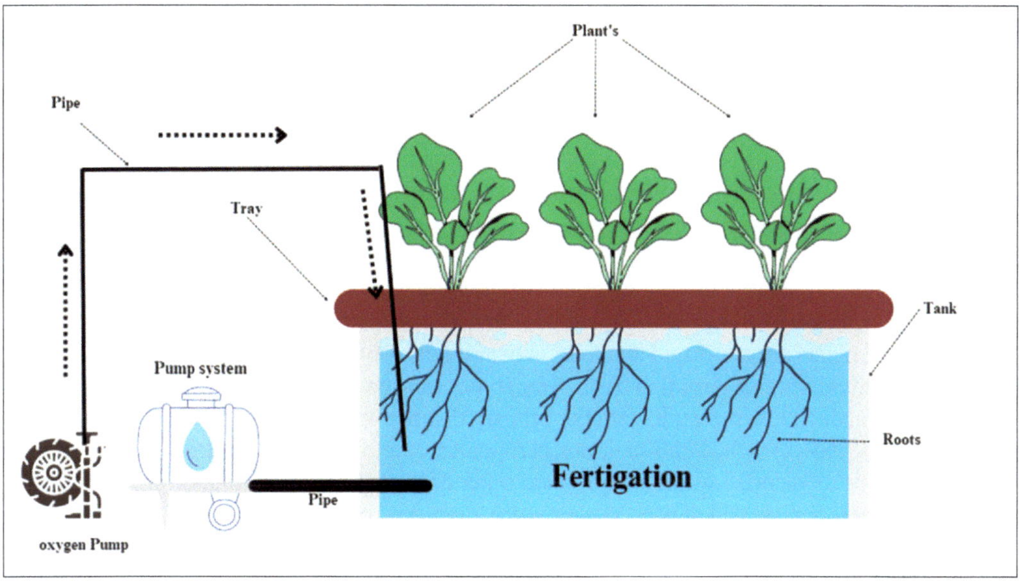

FIGURE 9.3 Representation of deep water culture (DWC).

from the grow tray. Up until the grow bed is flooded again and the procedure is repeated, the grow media is still damp from the flooding. The plants develop quickly and healthily because to the frequent flooding and drainage, which also feeds the soil with an abundance of oxygen and nutrients.

9.3.6 Deep Water Culture (DWC)

A hydroponic system known as a DWC or floating raft system suspends plants over a reservoir using a drifting raft, with the roots immersed in liquid fertilizer and ventilated by an air pump (Figure 9.3). The huge quantity of water carried in the DWC channels, in contrast to NFT systems, enables significant nutrient uptake by the plants. In NFT systems, the minerals in the narrow film of water running at root level soon become exhausted. The channels' length, which might be anything between one and ten meters, is consequently less of a concern.

Short green vegetables only need a depth of 10 cm or even less, while a depth of 30 cm or more is advised to provide appropriate plant root area. Every canal has a retention time (the length of time needed to completely replenish the water in a barrel) of between one and four hours and a flow rate of water entering each one that is rather low. The discharge rate, however, may not be intense enough to give adequate oxygen, necessitating further aeration. This culture is frequently used to produce some plants since they do well in water. For large commercial activities cultivating one particular crop (usually spinach, baby spinach, or basil), this method is usually the most popular approach since it lends itself well to mechanization. Lettuce plantation explored for energy use efficiency in nutrient film technique and deep-water culture hydroponic systems (Gillani et al., 2023).

9.3.7 Drip Irrigation System

In this setup, nutritional solution is added to the medium surrounding the plants by drip irrigating it. A meter regulates the spraying rate according to the demands of the crops. The roots are immediately fed using water and nutrients as the mixture circulates surrounding them. As the extra liquid drains, oxygen is drawn into the root system, which is crucial for plant development. The drainage system prevents flood damage, and that is beneficial for root growth. Good quality is produced as a result of the flexible watering interval, which also results in excellent hydration uptake.

9.3.8 Deep Flow Technique (DFT)

A variation of the NFT method, also known as the nutrient flow method, is the deep flow technique. The surrounding mixture is about 4 cm high, replacing the small mineral coating that was previously there. It follows the same fundamental process. Re-using is operating in the system. This particular hydroponic setup is safest since the roots are still supplied if the pump fails thanks to the deep flow approach. The technique is not used very frequently, though, as oxygen supplies to the plants might vary, leading to unequal plant development, specifically in extended or bigger systems.

9.3.9 Aeroponics and Fogponics

In aeroponics, plants are cultivated and fed via hanging its root systems in the air and misting them with nutritional mixture on a routine basis. Aeroponics may be divided into two basic categories: (i) high pressure and (ii) low pressure, with the primary distinction becoming the diameter of the spray particles employed in each. High-pressure aeroponics employs high-pressure (around 120 PSI), low-flow pumps to vaporize water and produce water droplets of at least 50 microns, as opposed to low-pressure aeroponics, which uses low-pressure, high-flow pumps (Figure 9.4).

Fogponics is the name used to describe exceedingly thin droplets, which mimics fog. Fogponics is a soil-less cultivation of plants suspending in air or fog to provide the nutrition to plants in the absence of soil (Uddin & Suliaman, 2021). Fogponics differs from other types of aeroponics in that it utilizes one of a variety of techniques (such as ultrasonic, pressurized air, or heating devices) to create a suspension of very tinier droplets (5–30 microns), or even steam. Aeroponics customarily provides nutrient-rich water to the roots, which use a mist (comparatively big atoms in air). Atoms between 1 and 50 microns in size are best absorbed by plants. It is presumable that the energy needed for growth and the rate of absorbance is negatively correlated to droplet size. Due to their extensive exposure to more oxygen, plants produced using an aeroponic system often grow more quickly compared to other types of hydroponic systems.

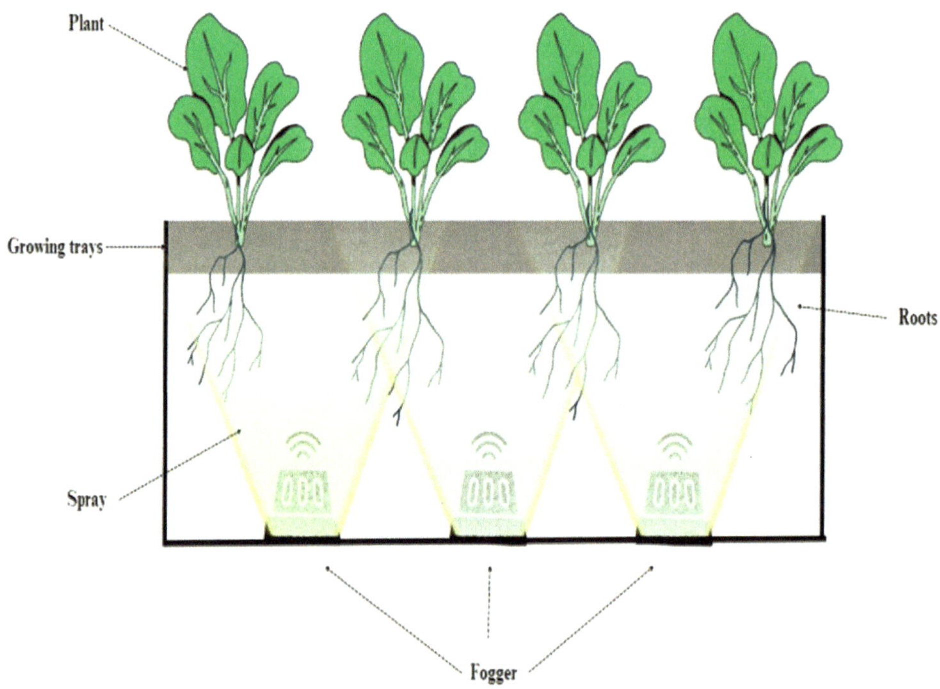

FIGURE 9.4 Representation of the aeroponics system.

9.4 GROWTH CONDITION REQUIREMENT FOR A HYDROPONICS SYSTEM

Optimum growth conditions are needed for cultivation of specific vegetables, herbs, and crops in a hydroponics system, which includes light, temperature, humidity, pH, air temperature, and suitable substrate.

9.4.1 Light

The proper amount and quality of light transmission is critical for effective photosynthesis, growth, and production. UVC 100–280 nanometres (nm), UVB 280–315 nm, UVA 315–400 nm, visual and photosynthesis process activity rays (PAR) 400–700 nm, far-red 700–800 nm, and infra-red 800–4000 nm are all produced by the sun. The visible spectrum's wavebands are also sub-divided into three color combinations: blues 400–500 nm, greenish 500–600 nm, and red 600–700 nm.

Chlorophyll is divided into four categories chlorophyll a, which is observed in all higher plants, algae, and cyanobacteria; chlorophyll b, which also is observed in higher plants and algae; chlorophyll c, which would be observed in diatoms, dinoflagellates, and brown algae; and chlorophyll d, which would be observed only in red algae.

The most prevalent photosynthetic pigment includes chlorophyll a, which captures blue, red, as well as purple frequencies mostly in the visible light spectrum. It mainly involves the photosynthesis process, wherein oxygenation is the primary byproduct. Chlorophyll b receives blue light predominantly and is utilized to supplement the maximum absorption of chlorophyll by broadening the wide range of frequencies that a photosynthetic organism may capture. Chlorophyll a and b function together to maximize transmittance mostly in blue to red range.

Plant light responses have developed to assist plants in adjusting to a wide range of light situations. Most plants behave better to moderate and intense light levels, although a few are suited to work most effectively in full sun, whereas others need less shadow. Plants breathe air and create CO_2 in the shadow. The physiological rate hikes even as frequency is increased, while at a specific light level (the efficiency curve point), the oxygen level equals the photosynthetic activity (no net uptake or loss of CO_2). The light source, in combination to its strength, determines the pace of photosynthesis. Crops may utilize frequencies ranging from 400 to 700 nm for photosynthesis.

The quantity of light accessible to plants varies greatly throughout the world and during the periods. At solar minimum heights, for comparison purposes, signal should transfer through some kind of greater amount of environment until reaching the land temperature, causing changes in the spectral region because as surroundings screen correspondingly more of the relatively short wavelengths of the light, so it screens so much more UV than blue, and much more blue than green or red.

9.4.2 Temperature

The best water temperature variation as with most crops is 14–22 °C, even though this completely depends on the variety of plant. Temperature of the water, instead of atmospheric temperature, does have the largest influence on crops overall, and other microorganisms have such a specific maximum temperature too. Nutrient uptake microorganisms, for example, demand a mean temperature of almost 20 °C.

9.4.3 pH

The water's pH level affects the amount of several micronutrients. Many crops tolerate pH levels ranging from 5.5 through 7.5. If indeed the pH decreases below this region, plants have developed nutritional lockup, which indicates that such minerals exist in the liquid but cannot be used by the crops. Iron, calcium, and magnesium seem to be very important. Conversely, there seems to be indications suggesting nutrition lockdown is just less frequent in adults in hydroponic systems.

9.4.4 Air Temperature

Average temperature is an affecting environment element that regulates crop plant cycles from either the earliest phases of development through bloom creation. Every plant species does have a different ideal mean temperature. Plants "seek" an appropriate temperature, which is determined by a combination of atmospheric temperature, moisture content, as well as lighting. When light levels are very high, crops warm up, producing in a thermal conduction between both the crop and the atmosphere. To absorb heat, the rate of evaporation of the crop should rise. Rising temperatures or decreases beyond or below the ideal range are known to change various biological processes in crops and harm plant cells, hence influencing growth.

9.4.5 Relative Humidity

The level of water droplets found in the air represented as just a percentage of the volume required for absorption at a given temperature is referred to at humidity levels. Humidity levels impact soil's water relations intrinsically and extrinsically, as well as leaf development, respiration, and disease prevalence.

When the relative humidity level is high, its evaporation rate dropped, when osmotic pressure increases, so plants proliferate. Whenever the relative humidity (RH) is poor, evaporation rises, generating water deficiencies in the crop and perhaps crop wilting. Water shortages cause the stomata to close partially or completely, preventing carbon dioxide from entering and reducing photosynthesis. High RH promotes the growth of insect pests and illnesses, as well as the easy development of fungal spores on leaf tissue.

9.4.6 Substrate: Support System (Foam, Cocopeat, Agar Gel)

The suitability of a varied substrate varies according to the hydroponic system. The medium might either be artificial, inorganic, or organic. It must have a sufficient surface area and be conductive to air and water in order to support bacterial development, water movement, and root metabolism. It must not be poisonous, have a pH of 7, and be mold-resistant. Additionally, it must not be so thin that it swims. The medium affects water holding strength, ventilation, and pH equilibrium, among other factors. Diameter, dimensions, and porosity all affect how much water is kept because it adheres to the surface of the particles and fills the pore space.

The bigger the surface area and pore space, the better the water absorption because of the tiny particle size. Compared to smooth, fine aggregates, oddly shaped molecules have a larger surface area and a greater water holding rate. Water may be stored in porous materials, hence water holding is large. The medium should be able to keep water well while also allowing for proper drainage. So, it is essential to refrain from using too small of a substance in addition to avoiding water holding and oxygen deficiency. Granular or fibrous substrates are two different types of substrates. Gravel, vermiculite, perlite, pumice, and light expanded clay aggregate are a few examples of granular substrates. Coconut fiber and rock wool are examples of fibrous substrates.

It is crucial to have a good balance of large and tiny pores. Coarse foundations have a high macroporosity (the amount of air they can hold) but a low microporosity (the amount of water they can hold), whereas fibrous substrates have a great microporosity but a poor macroporosity.

9.4.6.1 Volcanic Gravel (Tuff)

It is mostly chemically inert, with the exception of tiny discharges of micronutrients like iron and magnesium and the intake of phosphate and potassium ions in the initial months. It has an extremely good surface area to volume ratio that gives adequate room for bacteria to colonize. Volcanic gravel with a diameter of 8 to 20 mm is advised. Greater gravel is unable to provide the necessary contact area or plant support, whereas lesser gravel is more prone to become clogged with solid waste.

9.4.6.2 Light Expanded Clay Aggregate (LECA)

When relative to other media, it is quite lighter, that makes it suitable for rooftop aquaponics. It exists in a range of sizes; for aquaponics, the bigger ones with dimensions of 8–20 mm are advised. Even though microbes coat the surfaces, higher porosity gaps (macroporosity) provide for improved liquid infiltration via medium and improved air delivery. Nevertheless, this substrate possesses microscopic micropores, which limits its ability to store water.

9.4.6.3 Limestone Gravel

Despite occasionally being utilized as a substrate, limestone gravel is not advised. Compared to volcanic gravel, limestone has a lesser surface-to-volume ratio, is heavier, and is not inert. Calcium carbonate ($CaCO_3$), which combines with water, makes up the majority of limestone's chemical makeup. This will raise the pH, hence it should only be employed in waterbodies that are highly acidic or low in alkalinity. However, a tiny amount of limestone can assist in offsetting the nitrification bacteria's tendency to make things more acidic, reducing the necessity for routine water buffering.

9.4.6.4 Perlite

A siliceous substance with volcanic origins, perlite is extracted from lava flows. The temperature is increased to 760 °C, which causes the tiny portion of water to convert into vapor and causes the atoms to swell into tiny, sponge-like kernels. Perlite is incredibly thin and can hold up to four times its own weight in water. It includes a pH range of 6.0 to 8.0, is virtually balanced yet lacks buffering ability; also, in contrast to vermiculite, there is no cation exchange capacity and no trace amounts of micronutrients. To promote both drainage and aeration, it should not be used alone but instead combined with another substrate.

9.4.6.5 Vermiculite

A micaceous mineral called vermiculite spreads whenever warmed to temperatures exceeding 1000 °C. Water is converted to vapor, creating tiny, porous kernels that resemble sponges. Vermiculite is lightweight and has a high water absorption capacity. It is a hydrated magnesium-aluminum-iron silicate chemically. It may retain minerals in reserve and discharge them subsequently since it is balanced in reactions, has strong buffering capabilities, and has a comparatively large cation exchange capacity. Magnesium and potassium are also present, and they are accessible to plants.

9.4.6.6 Rock Wool

Basalt rock that has been heated to a molten state in flames at 1500 °C is used to make rock wool. The threaded, compacted wool packages made from the liquid basalt are then carved into chunks, bricks, or plugs. Over the past 20 years, rock wool culture has accounted for the majority of the greenhouse industry's explosive growth. Due to the fact that it does not decompose in landfills, worries have been lifted in recent times over how to dispose of it.

9.4.6.7 Coconut Fiber (or Coir)

Coir, often known as coconut fiber, is an organic medium made of pulverized and shredded coconut shells. It holds water and enables for a healthy quantity of oxygen for the roots and has a pH that is very near to balanced.

9.4.6.8 Pumice

Pumice and perlite are both siliceous volcanic materials with nearly identical characteristics. But nonetheless, because it has not been moistened, the crude ore is thicker and does not retain moisture as quickly because it has not been heated after grinding and screening.

9.5 NUTRIENT MANAGEMENT IN LIQUID SOLUTION TO IMPROVE PLANT PERFORMANCE

Growth parameters of hydroponics system can be altered/controlled easily, therefore nutrient solution management is quite easy according to the needs of the plant (Sathyanarayan et al., 2023). When it comes to nutrition management in hydroponics, plants also get effected by pH and it shows different effects at different level of pH like uptake of nutrition (Gillespie et al., 2021). To optimize nutrient absorption, hydroponic leafy green production necessitates careful regulation of pH and electrical conductivity (EC). The pH of nutrient mixtures is normally maintained within 5.5 and 6.5, but reducing the pH to a more acidic level may help to reduce dangerous waterborne infections. In order to evaluate a new hydroponic nutrition management method, "Corvair" spinach crops were raised in a growth chamber utilizing a deep-water culture system and pH levels of 4.0, 4.5, 5.0, and 5.5 of a hydroponic nutrient solution. Development of plants and appearance were normal but modest at pH 4.5 or 5.0. Plant tissue investigations revealed that when pH fell, N, P, K, Mg, S, Cu, Fe, Mn, and Zn concentrations declined. When the intensity of the nutritional mixture was raised threefold at a low pH of 4.5, spinach shoot and root fresh weight grew dramatically yet remained lesser than in the control. Increased nutrition formula optimization may boost spinach growth at low pH.

As a mineral fertilizer, phosphorus (P), a major nutrient, is utilized in agriculture. Hydroponically cultivating crops can maximize nutrient uptake, however irrigation water of poor quality hinders crop production. Different P and NaCl concentrations in the root area were used in two trials (Germano et al., 2022). A higher root/shoot ratio was the most important morphological plant response to the low P concentration. Up to a dose of 30 mol m^{-3}, sodium chloride did not significantly impair yield and quality. These findings supported the hypothesis that, with the right governance, basil may be cultivated hydroponically at reduced P and elevated NaCl concentrations. However, temperature, electrical conductivity, humidity, pH, and oxygen content are the main concerned factors to regulate plant growth and productivity. The quality of plants and metabolites may improve under sodium chloride (NaCl) treatment in a hydroponics system without affecting normal growth of plants as already explored in Chinese chives (*Allium tuberosum* Rottler) (Xie et al., 2023). The management approach of nutrient solution may enhance nutritional quality and flavor of Chinese chives through adding 10 mM NaCl in the hydroponics system (Xie et al., 2023).

Generally, the need for nutrients, is met with the help of liquid fertilizer. But research proves that, biofertilizer can cut down on the need for mineral fertilizers by up to 50% while also enhancing the grade of baby spinach cultivated hydroponically (Dasgan et al., 2023). The total phenolic, vitamin C, soluble solids, phosphorus, iron, SPAD-chlorophyll, and titrable acidity were produced by mycorrhiza, algae, and bacteria. The most potassium, calcium, magnesium, zinc, and manganese were produced by microalgae. Therefore, these results show that the use of biofertilizers has the ability to up to 50% lessen the dependence on mineral fertilizers. In addition, it is probable that baby spinach's value will get better due to a surge in components that are healthy.

The productivity of lettuce (*Lactuca sativa* L.) can be significantly impacted by a minor nutrient deficit. According to a factorial experiment, the fresh and dry weight, chlorophyll index, chlorophyll a and b, total chlorophyll, carotenoids content, total protein concentration, and leaf zinc concentration all jumped considerably as the content of zinc raised (Behtash et al., 2023). The levels of proline, malondialdehyde (MDA), guaiacol peroxidase (GPX), catalase (CAT), ascorbate peroxidase (APX), and boron concentration all rose considerably as boron content climbed. High boron contents had a negative impact on the aforementioned metrics, although injecting zinc to the mixture reduced this effect.

Three different nutrient-solution depth treatments were tested in hydroponics to evaluate the growth parameters, phytochemicals composition and antioxidant activity of kale grown. They found that a depth of 2 cm at 24 days exhibits maximum levels of plant height, total leaf area, stem diameter total root length, and root surface area, high fresh and dry weight of the shoot and root. In addition to that total chlorophyll accumulation, nutrient ions uptake, contents of carotenoid, soluble sugars, and vitamin C were also in good proportion in the depth of 2 cm and 3 cm (Tan et al., 2023). Agrobacterium-specific phages express potential as disinfectant of nutrient solution and helpful to manage hairy root disease (HRD) in hydroponic greenhouses (Fortuna et al., 2023).

9.6 ROOT MODELING AND BIOFORTIFICATION THROUGH ADVANCE TECHNIQUES

Biofortification is a significant field of study that aims to boost the nutrient levels in crops to enhance human health. Several approaches can be used to achieve this goal, including genetic engineering, breeding, and nutrient management (Kaur et al., 2020; Saini et al., 2020). One of the techniques used to accomplish biofortification is enhancing the nutrient uptake and usage effectiveness of plant roots. This can be done by modifying root architecture or enhancing the efficiency of nutrient transport through root membranes. Advanced technologies such as molecular breeding, genome editing, and biotechnology can be employed to increase nutrient content. For instance, scientists can use CRISPR-Cas9 gene editing techniques to modify genes that are responsible for nutrient uptake and transport, hence improving the efficiency of nutrient absorption and utilization. Advanced imaging techniques like CT and MRI can also be used to study root architecture in detail, providing valuable insights into root–soil interactions.

Precision agriculture, which utilizes drones, sensors, and other technologies to monitor crop growth and optimize yield and quality, is another advanced technique used in biofortification. With precision agriculture, farmers can apply the right amount of nutrients and water to crops, enhancing their nutritional content.

Food crops that were genetically diverse, conventional, or old-fashioned included nutrient-dense substances, which include micronutrients. However, in modern times, farmers have chosen to increase agricultural production in order to increase earnings, which has resulted in a drop in the amount of micronutrients and possibly the disappearance of some within food crops. Moreover two billion people worldwide suffer from nutritional deficiency with Asia and Africa being two of the most severely impacted continents. With traditional breeding, biotechnological, and genomic methods, biofortification aims to increase the amounts of essential nutrients, such as vitamins, minerals, and proteins, in the edible parts of plants.

Biofortified foods may help people consume iodine, which is necessary for human metabolism (Figueiredo et al., 2022). The purpose of the research was to raise the iodine content of lettuce and assess how it affected plant development and productivity. Two cultivars were harvested 40 days after transplantation and measured for fresh and dry matter weight, iodine content and levels in the leaves, and root volume. Although the results indicated phytotoxicity, all dosages encouraged increases in total iodine levels. As a result, it is feasible to raise the iodine concentration of lettuce leaves while still needing to modify the dosages that will be utilized.

Alternative dietary sources of this small amount must be incorporated into the human diet in order to prevent the negative effects of iodine shortage on health (Sularz et al., 2020). The biofortification of crops with minerals and vitamins is one of the most efficient techniques to improve nutrition. Crops were grown for the study in a hydroponic nutrient film technique (NFT) system. The findings of this study suggest that the ingestion of iodine compounds affects alterations in iodine and various chemical concentrations in the treated samples. Linolenic acid was the predominant fatty acid in lettuce; yet, iodine fertilization had little impact on plant fatty acid composition other than to increase myristic and arachidic acid contents. Additionally, we discovered that iodine enrichment increases the levels of several minor and major nutrients. In addition, 3.5-diISA treatment reduced nitrate levels in comparison to control and different treatments. Because lettuce may be a rich source of iodine and different components in the human diet, it may be argued that producing it with iodosalicylates added is worthy of attention. In one study, a biofortification method to raise the Zn tissue content of two distinct purslane kinds was put to the test (D'Imperio et al., 2022). The procedure of biofortification raised the Zn concentration in the NS by 1.3, 2.6, and 5.2 mg/L, respectively, increasing it by 1.8-, 2.3-, and 2.7-fold. Consuming hydroponically grown purslane that has been biofortified with zinc might help people get more zinc.

Advanced methodologies for plant root modeling and biofortification are major topics of research in modern agriculture. Plant roots are essential for nutrient absorption, water uptake, and plant attachment to the soil. Understanding the structure and function of plant roots is therefore critical for the creation of crops with higher yield and quality.

Computer simulations are one of the sophisticated approaches utilized in plant root modeling. Scientists may use computer simulations to generate models of plant roots and mimic their growth and development under various environmental circumstances. This can aid in determining the best root architecture for nutrient absorption and water utilization. Finally, sophisticated approaches for plant root modeling and biofortification have the capacity to revolutionize current agriculture. These strategies can aid in the development of crops with higher yield, quality, and nutrient content, therefore contributing to food security and improving human health. Overall, the combination of plant root modeling and biofortification through advanced techniques shows significant potential in enhancing the nutritional quality of crops and improving food security for the expanding population.

Transcriptomic and hormonal analysis was evaluated for roots of maize seedlings under hydroponic system at low temperature and found that two sensitive traits (cell cycle duration and cell elongation) have been regulated by temperature reduction (Friero et al., 2023). Biofortification of ten bread wheat genotypes (jala-2016, Galaxy-2013, Punjab-2011, Faisalabad-2008, SARC-I, SARC-VIII, Anaj-2017, Lasani-2008, Inqalab-91 and Sehar-2006) was investigated using varying concentration of Zn (0.2, 0.4 μM) and Fe (25, 50 μM) under saline and normal conditions in a hydroponics experiment (Naz et al., 2023). As a result, Faisalabad-2008 and Galaxy-2013 performed well and biofortified for Fe and Zn for tested traits (chlorophyll contents, root/shoot dry weight, Zn, Fe, K and Na concentration in root and shoot) under saline conditions.

9.7 APPLICATION OF INTEGRATED SYSTEM OF HYDROPONIC AND BIOFOULING FOR PROFICIENTLY PURIFIED WASTEWATER FROM AQUACULTURE

Hydroponic effluent can be considered as low-cost growth media due to comprising of judicious quantity of residual nutrients to grow microalgae mediated resource recovery and water recycling. Moreover, allelopathic effects of root exudates (mainly, benzoic acid) has been explored to evaluate the growth inhibition and stimulation of microalgae (Tiwari & Prajapati, 2023). Biofouling or biological fouling is a very common phenomenon occurring in nature due to the formation of biofilms. These biofilms are formed by access growth of micro or macro-organisms, leading to accumulation of biological matter on surfaces. When biofouling is caused due to accumulation of micro-organisms like bacteria and protists it is referred to as micro-biofouling, on the other hand, accumulation of macro-organisms like invertebrates and algae results in macro-biofouling (Wahl, 1989). Apart from fouling due to the formation of biofilms, inorganic fouling due to deposition from crystallization, corrosion, ice, oil and suspended particles is also prevalent in nature. Biofouling is prevalent in industry, marine environment, and medical fields. In all the three fields it poses significant health-related threats, deleterious impact on environment and resultant financial losses (Walker et al., 2000; Railkin, 2004; Hellio & Yebra, 2009; Vo-Dinh, 2007). With the nature and extent of fouling varying with immediate environment, organisms and deposition of inorganic compounds, it is also different in medical, industrial, and marine fields. While biofouling in the medical field includes only biofilm formation, industrial and marine biofouling includes a combination of biofilm, macrofouling, and inorganic deposits. Biofouling in the medical field can be seen in orthopaedic or dental implants resulting in removal due to infection and periodontal disease, respectively (Aminian et al., 2022). It is also common in biosensors, respirators, catheters, and heamodialysis. Biofouling is most common in the marine environment where it occurs on ship hulls, and engine and marine platforms, resulting in more fuel consumption, more stress from excess drag, and increase in load on the marine structure. In industrial biofouling observed in membranes it reduces flux, in the food, paper and paint industry along with metal cutting fluid, it results in spoilage of food, filter blockages, and poses health risks to workers.

Concerns arising from biofouling call for methods to prevent, reduce, and overcome effects and losses associated with it, by adopting antifouling strategies. These strategies can exert controls by both physical and chemical means and even a combination of both (Bhushan, 2009; Bar-Cohen, 2011). Physical control may be imposed by altering wettability, reducing adhesion and drag, microtexture and several miscellaneous behaviors, whereas the secretions form a dominant part of chemical controls. Antifouling methods can be learnt and adapted from flora and fauna in nature as well. For example, the hydrophobic wax covered wings of dragonfly (*Libellula)*, super hydrophobic feathers of dove (*Zenaida*), pigeon (*Columba)* and duck (*Anas) and* superhydrophobic self-cleaning surface of lotus (*Nelumbo nucifera)* and taro (*Colocasia esculanta*). These natural mechanisms existing in flora and fauna can be mimicked for developing antifouling techniques. Apart from these coatings are also used to control biofouling. Some of these are polyethylene glycol coatings and zwitterionic polymers used on medical devices to deter adsorption and adhesion of proteins. To prevent antifouling in marine and industrial equipments metal coatings, silicone elastomer, self-polishing copolymer are applied such that some of these act by either lowering surface energy thus promoting low adhesion or form toxic coated surface. Even enzymes and hormones are coated on surfaces as they lower bio-adhesion, causing bacterial lysis and premature metamorphosis of larva.

9.7.1 Biofilms: A Prerequisite to Biofouling

In nature, microorganisms and their communities attach to surfaces by means of polymer matrix forming complex structures in a dynamic manner called biofilms. However, a biofilm can consist of a single or many different species of microorganisms that attach to each other in a self-synthesized polymer matrix made of proteins, polysaccharides and extracellular DNAs (Priyadarshanee & Das, 2023). Bacteria, algae, archaea, protozoa, and fungi are some biofilm-forming species. Formation of bacterial biofilm generally depends upon the interaction between bacterial cells, the surrounding media and the substrates (Watson et al., 2023). Furthermore, biofilm formation is a multi-step process classified into five phases as follows (Figure 9.5).

Reversible attachment (stage 1): the first step of biofilm formation, induced by some favorable interactions between the planktonic cells and substrate surfaces. While the bacterial movement towards the surfaces may occur by convection, sedimentation, or Brownian movement (Zhao et al., 2023), bacterial growth on surfaces is made possible by chemotaxis, which further enable cell-surface interactions.

Attachment to substrate or host surfaces is triggered by weak interactions like van der Waals forces, hydrophobic interactions and electrostatic interactions between cells and substrate surfaces together with shear forces within the cell and is further driven by locomotive appendages – flagella (Yin et al., 2019; Beitelshees et al., 2018).

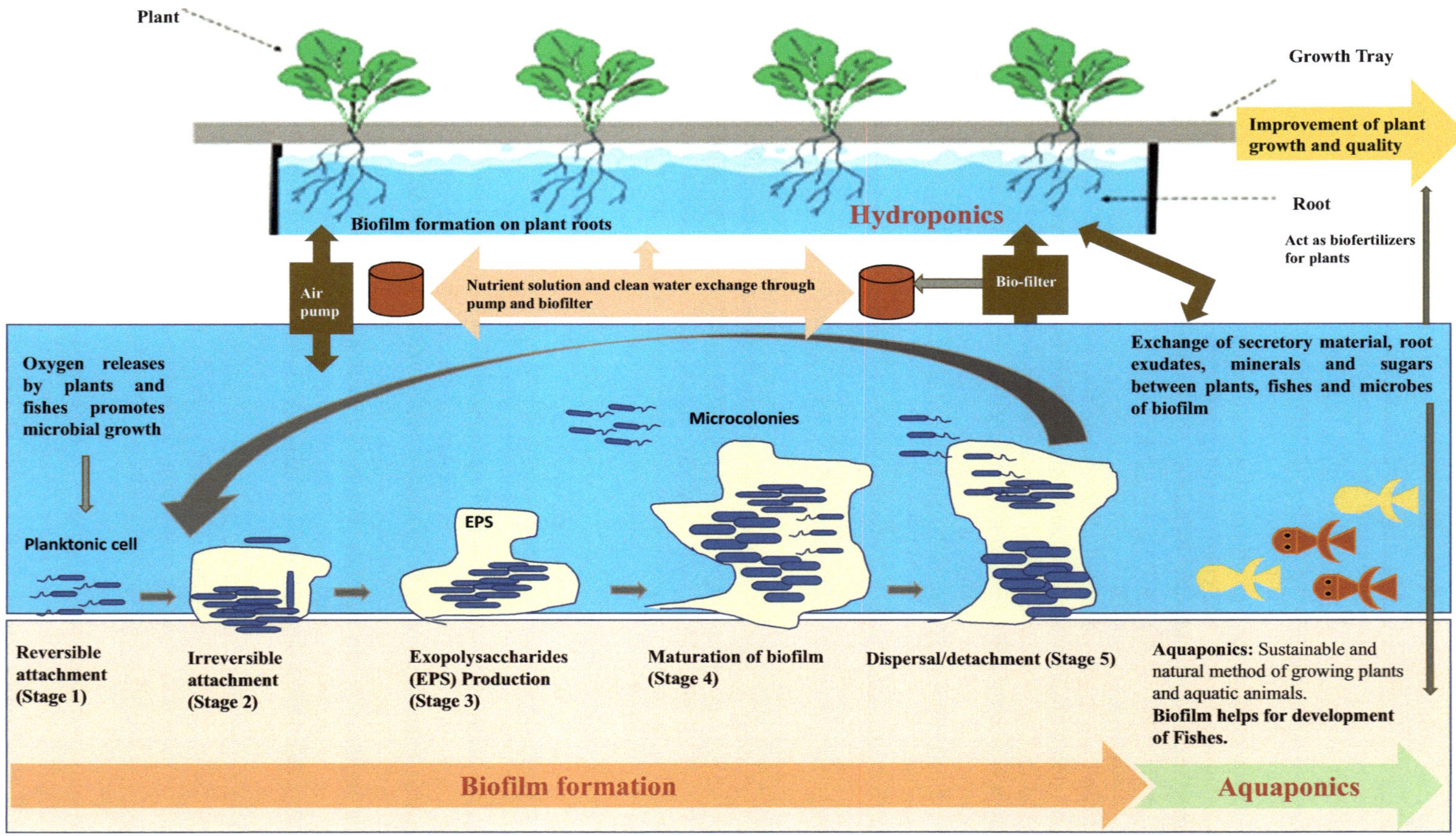

FIGURE 9.5 Integrated system of hydroponics, aquaponics, and biofilm to maintain and improve ecosystem for growth and development of plants, fishes, and microbes in liquid medium.

Irreversible attachment (stage 2): at this stage the interactions are mediated by short-range interactions including hydrophobic, dipole–dipole interactions, formations of hydrogen, covalent and ionic bonds in conjunction with adhesion of bacterial structures. Presence of adhesion structures like pili, flagella, fimbriae on the surface of bacteria also play a pivotal role in the formation of biofilms as they mediate physical contact with the surface (Berne et al., 2015). Apart from surface adhesion structures cell-to-cell communication called quorum sensing also mediates co-ordination between individual cells to form biofilms (Lamret et al., 2023). Via quorum sensing, the bacteria produce chemical messengers like autoinducers (AIs) and acyl homoserine lactones (AHLs, primarily in Gram-positive bacteria) for cell-to-cell communication. The stimulus for transition from reversible to irreversible adhesion of cells is received from c-di-GMP, a secondary messenger and other cell surface structures.

Biofilm bioadhesion mechanism through exopolysaccharides (EPS) production (stage 3): EPS is composed of extracellular DNAs (eDNAs), proteins, lipids, polysaccharides, and several other polymers, which may vary with bacterial species and environmental conditions (Costa et al., 2018). It forms a vital constituent of the biofilm extracellular matrix while the synthesis and secretion is regulated by residing bacterial cells through quorum sensing. Cohesion among the bacterial cells within the biofilm and adhesion of biofilms to surfaces is possible because of EPS production by formation of hydrophobic and ion bridging interactions (Costa et al., 2018). Apart from adherence to surfaces and biofilm formation, other essential attributes of EPS include protection of cells, cell–cell recognition, retention of water and nutrients, association with plants and genetic exchange.

Maturation of biofilm (stage 4): up to this stage the bacterial cells multiply to increase in population using AI signals, within the EPS matrix. This culminates in formation of microcolonies and further maturation of biofilms (Toyofuku et al., 2016). Also, the genes involved in EPS accumulation are induced once the level of AIs cross a certain threshold. 15 gene-long *epsA-O* cluster for biofilm formation in *Bacillus subtilis* is one such example. The alteration in expression of genes induced as a consequence of microcolony formation results in products that are further utilized for EPS production that works to bind the embedded bacterial cells within the matrix acting as "biological glue" (Fathima et al., 2023). Upon maturation, the cells are confined for movement within the microcolonies because gene expression of sessile cells significantly differs from planktonic cells leading to inhibition of formation of cell surface structures. AIs are crucial for maintenance of biofilms.

Dispersal/detachment (stage 5): this marks the termination of biofilm formation/development beyond which the bacterial cells move forward by leaving the existing biofilm to start a new cycle of biofilm formation (Liaqat et al., 2023). The complex phase of bacterial cell dispersal is regulated by factors such as environmental signals, signal transduction pathways, and effector molecules (Kaplan, 2010). Dispersal is achieved in three stages wherein the cells in the microcolony first detach (stage 1), followed by movement to a fresh substrate (stage 2) and finally adhesion to the new substrate (Kaplan, 2010). This stage witnesses the upregulation of genes for synthesis of motility structures like flagella and EPS degradation, while downregulation of genes involved in attachment, EPS (or polysaccharide) and fimbriae synthesis. Blocking c-di-GMP signaling pathways leading to reduction in intracellular c-di-GMP levels can effectively inhibit either biofilm development or facilitate its dispersal (Kaplan, 2010).

Fouling encompasses both biofouling as well as inorganic fouling. While inorganic fouling occurs due to depositions from suspended particles, oil, ice, corrosion and crystallization, biofouling is caused by agglomeration of unwanted biological material, which may be organic or inorganic. Formation of biofilm on a surface, followed by colonization by micro- and macro-organisms resulting in deterioration of surface is called biofouling. The extent and composition of fouling varies significantly with environment, organisms involved and inorganic deposits from surroundings. Depending on the thickness, density, bioadhesive strength, structure and weight of fouling organism, the appearance may be characterized. Since the causal organism can range from micro- to macro-organisms, the properties can be studied using various microscopic techniques like light, transmission electron (TEM), scanning electron (SEM), atomic force or fluorescence microscopy as well as weight scales. Both biofouling and inorganic fouling are influenced by surface factors including colors, contours, wettability, and microtexture. Depending on the surface properties, the fouling organisms may vary. For example, larvae of mussel and bryozoans colonize hydrophobic surfaces while barnacles prefer surfaces either with light color or convex contours. Variation in the extent of biofouling is also observed in susceptible fields like marine, medical, industry, etc. Due to the occurrence of biofouling these fields experience significant health risks and financial losses simultaneously. While biofouling in the medical field may be only due to the formation of biofilms, it is a combination of biofilm, inorganic fouling, and macro-fouling in marine and other industries.

9.7.2 Strategies for Prevention of Biofilm Formation or Antifouling: Prevention and Control Aspects

The economic losses incurred due to biofouling in various fields discussed above, necessitates strategies for control and prevention of biofouling. Such strategies may employ physical and chemical control or a combination of both, as suitable depending upon the complexity of fouling organism. As biofouling is a natural phenomenon, likewise nature offers several antifouling lessons or mechanisms that can be adapted from flora and fauna and mimicked for formulating effective mitigation techniques. Some of these primary mechanisms include altering surface wettability, low adhesion, low drag, grooming, microtexture, sloughing and chemical secretions (Bhushan, 2009). For example, shark skin exhibits antifouling properties due to its low drag, riblet microtexture, mucous layers, and flexion of scales. Mimicking the microtexture of shark skin on marine structures can prevent settling down of microorganisms as definite width and depths are preferred by certain microorganisms (Murthy et al., 2023). Similarly, low drag significantly decreases the time of microorganism settlement on surfaces and washes them away (Gizer et al., 2023).

Several other antifouling examples from nature harness surface wettability properties. These include low adhesion super-hydrophobic surfaces as found in leaves of lotus, broccoli, feathers of pigeon and wings of butterfly and superoleophobic surfaces of fish scales. Such surfaces withstand fouling as water enhances self-cleaning action (Parambath et al., 2023). While grooming involves physical removal of biofouling microorganisms from host surfaces, sloughing refers to the gradual dispersal of the outermost layer of an organism as in the case of seaweed, crustaceans, and stonefish. However, grooming can help control both fast and slow growing biofouling, only slow growing biofouling can be controlled by sloughing. Removal of fouling organisms from gills of crayfish by *Branchiobdellid annelids* by feeding on them is an example of grooming. Further, chemical secretions act by either preventing or removing biofouling organisms, as can be seen in bacteria *Roseobacter gallaeciensis*, which produces antifouling chemicals (Thirunavukkarasu et al., 2023), a red seaweed – *Delisea pulchra* manipulates the attraction message of colonizing bacteria by using halogenated furanone (Pinheiro et al., 2023). Such lessons from nature may serve as the basis for designing techniques for both physical and chemical control of biofouling. Antifouling coatings like tributyltin coatings used to control marine biofouling have deleterious effects on marine environment, paving way for diverting towards non-toxic antifouling and foul release coatings.

9.7.2 Biofouling Prevalence in Nature and Implementation for Ecosystem Improvement

9.7.2.1 Marine Environment

Biofouling is probably considered to be most prevalent in the marine environment. This is because water currents act as carriers of nutrients and cleaning agents of wastes creating an ideal condition for establishment by sessile and planktonic organisms. In marine environments, the initial biofilm formation is followed by attraction of macrofoulers (Sousa-Cardoso et al., 2023). Both micro and macrofoulers establish themselves and colonize ships hulls and ballast tanks, sonar devices, marinas, underwater acoustic instruments and cables, buoys, oil installations and pontoons. This results in reduced speed due to increased drag, which further increases the consumption of fuels. The acid producing bacteria in the biofilm lead to biological corrosion, whereas the slippery coating and structures formed by algal species create environment and safety concerns. Biofouling causes increased loading, which may damage fishing nets and sink buoys. In marine environment, biofouling is commonly caused by hard-shelled barnacles (*Tetraclitella purpurescens*), green algae (*Ulva australis*), and brown algae (*Ectocarpus*).

9.7.2.2 Medical Industry

Biofouling in the medical sector encompasses biofilm formation in orthopaedic, prosthetic and dental implants, dental, biosensors, catheters, contact lens and devices or equipments like respirator or haemodialysis machines used for diagnostic or support purposes. As a consequence, rejection of implants, dysfunctioning of biosensors and equipments, and spread of infectious diseases may occur (Latif et al., 2023). A major proportion of hospital-transmitted infections are attributed to biofilm formation on medical devices. Biofouling in medical devices or implants occurs due to adhesion of proteins or microorganisms, which often results in biocompatibility-related rejection, consequently leading to medical device failure. The rectification may either involve surgical replacement of the medical device that poses a threat for antibiotic resistance and even the life of the patient. However, biofilm formation in devices like haemodialysis machines with most advanced purification systems also, could not be prevented. Such infectious biofilms are lethal for sick patients and also difficult to disinfect and clean (Rowan et al., 2023).

9.7.2.3 Industries (Waste Water Treatment Plants)

Biofouling due to biofilm formation can be seen in industries such as dairy industry, nuclear or other power plants, water supplies, food production units and even membranes used for purification and desalination purposes. Similar to impediments posed due to biofilm formation in marine environment, in industries also biofilm formation results in more friction, decline in heat-transfer efficiency, pipe pressure and energy requirements (Latif et al., 2023). Biofouling in membranes is evident in bioreactors, reverse osmosis, and nano/microfiltration membranes as well as virus removal. This results due to deposition of organic and inorganic material together with microorganisms, which consequently reduces membrane flux and increases hydraulic resistance and osmotic pressure.

9.7.2.4 Crop Production/Agriculture/Aquaculture/Plant World

Aquaculture, also known as aquatic or underwater agriculture is the cultivation of aquatic plants and animals under controlled or semi-controlled conditions. It is considered to be the most potential and sustainable source for production of animal protein (Jin et al., 2023). According to the United Nation's Food and Agriculture Organization (FAO), it is the most rapidly growing animal-protein production industry/sector across the globe.

The framework of equipment required for cultivation of aquatic organisms mainly like marine shellfish and marine finfish includes a conglomeration of components, such as nets, cages ropes, and floats. The surfaces of these components, shells in shellfish culture are the target sites for settling down of fouling organisms like algae, tubeworms, barnacles, molluscs, and other organisms resulting in biofouling. This poses significant economic impacts on aquaculture production chain resulting from detrimental effects on framework as well as eradication of biofouling. Further, agriculture irrigation systems also witness biofilm formation subsequently leading to biofouling. This leads to clogging and reduced efficiency of irrigation systems (Oliver et al., 2014; Zhou et al., 2017) affecting gross food productivity (Ngan & Habimana, 2020).

ACKNOWLEDGEMENT

The authors gratefully acknowledge the Department of Biotechnology, Sharda School of Engineering and Technology andSharda School of Agricultural Sciences, Sharda University, Greater Noida, Uttar Pradesh, India.

REFERENCES

Aminian, A., Fathi, A., Gerami, M.H., Arsan, M., Forutan Mirhosseini, A., & Seyed Mohammad, T. (2022). Nanoparticles to overcome bacterial resistance in orthopedic and dental implants. *Nanomedicine Research Journal, 7*, pp.107–123.

Arnon, D.I., & Hoagland, D.R. (1940). Crop production in artificial culture solutions and in soils with special reference to factors influencing yields and absorption of inorganic nutrients. *Soil Science, 50*, pp.463–485.

Arshad, M.A., & Martin, S. (2002). Identifying critical limits for soil quality indicators in agro-ecosystems. *Agriculture, Ecosystems & Environment, 88*, pp.153–160.

Ashraf, M., Ahmad, R., Bhatti, A.S., Afzal, M., Sarwar, A., Maqsood, M.A., & Kanwal, S. (2010). Amelioration of salt stress in sugarcane (*Saccharum officinarum* L.) by supplying potassium and silicon in hydroponics. *Pedosphere, 20*, pp.153–162.

Bar-Cohen, Y. (2011). *Biomimetics: nature based innovation*. Boca Raton, FL: CRC Press.

Behtash, F., Fakhrghazi, H., Hajizadeh, H.S., Aghaee, A., & Pourmohammad, A. (2023). Effect of zinc and boron on yield, quality attributes and physiological traits of lettuce under hydroponic culture. *Pakistan Journal of Botany, 554*, pp.1267–1276.

Beitelshees, M., Hill, A., Jones, C.H., & Pfeifer, B.A. (2018). Phenotypic variation during biofilm formation: implications for anti-biofilm therapeutic design. *Materials, 11*, p.108.

Berne, C., Ducret, A., Hardy, G.G., & Brun, Y.V. (2015). Adhesins involved in attachment to abiotic surfaces by Gram-negative bacteria. *Microbial Biofilms, 3*, pp.163–199.

Bhushan, B. (2009). Biomimetics: lessons from nature–an overview. *Philosophical Transactions of the Royal Society A: Mathematical, Physical and Engineering Sciences, 367*, pp.1445–1486.

Brady, N.C., Weil, R.R., & Weil, R.R. (2008). The soil around us. *The nature and properties of soils.* Nyle C. Brady and ray R. Weil (Eds.) 14th edition. *13*, pp.662–710. Upper Saddle River, NJ: Prentice Hall.

Costa, O.Y., Raaijmakers, J.M., & Kuramae, E.E. (2018). Microbial extracellular polymeric substances: ecological function and impact on soil aggregation. *Frontiers in Microbiology, 9*, p.1636.

Dasgan, H.Y., Kacmaz, S., Arpaci, B.B., İkiz, B., & Gruda, N.S. (2023). Biofertilizers improve the leaf quality of hydroponically grown baby spinach (*Spinacia oleracea* L.). *Agronomy, 13*, p.575.

de Carvalho, R.O., Machado, M.B., Scherer, V.S., Fuentes, G.C., da Luz, C.A.S., & da Luz, M.L.G.S. (2015). Hydroponic lettuce production and minimally processed lettuce. *Agricultural Engineering International: CIGR Journal,* 2015, 290–293.

D'Imperio, M., Durante, M., Gonnella, M., Renna, M., Montesano, F.F., Parente, A., Mita, G., & Serio, F. (2022). Enhancing the nutritional value of *Portulaca oleracea* L. by using soilless agronomic biofortification with zinc. *Food Research International, 155*, p.111057.

Domingues, D.S., Takahashi, H.W., Camara, C.A., & Nixdorf, S.L. (2012). Automated system developed to control pH and concentration of nutrient solution evaluated in hydroponic lettuce production. *Computers and Electronics in Agriculture, 84*, pp.53–6.

Doran, J.W. (1996). Soil health and sustainability. *Advances in Agronomy, 56*, pp.1–54.

Fathima, A., Arafath, Y., Hassan, S., Prathiviraj, R., Kiran, G.S., & Selvin, J. (2023). Microbial biofilms: a persisting public health challenge. In *Understanding Microbial Biofilms,* Surajit Das, Neelam

Amit Kungwani (Eds.) pp.291–314. Elsevier Academic Press: Amsterdam, The Netherlands. ISBN 9780323999779.

Figueiredo, M.N.D., Martinez, H.E.P., Fontes, E.A.F., Milagres, C.D.C., & Silva, J.M.D. (2022). Agronomic biofortification with iodine in lettuce plants cultivated in floating hydroponic system. *Revista Ceres, 69*, pp.210–217.

Fortuna, K.J., Holtappels, D., Venneman, J., Baeyen, S., Vallino, M., Verwilt, P., Rediers, H., De Coninck, B., Maes, M., Van Vaerenbergh, J., & Lavigne, R. (2023). Back to the roots: *Agrobacterium*-specific phages show potential to disinfect nutrient solution from hydroponic greenhouses. *Applied and Environmental Microbiology*, *89*(4), pp.e00215–e00223.

French, S., Levy-Booth, D., Samarajeewa, A., Shannon, K.E., Smith, J., & Trevors, J.T. (2009). Elevated temperatures and carbon dioxide concentrations: effects on selected microbial activities in temperate agricultural soils. *World Journal of Microbiology and Biotechnology*, *25*(11), pp.1887–1900.

Friero, I., Larriba, E., Martínez-Melgarejo, P.A., Justamante, M.S., Alarcón, M.V., Albacete, A., Salguero, J., & Pérez-Pérez, J.M. (2023). Transcriptomic and hormonal analysis of the roots of maize seedlings grown hydroponically at low temperature. *Plant Science, 326*, p.111525.

Gao, W., He, D., Ji, F., Zhang, S., & Zheng, J. (2020). Effects of daily light integral and LED spectrum on growth and nutritional quality of hydroponic spinach. *Agronomy, 10*, p.1082.

Germano, R.P., Melito, S., Cacini, S., Carmassi, G., Leoni, F., Maggini, R., Montesano, F.F., Pardossi, A., & Massa, D. (2022). Sweet basil can be grown hydroponically at low phosphorus and high sodium chloride concentration: effect on plant and nutrient solution management. *Scientia Horticulturae, 304*, p.111324.

Gillani, S.A., Abbasi, R., Martinez, P., & Ahmad, R. (2023). Comparison of energy-use efficiency for lettuce plantation under nutrient film technique and deep-water culture hydroponic systems. *Procedia Computer Science, 217*, pp.11–19.

Gillespie, D.P., Papio, G., & Kubota, C. (2021). High nutrient concentrations of hydroponic solution can improve growth and nutrient uptake of spinach (*Spinacia oleracea* L.) *grown in acidic nutrient solution. HortScience, 56*, pp.687–694.

Gizer, G., Önal, U., Ram, M., & ŞAHİNER, N. (2023). Biofouling and mitigation methods: a review. *Biointerface Research in Applied Chemistry, 13*, 1–25.

Glanz, J. (1995). *Saving our soil: solutions for sustaining earth's vital resource*. Johnson Books.

Hartemink, A.E. (2006). Soil fertility decline: definitions and assessment. *Encyclopedia of Soil Science, 2*, pp.1618–1621.

Hazelton, P., & Murphy, B. (2016). *Interpreting soil test results: what do all the numbers mean?*. Australia: CSIRO Publishing.

Hellio, C., & Yebra, D. (2009). *Advances in marine antifouling coatings and technologies*. Boca Raton, FL: CRC Press.

Jayachandran, A., Shikha, J., Shikha, S., Poonam, M., Subhashree, S., Kuldeep, K., & Bhargav, K. (2022). Hydroponics: an art of soil less farming. *The Pharma Innovation Journal, 11*, pp.1049–1053.

Jin, S., Kong, Q., John, C.K., Wang, Z., Zhang, T., Li, X., Zhu, X., Li, J., Luo, Y., Qian, M., & Chen, F. (2023). Natural biota's contribution to cultured aquatic animals' growth in aquaculture cannot be ignored. *Aquaculture Research, 2023*, pp. 1–25.

Kaplan, J.Á. (2010). Biofilm dispersal: mechanisms, clinical implications, and potential therapeutic uses. *Journal of Dental Research, 89*, pp.205–218.

Karlen, D.L., Mausbach, M.J., Doran, J.W., Cline, R.G., Harris, R.F., & Schuman, G.E. (1997). Soil quality: a concept, definition, and framework for evaluation (a guest editorial). *Soil Science Society of America Journal, 61*, pp.4–10.

Kaur, T., Rana, K.L., Kour, D., Sheikh, I., Yadav, N., Kumar, V., Yadav, A.N., Dhaliwal, H.S., & Saxena, A.K. (2020). Microbe-mediated biofortification for micronutrients: present status and future challenges. In *New and future developments in microbial biotechnology and bioengineering* Ali Asghar Rastegari, Ajar Nath Yadav, Neelam Yadav (Eds.), pp.1–17. Amsterdam, The Netherlands: Elsevier.

Knop, W. (1865). Quantitative Untersuchungen uber die Ernah rungsprozesse der Pflanze. *Die Landwirtschaftlichen Versuchs-Stationen, 7*, pp.93–107.

Lamret, F., Lemaire, A., Lagoutte, M., Varin-Simon, J., Abraham, L., Colin, M., Braux, J., Velard, F., Gangloff, S.C., & Reffuveille, F. (2023). Approaching prosthesis infection environment: development of an innovative *in vitro Staphylococcus aureus* biofilm model. *Biofilm, 5*, p.100120.

Latif, S., Imran, M., Siddique, M.H., & Hussain, N. (2023). Biofouling: current status and challenges. *Advances in Nanotechnology for Marine Antifouling, 2023*, pp.1–13.

Lehmann, J., Pereira da Silva, J., Steiner, C., Nehls, T., Zech, W., & Glaser, B. (2003). Nutrient availability and leaching in an archaeological Anthrosol and a Ferralsol of the Central Amazon basin: fertilizer, manure and charcoal amendments. *Plant and Soil, 249*, pp.343–357.

Liaqat, M.A., Mahmood, M.S., Akram, M.N., Arshad, M.I., Aslam, R., Khan, M.K., Khan, M.U.Z., Fatima, N., & Ali, S. (2023). Bacterial biofilms and their regulatory approaches. *One Health Triad, Unique Scientific Publishers, Faisalabad, Pakistan, 2*, pp.8–17.

Maharana, L., & Koul, D.N. (2011). The emergence of Hydroponics. *Yojana, 55*, pp.39–40.

Malik, A.M., Mughal, K.M., Mian, S.A., & Khan, M.A.U. (2018). Hydroponic tomato production and productivity improvement in Pakistan. *Pakistan Journal of Agriculture Research, 31*, pp.133–144.

Maluin, F.N., Hussein, M.Z., Nik Ibrahim, N.N.L., Wayayok, A., & Hashim, N. (2021). Some emerging opportunities of nanotechnology development for soilless and microgreen farming. *Agronomy, 11*, p.1213.

Mamatha, V., & Kavitha, J.C. (2023). Machine learning based crop growth management in greenhouse environment using hydroponics farming techniques. *Measurement: Sensors, 25*, p.100665.

Maneejantra, N., Tsukagoshi, S., Lu, N., Supoaibulwatana, K., Takagaki, M., & Yamori, W. (2016). A quantitative analysis of nutrient requirements for hydroponic spinach (*Spinacia oleracea* L.) production under artificial light in a plant factory. *Journal of Fertilizers & Pesticides*, *7*(2). https://newatlas.com/aerofarms-urban-agriculture/15371/

Marginson, S. (2010). *Aerofarms urban agriculture system: less space, less water, and no pesticides*. www.gizmag.com/aerofarms-urban-agriculture/15371/

Maucieri, C., Nicoletto, C., Junge, R., Schmautz, Z., Sambo, P., & Borin, M. (2018). Hydroponic systems and water management in aquaponics: a review. *Italian Journal of Agronomy, 13*, pp.1–11.

Mengel, K., Kirkby, E.A., Kosegarten, H., & Appel, T. (2001). The soil as a plant nutrient medium. In Mengel, K., Kirkby, E.A., Kosegarten, H., Appel, T. (eds). *Principles of plant nutrition*, pp.15–110. Springer, Dordrecht.

Murthy, P.S., Nanchariah, Y.V., Venkatnarayanan, S., Das, A., & Mohan, T.K. (2023). Antifouling mechanisms in and beyond nature: leverages in realization of bioinspired biomimetic antifouling coatings. In *Advances in nanotechnology for marine antifouling* Ram K. Gupta, Ashok Kumar Nadda, Tuan Anh Nguyen, Swati Sharma, Muhammad Bilal (Eds.), pp.329–362. Amsterdam, The Netherlands: Elsevier.

Naz, T., Iqbal, M.M., Akhtar, J., & Saqib, M. (2023). Baseline hydroponic study for biofortification of bread wheat genotypes with iron and zinc under salinity: growth, ionic, physiological and biochemical adjustments. *Journal of Plant Nutrition, 46*, pp.743–764.

Ng, J.F., Ahmed, O.H., Jalloh, M.B., Omar, L., Kwan, Y.M., Musah, A.A., & Poong, K.H. (2022). Soil nutrient retention and ph buffering capacity are enhanced by calciprill and sodium silicate. *Agronomy, 12*, p.219.

Ngan, W.Y., & Habimana, O. (2020). From farm-scale to lab-scale: the characterization of engineered irrigation water distribution system biofilm models using an artificial freshwater source. *Science of The Total Environment, 698*, p.134025.

Nguyen, N.T., McInturf, S.A., & Mendoza-Cózatl, D.G. (2016). Hydroponics: a versatile system to study nutrient allocation and plant responses to nutrient availability and exposure to toxic elements. *Journal of Visualized Experiments (JoVE), 113*, p.e54317.

Oliver, M.M.H., Hewa, G.A., & Pezzaniti, D. (2014). Bio-fouling of subsurface type drip emitters applying reclaimed water under medium soil thermal variation. *Agricultural Water Management, 133*, pp.12–23.

Orji, U.E., & Obasi, S.N. (2012). Properties and classification of erosion prone soils of Ukpor, Nnewi South LGA, Anambra State, Nigeria. *International Journal of Agriculture and Rural Development, 15*, pp.1079–1084.

Parambath, J.B., Arooj, M., & Mohamed, A.A. (2023). Biomedical antifouling polymer nanocomposites. In *Advances in biomedical polymers and composites* Kunal Pal, Sarika Verma, Pallab Datta, Ananya Barui, S.A.R. Hashmi, Avanish Kumar Srivastava (Eds.), pp.563–589. Amsterdam, The Netherlands: Elsevier.

Pinheiro, J., Lyons, T., Heras, V.L., Recio, M.V., Gahan, C.G., & O'Sullivan, T.P. (2023). Investigation of halogenated furanones as inhibitors of quorum sensing-regulated bioluminescence in *Vibrio harveyi*. *Future Medicinal Chemistry, 15*, pp.317–332.

Priyadarshanee, M., & Das, S. (2023). Bacterial extracellular polymeric substances: biosynthesis and interaction with environmental pollutants. *Chemosphere, 332*, p.138876.

Railkin, A.I. (2004). *Marine biofouling colonization processes and defenses*. Boca Raton, FL: CRC Press.

Ramirez-Arias, J.A., Hernández-Ibarra, U., Pineda-Pineda, J., & Fitz-Rodríguez, E. (2017). Horizontal and vertical hydroponic systems for strawberry production at high densities. In *International Symposium on New Technologies for Environment Control, Energy-Saving and Crop Production in Greenhouse and Plant* Qichang Yang and Tao Li (Eds.), 1227, pp.331–338.

Rayment, G.E., & Higginson, F.R. (1992). *Australian laboratory handbook of soil and water chemical methods*. Australia: Inkata Press.

Rengasamy, P., & Churchman, G.J. (1999). Cation exchange capacity, exchangeable cations and sodicity. In Peverill, K.I., Sparrow, L.A. and Reuter, D.J. (eds). *Soil analysis an interpretation manual*. CSIRO: Melbourne.

Roba, T.B. (2018). Review on: The effect of mixing organic and inorganic fertilizer on productivity and soil fertility. *Open Access Library Journal, 5*(06), p.1.

Rosa-Rodríguez, R.D.L., Lara-Herrera, A., Trejo-Téllez, L.I., Padilla-Bernal, L.E., Solis-Sánchez, L.O., & Ortiz-Rodríguez, J.M. (2020). Water and fertilizers use efficiency in two hydroponic systems for tomato production. *Horticultura Brasileira, 38*, pp.47–52.

Rowan, N.J., Kremer, T., & McDonnell, G. (2023). A review of Spaulding's classification system for effective cleaning, disinfection and sterilization of reusable medical devices: viewed through a modern-day lens that will inform and enable future sustainability. *Science of the Total Environment, 878*, p.162976.

Saini, D.K., Devi, P., & Kaushik, P. (2020). Advances in genomic interventions for wheat biofortification: a review. *Agronomy, 10*, p.62.

Sathyanarayan, S.R., Warke, V.G., Mahajan, G.B., & Annapure, U.S. (2023). Soil free nutrient availability to plants. *Journal of Plant Nutrition, 46*, pp.801–814.

Somerville, C., Cohen, M., Pantanella, E., Stankus, A., & Lovatelli, A. (2014). Small-scale aquaponic food production: integrated fish and plant farming. *FAO Fisheries and Aquaculture Technical Paper, 589*, p.I.

Sousa-Cardoso, F., Teixeira-Santos, R., Campos, A.F., Lima, M., Gomes, L.C., Soares, O.S., & Mergulhão, F.J. (2023). Graphene-based coating to mitigate biofilm development in maeine Environments. *Nanomaterials, 13*, p.381.

Sularz, O., Smoleń, S., Koronowicz, A., Kowalska, I., & Leszczyńska, T. (2020). Chemical composition of lettuce (*Lactuca sativa* L.) biofortified with iodine by KIO3, 5-Iodo-, and 3.5-diiodosalicylic acid in a hydroponic cultivation. *Agronomy, 10*, p.1022.

Suma, A., Francies, R.M., Radhakrishnan, V.V., Sureshkumar, P., & Varghese, E. (2022). Elucidating the response of rice genotypes to iron toxicity through hydroponics. *Indian Journal of Plant Genetic Resources, 35*, pp.66–72.

Tan, J., Jiang, H., Li, Y., He, R., Liu, K., Chen, Y., He, X., Liu, X., & Liu, H. (2023). Growth, phytochemicals, and antioxidant activity of kale grown under different nutrient-solution depths in hydroponic. *Horticulturae, 9*, p.53.

Thirunavukkarasu, R., Pushparaj, S., Arokiarajan, M.S., Pandi, P., Subramanian, K., & Joseph, J. (2023). Microbial symbionts from Algae. *Microbial Symbionts, 1*, p.311.

Tiwari, H., & Prajapati, S.K. (2023). Allelopathic effect of benzoic acid (hydroponics root exudate) on microalgae growth. *Environmental Research, 219*, p.115020.

Tong, R.C., Whitehead, C.S., & Fawole, O.A. (2021). Effects of conventional and bokashi hydroponics on vegetative growth, yield and quality attributes of bell peppers. *Plants, 10*, p.1281.

Toyofuku, M., Inaba, T., Kiyokawa, T., Obana, N., Yawata, Y., & Nomura, N. (2016). Environmental factors that shape biofilm formation. *Bioscience, Biotechnology, and Biochemistry, 80*, pp.7–12.

Uddin, M.R., & Suliaman, M.F. (2021). Energy efficient smart indoor fogponics farming system. In *IOP Conference Series: Earth and Environmental Science, 673*(1), p.012012. Glassfields, Avon Street, Bristol, BS2 0GR, UK: IOP Publishing.

van Reeuwijk, L.P. (ed.) (2002). *Procedures for Soil Analysis*. 6th edition. Technical Paper/International Soil Reference and Information Centre. Wageningen, The Netherlands.

Velazquez-Gonzalez, R.S., Garcia-Garcia, A.L., Ventura-Zapata, E., Barceinas-Sanchez, J.D.O., & Sosa-Savedra, J.C. (2022). A review on hydroponics and the technologies associated for medium-and small-scale operations. *Agriculture, 12*, p.646.

Vigneshwaran, R., Baskar, K., Mahendran, P.P., Murugan, E., & Ragavan, T. (2022). *Screening of saline tolerant sorghum genotypes by hydroponics method,* The *Pharma Innovation Journal, 11*(2): 2243–2249.

Vo-Dinh, T. (ed.) (2007). *Nanotechnology in biology and medicine*. Boca Raton, FL: CRC Press.

Wahl, M. (1989). Marine epibiosis. I. Fouling & antifouling: some basic aspects. *Marine Ecology Progress Series, 58*, pp.175–189. doi:10.3354/meps058175.

Walker, J., Surman, S., & Jass, J. (2000). *Industrial biofouling detection, prevention and control.* New York, NY: Wiley.

Watson, F., Swann, M., Zimmer, A., Chen, R., Foulkes, L., & Percival, S.L. (2023). Aspects of biofilms on medical devices. In *Understanding microbial biofilms,* Surajit Das, Neelam Amit Kungwani (Eds.) pp.91–105. Amsterdam, The Netherlands: Academic Press.

Xie, B., Xiao, X., Li, H., Wei, S., Li, J., Gao, Y., & Yu, J. (2023). Moderate salinity of nutrient solution improved the nutritional quality and flavor of hydroponic chinese chives (*Allium tuberosum* Rottler). *Foods, 12*, p.204.

Yin, W., Wang, Y., Liu, L., & He, J. (2019). Biofilms: the microbial "protective clothing" in extreme environments. *International Journal of Molecular Sciences, 20*, p.3423.

Zhao, A., Sun, J., & Liu, Y. (2023). Understanding bacterial biofilms: from definition to treatment strategies. *Frontiers in Cellular and Infection Microbiology, 13*, 1–25.

Zhou, B., Wang, T., Li, Y., & Bralts, V. (2017). Effects of microbial community variation on bio-clogging in drip irrigation emitters using reclaimed water. *Agricultural Water Management, 194*, pp.139–149.

Zhu, Z., Li, D., Cong, L., & Lu, X. (2021). Identification of microRNAs involved in crosstalk between nitrogen, phosphorus and potassium under multiple nutrient deficiency in sorghum. *The Crop Journal, 9*, pp.465–475.

Zhu, Z., Li, D., Wang, P., Li, J., & Lu, X. (2020). Transcriptome and ionome analysis of nitrogen, phosphorus and potassium interactions in sorghum seedlings. *Theoretical and Experimental Plant Physiology, 32*, pp.271–285.

10 Emerging Research in Nanotechnology, Metagenomics, and Other Omics-Based Technologies for Rhizosphere Management and Increased Plant Growth and Productivity

Priyanka Yadav, Praveen Kumar Oraon*, Garima Malik, Surendra Pratap Singh and Astha Gupta*
* Equal contribution.

10.1 INTRODUCTION

According to the World Population Prospects 2022, it is projected that the global population will reach 8.5 billion by 2030 and 9.7 billion by 2050 (World Population Prospects 2022). The growing world population will always drive the demand for agricultural growth. Agricultural demands depend on various factors such as income growth projection, cultural preferences in diets and nutrition and differences in income levels, which eventually defines the patterns of consumption between countries. However, the Global Report on Food Crises (GRFC 2022) warned about 193 million people falling into IPC/CH (Integrated Food Security Phase Classification/Cadre Harmonisé) Phase 3 or above, which indicates a worse situation of crisis and a total of 53 countries/territories across the globe fall into crisis zones. Major players, which lead to such crisis conditions, are extreme weather, economic shock due to situations like COVID-19 and food security. Along with this, plant stress, i.e., biotic, and abiotic can severely affect plant growth, yield and survival. Therefore, to feed the ever-expanding world population with different patterns of diet and consumption along with minimizing such crisis situations, it is imperative to enhance food production. Increasing food production can be targeted via rhizosphere management in a planned manner, which can effectively enhance crop yield and agricultural outputs (Fasusi et al., 2021). Significant advancements in molecular biology and the advent of next-generation sequencing technologies have led to advancements in nanotechnology, metagenomics, meta-transcriptomics, and meta-proteomics (Knief 2014). Thereby, opening the window in plant rhizosphere research to increase plant growth and productivity would be helpful. Rhizosphere originates from the Greek word "rhiza" meaning roots is an area/region around the root at 1–3 mm where microbiota resides acting

DOI: 10.1201/9781003354109-10

as an extension of roots, enabling the plants for better utilization of nutrients available in the soil, establishing the rhizosphere as one of the most dynamic interfaces on the planet (Hartmann et al., 2008). Dynamic alterations in the physical, chemical, and biological characteristics of rhizospheric soil significantly influence plant growth. Colonization of microbial population in plant rhizosphere results in plant microbe and microbe–microbe interplay, which influences the process, such as seed germination, plant growth, disease resistance, and productivity. Majorly residing microbes in the rhizosphere are fungi, bacteria, nematodes, algae, and worms (Bonkowski et al., 2009). Interaction between plants and microbes depends on numerous factors such as climatic factors, type of soil, root exudates, plant species, developmental stages, and root architecture (Dastogeer et al., 2020). Plants provide shelter to numerous microbes during the entire course of their lifecycle and microbes inhabiting in the rhizosphere can be good or bad. Plants are benefited via direct or indirect modes of association with microbes by providing available nutrients and minerals present in the soil acting as extensions of roots and through the secretion of phytohormones. Plant growth-promoting rhizobacteria (PGPR) are a group of microbes borne and found in the rhizosphere known to play a role in critical processes such as improvement of soil fertility, nitrogen fixing, antagonistic microbial effects, absorption of heavy metals, enhancing chlorophyll content, carbon assimilation, and photosynthetic efficiency (Hassan et al., 2019). Improvement in sustainable agriculture methods using biological control is advancing due to the role of microbes in plant defense, abiotic stresses, pests, and plant growth, which is becoming evident and established with more and more studies (Bhardwaj et al., 2014; Rahman et al., 2018). Products formed due to interaction between plants and microbes and the microbiota niche present in the soil are being utilized as biofertilizers, biopesticides, and biostimulators to gain better insights into the rhizosphere biology to understand, utilize, exploit, and manage the rhizosphere in a planned manner to take advantage in the agricultural sector to enhance food production and security (Trivedi et al., 2021). Metagenomic is one of the approaches for gaining insights into the global picture for studying the plant microbe and microbe–microbe interaction belonging to different geographical locations, different plant species, existing microbial genetic diversity, and identification of novel genes metabolic pathways, which can also be utilized in industrial sectors (Rosselli et al., 2015). Microbiology has a long history starting from way back to 1676 when Leeuwenhoek first reported oral microbiota. In 1888, Rober Koch isolated microorganisms. In the year 1977, Carl Woes suggested rRNA as a suitable marker for taxonomy-based studies and Frederick Sanger developed DNA sequencing chemistry. Giovannoni et al., (1996) conducted and reported the first study on microbes at the community level utilizing 16s rRNA libraries. Handelsman coined the term metagenomics in 1998 and since then, NGS technology came into light with subsequent discoveries and improvements. In 2005, the first NGS machine by Roche came into light followed by Illumina, PacBio, Genomics10X, Nanopore. These advancements in NGS have made understanding complex microbial communities possible. Metagenomic approaches are mainly opted to explore the species diversity and richness (Fadiji et al., 2020). The number of species present in a particular sample is referred to as species richness whereas species differential abundance is the quantification of species unevenness present in a sample. Species richness is the measurement of different species that exist in a given ecosystem. For a single ecological community or micro-habitat, it can be used as a measure of species diversity (Brown et al., 2007) but it cannot be solely used as a parameter to describe species diversity in a particular microbial community since, if two communities having the same number of species can differ in their differential abundance of species then the community with lower abundance will be considered as more diverse than the other (Ramette et al., 2005). To study and compare microbial communities in a more comprehensive manner, few metrics are also used in metagenomics, i.e., alpha (α), which is the measure of local diversity of a community whereas gamma (γ) measures regional diversity involving multiple community's beta metrics measures and links how alpha, and beta (β) communities are present in an area (Ghosh et al., 2021; Dhondge et al., 2022). Statistical approaches have also been opted to study species diversity in a quantitative manner and non-parametric estimators are used as they consider low abundance species present in microbial communities. Two mostly used non-parametric estimators used in metagenomic studies are Simpson's diversity index (*D*), which measures the probability of assigning two independent individuals sampled randomly from the microbial community into the same species. Shannon–Wiener function (Shannon–Weaver species diversity) is the measurement of entropy and it increases with increase in number of species in a sample (Yousuf et al., 2012). Metagenomic studies have shown the microbial diversity estimated using these indices, which have further demonstrated that molecular markers can be used to compare the diversity of microbial communities (Soni et al., 2017). However, there are instances where species diversity estimators were unsuccessful in complex microbial communities. Metagenomics takes care of culture-dependent (i.e., can be grown under laboratory conditions) and independent (i.e., cannot be grown under laboratory conditions) samples using two approaches: amplicon sequencing and whole genome sequencing (Tracanna et al., 2021). By employing these two approaches, we can delve into the exploration and identification of new genes, proteins and genomes of non-culturable species in a more efficient and precise manner compared to traditional microbiology methods. Metagenomics offers valuable insights by providing information on the genome and the state and composition of the microbial community during the time of sample collection, offering a snapshot of the microbial ecosystem (Vieira et al., 2021). However, when we only deal with the genic/functional counterpart of a microbial community from collected samples and study the active transcripts within and between microbial communities, they are referred as meta-transcriptome (Kumar et al., 2019). Similarly, when dealing with complete sets of protein, they are known as metaproteome. A study of all the metabolites from a particular site or ecosystem comes under the field of metabolome (Mhlongo et al., 2018). Metaepigenome deals with epigenetic modification in the genome known to regulate expression of genes (Allen et al., 2017). Similarly, when dealing with modifications occurring in proteins such as post-translational modifications in protein (phosphorylation) comes under metaepiproteome (Ijaz et al., 2021). All the studies combined fall under the umbrella of the omics-based approach, which aims to comprehend the intricate nature of the entire rhizosphere as a system biology (Pinu et al., 2019). This chapter deals with the recent advancement of emerging technologies in the field of nanotechnology, microbiome biology, and their significance in enhancing plant growth and productivity.

10.2 NANOTECHNOLOGY: THE BOON FOR PLANT GROWTH AND DEVELOPMENT

In the quest to meet the increasing demands of growing global population, the responsibility falls on the shoulders of plant scientists and breeders to seek innovative methods to boost crop production. Classical/conventional breeding, molecular breeding, and biotechnology sciences have been helping in crop production improvement since long ago but, will nanotechnology be able to contribute to this endeavor? Will the emergence of "nanobiotechnology" prove to be a game-changer for plant growth and development? This section aims to shed light on these inquiries and explore the potential of nanobiotechnology as a catalyst for enhancing crop production.

As per the standards set by the American Society for Testing and Materials (ASTM), nanomaterials (NMs) are defined as natural or manufactured materials with sizes ranging from 1 to 100 nm (Astm E2456 - 06, 2012). Nanoparticles possess distinctive physicochemical properties characterized by their small size and a high surface-to-volume ratio (Roduner 2006). The field of nanotechnology is gaining significant attention in plant science due to the growing demand for an efficient, miniaturized system that can enhance seed germination, growth, and provide protection against abiotic and biotic stresses, thereby promoting improved plant growth and development (Wang et al., 2014). Agriculture has been a vital and stable sector, providing essential raw materials for food and various industries. However, the limitation imposed by scarce natural resources, such as land, water, and soil, coupled with a burgeoning population, necessitates the development of agriculture in an economically and environmentally sustainable manner. To address these challenges, innovative techniques like nanotechnology have been embraced on agriculture, giving rise to the field of agronanotechnology. This collaborative approach focuses on leveraging the potential of nanotechnology for the growth and development of plants, aiming to achieve sustainable agricultural practices that can combat poverty, hunger, and drive economic growth.

10.2.1 Types of Nanoparticles

Based on their origin there are two types of nanoparticles:

(a) Natural nanoparticles
  Naturally occurring particles that existed even before the beginning of life on earth. For instance-soil organic colloids, polysaccharides, humic materials and peptidoglycan, clay, ocean salt, silicate clay minerals, and sesquioxides.
(b) Anthropogenic/man-made nanoparticles
  i. Engineered nanoparticles: these are prepared by human beings, and they do possess specific physicochemical properties.
  ii. Accidental nanoparticles: these are produced accidentally during the manufacturing and several anthropogenic processes, such as electricity generation, diesel-burning, welding, etc. (Kah et al., 2019).

Based on their components there are three types of nanoparticles:

(a) Organic: organic nanoparticles exhibit desirable properties such as biodegradability, non-toxicity, and sensitivity to thermal and electromagnetic radiations, including heat and light particles. Examples: micelles, ferritin, and liposomes.
(b) Inorganic: these particles are made of metals and metal oxides (Shanmugam et al., 2020).
  i. Metal-based: metal-based materials are produced to nanometric sizes from metals, either destructively or creatively. Commonly employed metals for nanoparticle synthesis include Fe, Cd, Ag, Co, Pb, Au, Al, and Zn (Salavati et al., 2008).
  ii. Metal oxide-based: these nanoparticles are designed to modify and enhance the properties of their metal counterparts, often leading to increased reactivity and efficiency. Some commonly synthesized nanomaterials in these categories include aluminium oxide, zinc oxide, silicon dioxide, iron oxide, and cerium oxide (Tai et al., 2007).
(c) Carbon-based nanoparticles: they are exclusively formed from carbon atoms and can be classified into several categories, including fullerenes, graphene, carbon nanotubes, carbon nanofibers, and carbon black (Liu et al., 2020).

10.2.2 Effects of Nanoparticles on Plant Growth and Development

The interaction between nanoparticles and plants can induce various morphological and physiological changes, which are influenced by the nanoparticle's properties, such as chemical composition, size, surface coating, reactivity, and most importantly, the dosage at which they are applied (Mirkasymov et al., 2021). Researchers have observed both positive and negative effects of engineered nanoparticles (ENPs) in plant growth and development. The impact of ENPs on plants is dependent on factors such as composition, concentration, size, and physical and chemical properties of the nanoparticles, as well as specific to plant species (Ma et al., 2010). The efficacy of nanoparticles varies depending on their concentration and differs among plant species. However, nanoparticles have been found to influence seed germination, root development, plant growth, and photosynthesis. The following studies provide insights into the impact of nanoparticles (NPs) on plant growth and development.

10.3 IMPROVED PLANT GROWTH UNDER BIOTIC STRESS: NANOPESTICIDES

Two important biotic stresses are pathogen infection and herbivore attacks. The utilization of conventional pesticides has contributed to the improvement of agricultural production and overall productivity, making it one of the primary factors influencing crop yield. On the other side, it was also responsible for the ill effects on the environment and health. According to researchers "nanopesticides" are defined as any formulation of pesticide or product in which the active ingredients are engineered nanomaterials possessing biocidal properties, either as a whole or as a component of the engineered structure (Adisa et al., 2019). AgNP, CuNPs, and AuNPs, are several NPs, that possess antibacterial and pest-control functions (Gogos et al., 2012; Cao et al., 2018). Generally, polymers are also used in the nanoencapsulated pesticide formulations along with nanopesticides. Some other excellent publications (Servin et al., 2015; Kah et al., 2014) are also available on nanopesticides. Based on previous studies, metal nanomaterials such as CuNPs, AgNPs, MgNPs, CeNPs, and even carbon nanoparticles and silicon-based nanoparticles shield plants from the attack of insects, pests, and pathogens.

10.3.1 Copper (Cu) Nanoparticles

The roles of Cu ions are well established for antibacterial as well as antifungal action. The commercial pesticide Kocide 3000 has Cu oxide NPs as an active ingredient. *Escherichia coli* and *Bacillus subtilis* are the main bacterial targets and *A. alternata*, *F. oxysporum*, and *C. lunata* are the most common targets of CuNPs (Kanhed et al., 2014; Yoon et al., 2007). Borgatta et al., (2018) conducted a study on the fungal root disease caused by *Fusarium oxysporum f.* sp. *niveum* in watermelon (*Citrullus lanatus*). They compared copper oxide NPs and nanosheets of $Cu_3(PO_4)_2{\cdot}3H_2O$ ability to repress the caused infection. Greenhouse experiment, nanosheets of $Cu_3(PO_4)_2{\cdot}3H_2O$. (10 mg/L) was able to control the fungal infection significantly (58% decrease) in disease spread while CuONPs were significantly effective at higher concentration (1000 mg/L). This has indicated that dissolution profile of NPs defines their performance. Recently, Cumplido-Nájera (2019) investigated the effects of CuNPs and K_2SiO_3NPs on tomato plants that were cultivated hydroponically. Plants were exposed to the *Clavibacter michiganensis* pathogen and different concentrations and a combination of NPs was applied. For instance 50 mg/L CuNPs + 184 mg/L Si; 50 mg/L CuNPs + 460 mg/L Si; 250 mg/L CuNPs + 184 mg/L Si; and 250 mg/L CuNPs + 460 mg/L Si were applied. The conclusions of the study indicated that both NPs are efficient in lowering the effect of infection of *C. michiganensis*.

Furthermore, the application of NPs in combination was observed to lead to an increase in the levels of both non-enzymatic and enzymatic metabolites, which play a crucial role in plant defence mechanisms. Van et al. (2016) concluded that copper oxide NPs exhibit insecticidal effects on conventional cotton and Bt-transgenic cotton (Figure 10.1).

10.3.2 Silver (Ag) NPs Nanoparticles

The increasing number of chemical pesticides to which pests and fungi have become resistant has emphasized the requirement for new ways of crop protection. AgNPs are considered as potential nanopesticide because of the antibacterial activities at wide scale. Ocsoy et al. (2013) synthesized DNA-directed AgNPs grown on graphene oxide; Ag@dsDNA@GO DNA-directed silver (Ag)

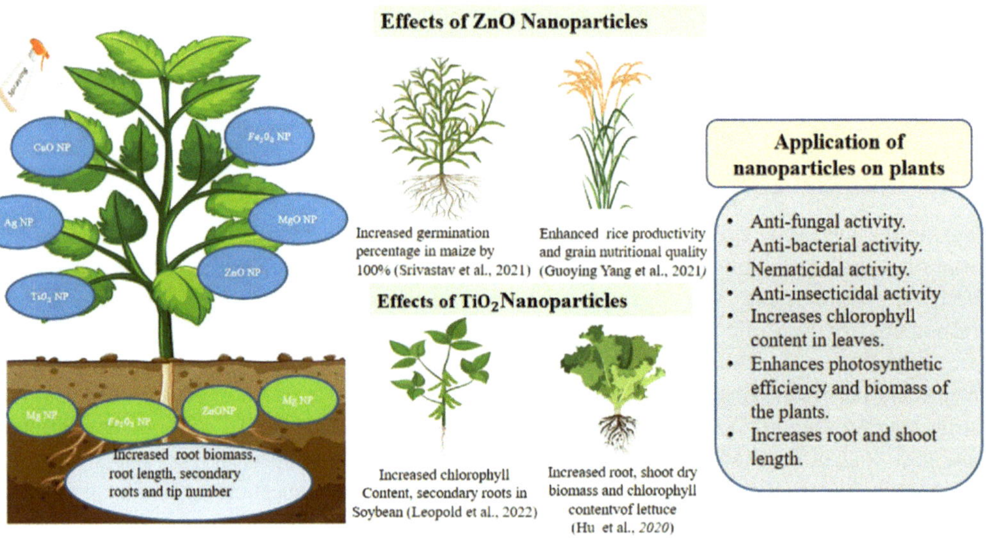

FIGURE 10.1 Generalized application of nanoparticles on plant growth, productivity, and microbial communities.

nanoparticles (NPs) grown on graphene oxide (GO). They found that at 16 mg/L concentration of Ag@dsDNA@GO the infection of *Xanthomonas perforans* which infects tomatoes with bacterial spots was decreased significantly. Moreover, AgNPs proved to be potential against nematodes as well (Ahn et al., 2014). Cromwell et al. (2014) showed that AgNPs (30–150 mg/mL) was successful in reducing 99% of *Meloidogyne* spp. within 6 hr. In a field experiment on day 2 and day 4, AgNPs at 150 mg/mL were observed to reduce the number of nematodes by 82% and 92%, respectively.

Green-synthesized AgNPs are environment friendly as compared to chemical pesticides. Various plant or bacterial extracts have been identified as effective capping and reducing agents in the fabrication of AgNPs. Mishra et al. (2014) utilized *Serratia* sp., a plant-growth-promoting rhizobacterium, to biosynthesize AgNPs with remarkable antifungal activity against *Bipolaris sorokiniana*, which infects wheat with spot blotch. Narayanan and Park (2014) employed turnip leaf extract to green-synthesize AgNPs with a size of approximately 16 nm and a zeta potential of −14.8 mV. These AgNPs exhibited a broad spectrum of antifungal activity against fungi (wood-degrading) such as *Phanerochaete sordida, Gloeophyllum abietinum*, and *Chaetomium globosum. Artemisia absinthium* extract was utilized for the environmentally friendly production of AgNPs. When these AgNPs, synthesized through green synthesis, were applied to plants through foliar application at a concentration of 10 mg/L, they demonstrated effective suppression of infection (Ali et al., 2015). Additionally, the application of these AgNPs resulted in improved plant growth and development, showcasing their potential as a promising approach for plant disease management and enhancing overall plant health.

It is also necessary to address concerns regarding AgNPs potential toxicity to plants. Zhang et al. (2018) conducted a study to evaluate the toxicity of commercial AgNPs at both molecular and physiological levels. Their findings revealed that foliar application of AgNPs at a concentration of 0.4 mg per plant induced oxidative stress in cucumber leaves. It is important to conduct further investigations to determine the concentration response of plants to AgNPs and assess their biological impact. These studies will help better understanding the potential risks and benefits of AgNPs and enable their safe and effective utilization in plant-related applications.

10.3.3 Carbon (C) Nanoparticles

In their study, Wang et al. (2014) examined the antifungal behavior of various carbon based nanomaterials (62.5, 125, 250, and 500 mg/L) against two highly pathogenic fungi, *Fusarium graminearum* and *Fusarium poae*. The antifungal tested included single-walled carbon nanotubes (SWCNTs), multiwalled carbon nanotubes (MWCNTs), graphene oxide (GO), reduced graphene oxide (rGO), fullerene (C60), and activated carbon (AC). The fungi were coincubated with the nanoparticles for 5 hrs and 12 hrs away from light. Results showed that SWCNTs exhibited the strongest antifungal activity at a concentration of 500 mg/L, followed by MWCNTs, GO, and rGO. On the other hand, activated carbon at the tested concentration range did not show any antifungal activity. Spore germination of *F. graminearum* was inhibited by C60 at 500 mg/L but the same was not observed with *F. poae*. CNMs inhibit water uptake and cause plasmolysis in pathogens, which makes them successful antifungal agents.

10.3.4 Titanium (Ti) Nanoparticles

Antimicrobial properties of photochemically active TiO_2NPs makes them a suitable material for nanopesticides to improve plant growth and productivity. Paret et al. (2013) found out *X. perforans* induces spot disease in tomato plants and photocatalytic TiO_2NPs was observed to show the antibacterial activity against *X. perforans*. In controlled greenhouse conditions, TiO_2 significantly lowered down the spot infection at concentration of 500–800 mg/L in comparison to untreated and Cu-treated controls.

10.3.5 Magnesium (Mg) Nanoparticles

Imada et al. (2016) showed the MgO NP effects on tomato plants that conferred resistance against *Ralstonia solanacearum*. Pretreatment of tomato roots with MgO nanoparticles (NPs) at various concentrations (0.1%, 0.5%, 0.7%, or 1%) resulted in a significant reduction in the penetration level of infection. The observed decrease in infection level could be attributed to the increased concentration of salicylic acid, jasmonic acid, and ethylene. Cai et al. (2018) demonstrated that the application of MgO NPs at concentrations of or 250 mg/L effectively suppressed the growth of *R. solanacearum*, the causative agent of bacterial wilt in tobacco. Huang et al. (2018) compared the antibacterial activities of synthesized $Mg(OH)_2$ NPs with the commercial pesticide Kocide 3000, which contains $Cu(OH)_2$ as an active ingredient. The study found that $Mg(OH)_2$ NPs effectively inhibited the growth of bacteria such as *X. alfalfae*, *Pseudomonas syringae*, and *E. coli* within a 4 hr period. The efficiency of magnesium hydroxide NPs was in equal proportion with Kocide 3000, which indicates the replacement of toxic chemicals.

10.4 IMPROVED PLANT GROWTH THROUGH NUTRIENT TRANSFER: NANOFERTILIZERS

Micronutrients play a crucial role in the growth, development, and defense mechanisms of plants against pests and diseases. Servin et al. (2015) suggested that directly killing of pathogens to increase stress tolerance has become an outdated strategy, instead we need to upgrade the nutritional level of plants to fight against diseases and pathogens. The use of NPs can provide such nutritional enrichment to plants. Elmer and White (2016) conducted research on the development of micronutrients in the form of metal oxide NP (Al_2O_3, CuO, FeO, MnO, NiO, and ZnO) for foliar sprays. Testing of the foliar spray was done on tomato and eggplant, which were pathogen infected. Enhancement in the production of both tomato and eggplant has been observed despite being infected with *Fusarium* wilt fungus. CuO, MnO, and ZnO have been found to effectively reduce the level of infection in plants, thereby enhancing their growth and ultimately increasing yield (Elmer and White, 2016). Kah et al. (2019) showed that there was 20–30% gain in production level due to nanofertilizers as compared to conventional products. Several NPs are used in promoting plant growth and some are listed here.

10.4.1 Iron (Fe) Nanoparticles

The most crucial physiological pathways happening in plants are the biosynthesis of pigments involved in photosynthesis and the process of photosynthesis, and respiration. All these processes have an important role of iron as it is involved in electron transport during photosynthesis and respiration and is also important in chlorophyll synthesis (Clarkson, 1995). Many studies have pointed toward the role of NPs based on Fe in plant growth enhancement under non-stress conditions. Lettuce (*Lactuca sativa*) seedlings were grown in a liquid medium by Liu et al. (2016). They provided different concentrations of FeOX NPs and 5–20 mg/L concentration has increased the length of shoot in lettuce, by 12–26%. Ghafariyan et al. (2013) showed increased content of chlorophyll in hydroponic soybean (*Glycine max* L.) through Fe_2O_3 NPs at concentration of 30–60 mg/L. Recently, Rui et al. (2016) confirmed that Fe_2O_3 is more effective than the Fe-based traditional fertilizers. Alidoust and Isoda (2013) showed the elevated root length and the photosynthetic efficiency of soybean seedlings after a foliar spray of nano-Fe_2O_3, which was citric-acid-coated.

10.4.2 Zinc (Zn) Nanoparticles

Zn is an important component in maintaining enzyme activity and structure. There are several studies that prove ZnO NPs as plant growth promoters. For instance, Kareem et al. (2022) and Dhoke et al. (2013) concluded that when ZnO NP particles were applied on foliar leaves they significantly enhanced the root and above-ground biomass of the hydroponically grown *Vigna radiata* seedlings. The effect of ZnO NPs was studied on sorghum through the soil and foliar exposure. The use of macronutrient efficiency, Zn grain enrichment, and yield were observed. They finally concluded that ZnO significantly enhanced the productivity and nutritional quality of grains in sorghum under both high and low concentrations of NPK (Dimkpa et al., 2017). Subbaiah et al. (2016) concluded to the point that 1500 ppm ZnO NPs increase the percent germination and the vigor of seedlings in maize. Along with its additional role in plants under stress-free conditions, ZnO nanoparticles also enhanced plant efficiency under salinity stress (Alharby et al., 2016), drought stress (Dimkpa et al., 2019) and Cd stress (Rizwan et al., 2019).

10.4.3 Magnesium (Mg) Nanoparticles

Magnesium is important for a number of physiological and chemical processes and is an important micronutrient responsible for plant growth (Li et al., 2001). Black-eyed pea was treated with foliar spray of MgNPs. Photosynthetic machinery and the plant biomass were enhanced by a concentration of 0.5 g/L (Delfani et al., 2014). Rathore and Tarafdar (2015) did foliar spray on wheat plants with biosynthesized MgNPs (<5.9 nm, conc.20 mg/L). Root biomass, length of roots, and the number of tips seemed to be enhanced. Magnesium is important for increasing the activities of antioxidant enzymes. MgNPs can be used to suppress the symptoms of deficiency of Mg in plants and can be helpful in fighting against stress. For instance, in comparison to bulk MgO, the application of MgO NPs (150 and 250 mg/L) reduced *Ralstonia solanacearum* bacteria to a significant level in soil-grown tobacco (Cai et al., 2018). A list of nanomaterials and their impact on growth and development of different plant species has been shown in Table 10.1.

TABLE 10.1
List of Nanomaterials and Their Impact on Growth and Development of Different Plant Species

Nanomaterial	Plant Species	Impact	Reference
CuO NP	*Citrullus lanatus*	Anti-fungal action	Borgatta et al., (2018)
CuO NP	Conventional cotton and Bt-transgenic cotton	Anti-insecticidal action	Van et al., (2016)
Cu NPs and K_2SiO_3 NPs	Hydroponically cultivated tomato	Anti-bacterial action	Cumplido-Nájera (2019)
Ag@dsDNA@GO	Tomato	Anti-bacterial action	Ocsoy et al., (2013)
AgNP	Bermuda grass	Nematicidal effect	Cromwell et al., (2014)
AgNP	Wheat	Anti-fungal activity	Mishra et al., (2014)
TiO_2NPs	Tomato	Anti-bacterial potential	Paret et al., (2013)
MgO NPs	Tomato	Anti-bacterial activity	Imada et al., (2016)
MgO NPs	Tobacco	Anti-bacterial activity	Cai et al., (2018)
MgNPs	Black-eyed pea	Enhanced Photosynthesis and plant biomass	Delfani et al., (2014)
MgNPs	Wheat	Increased root biomass, root length and tip number	Rathore and Tarafdar (2015)
FeOX NPs	*Lactuca sativa*	Increased shoot length of lettuce	Liu et al., (2016)
Fe_2O_3 NPs	Hydroponic soybean	Increased chlorophyll content	Ghafariyan et al.,(2013)
Fe_2O_3 NPs coated NPS with citric acid	Soybean seedlings	Increased photosynthetic efficiency and root length.	Alidoust and Isoda (2013)
ZnO NPs	*Vigna radiata* seedlings.	Enhanced the biomass of roots and above-ground tissues	Dhoke et al., (2013)
ZnO NPs	Sorghum	Increase in nutritional quality of grains and plant productivity	Dimkpa et al., (2017)
ZnO NPs	Maize	Seed vigor and germination percentage was increased	Subbaiah et al., (2016)

10.5 INFLUENCE OF NANOMATERIALS ON THE RHIZOSPHERE ENVIRONMENT

The influence of specific metal nanoparticles on soil microflora has not been studied well. To a minor extent, studies had been done to know the influence of nanomaterials on the rhizosphere (Rajput et al., 2023; Wang et al., 2022). Nanomaterials are materials with at least one dimension at the nanoscale, which is 1 to 100 nm. These are made up of metals or polymers and can be used in the form of nanoparticles, nanofilms, and nanotubes (Baig et al., 2021). Nanoparticles can be present inside or as an external coating of nanomaterials, which possibly makes the direct application of nanomaterials to the soil (Abid et al., 2022). Due to their small size, they can directly interact with soil properties and soil microbes. The application of nanomaterials on land leads to direct contact between existing NPs in nanomaterial and the soil microbes. Nanomaterials are efficiently adsorbed and accumulated by microbes, which are responsible for nanomaterial mobilization through food chains. The presence of nanoparticles has the potential to disrupt the communities within food webs, which consist of various populations such as plants, fish, and bacteria. This alteration can have implications for the interaction and dynamics of these populations, as observed by Holden et al. (2013). NPs affect the plant roots and rhizospheric microorganisms in both a positive and negative manner. A study conducted by Zeliadt et al. (2010) revealed that the application of microbicidal silver nanoparticles (AgNPs) can have a negative impact on plant growth and the soil microbes that play a crucial role in nutrient extraction from the soil. This suggests that the use of AgNPs may disrupt the beneficial symbiotic relationship between plants and soil bacteria and fungi, which are essential for nutrient uptake and overall plant health (Zeliadt et al., 2010). Haris and Ahmad (2017) provide evidence of changes in the bioavailability of plant toxins and the activity of soil microorganisms. In a study by Shcherbakova et al. (2016), the effect of molybdenum nanoparticles on the root exudate composition of chickpea plant rhizosphere was investigated. The researcher observed that co-inoculation of *Mesorhizobium ciceri* strain ST282 with Mo NPs and *Bacillus subtilis* a helper bacteria resulted in improved symbiotic effectiveness. Additionally, the changes in the root exudate composition led to an increase in the structural diversity of the microbial community in the rhizosphere. In a research carried out by Moll et al. (2017), the influence of two different varieties of titanium dioxide nanoparticles TiO_2 NPs namely P25 and E171, as well as bulk TiO_2 particles with a size greater than 100nm, on the variety and composition of soil microorganisms was examined. They utilized short read illumine sequencers for sequencing and assessing microbial diversity based on ribosomal markers. The findings revealed that the structure of the prokaryotic community, rather than the fungal community, was altered by the presence of TiO_2NP treatments depending on the microbial strain. In a study conducted be Wang et al. (2017), the impact of different concentrations of silver nanoparticles Ag-NPs on the soil microbial community was investigated over a period of 7 days. The study revealed that there is a negative correlation between the concentration of Ag-NPs NPs (10, 50, and 100 mg kg^{-1}) and the abundance of acidobacteria, actinobacteria, and cyanobacteria, indicating a decrease in their number with increase in AgNP concentrations. Conversely, other phyla such as Proteobacteria and Planctomycetes, showed an increase in abundance with higher concentrations of AgNPs. In another study by Shan et al. (2015), the effect of biochar, activated carbon (AC), SWCNTs, and MWCNTs at various concentrations (0, 0.2, 20, and 2000 mg kg^{-1} dry soil) on the level of 14C-catechol and the soil microbial community examined. The result indicated that SWCNTs at 2000 mg kg^{-1} significantly altered mineralization. Furthermore, both AC and SWCNTs inhibited soil microbial activities and influenced the microbial communities. However, MWCNTs at a concentration of 0.2 mg kg^{-1} significantly enhanced mineralization and stimulated microbial activity.

10.5.1 Nano Herbicides

As nanotechnology can be used to improve the performance of many existing herbicides or to create an alternative tool that is a quick, efficient, and affordable means and exceeds the original product, nano herbicides are being developed to address the issues of all types of weeds (Manjunatha et al., 2016). Therefore, weeds are undesirable plants that grow in the field alongside the primary crops. Herbicides are applied in an effort to get rid of certain weeds. However, when non-original herbicides are applied, it affects the main crop too, and because of this, there are huge losses in crop yield. By using nano herbicides with a range of 1 to 100 nm, they attempt to incorporate into the soil and eradicate the weed plant from its root without incorporating it with the main crop, which induces a high yield (Prasad et al., 2017). And by lowering herbicide inputs and giving users control over where and when an active component is delivered, nano herbicides serve as a "smart delivery system" that promotes environmental sustainability (Pérez-de-Luque and Rubiales, 2009). The most commonly used herbicides; atrazine and triazine were nano-compressed to get an efficient release to the plants. Atrazine is used worldwide to kill weeds (Camilo 2016). Atrazine use on a regular basis causes the soil to lose all of its nutrients and renders it resistive to the main crop. As a result, using modified silver with nanoparticles and carboxymethyl cellulose facilitates the removal of herbicides (Mahto et al., 2019).

10.5.2 Nanopesticides

Pesticides known as nanopesticides use nanotechnology to increase their efficacy and efficiency. These insecticides are created to be more effective at controlling pests, need fewer applications, and use less active ingredients overall (Kah et al., 2014). By adding nanomaterials, such as nano capsules or nanoparticles, into the pesticide formulation, nanopesticides can be produced. These nanoparticles can make the pesticide more soluble and stable, which will enhance its absorption and efficiency. Additionally, nanopesticides can increase the bioavailability of the active component, which can lower the amount of pesticide required to get the desired result (Balaure et al., 2017). And the pesticides are subdivided into four categories and those are herbicides that act on unwanted plants, insecticides on multiple insects, fungicides that control moulds, and disinfectants and algae for the controlling of bacteria. If we make these pesticides in a closed environment, they will be slow and efficient in their release to the plants. Solubility will be increased, and runoff of the soil decreased (Chaud et al., 2021). Certain pesticides like Zineb and Mancozeb were held together with multiwall carbon nanotubes, posing major toxicity to the Alternaria alternative fungi (Kumar et al., 2019).

10.6 METAGENOME A WAY FORWARD IN MICROBIOME RESEARCH

Understanding the microbiome diversity in the soil has always been the goal of researchers to gain insights into how the plant–microbe and microbe–microbe interaction works (in the rhizosphere) (Table 10.2). Microbial community present in the rhizosphere differs from the microbial community found in bulk soil. Various studies have shown that host genetics governs the rhizospheric microbial community. Even plant microbes' associations differ between host genotype to genotype and between different soil types. Metagenomic studies are mainly conducted in two ways

10.6.1 Metataxonomy/Metaprofiling

Marker based which is also known as meta-taxonomy or metaprofiling, this approach utilizes sequencing of already known and highly conserved markers, for example, 16s rRNA, small subunit rRNA, and intergenic transcribed spacers (ITS) markers. With the advancement in sequencing technology and reduction in cost, microbiome research has also taken a leap forward via sequencing amplicons using high throughput sequencing. Initially, the amplification from the samples was performed using universally phylogenetic conserved markers like 16s rRNA gene in the case of prokaryotes and ITS when dealing with fungal samples. Amplified products were earlier used to be sequenced with Sanger sequencing but now with advancement and evolution in second-generation NGS has made it more cost efficient plus generating high throughput data with less computing time involved in analysing the data. In brief, raw data generated from amplicon sequencing undergoes sequential steps comprising of quality check, quality filtering, trimming, merging of forward and reverse read pairs, and removal of chimeric reads to obtain a high-quality clean read. Then, cleaned and high-quality reads are subjected to reference databases for alignment and as a result, it matches with highest alignment similarity and lengths are clustered into "operational taxonomic units" (OTU), which means clustered reads belong to the same species. The drawbacks of marker-based quantification of the microbiome community present in the sample are PCR-based amplification of markers and sequencing errors, which are prone to show biases in the results. The classification step is limited at the genus level due to the database selected as the reference and classifier used, which also provides

TABLE 10.2
List of Commonly Used Tools in Metagenome Data Analysis

Function	Tools	References
Quality control	FastQC	https://www.bioinformatics.babraham.ac.uk/projects/fastqc/
	MultiQC	https://doi.org/10.1093/bioinformatics/btw354
	BBTools	http://jgi.doe.gov/data-and-tools/bbtools/bb-tools-user-guide/
Trimming	Trimmomatic	https://doi.org/10.1093/bioinformatics/btu170
	Cutadapt	https://doi.org/10.14806/ej.17.1.200
Assembly	Meta-IDB	https://doi.org/10.1093/bioinformatics/btr216
	MetaVelvet	https://doi.org/10.1093/nar/gks678
	SOAP	https://doi.org/10.1093/bioinformatics/btn025
	MetaSPAdes	https://doi.org/10.1101/gr.213959.116
	MetAMOS	https://doi.org/10.1186/gb-2013-14-1-r2
Alignment	Bowtie2	https://doi.org/10.1038/nmeth.1923
	HMMER	https://doi.org/10.1371/journal.pcbi.1002195
	DIAMOND	https://doi.org/10.1038/nmeth.3176
Binning/clustering	IMG/M	https://doi.org/10.1093/nar/gkm869
	MG-RAST	https://doi.org/10.1101/pdb.prot5368
	COCACOLA	https://doi.org/10.1093/bioinformatics/btw290
	MEGAN	https://doi.org/10.1101/gr.5969107
	MetaPhyler	https://doi.org/10.1186/1471-2164-12-s2-s4
	PCAHIER	https://doi.org/10.1142/s0219720010005051
	MaxBin	https://doi.org/10.1093/bioinformatics/btv638
Taxonomic classification	SILVA	https://doi.org/10.1093/nar/gkm864
	RDP	https://doi.org/10.1093/nar/gkn879
	GreenGe ne	https://doi.org/10.1128/aem.03006-05
	Kaiju	https://doi.org/10.1038/ncomms11257
	Centrifuge	https://doi.org/10.1101/gr.210641.116
Annotation	MetaGene	https://doi.org/10.1093/dnares/dsn027
	Orphelia	https://doi.org/10.1093/nar/gkp327
	Prodigal	https://doi.org/10.1186/1471-2105-11-119
Statistical analysis	Primer-E	https://doi.org/10.1111/j.1442-9993.1993.tb00438.x
	ShotgunFunctionalizeR	https://doi.org/10.1093/bioinformatics/btp508

limited information about the functionality. The use of only one phylogenetic marker 16s rRNA in prokaryotes and ITS in fungus can introduce technical differences in the microbial community which raises questions and concerns about using these markers for targeted sequencing. Therefore, there is a scope of improvement in amplicon-based metagenomic studies to develop more markers to be included with reduction of errors incorporated during library preparation methods and sequencing. Only classification of known species can be done using marker-based study as novel species will not be present in the database, as a result, it will not get aligned and will remain unannotated (Escobar-Zepeda et al.,2015; Puntillo et al., 2022).

10.6.2 Shotgun Whole Genome Sequencing

The need for whole genome sequencing (WGS) arises to circumvent the drawbacks of marker-based. It also provides complete insights into the microbiome present in a particular environment. The WGS method reveals everything present at the genomic level in the species and allows investigation of the coding, non-coding, conserved regions, and transposable elements. The functional aspect of the species present in the sample will enable us to draw the relationship between the host plant and a particular microbe. Novel microbes can be captured using shotgun sequencing, which was not possible in the case of marker-based metagenomics. Sequencing platforms, such as 454 pyrosequencing, ion torrent PGM, Illumina, and PacBio are majorly opted for studying microbial diversity. However, third-generation sequencers such as PacBio and Nanopore with their advancements can generate data with fewer errors and more completeness, correctness, and better contiguity than second-generation sequencers. Structural and functional diversity can be studied extensively to draw lines of conclusions about how microbe–plant and microbe–microbe interacts and exhibit positive or negative relationships and their role in plant growth and development. Due to higher depth and better coverage of sequencing data generated using shotgun sequencing, even the least abundant species present in the sample can be captured at a better resolution. Handling of novel species using the de novo assembly pipeline is also feasible as it includes both a reference database and de novo-based procedures to detect all the species present in the samples (Rounsley et al., 2010; Kumar et al., 2023). A very general pipeline for analysis of metagenome is depicted in Figure 10.2.

Based on the goal of studying and exploring different aspects of the rhizosphere, researchers use different approaches of metagenomics to understand the rhizosphere biology and how it can play its role in plant growth and productivity. A cropping system selected/planned during a season can significantly affect the health and fertility of the soil. Sun et al. (2021) showed a sugarcane-peanut intercropping system can control pests and diseases, enhance/increase the cane stalk height and weight, stalk number and edaphic factors such as pH, phosphorus content in both the rhizosphere and bulk soil. Rhizosphere management in this way compared to monocropping sugarcane significantly increases bacterial diversity and abundance without having any detrimental effect on soil. Another study on sugarcane showed detrimental effects of continuous cropping on sugarcane yield, sugar content and soil physiochemical properties, such as pH, organic matter, total nitrogen, total sulfur, total potassium, and available nitrogen contents, which decreases as continuous cropping years increases. Along with detrimental effects, it also reduces beneficial rhizosphere bacterial and fungal communities from the samples where continuous cropping was done for as long as 10 to 30 years. Short-read sequencing using Illumina was used in order to identify the presence of bacterial and fungal composition and abundance in the soil affected by continuous cropping. Analysis revealed variation in communities is mainly caused by soil pH, total nitrogen, and total sulfur. Also, continuous cropping caused declination of genera involved in nitrogen and sulfur cycle and inclination of harmful microbial communities that might be one of the causal reasons for decreasing sugarcane quality and yield (Wu et al., 2022). Moreover, Illumina based 16s rRNA sequencing revealed the majority of bacteria profoundly found in the root endosphere of the sugarcane. PCR-based amplification and sequencing of 16s rRNA and phylogenetic analysis revealed 62 rhizospheric and 28 root endospheric bacterial isolates that were identified from wheat-rice intercropping system, which showed antagonistic effects against one or multiple root rot pathogens of wheat (*Fusarium oxysporum, Fusarium moniliforme, Macrophomina phaseolina,* and *Rhizoctonia solani*). Rhizosphere management in this manner can lead to a more ecofriendly way of increasing plant growth and productivity and biological controls (Zia et al., 2021). Wu et al. (2022) utilized amplification of ITS1 for fungal species and 16s rRNA for bacterial species identification followed by Illumina-based sequencing. Their study showed that monocropping or intercropping tomato with onion in conjugation with applying biochar significantly increases plant height, ability of tomato to absorb N, P, K, dry weight, and improves physiochemical state of soil such as increase in pH and decrease in bulk density and NO_3. Abundance of bacteria and fungi has increased with the increase in biochar concentration, but the fungal diversity has reduced and bacterial diversity remained unaffected in the soil (Wang et al., 2022). Continuous use of agricultural land put soil in a state of compaction and added constraints to microbial functioning and its diversity, additionally, affecting soil physiochemical properties. Soil compaction is known to reduce and delay germination in crops which can result in yield loss upto 60% in the case of pea and 40% in wheat (Longepierre et al., 2022). A study revealed that differences in soil moisture

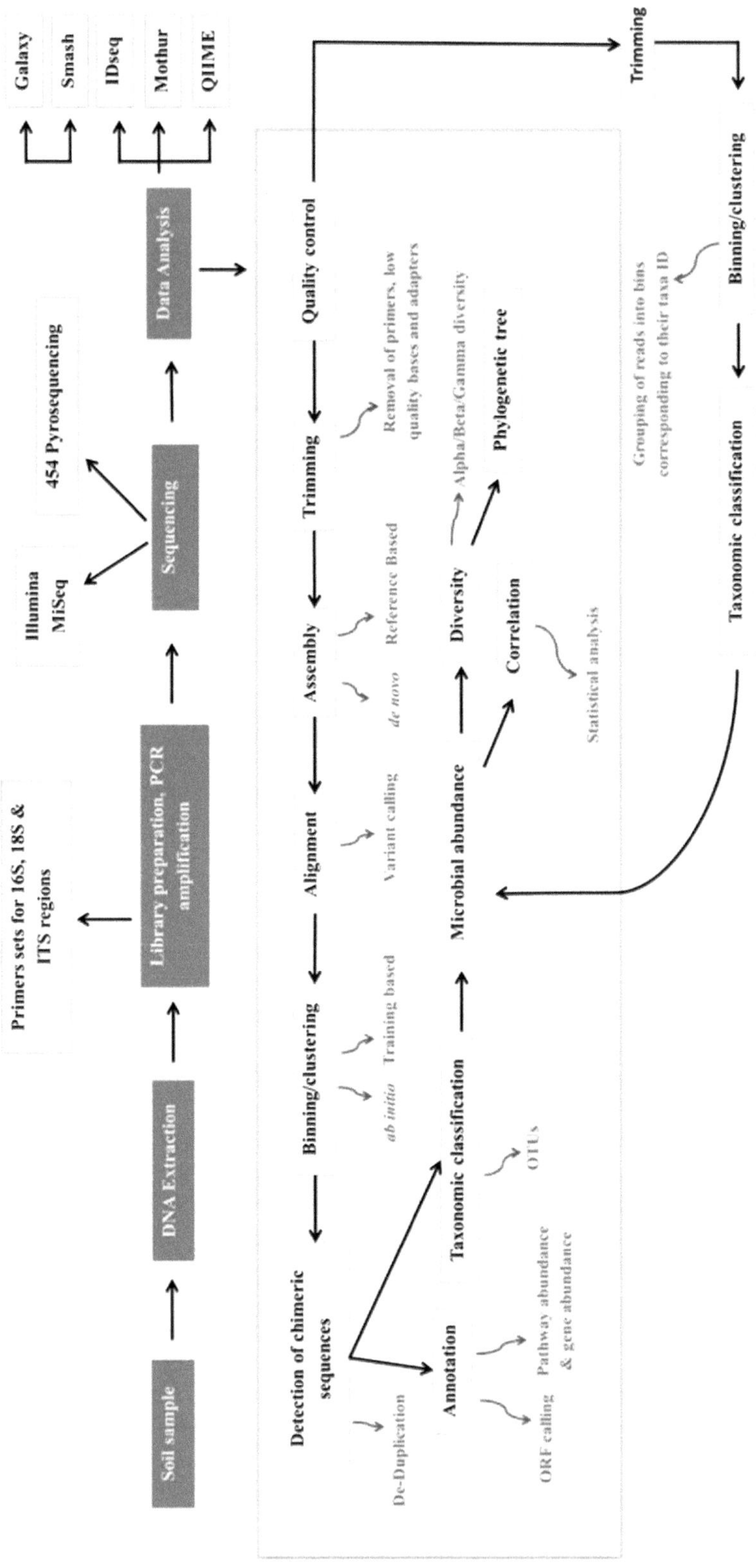

FIGURE 10.2 A general metagenomic analysis pipeline.

content can drive the establishment of crop-specific microbial communities (aerobic and anaerobic shifts) in compact soil. Therefore, soil moisture can be served as a good indicator for sustainable agricultural practices in these conditions (Hartmann et al., 2022). PCR amplification of 16s rRNA was done followed by transformation of purified product into *E. coli* and sequencing of clones and the resulting sequencing data was analyzed in order to identify PSB strains. Novel phosphate solubilizing bacteria *Acinetobacter* sp. *Ac-14* was identified as a stable, efficient, and sustainable candidate for phosphate solubilization and a plant growth, promoting ability by producing Indole 3 acetic acid and ammonia and by dissolving insoluble phosphate into 23 organic compounds. This study identified a bacterial strain, *Acinetobacter* sp. *Ac-14*, which might have adapted high calcium environment and its ability to improve soil fertility can lead towards developing sustainable agriculture in phosphate deficient lands (Xie et al., 2021). Agriculture management in an efficient way can lead to better crop yield and less loss due to pathogens whereas, maintaining beneficial rhizosphere microbial communities and reducing harmful microbial communities improved soil fertility and physiochemical property suited for crops. Crop rotation rather than monoculture has shown to be a better way of agriculture management as Bian et al. (2022) showed that ginseng cultivation followed by *Allium fstulosum* and *Brassica napus* drastically improves soil available phosphorus and potassium where the shifts in microbial diversity made by both the crops increased available nutrients present in the soil and reduced salinization and pathogen fungus *Ilyonectria* in ginseng cultivation. Continuous watermelon cropping revealed that due to soil physiochemical properties such as moisture and pH can cause shifts in microbial communities (bacterial and fungus), which later contribute to yield reduction (Gu et al., 2022). A list of metagenomic studies on different plant species is represented in Table 10.3

TABLE 10.3
List of Metagenomic Studies on Different Plant Species

S. No.	Plant Species	Inference	Reference
1.	*Sorghum bicolor*	The plant exhibits increased root and shoot biomass	Kuramae et al., 2020
2.	Banana	Metagenome analysis revealed the inhibitory effect of plant growth-promoting rhizobacteria on *Fusarium oxysporum* f.sp. *cubense* (Foc)	Effendi et al., 2021
3.	*Paspalum scrobiculatum*	Metagenomic analysis of the rhizosphere concluded that microorganisms present in this zone help the plant in its growth and development under nutrient-deprived, semi-arid, and dry biotic conditions. It happens through various secondary metabolite pathways which are involved in the synthesis of bacteriocins or antibacterial peptides, as well as the development of resistance against antimicrobial comounds.	Prabha et al., 2019
4.	*Beta vulgaris* L.	Metagenome analysis revealed that plant growth-promoting traits PGPTs play a crucial role in the growth of sugar beet. Conversely, the analysis did not reveal any genes involved in IAA production and N2 fixation.	Tsurumaru et al., (2015)
5.	Plantation Crops such as Coconut, Cocoa and Arecanut	Various genes involved in mineral phosphate solubilization, siderophores, acetoin, butanediol, 1-aminocyclopropane-1-carboxylate (ACC) deaminase, and various PGPTs were found supportive in plant growth and development.	Gupta et al., (2014)

(*continued*)

TABLE 10.3 (Continued)
List of Metagenomic Studies on Different Plant Species

S. No.	Plant Species	Inference	Reference
6.	*Panax ginseng*	The effect of endophytic bacteria which is present in the roots of *P. ginseng* was observed on the growth and development of this plant through phosphate and nitrogen metabolism and this effect is age dependent. Bacteria show their maximum effect when the plant is 3 years old.	Hong et al., (2019)
7.	*Emilia sonchifolia* (Linn.) DC	Genes involved in plant growth promotion such as nitrogen metabolism (nif) and ACC deaminase coding genes were found to be regulated positively for host plants through metagenome studies.	Urumbil et al., (2021)
8.	Cabbage	The metagenomic library screening identified chitinase as an antagonistic agent against various fungal pathogens.	Hjort et al., (2014)
9.	Maize	Genes crucial in nitrogen fixation phosphate solubilization, siderophore production, disease control, and stress resistance were found to improve plant growth.	Akinola et al., (2021)
10.	*Oryza sativa*	The presence of *Bacillus* as the dominant endophytic genera in rice roots is believed to have a significant impact on nitrogen fixation, leading to improved plant growth and development.	Sengupta et al., (2017)
11.	*Radix Angelica sinensis*	The microbial community in the rhizosphere plays a vital role in promoting the branching and growth of the root system in *Angelica* sp.	Yan et al., (2021)
12.	Barley	A functional metagenomic approach revealed that mineral phosphate solubilizing microorganisms plays an important role in plant growth and they were found to be positively related to plant growth.	Chhabra et al., (2013)
13.	Sugarbeet	Metagenome analysis identified the upregulation of stress-related genes in response to *Rhizoctonia solani,* the fungal root pathogen, leading to disease suppression.	Chapelle et al., (2016)

10.7 UNDERSTANDING OF PLANT GROWTH AND THE RHIZOSPHERE THROUGH OMICS LENS

Omics-based approaches to understand rhizosphere biology are an emerging branch of system biology. Integrating multiple omics approaches can reveal major insights into the biology of plant–microbe interaction. Currently rhizosphere studies are mainly genome-centric, i.e., metagenome based however, with gaining advancements into the sequencing technologies and techniques, studies such as meta-transcriptome, meta-proteome, meta-epigenome, and meta-epiproteome are now being conducted to understand different planes of rhizosphere and its association with plant growth to understand the larger picture of the whole system biology (White et al., 2017). Soil-borne pathogen *Rhizoctonia solani,* a fungus, was studied between two wheat grown fields, i.e., suppressive, and non-suppressive against the disease to identify DEGs and annotate genes that are causal differences in plant–microbe–pathogen interaction and how microbe mediates disease suppressiveness (Hayden et al., 2018). A study revealed DEGs belonging to non-suppressive samples were dominated by

Arthrobacter spp., *Pseudomonas* spp., *Stenotrophomonas* spp., and *Buttiauxella* spp. were dominant suppressive samples and can be applied in future wheat crop cultivation and rhizosphere management against the disease (Hayden et al., 2018). Metaproteome approaches can be utilized to characterize bacterial communities associated with *Vitis vinifera* rhizosphere (Bona et al., 2019). Extracted proteins from soil samples followed by mass spectrometry analysis resulting protein sequence undergo annotation against NCBI non-redundant database and undergo functional annotation using blast2GO. *Vitis vinifera* rhizosphere is majorly colonized via *Streptomyces*, *Bacillus*, *Bradyrhizobium*, *Burkholderia*, and *Pseudomonas* genera based on protein expression evidence (Bona et al., 2019).

Although single omics can provide a lot of information on the rhizosphere biology moreover, a multi-omics approach can give a better understanding of the different processes such as gene expression, transcription, translation, and metabolite production at system biology level. Multi-omics techniques are used in understanding all the possible interaction networks of microbe–plant–pathogen, biogeochemical properties of soil and chemical diversity, in the rhizosphere and to identify the potential candidate genes and their pathways. Integrating metagenomics and metatranscriptomics with viromics revealed 1059 dsDNA and 16541 ssRNA populations from a crop-rotated field of *B. napus* L. for 3 years. The study revealed that viral diversity and its composition play an important role in plant–microbe interactions in the rhizosphere via rewiring host bacterial metabolism. Muscatt et al. (2022) and Zhong et al. (2022) carried out their study on the plant *Glycyrrhiza uralensis*, which has medicinal importance. The studies showed how the multi-omics approach can be utilized to decipher deferentially expressed genes (DEGs) involved in liquiritin and glycyrrhizic acid biosynthesis pathways. Also, the role of microbial community in liquiritin and glycyrrhizic accumulation has been shown. A combination of root metabolome, root transcriptome, and metagenome helped in resolving how microbial community and different plant types, i.e., wild type, cultivated for 1 year and 3 years have different levels of accumulation of important biosynthetic compounds and DEGs (Zhong et al., 2022). Multi-omics approaches can also shed light on the effect of *Bacillus thuringiensis* (Bt) crops and its adverse impact on environment and how Bt crops can modulate soil microbiome. However, Bt rice has been explored using two layers of information from metabolome profiling of roots and soil and surface water metagenome (Li et al., 2023). Results showed that Bt rice has no adverse impact on soil microbiome but can alter soil carbon and nitrogen and can selectively modulate the assemblage of specific microbiota via altering metabolic composition of root exudates (Li et al., 2023). Management of agricultural practices, i.e., the input and output in the form of plant phenotype is a very critical point to be considered during cultivation of any crop. Multi-omics approach was applied to *Brassica rapa* to demonstrate the role of overall different management agricultural practices and their impact on the association of plant traits heterogeneity with soil metabolites, minerals, and microbes. Application of soil solarization, a well-known environment-friendly method to deal with pests and weeds showed an increase in shoot biomass and crop yield by induction of organic nitrogen, regardless of the fertilizers applied in conjugation (Ichihashi et al., 2020). Integration of NMR-based metabolomics, ICP-OES-based ionomics, along with 16s rRNA based metagenome using Illumina opened the multilayer complicated network interaction in the agroecosystem. Soil solarization induces the increase in growth response via organic nitrogen availability and important soil rhizosphere bacteria known to play a role in better crop production indicating that solarization of soil not only allows colonization of bacteria, which helps in decomposition of inorganic nitrogen (Ichihashi et al., 2020). Xu et al. (2018) performed a large-scale field evaluation on sorghum and its rhizome to study drought stress using a multi-omics approach. The outcome of the study showed, drought delays the development of early sorghum rhizome and causes increased abundance and activity of monoderm bacteria by using 16S rRNA amplicon and metatranscriptome sequencing, and root metabolomics. Increased activity of bacterial ATP-binding cassette (ABC) transporter genes and drought has altered plant metabolism. Overall individual omics study is centric towards the goal of identification and characterization of rhizosphere microbiome associated with the plant species under study. Whereas, holistic-omics-based studies are mainly focused on understanding the complex interactions in all possible ways, which can portray the true picture of system biology and can aid in sustainable crop productivity and agroecological and rhizosphere management.

10.8 CONCLUSION

To satisfy the growing need for food by an elevating population it has become necessary to enhance plant growth and yield. There are several traditional techniques and approaches for enhancing plant productivity but unfortunately, the proportion to which these approaches enhance the production and the required proportion of enhancement do not match well. Nanotechnology is in bloom and this is being considered to increase plant growth, development, and yield. Nanomaterials have been used as crop plant nutrients in the form of nano fertilizers and as crop protection agents in the form of nanopesticides and nano herbicides. The application of nanomaterials has played an important role in enhancing plant production. Nanomaterials are being used as "smart" delivery systems in medical sciences, similarly, it has been shown in recent times that nanoparticles can be used as chemical delivery agents to target molecules such as genes/DNA or specifically some cellular organelles such as nuclei in plants. The function and use of nanoparticles depend on their concentration and structure, so the major interest is optimizing their shape, size, and concentration to manipulate the catalytic, magnetic, and mechanical properties so that they can interact with living systems in the best way. It was in bloom in the last decade however more research is needed to understand the biological, physiological, and molecular mechanisms of plants for nanoparticles. The mode of action of nanoparticles, their interaction with the biomolecules, and their effect on gene expression regulation in plants are still under investigation. Sometimes at higher concentrations nanoparticles lead to mitotic arrest, cell wall breakage, metaphase disturbance, inhibition of gene expression, and chromatin condensation, and some metal oxide nanoparticles negatively affect the beneficial microbial systems that are present in the soil rhizosphere. More research is required to study the phytotoxic effects caused by different nanoparticles and the mechanisms involved in phytotoxicity. Along with the growing field of nanotechnology, multi-omics technology in recent times has provided a deep understanding of rhizosphere biology and significantly plays its role in plant growth and development. Proteomics, genomics, metabolomics, and transcriptomics have provided their contributions to the discovery of new strains and substances, to study the gene–gene and protein–protein interactions, and to provide the evolutionary relationship between the strains. Metagenomic studies done in plant species *Beta vulgaris*, *Paspalum scrobiculatum*, *Panax ginseng*, *Oryza sativa*, and *Zea mays* showed the role of plant growth-promoting traits (PGPTs) and endophytic bacteria, which are responsible for the positive regulation of genes involved in mineral phosphate solubilization, plant growth promotion such as nitrogen metabolism (nif) and ACC deaminase coding genes. However, few studies have been carried out using multi-omics technology. With the advancements in sequencing techniques, there is room to progress in effectively linking DNA sequencing with RNA sequencing and also the other omics data that will be helpful in giving a broader and more clear picture of the rhizosphere community and its effect on overall physiological growth of the plant. The progress in understanding the functional aspect of the soil is slow because of various interfering molecules that are present in the soil samples and the insufficiency of databases of soil. Also, there is a lack of analytical techniques which is a major hurdle in analyzing and the management of the huge multi-omics data. Apart from this, non-professionals and new users find it difficult to use the available software and are unable to reproduce results. So, there is a great need to develop robust bioinformatic tools for analyzing complex multi-omics data.

ACKNOWLEDGEMENTS

The authors gratefully acknowledge the Department of Biotechnology, Sharda School of Engineering and Technology, Sharda University, Greater Noida, Uttar Pradesh, India, Department of Botany, University of Delhi, New Delhi, India, Plant Molecular Biology Laboratory, Department of Botany, Dayanand Anglo-Vedic (PG) College, Chhatrapati Shahu Ji Maharaj University, Kanpur, India and Department of Life Sciences, Presidency University, College Street, Kolkata, West Bengal, India.

REFERENCES

Ab Rahman, S.F.S., Singh, E., Pieterse, C.M. and Schenk, P.M., 2018. Emerging microbial biocontrol strategies for plant pathogens. *Plant Science, 267*, pp.102–111.

Abid, N., Khan, A.M., Shujait, S., Chaudhary, K., Ikram, M., Imran, M., Haider, J., Khan, M., Khan, Q. and Maqbool, M., 2022. Synthesis of nanomaterials using various top-down and bottom-up approaches, influencing factors, advantages, and disadvantages: a review. *Advances in Colloid and Interface Science, 300*, p.102597.

Adisa, I.O., Pullagurala, V.L.R., Peralta-Videa, J.R., Dimkpa, C.O., Elmer, W.H., Gardea-Torresdey, J.L. and White, J.C., 2019. Recent advances in nano-enabled fertilizers and pesticides: a critical review of mechanisms of action. *Environmental Science: Nano*, *6*(7), pp.2002–2030.

Ahn, J.M., Eom, H.J., Yang, X., Meyer, J.N. and Choi, J., 2014. Comparative toxicity of silver nanoparticles on oxidative stress and DNA damage in the nematode, Caenorhabditis elegans. *Chemosphere, 108*, pp.343–352.

Akinola, S.A., Ayangbenro, A.S. and Babalola, O.O., 2021. The diverse functional genes of maize rhizosphere microbiota assessed using shotgun metagenomics. *Journal of the Science of Food and Agriculture, 101*(8), pp.3193–3201.

Alharby, H.F., Metwali, E.M., Fuller, M.P. and Aldhebiani, A.Y., 2016. The alteration of mRNA expression of SOD and GPX genes, and proteins in tomato (Lycopersicon esculentum Mill) under stress of NaCl and/or ZnO nanoparticles. *Saudi Journal of Biological Sciences*, *23*(6), pp.773–781.

Ali, M., Kim, B., Belfield, K.D., Norman, D., Brennan, M. and Ali, G.S., 2015. Inhibition of Phytophthora parasitica and P. capsici by silver nanoparticles synthesized using aqueous extract of Artemisia absinthium. *Phytopathology*, *105*(9), pp.1183–1190.

Alidoust, D. and Isoda, A., 2013. Effect of γFe 2 O 3 nanoparticles on photosynthetic characteristic of soybean (Glycine max (L.) Merr.): foliar spray versus soil amendment. *Acta Physiologiae Plantarum, 35*, pp.3365–3375.

Allen, B., Pezone, A., Porcellini, A., Muller, M.T., and Masternak, M.M. (2017). Non-homologous end joining induced alterations in DNA methylation: A source of permanent epigenetic change. *Oncotarget*, *8*(25), p.40359.

Baig, N., Kammakakam, I. and Falath, W., 2021. Nanomaterials: a review of synthesis methods, properties, recent progress, and challenges. *Materials Advances*, *2*(6), pp.1821–1871.

Balaure, P.C., Gudovan, D. and Gudovan, I., 2017. Nanopesticides: a new paradigm in crop protection. *In* Grumezescu, A.M. (ed) *New pesticides and soil sensors* (pp.129–192). Academic Press.

Bhardwaj, D., Ansari, M.W., Sahoo, R.K. and Tuteja, N., 2014. Biofertilizers function as key player in sustainable agriculture by improving soil fertility, plant tolerance and crop productivity. *Microbial Cell Factories, 13*, pp.1–10.

Bian, X., Yang, X., Li, Q., and Sun, X., 2022. Effects of planting of two common crops, *Allium fistulosum* and *Brassica napus*, on soil properties and microbial communities of ginseng cultivation in northeast China. *BMC Microbiology*, *22*(1), p.182

Bona, E., Massa, N., Novello, G., Boatti, L., Cesaro, P., Todeschini, V., Magnelli, V., Manfredi, M., Marengo, E., Mignone, F. and Berta, G., 2019. Metaproteomic characterization of the Vitis vinifera rhizosphere. *FEMS Microbiology Ecology*, *95*(1), p.fiy204.

Bonkowski, M., Villenave, C. and Griffiths, B., 2009. Rhizosphere fauna: the functional and structural diversity of intimate interactions of soil fauna with plant roots. *Plant Soil*, 321, pp. 213–233. https://doi.org/10.1007/s11104-009-0013-2

Borgatta, J., Ma, C., Hudson-Smith, N., Elmer, W., Plaza Perez, C.D., De La Torre-Roche, R., Zuverza-Mena, N., Haynes, C.L., White, J.C. and Hamers, R.J., 2018. Copper based nanomaterials suppress root fungal disease in watermelon (Citrullus lanatus): role of particle morphology, composition and dissolution behavior. *ACS Sustainable Chemistry & Engineering, 6*(11), pp.14847–14856.

Brown, R.L., Jacobs, L.A. and Peet, R.K., 2007. *Species richness: small scale. eLS.*

Cai, L., Chen, J., Liu, Z., Wang, H., Yang, H. and Ding, W., 2018. Magnesium oxide nanoparticles: effective agricultural antibacterial agent against Ralstonia solanacearum. *Frontiers in Microbiology, 9*, p.790.

Camilo, B.K., 2016. *Potential of bioreactors for attenuation of herbicides and nitrate in drainage waters.* Technische Universitaet Berlin (Germany).

Cao, L., Zhou, Z., Niu, S., Cao, C., Li, X., Shan, Y. and Huang, Q., 2017. Positive-charge functionalized mesoporous silica nanoparticles as nanocarriers for controlled 2, 4-dichlorophenoxy acetic acid sodium salt release. *Journal of Agricultural and Food Chemistry, 66*(26), pp.6594–6603.

Chapelle, E., Mendes, R., Bakker, P.A. and Raaijmakers, J.M., 2016. Fungal invasion of the rhizosphere microbiome. *The ISME Journal, 10*(1), pp.265–268. https://doi.org/10.1038/ismej.2015.82

Chaud, M., Souto, E.B., Zielinska, A., Severino, P., Batain, F., Oliveira-Junior, J. and Alves, T., 2021. Nanopesticides in agriculture: benefits and challenge in agricultural productivity, toxicological risks to human health and environment. *Toxics*, 9(6), p.131.

Chhabra, S., Brazil, D., Morrissey, J., Burke, J.I., O'Gara, F. and Dowling, N.D., 2013. Characterization of mineral phosphate solubilization traits from a barley rhizosphere soil functional metagenome. *Microbiologyopen, 2*(5), pp.717–724.

Clarkson, D.T. and Marschner, H., 1995. Mineral nutrition of higher plants. second edition. 889 pp. London: Academic Press, £29.95 (paperback). *Annals of Botany, 78*(4), pp.527–528.

Cromwell, W.A., Yang, J., Starr, J.L. and Jo, Y.K., 2014. Nematicidal effects of silver nanoparticles on root-knot nematode in bermudagrass. *Journal of Nematology, 46*(3), p.261.

Cumplido-Nájera, C.F., González-Morales, S., Ortega-Ortíz, H., Cadenas-Pliego, G., Benavides-Mendoza, A. and Juárez-Maldonado, A., 2019. The application of copper nanoparticles and potassium silicate stimulate the tolerance to Clavibacter michiganensis in tomato plants. *Scientia Horticulturae, 245*, pp.82–89.

Dastogeer, K.M., Tumpa, F.H., Sultana, A., Akter, M.A. and Chakraborty, A., 2020. Plant microbiome–an account of the factors that shape community composition and diversity. *Current Plant Biology, 23*, p.100161.

Delfani, M., Baradarn Firouzabadi, M., Farrokhi, N. and Makarian, H., 2014. Some physiological responses of black-eyed pea to iron and magnesium nanofertilizers. *Communications in Soil Science and Plant Analysis, 45*(4), pp.530–540.

Dhoke, S.K., Mahajan, P., Kamble, R. and Khanna, A., 2013. Effect of nanoparticles suspension on the growth of mung (Vigna radiata) seedlings by foliar spray method. *Nanotechnology Development, 3*(1), pp.e1–e1.

Dhondge, H.V., Barvkar, V.T., Paul, D., Dastager, S.G., Pable, A.A. and Nadaf, A.B., 2022. Exploring the core microbiota in scented rice (Oryza sativa L.) rhizosphere through metagenomics approach. *Microbiological Research, 263*, p.127157.

Dimkpa, C.O., Singh, U., Bindraban, P.S., Elmer, W.H., Gardea-Torresdey, J.L. and White, J.C., 2019. Zinc oxide nanoparticles alleviate drought-induced alterations in sorghum performance, nutrient acquisition, and grain fortification. *Science of the Total Environment, 688*, pp.926–934.

Dimkpa, C.O., White, J.C., Elmer, W.H. and Gardea-Torresdey, J., 2017. Nanoparticle and ionic Zn promote nutrient loading of sorghum grain under low NPK fertilization. *Journal of Agricultural and Food Chemistry, 65*(39), pp.8552–8559.

Effendi, Y., and Pancoro, A. (2021). Revealing the role of Plant Growth Promoting Rhizobacteria in suppressive soils against *Fusarium oxysporum* f. sp. cubense based on metagenomic analysis. *Heliyon, 7*(8), p.e07636.

Elmer, W.H. and White, J.C., 2016. The use of metallic oxide nanoparticles to enhance growth of tomatoes and eggplants in disease infested soil or soilless medium. *Environmental Science: Nano, 3*(5), pp.1072–1079.

Escobar-Zepeda, A., Vera-Ponce de León, A. and Sanchez-Flores, A., 2015. The road to metagenomics: from microbiology to DNA sequencing technologies and bioinformatics. *Frontiers in Genetics, 6*, p.348.

Fadiji, A.E. and Babalola, O.O., 2020. Metagenomics methods for the study of plant-associated microbial communities: a review. *Journal of Microbiological Methods, 170*, p.105860.

Fasusi, O.A., Cruz, C. and Babalola, O.O., 2021. Agricultural sustainability: microbial biofertilizers in rhizosphere management. *Agriculture*, *11*(2), p.163.

Ghafariyan, M.H., Malakouti, M.J., Dadpour, M.R., Stroeve, P. and Mahmoudi, M., 2013. Effects of magnetite nanoparticles on soybean chlorophyll. *Environmental Science & Technology*, *47*(18), pp.10645–10652.

Ghosh, A., Rathore, A. and Jayanth, B., 2021. Metagenomics of plant rhizosphere and endophytic association: concepts and applications. In Dubey, S.K. and Verma, S.K. (eds) *Plant, soil and microbes in tropical ecosystems* (pp.275–291). Springer Nature.

Global Report on Food Crises. 2022. www.wfp.org/publications/global-report-food-crises-2022.

Gogos, A., Knauer, K. and Bucheli, T.D., 2012. Nanomaterials in plant protection and fertilization: current state, foreseen applications, and research priorities. *Journal of Agricultural and Food Chemistry*, *60*(39), pp.9781–9792.

Gu, X., Yang, N., Zhao, Y., Liu, W. and Li, T., 2022. Long-term watermelon continuous cropping leads to drastic shifts in soil bacterial and fungal community composition across gravel mulch fields. *BMC Microbiology*, *22*(1), p.189.

Gupta, A., Gopal, M., Thomas, G.V., Manikandan, V., Gajewski, J., Thomas, G., Seshagiri, S., Schuster, S.C., Rajesh, P. and Gupta, R., 2014. Whole genome sequencing and analysis of plant growth promoting bacteria isolated from the rhizosphere of plantation crops coconut, cocoa and arecanut. *PLoS One*, *9*(8), p.e104259.

Haris, Z. and Ahmad, I., 2017. Impact of metal oxide nanoparticles on beneficial soil microorganisms and their secondary metabolites. *International Journal of Life-Science Research*, *3*(April), pp.1020–1030.

Hartmann, A., Rothballer, M. and Schmid, M., 2008. Lorenz Hiltner, a pioneer in rhizosphere microbial ecology and soil bacteriology research. *Plant and Soil, 312*, pp.7–14.

Hassan, M.K., McInroy, J.A. and Kloepper, J.W., 2019. The interactions of rhizodeposits with plant growth-promoting rhizobacteria in the rhizosphere: a review. *Agriculture*, *9*(7), p.142.

Hayden, H.L., Savin, K.W., Wadeson, J., Gupta, V.V. and Mele, P.M., 2018. Comparative metatranscriptomics of wheat rhizosphere microbiomes in disease suppressive and non-suppressive soils for Rhizoctonia solani AG8. *Frontiers in Microbiology, 9*, p.859.

Hjort, K., Presti, I., Elväng, A., Marinelli, F. and Sjöling, S., 2014. Bacterial chitinase with phytopathogen control capacity from suppressive soil revealed by functional metagenomics. *Applied Microbiology and Biotechnology*, *98*(6), 2819–2828. https://doi.org/10.1007/s00253-013-5287-x

Holden, N.J., Pritchard, L., Wright, K. and Toth, I.K., 2013. Mechanisms of plant colonization by human pathogenic bacteria: an emphasis on the roots and rhizosphere. *Molecular Microbial Ecology of the Rhizosphere, 1*, pp.1217–1226.

Hong, C.E., Kim, J.U., Lee, J.W., Bang, K.H. and Jo, I.H., 2019. Metagenomic analysis of bacterial endophyte community structure and functions in Panax ginseng at different ages. *3 Biotech*, *9*(8), 1–8. www.un.org/development/desa/pd/sites/www.un.org.development.desa.pd/files/undesa_pd_2022_wpp_key-messages.pdf

Huang, Z., Rajasekaran, P., Ozcan, A. and Santra, S., 2018. Antimicrobial magnesium hydroxide nanoparticles as an alternative to Cu biocide for crop protection. *Journal of Agricultural and Food Chemistry*, *66*(33), pp.8679–8686.

Ichihashi, Y., Date, Y., Shino, A., Shimizu, T., Shibata, A., Kumaishi, K., Funahashi, F., Wakayama, K., Yamazaki, K., Umezawa, A. and Sato, T., 2020. Multi-omics analysis on an agroecosystem reveals the significant role of organic nitrogen to increase agricultural crop yield. *Proceedings of the National Academy of Sciences*, *117*(25), pp.14552–14560.

Ijaz, M., Iqbal, M., Rasool, B., Zubair, M., Umirbekovna, I.A. and Bukhari, S.A., 2021. Rhizosphere dynamics: an OMICS perspective. In Pudake, R.N., Sahu, B.B, Kumari, M. and Sharma, A.K. (eds), *Omics science for rhizosphere biology* (pp.73–88). Springer Nature.

Imada, K., Sakai, S., Kajihara, H., Tanaka, S. and Ito, S., 2016. Magnesium oxide nanoparticles induce systemic resistance in tomato against bacterial wilt disease. *Plant Pathology*, *65*(4), pp.551–560.

Kah, M. and Hofmann, T., 2014. Nanopesticide research: current trends and future priorities. *Environment International, 63*, pp.224–235.

Kah, M., Tufenkji, N. and White, J.C., 2019. Nano-enabled strategies to enhance crop nutrition and protection. *Nature Nanotechnology*, *14*(6), pp.532–540.

Kanhed, P., Birla, S., Gaikwad, S., Gade, A., Seabra, A.B., Rubilar, O., Duran, N. and Rai, M., 2014. In vitro antifungal efficacy of copper nanoparticles against selected crop pathogenic fungi. *Materials Letters, 115*, pp.13–17.

Kareem, H.A., Saleem, M.F., Saleem, S., Rather, S.A., Wani, S.H., Siddiqui, M.H., Alamri, S., Kumar, R., Gaikwad, N.B., Guo, Z. and Niu, J., 2022. Zinc oxide nanoparticles interplay with physiological and biochemical attributes in terminal heat stress alleviation in mungbean (Vigna radiata L.). *Frontiers in Plant Science, 13*, p.101.

Knief, C., 2014. Analysis of plant microbe interactions in the era of next generation sequencing technologies. *Frontiers in Plant Science, 5*, p.216.

Kumar, D., Patel, Z., Pandit, P.R., Pandit, R., Puvar, A., Patel, A.K., Joshi, M. and Joshi, C.G., 2023. Textile industry wastewater microbiome: recovery of metagenome assembled genomes (MAGs) using shotgun sequencing approach from Jetpur, Gujarat, India. *Ecological Genetics and Genomics, 26*, p.100155.

Kumar, S., Nehra, M., Dilbaghi, N., Marrazza, G., Hassan, A.A. and Kim, K.H., 2019. Nano-based smart pesticide formulations: emerging opportunities for agriculture. *Journal of Controlled Release,* 294, pp.131–153.

Kuramae, E.E., Derksen, S., Schlemper, T.R., Dimitrov, M.R., Costa, O.Y. and da Silveira, A.P., 2020. Sorghum growth promotion by Paraburkholderia tropica and Herbaspirillum frisingense: putative mechanisms revealed by genomics and metagenomics. *Microorganisms*, *8*(5), p.725.

Le Van, N., Ma, C., Shang, J., Rui, Y., Liu, S. and Xing, B., 2016. Effects of CuO nanoparticles on insecticidal activity and phytotoxicity in conventional and transgenic cotton. *Chemosphere, 144*, pp.661–670.

Li, L., Tutone, A.F., Drummond, R.S., Gardner, R.C. and Luan, S., 2001. A novel family of magnesium transport genes in Arabidopsis. *The Plant Cell, 13*(12), pp.2761–2775.

Li, P., Ye, S., Chen, J., Wang, L., Li, Y., Ge, L., Wu, G., Song, L., Wang, C., Sun, Y. and Wang, J., 2023. Combined metagenomic and metabolomic analyses reveal that Bt rice planting alters soil CN metabolism. *ISME Communications*, *3*(1), p.4.

Liu, J., Li, R. and Yang, B., 2020. Carbon dots: a new type of carbon-based nanomaterial with wide applications. *ACS Central Science*, *6*(12), pp.2179–2195.

Liu, R., Zhang, H. and Lal, R., 2016. Effects of stabilized nanoparticles of copper, zinc, manganese, and iron oxides in low concentrations on lettuce (Lactuca sativa) seed germination: nanotoxicants or nanonutrients? *Water, Air, & Soil Pollution, 227*, pp.1–14.

Longepierre, M., Feola Conz, R., Barthel, M., Bru, D., Philippot, L., Six, J., and Hartmann, M., 2022. Mixed effects of soil compaction on the nitrogen cycle under pea and wheat. *Frontiers in Microbiology*, 12, p.822487.

Ma, X., Geiser-Lee, J., Deng, Y. and Kolmakov, A., 2010. Interactions between engineered nanoparticles (ENPs) and plants: phytotoxicity, uptake and accumulation. *Science of the Total Environment*, *408*(16), pp.3053–3061.

Mahto, R., Chatterjee, N., Priya, T. and Singh, R.K., 2019. Nanotechnology and its role in agronomic crops. *In Hasanuzzaman, M. (ed) Agronomic crops* (pp.605–636). Springer.

Manjunatha, S.B., Biradar, D.P. and Aladakatti, Y.R., 2016. Nanotechnology and its applications in agriculture: a review. *Journal of Farm Sciences*, *29*(1), pp.1–13.

Mhlongo, M.I., Piater, L.A., Madala, N.E., Labuschagne, N. and Dubery, I.A., 2018. The chemistry of plant–microbe interactions in the rhizosphere and the potential for metabolomics to reveal signaling related to defense priming and induced systemic resistance. *Frontiers in Plant Science, 9*, p.112.

Mirkasymov, A.B., Zelepukin, I.V., Nikitin, P.I., Nikitin, M.P. and Deyev, S.M., 2021. In vivo blockade of mononuclear phagocyte system with solid nanoparticles: efficiency and affecting factors. *Journal of Controlled Release, 330*, pp.111–118.

Mishra, S., Singh, B.R., Singh, A., Keswani, C., Naqvi, A.H. and Singh, H.B., 2014. Biofabricated silver nanoparticles act as a strong fungicide against Bipolaris sorokiniana causing spot blotch disease in wheat. *Plos One*, *9*(5), p.e97881.

Moll, J., Klingenfuss, F., Widmer, F., Gogos, A., Bucheli, T.D., Hartmann, M. and van der Heijden, M.G., 2017. Effects of titanium dioxide nanoparticles on soil microbial communities and wheat biomass. *Soil Biology and Biochemistry, 111*, pp.85–93.

Muscatt, G., Hilton, S., Raguideau, S., Teakle, G., Lidbury, I.D., Wellington, E.M., Quince, C., Millard, A., Bending, G.D. and Jameson, E., 2022. Crop management shapes the diversity and activity of DNA and RNA viruses in the rhizosphere. *Microbiome*, *10*(1), p.181.

Narayanan, K.B. and Park, H.H., 2014. Antifungal activity of silver nanoparticles synthesized using turnip leaf extract (Brassica rapa L.) against wood rotting pathogens. *European Journal of Plant Pathology, 140*, pp.185–192.

Ocsoy, I., Paret, M.L., Ocsoy, M.A., Kunwar, S., Chen, T., You, M. and Tan, W., 2013. Nanotechnology in plant disease management: DNA-directed silver nanoparticles on graphene oxide as an antibacterial against Xanthomonas perforans. *ACS Nano*, *7*(10), pp.8972–8980.

Paret, M.L., Vallad, G.E., Averett, D.R., Jones, J.B. and Olson, S.M., 2013. Photocatalysis: effect of light-activated nanoscale formulations of TiO2 on Xanthomonas perforans and control of bacterial spot of tomato. *Phytopathology*, *103*(3), pp.228–236.

Pérez-de-Luque, A. and Rubiales, D., 2009. Nanotechnology for parasitic plant control. *Pest Management Science: Formerly Pesticide Science*, *65*(5), pp.540–545.

Pinu, F.R., Beale, D.J., Paten, A.M., Kouremenos, K., Swarup, S., Schirra, H.J. and Wishart, D., 2019. Systems biology and multi-omics integration: viewpoints from the metabolomics research community. *Metabolites*, *9*(4), p.76.

Prabha, R., Singh, D.P., Gupta, S., Gupta, V.K., El-Enshasy, H.A. and Verma, M.K., 2019. Rhizosphere metagenomics of Paspalum scrobiculatum l.(kodo millet) reveals rhizobiome multifunctionalities. *Microorganisms*, *7*(12), p.608.

Prasad, R., Bhattacharyya, A. and Nguyen, Q.D., 2017. Nanotechnology in sustainable agriculture: recent developments, challenges, and perspectives. *Frontiers in Microbiology, 8*, p.1014.

Puntillo, M., Peralta, G.H., Bürgi, M.D.M., Huber, P., Gaggiotti, M., Binetti, A.G. and Vinderola, G., 2022. Metaprofiling of the bacterial community in sorghum silages inoculated with lactic acid bacteria. *Journal of Applied Microbiology*, *133*(4), pp.2375–2389.

Rajput, V.D., Kumari, A., Upadhyay, S.K., Minkina, T., Mandzhieva, S., Ranjan, A., Sushkova, S., Burachevskaya, M., Rajput, P., Konstantinova, E. and Singh, J., 2023. Can nanomaterials improve the soil microbiome and crop productivity? *Agriculture, 13*(2), p.231.

Ramette, A., LiPuma, J.J., and Tiedje, J.M., 2005. Species abundance and diversity of Burkholderia cepacia complex in the environment. *Applied and Environmental Microbiology*, 71(3), 1193–1201.

Rathore, I. and Tarafdar, J.C., 2015. Perspectives of biosynthesized magnesium nanoparticles in foliar application of wheat plant. *Journal of Bionanoscience*, *9*(3), pp.209–214.

Rizwan, M., Ali, S., ur Rehman, M.Z., Adrees, M., Arshad, M., Qayyum, M.F., Ali, L., Hussain, A., Chatha, S.A.S. and Imran, M., 2019. Alleviation of cadmium accumulation in maize (Zea mays L.) by foliar spray of zinc oxide nanoparticles and biochar to contaminated soil. *Environmental Pollution, 248*, pp.358–367.

Roduner, E., 2006. Size matters: why nanomaterials are different. *Chemical Society Reviews*, *35*(7), pp.583–592.

Rosselli, R. and Squartini, A., 2015. Metagenomics of plant–microbe interactions. In Sablok, G., Kumar, S., Ueno, S., Kuo, J., and Varotto, C. (eds) *Advances in the understanding of biological sciences using next generation sequencing (NGS) approaches* (pp.135–153). Springer Nature.

Rounsley, S.D. and Last, R.L., 2010. Shotguns and SNPs: how fast and cheap sequencing is revolutionizing plant biology. *The Plant Journal*, *61*(6), pp.922–927.

Rui, M., Ma, C., Hao, Y., Guo, J., Rui, Y., Tang, X., Zhao, Q., Fan, X., Zhang, Z., Hou, T. and Zhu, S., 2016. Iron oxide nanoparticles as a potential iron fertilizer for peanut (Arachis hypogaea). *Frontiers in Plant Science, 7*, p.815.

Salavati-Niasari, M., Davar, F. and Mir, N., 2008. Synthesis and characterization of metallic copper nanoparticles via thermal decomposition. *Polyhedron*, *27*(17), pp.3514–3518.

Sengupta, S., Ganguli, S. and Singh, P.K., 2017. Metagenome analysis of the root endophytic microbial community of Indian rice (O. sativa L.). *Genomics Data, 12*, pp.41–43.

Servin, A., Elmer, W., Mukherjee, A., De la Torre-Roche, R., Hamdi, H., White, J.C., Bindraban, P. and Dimkpa, C., 2015. A review of the use of engineered nanomaterials to suppress plant disease and enhance crop yield. *Journal of Nanoparticle Research, 17*, pp.1–21.

Shan, J., Ji, R., Yu, Y., Xie, Z. and Yan, X., 2015. Biochar, activated carbon and carbon nanotubes have different effects on fate of 14C-catechol and microbial community in soil. *Scientific Reports*, *5*(1), pp.1–11.

Shanmugam, S., Hari, A., Pandey, A., Mathimani, T., Felix, L., & Pugazhendhi, A. (2020). Comprehensive review on the application of inorganic and organic nanoparticles for enhancing biohydrogen production. *Fuel, 270*, 117453.

Shcherbakova, E.N., Shcherbakov, A.V., Andronov, E.E., Gonchar, L.N., Kalenskaya, S.M. and Chebotar, V.K., 2017. Combined pre-seed treatment with microbial inoculants and Mo nanoparticles changes composition of root exudates and rhizosphere microbiome structure of chickpea (Cicer arietinum L.) plants. *Symbiosis, 73*, pp.57–69.

Soni, R., Kumar, V., Suyal, D.C., Jain, L. and Goel, R., 2017. Metagenomics of plant rhizosphere microbiome. *In Singh, R.P., Kothari, R., Koringa, P.G., and Singh, S.P. (eds) Understanding host-microbiome interactions-an omics approach: omics of host-microbiome association* (pp.193–205). Springer Nature.

Subbaiah, L.V., Prasad, T.N.V.K.V., Krishna, T.G., Sudhakar, P., Reddy, B.R. and Pradeep, T., 2016. Novel effects of nanoparticulate delivery of zinc on growth, productivity, and zinc biofortification in maize (Zea mays L.). *Journal of Agricultural and Food Chemistry*, *64*(19), pp.3778–3788.

Sun, H., Jiang, S., Jiang, C., Wu, C., Gao, M. and Wang, Q., 2021. A review of root exudates and rhizosphere microbiome for crop production. *Environmental Science and Pollution Research, 28*, pp.54497–54510.

Suzuki, M.T. and Giovannoni, S.J., 1996. Bias caused by template annealing in the amplification of mixtures of 16S rRNA genes by PCR. *Applied and Environmental Microbiology*, *62*(2), pp.625–630.

Tai, C.Y., Tai, C.T., Chang, M.H. and Liu, H.S., 2007. Synthesis of magnesium hydroxide and oxide nanoparticles using a spinning disk reactor. *Industrial & Engineering Chemistry Research*, *46*(17), pp.5536–5541.

Tracanna, V., Ossowicki, A., Petrus, M.L., Overduin, S., Terlouw, B.R., Lund, G., Robinson, S.L., Warris, S., Schijlen, E.G., Van Wezel, G.P. and Raaijmakers, J.M., 2021. Dissecting disease-suppressive rhizosphere microbiomes by functional amplicon sequencing and 10× metagenomics. *MSystems, 6*(3), pp.e01116–20.

Trivedi, P., Mattupalli, C., Eversole, K. and Leach, J.E., 2021. Enabling sustainable agriculture through understanding and enhancement of microbiomes. *New Phytologist*, *230*(6), pp.2129–2147.

Tsurumaru, H., Okubo, T., Okazaki, K., Hashimoto, M., Kakizaki, K., Hanzawa, E., Takahashi, H., Asanome, N., Tanaka, F., Sekiyama, Y. and Ikeda, S., 2015. Metagenomic analysis of the bacterial community associated with the taproot of sugar beet. *Microbes and Environments*, *30*(1), pp.63–69.

Urumbil, S.K. and Anilkumar, M., 2021. Metagenomic insights into plant growth promoting genes inherent in bacterial endophytes of Emilia sonchifolia (Linn.) DC. *Plant Science Today*, *8*(sp1), pp.6–16.

Vieira, A.F., Moura, M. and Silva, L., 2021. Soil metagenomics in grasslands and forests–A review and bibliometric analysis. *Applied Soil Ecology, 167*, p.104047.

Wang, C., Yue, L., Cheng, B., Chen, F., Zhao, X., Wang, Z. and Xing, B., 2022. Mechanisms of growth-promotion and Se-enrichment in Brassica chinensis L. by selenium nanomaterials: beneficial rhizosphere microorganisms, nutrient availability, and photosynthesis. *Environmental Science: Nano*, *9*(1), pp.302–312.

Wang, H., Zhang, R., Zhao, Y., Shi, H. and Liu, G., 2022. Effect of biochar on rhizosphere soil microbial diversity and metabolism in tobacco-growing soil. *Ecologies*, *3*(4), pp.539–556

Wang, L., Hu, C. and Shao, L., 2017. The antimicrobial activity of nanoparticles: present situation and prospects for the future. *International Journal of Nanomedicine, 12*, p.1227.

Wang, X., Liu, X., Chen, J., Han, H. and Yuan, Z., 2014. Evaluation and mechanism of antifungal effects of carbon nanomaterials in controlling plant fungal pathogen. *Carbon, 68*, pp.798–806.

White, R.A., Borkum, M.I., Rivas-Ubach, A., Bilbao, A., Wendler, J.P., Colby, S.M., Köberl, M. and Jansson, C., 2017. From data to knowledge: the future of multi-omics data analysis for the rhizosphere. *Rhizosphere, 3*, pp.222–229.

World Population Prospects. 2022. *Summary of Results.*

Wu, H., Fang, C., Malacrinò, A., Winkelmann, T. and Xiong, W., 2022. Rhizosphere conversation among the plant-plant microbiome-soil under consecutive monoculture regimes. *Frontiers in Microbiology, 13*, p.1061427.

Xie, J., Yan, Z., Wang, G., Xue, W., Li, C., Chen, X. and Chen, D., 2021. A bacterium isolated from soil in a Karst rocky desertification region has efficient phosphate-solubilizing and plant growth-promoting ability. *Frontiers in Microbiology, 11*, p.625450.

Xu, L., Naylor, D., Dong, Z., Simmons, T., Pierroz, G., Hixson, K. K., ... and Coleman-Derr, D., 2018. Drought delays development of the sorghum root microbiome and enriches for monoderm bacteria. *Proceedings of the National Academy of Sciences*, *115*(18), pp. E4284–E4293.

Yan, H., Zhu, L., Wang, Y., Zhang, S., Liu, P., Dong, T.T., Wu, Q. and Duan, J.A., 2021. Comparative metagenomics analysis of the rhizosphere microbiota influence on Radix Angelica sinensis in different growth soil environments in China. *Food Science and Technology, 41*, pp.775–784.

Yoon, K.Y., Byeon, J.H., Park, J.H. and Hwang, J., 2007. Susceptibility constants of Escherichia coli and Bacillus subtilis to silver and copper nanoparticles. *Science of the Total Environment*, *373*(2–3), pp.572–575.

Yousuf, B., Keshri, J., Mishra, A. and Jha, B., 2012. Application of targeted metagenomics to explore abundance and diversity of CO2-fixing bacterial community using cbbL gene from the rhizosphere of Arachis hypogaea. *Gene*, *506*(1), pp.18–24.

Zeliadt, N., 2010. Silver beware: antimicrobial nanoparticles in soil may harm plant life. *Scientific American.* www. scientificamerican. com/article. cfm.

Zhang, H., Du, W., Peralta-Videa, J.R., Gardea-Torresdey, J.L., White, J.C., Keller, A., Guo, H., Ji, R. and Zhao, L., 2018. Metabolomics reveals how cucumber (Cucumis sativus) reprograms metabolites to cope with silver ions and silver nanoparticle-induced oxidative stress. *Environmental Science & Technology*, *52*(14), pp.8016–8026.

Zhong, C., Chen, C., Gao, X., Tan, C., Bai, H. and Ning, K., 2022. Multi-omics profiling reveals comprehensive microbe–plant–metabolite regulation patterns for medicinal plant Glycyrrhiza uralensis Fisch. *Plant Biotechnology Journal*, *20*(10), pp.1874–1887.

Zia, R., Nawaz, M.S., Siddique, M.J., Hakim, S. and Imran, A., 2021. Plant survival under drought stress: implications, adaptive responses, and integrated rhizosphere management strategy for stress mitigation. *Microbiological Research, 242*, p.126626.

11 Nanofertilization and Root Growth Regulation

Ahmed Abou El-Yazied, Mahmoud A.M. Abd El-Hady, Mahmoud Adel Ahmed Ali, Hany G. Abd El-Gawad and Mohamed F.M. Ibrahim

11.1 INTRODUCTION

Nowadays, agriculture is facing a bunch of problems. On one hand, the population is increasing with a demand for a huge supply of foods; on the other hand, input use efficiency is decreasing, food and land quality is degrading with a subsequent environmental misbalance. The agricultural industry is dealing with a wide range of issues, including crop yield regression, poor nutrient use efficiency, declining organic matter in the soil, nutrient deficiencies, climate change, decreasing arable land, and water availability. To solve these issues, it is necessary to investigate promising technologies like "nanotechnology," which can accurately detect and administer the right amount of nutrients and pesticides while maintaining environmental safety and greater usage efficiency (Basit et al., 2022). In this scenario, adoption of more efficient tools, which could alleviate the drawbacks and lead the agriculture in a sustainable way is the need of the hour. Fertilizers are crucial for increasing soil fertility and crop yield. One third of crop productivity is influenced by fertilizers, which also have an impact on how effectively other agri-inputs are used. Despite our ongoing efforts over the previous decades, crop nutrients' utilization efficiency remained steady. Nutrients remaining in the soil can infiltrate the aquatic environment and cause eutrophication. Furthermore, to low nutrient efficiency, agriculture in developing areas, is bothered by low organic matter, uneven fertilization, and low fertilizer feedback, all of which have resulted in lack of crop production (Ditta et al., 2020a,b; Reeve and Fallowfield, 2018).

Nanofertilizers are among the advanced inputs that have been discovered in the field of agriculture and proven to be effective (Mejías et al., 2021). Using nanofertilizers that differ from traditional fertilizers, can feed plants in a gradual and regulated manner is one of the most significant applications of nanotechnology in agriculture today. As well as being easily absorbed, nanofertilizers are released slowly and are required in small amounts, minimizing the use of fertilizers and their unfavorable impact on the environment (Gómez-Merino and Trejo-Téllez, 2018). Chemical fertilizers can be replaced by nanofertilizers (NFs), which are effective, cutting-edge, and environmentally friendly alternatives that should maximize benefits and reduce risks (Kalwani et al., 2022). Nanofertilizers have a high use efficiency and can be transported to a rhizosphere target in a timely manner (Medici et al., 2023; Sayed et al., 2022).

The science of plant nanonutrition is concerned with the uptake, translocation, metabolism, bioavailability of nanonutrients in the rhizosphere for plant growth and development, or applications of nanotechnology to provide nanonutrients for crop production (Ditta et al., 2015). The phytotoxicity of nanoparticles, the interactions between nanonutrients and plants (Reddy et al., 2016), soil (He et al., 2016), soil biota (Ibrahim et al., 2016), and the entire agroecosystem (Servin and White, 2016) are also included in this branch of plant nutrition. As a result, this review concentrates on plant nanonutrients, including how plants and soil react to them.

 DOI: 10.1201/9781003354109-11

11.2 SYNTHESIS OF NANOFERTILIZERS

Nanofertilizers can be synthesized via a variety of techniques, including physical, chemical, and biological processes. The method chosen is determined by the type of nanofertilizer, its qualities, and the desired application.

The methodology of nanotechnology fertilizers includes the use of nutrient-containing nanoparticles. Both physical (top-down) and chemical (bottom-up) processes can be used to create the nanomaterial. These nanoparticles are employed clearly to boost plant growth and production. The top-down technique involves applying physical or chemical treatments to reduce bulk materials to the nanoscale. Grinding, ball milling, and sonication are examples of physical processes. Bottom-up nanoscale is the self-assembly and self-organization of smaller material building components to produce useful nanoscale materials. The resulting nanoscale fertilizer provides high production, nutritious food, conserves natural resources, improves soil biological health, increases microbial diversity, improves nutrient mobilization, and reduces fertilizer consumption (An et al., 2022; Kalia et al., 2020; Raliya et al., 2018; Solanki et al., 2015). Top-down and bottom-up synthesis methods can be further categorized as chemical, physical, and biological synthesis methods. Bottom-up methods mostly include chemical synthesis of nanoparticles, whereas top-down methods primarily involve physical and biological synthesis (Bahrulolum et al., 2021; Nagar and Devra, 2018; Najafi et al., 2020; Salih et al., 2021). For nanofertilizers, the broad categories of nonmaterial synthesis routes such as top-down and bottom-up work well. Manganese, zinc, and iron nanofertilizers with particle sizes ranging from 20 to 60 nm were created utilizing a green microwave aided hydrothermal technique using nitrate precursors (Shebl et al., 2019). The following phase in nanofertilizer manufacturing is to consider novel dosage mechanisms such as gradual release and controlled release. Silicon encapsulating chitosan nanofertilizers added to wheat improved salt tolerance in the wheat plant (Hajihashemi and Kazemi, 2022). Plant extracts like stem, leaf, root, and seed are applied for the synthesis of nanoparticles as they perform as stabilizing or reduction agents. Using biogenic synthesis, the potassium nanofertilizer was created from the plant leaves. Higher growth and yield were seen in the wheat crop when the well-characterized nanophase fertilizer was used. The dosage through spraying showed the simplicity of absorption and decrease in leaching (Sheoran et al., 2021).

Chemical synthesis methods use molecular or atomic-level processes and are referred to as bottom-up techniques since they start at the level of the atom and use chemical reactions to create nanoparticles. Various chemical processes increase up atoms to nanoparticles. These techniques offer more precise control of particle size, lower impurities, low temperature processes, and controlled synthesis (Buzea and Pacheco, 2017; Shalaby et al., 2016; Toksha et al., 2021). Physical synthesis can be done using a variety of methods, including the gas condensation method, attrition ball mills, planetary ball mills, vibrating ball mills, low energy tumble mills, and high energy (Basit et al., 2022; Toksha et al., 2021). Physical synthesis techniques are typically employed as the primary method in the creation of nanofertilizers due to their simplicity, reduced time requirements, and potential for production on a large scale (Ditta et al., 2020a; Toksha et al., 2021). Biological synthesis, microorganisms like bacteria and fungus are currently being used as prospective biomanufacturing units to perform nanosynthesis, which is the present emphasis of nanobiotechnology. Different plant extracts or microorganisms are primarily used in the production of nanoparticles. These environmentally friendly reagent-based biosynthetic pathways lower pollution risk at the source stage. Nanoparticles made from plants can be created or extracted from easily accessible plant material due to the plant's benign nature. The manufacture of nanoparticles or nanomaterials for the creation of nanofertilizers uses a variety of plant parts, including stems, leaves, roots, fruits, and their extracts (Sheoran et al., 2021).

11.3 MECHANISMS OF NANOFERTILIZERS

The two ways that plants are exposed to nanoparticles through foliar and root systems. The cuticle is believed to serve as a physical barrier to prevent nanoparticles from entering, higher plants' leaves are shielded against water loss and unregulated exchange of other solutes by their waxy cuticles (Philippe et al., 2020). Different modes are used to regulate the entrance and absorption of nanofertilizers via plant cells. The nanofertilizers bind to carrier proteins, and/or enter through aquaporins, ion channels, or endocytosis, by developing required pores or by binding to organic chemicals in the environment. Nanofertilizers link to carrier proteins, and/or submit through aquaporins, ion channels, endocytosis, through using necessary pores or via connecting to organic elements in the environment (Ma et al., 2018; Pradhan and Mailapalli, 2017). One of the crucial factors being thought about is designing and creating nanofertilizers for controlled release, which will lead to increased nutritional absorption rates. Slow and controlled release fertilizers are being developed because of nanofertilizer technology research for sustainable and precise agriculture. Less frequent administration, lower labor costs, assuring minimal input consumption, and use in the distribution of insecticides, food, and drugs are some of the uses and aims of controlled release (Jampílek and Kráľová, 2017; Shafiq et al., 2020; Tymoszuk, 2021; Yadav et al., 2022). As a result, fertilizers will eventually be used more effectively, which will reduce nutrient losses to the environment and make fertilizers more environmentally friendly. Coatings or encapsulation of inorganic or organic components are a characteristic of controlled release fertilizers. This characteristic regulates the ratio, pattern, and time of plant nutrient release (Su et al., 2019). By using semi-permeable coatings or gradually hydrolysis of water soluble, low molecular weight formulations, the just-needed release is accomplished. It was recommended to employ nano-regime fertilizers, and the crucial difficulties involved, such as societal acceptance and use as nano formulations for regulatory bodies. Since traditional fertilizers suffer from poor uptake efficiency, they must be utilized in large quantities. Nanocomposites of urea and hydroxyapatite show mature administration, lower volatilization, and continuous availability of N-P-K compared to conventional fertilizers, which phosphate fertilizers are lost by chemical bonding in soil and were unavailable to plants and nitrogen lost by leaching (Madusanka et al., 2017; Rop et al., 2018). In the ideal scenario, nanofertilizers would be able to release nutrients when plants needed them to, preventing the loss of extra fertilizers through gas phase or via leaching. To accomplish this, the release of nutrients is supposed to be dependent on plant signaling (Dimkpa and Bindraban, 2018; Marchiol et al., 2020).

A significant advancement toward controlled release nanofertilizers is the mechanism using stimuli-based release, which has the potential to offer more intelligent and active delivery systems than the current diffusion and ion exchange pathways. The synthesis, characterization, and demonstration of chitosan gated $FeSO_4$ power loaded mesoporous silica showed three benefits in fertilizer delivery, as shown below. First, only in the presence of an acidic root zone does this gate molecule exhibit selective release of loaded $FeSO_4$. Second, chitosan covering acts as a gate to stop carbonate and bicarbonate access, which would otherwise cause the iron to be converted into unavailable forms in the soil, such as oxyhydroxyl. Third, in comparison to the negatively charged pectin surface, the chitosan also improves the binding of the $FeSO_4$- mesoporous silica- chitosan composite to the root surface. When compared to a plant treated with uncoated composite, the benefits decreased the iron deficiency parameter FCR by about 20% in $FeSO_4$- mesoporous silica- chitosan treated plants (Bindra et al., 2019). Such stimuli, for instance, ethylene synthesis via roots and acidification of the rhizosphere released in response to P or K deficit in soils, are influenced by ethylene production with Arabidopsis roots (Sarmast and Salehi, 2016; Syu et al., 2014). It is reasonable to assume from studies that developed nanofertilizers may considerably decrease the quantity of fertilizer used improving their efficiency and lowering release into the soil as compared to conventional formulations. These nanofertilizers release measured amounts of necessary and adequate quantities of the essential ingredients over an extended period and to obtain full biological ability while minimizing the loss and harmful effects (Jampílek and Kráľová, 2017; Shafiq et al., 2020; Yadav et al., 2022).

Conventional fertilizers that contain nanoscale additives are referred to as nanoscale additive fertilizers. Crop rhizospheres may contain nanomaterial as an additive to promote plant growth, rather than necessarily as the nutrition itself. Study of treatment with nanostructured silica and its impact on select tree seedlings' growth (Bao-shan et al., 2004). They discovered that following absorption, nanostructured silica creates a protective coating at the cell wall that is expected to increase plant stress tolerance. They immersed the roots of seedlings in nanostructured silica solutions for their studies. A statistically significant increase in height, main root length, root diameter, and number of lateral roots was discovered at 500 μL of silica/L of treatment. Additionally, the chlorophyll content was higher after this treatment than it was in the controls (Bao-shan et al., 2004). Additionally, nano silica has been used as a water and mineral adsorbent in fertilizer treatments (Aminzare et al., 2013; Fernando et al., 2015). According to a study, adding nanoscale Se particles (less than 400 nm in size), created by milling, increased the amount of selenium in green tea, resulting in nanoscale selenium rich green tea with anticancer activity (El-Ramady et al., 2015). It is also worthwhile to note that the physiological circumstances are improved by the nanoadditives, the addition of ingredients to fertilizers in order to increase the nutrient uptake (Gámez-Espinosa et al., 2020). The synthesis of nanoscale coating fertilizers involves coating or loading nanoparticles onto conventional fertilizers. The modifications could consist of: first, encapsulating nutrients of interest in nanophases; second, coatings with thin polymer film particles; third, coating with nanoscale-sized emulsions. It is advantageous to encapsulate bacterial or fungal germs because this increases the availability of N-P-K in the root zone, which promotes plant growth (Ditta et al., 2020b; Mukhopadhyay and Kaur, 2016). Different dimensions at the nanoscale, such as coatings, nanothin films, or nanoporous materials used to contain fertilizers, could be utilized to expel nutrients from crops (Pitambara et al., 2019; Zulfiqar et al., 2019). The excessive use of chemical fertilizers can cause harm to human health due to trace elements that enter the body. Controlled release mechanisms, such as nanofertilizers, have well-defined release patterns and rates, making them a promising option. The use of free pesticide dosage has a hazardous effect on soil microorganisms, while slow-release mechanisms are more effective in preserving their bioremediation characteristics. Experimental results show that slow-release mechanisms can preserve the degradation potential of microflora and contribute towards removing harmful substances from the environment (Iavicoli et al., 2017; Lu et al., 2022; Vejan et al., 2021). The study by Sharma et al. (2020) utilized foliar spray and seed treatment modes with a chitosan nanofertilizer that released co-encapsulated copper and salicylic acid in a slow-release mode. This method allowed for store food remobilization with a porous nanofertilizer.

11.4 APPLICATION OF NANOFERTILIZATION IN PLANTS

Nanotechnology has magnificent implementations in agricultural sciences. Nanofertilizers enhance the efficiency of nutrients which reflect on high crop productivity (Tarafdar et al., 2013; Ditta et al., 2015; Panpatte et al., 2016). Nanoparticles have various features in reinforcing the agricultural productivity through promoting seed germination and growth (Khan et al., 2017), enhancing the efficacy of water and fertilizer doses. Many researchers reported the effect of nanofertilizers for plant nutrition (Liu et al., 2016; Panpatte et al., 2016), nanosmart delivery systems (Ditta et al., 2015), nanocapsules as carriers (Meredith et al., 2016). Several agricultural practices can be done for improving plant production, saving the time and costs by using various applications of nanofertilizers (Subramanian et al., 2016).

Nanomaterials in samll size assist plants easily penetrate and regulate water channels and improve adsorption and allocation substances, which reflect on plant growth (Khan et al., 2017). Regarding the uptake path by plants and nanoparticle translocation, two paths in plant system have been reported either from roots to leaves (soil application) or from leaves to roots (foliar application)

(Ma et al., 2015; Du et al., 2017). Various physiological, agronomical, photosynthetic parameters have been recorded due to the interaction between nanoparticles and plant system (Du et al., 2017; Reddy et al., 2016; Rizwan et al., 2017; Zuverza-Mena et al., 2017). According to reports, the tiny NPs enter plant roots through the root epidermal cells, capillary forces, or osmotic pressure (Lin and Xing, 2008; Du et al., 2011) (Figure 11.1). The semipermeable root cell wall's epidermal cells possess microscopic pores that limit the movement of big NPs. Some NPs caused the epidermal cell wall to develop additional pores, which made it easier for it to enter.

Mineral fertilizers are important in increasing plant production, but it pollutes the environment in addition to low use efficiency from plants (30–40%) as well as other amounts lost by leaching (Panpatte et al., 2016). Therefore, using nanofertilizers can enhance nutrient-use efficiency in addition to decreasing environmental pollution using proper amounts of nanofertilizers. On the other hand, nanofertilizers are mostly made by encapsulating fertilizers inside of nanoparticles. Rai et al. (2012) showed that the major techniques encapsulating fertilizers with nanoparticles either encapsulate fertilizers in nanoporous materials or coat with fluffy polymer film in addition to transferring nutrients in nanoscale dimensions (Panpatte et al., 2016).

Nanofertilizers are fertilizers consisting of nanostructured formulations that can be transported to targeted sites and allowed to release active ingredients, which keep demands of plant nutrients (Wang et al., 2016a, b). Therefore, nanofertilizers can be deemed promising to increase growth, nutrition, and plant productivity and regulate nutrient releases as well as improve nutrient-use efficiency under controlled environmental conditions (Yuvaraj and Subramanian, 2015; Subramanian et al., 2015; 2016).

Nanomembranes that permit the slow and consistent release of nutrients can be applied to fertilizer particles. This procedure increases crops' ability to utilize fertilizers while reducing nutrient loss. The low-release and super sorbent nitrogenous and phosphatic fertilizers can be used to deliver nanofertilizers in a timely manner to a rhizospheric target with high use efficiency. As

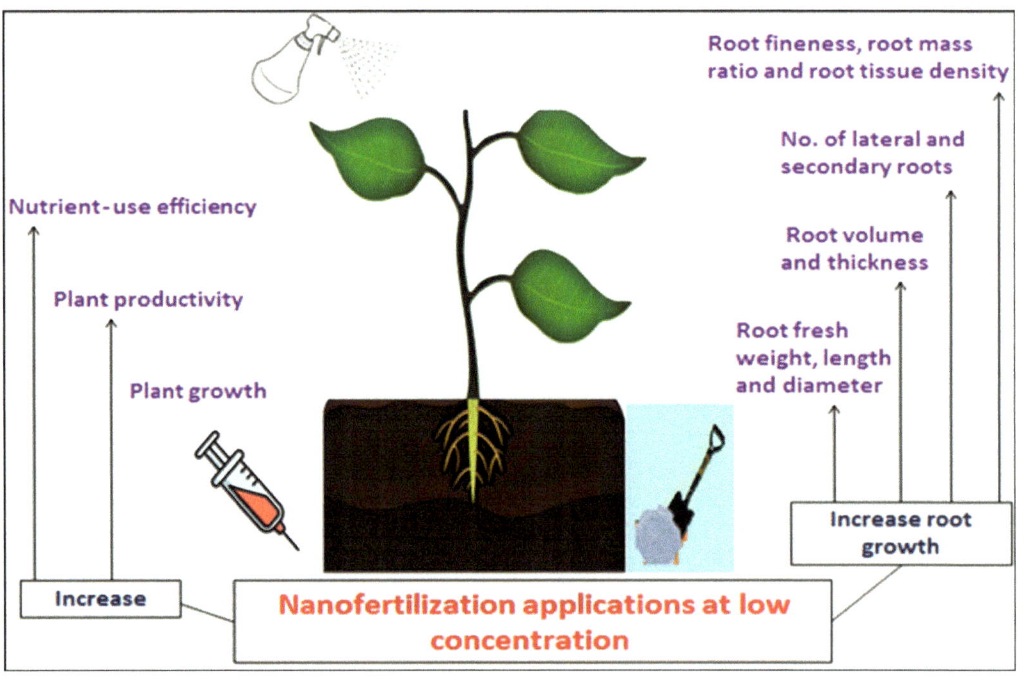

FIGURE 11.1 Schematic representation of different nanoparticle applications in plants showing the different effects on plants.

well, controlled-release fertilizers can ameliorate soil by lessening toxic impacts connected with high addition of mineral fertilizers (Suman et al., 2010). Nanoscale carriers could be applied for the effective delivery of fertilizers. These can ameliorate their stabilization against degradation in the environment then decrease an applied amount, which decreases chemical runoff and mitigates environmental problems. In this regard, Kottegoda et al. (2011) studied the effect of utilizing urea-modified hydroxyapatite nanoparticles that were encapsulated under pressure to sustain the release of nitrogen into soil, and showed that the release of N from the nanofertilizer followed a sequence of two-step operation: an initial explosion then slowly releases for up to 2 months.

Several plants have been fertilized with nanofertilizers as presented in Table 11.1. These researches contain fate and behavior of nanofertilizers in various agroecosystem compartments, and evaluation the possibility mechanism of nanofertilizer toxicity and their tolerance (Anjum et al., 2015). It has been found that synthetic nanoparticles are not constantly more harmful than chemical species that contain the same elements. For example, CuO nanoparticles are more toxic than Cu ions, while ZnO nanoparticles are as toxic as Zn ions, whereas MnO and FeO nanoparticles had lower toxicity than Mn and Fe ions in addition to an increase in lettuce growth (Liu et al., 2016). Thus, the application of nanofertilizers is new strategies could be performed in agricultural production to save the universal food production, which suffers from diverse challenges as pollution and low soil fertility.

By speeding up the reaction or synthesis process in the plant system, nanofertilizers improve the crop plant's surface area and nutrient availability, which contributes to a rise in the quality parameters of the plant (such as protein, oil, and sugar content). Several studies focused on the effect of nanofertilizers on yield quality of plants, that become one of the most paramount issues in plant nanofertilizers, such as Rico et al. (2014) on wheat, Rico et al. (2015) on barley, Lopez-Moreno et al. (2010); Hernandez-Viezcas et al. (2013); Peralta-Videa et al. (2014) on soybean, Wang et al. (2013); Barrios et al. (2016) on tomato, and Zhao et al. (2014) on cucumber.

Several studies cleared that nanofertilizers significantly affected the seed germination and seedling growth by easily penetrating the seed and enhancing nutrient availability, which causes extra shoot and root length. However, if concentration exceeds the ideal level, it may have an inhibiting effect on plant germination and seedling growth. ZnO nanoparticles' toxicity ahs an effect on the development of garlic roots (Talgar et al., 2011).

TABLE 11.1
List of Some Researches on Plant Nanofertilizers

Nanofertilizers	Plant Species	References
Ag nanoparticles	Potato	Tahmasbi et al. (2011)
CuO nanoparticles	Wheat	Hafeez et al. (2015)
TiO_2 nanoparticles	Soybean	Singh and Lee (2016)
nano-silicon	Rice	Wang et al. (2016c)
Nano NPK	Potato	Al-juthery and Al-Shami (2019)
Copper oxide nanoparticles	Tomato	Ashraf et al. (2021)
Zinc oxide nanoparticles	Maize	Abbas et al. (2021)
Nano-selenium	Cucumber	Shalaby et al. (2021)
Iron-oxide nanoparticles	Rice	Afzal et al. (2021)
Nano-nitrogen	Lettuce	Sharaf-Eldin et al. (2022)
Hydroxyapatite nanofertilizer	Rosemary	Elsayed et al. (2022)
Zinc oxide nanoparticles	Wheat	Lalarukh et al. (2022)
Zinc oxide nanoparticles	Cherry tomato	Almendros et al. (2022)
Chitosan nanoparticles	Maize	Sharma et al. (2022)
Nano-selenium and nano-copper	Strawberry	El-Bialy et al. (2023)

The majority of nanoparticles applied to seedlings or plants as a whole are absorbed through the roots, which introduces a bias for assessing their effects. This judgment is speculative primarily because, despite the existence of a few reports, the translocation of nanoparticles across plant tissues is not well understood. Overall, the impacts are related to how the nanoparticle interacts with roots, boosting or inhibiting nutrient availability and thus restricting translocation to higher tissues.

11.5 EFFECTS OF NANOPARTICLES ON ROOT GROWTH AND DEVELOPMENT

Nanotechnology ameliorates nutrient-use efficiency and reduces the costs of environment safeguard. Study findings have confirmed that plant roots can directly elevate nutrient ions from minerals' solid phase (Mukhopadhyay and Brar, 2006) then plants used ions for growth and development, which is reflected on the root system. Nanoparticles have the possibility to effect plant growth directly by inhibiting root growth and indirectly by compromising root functions, such as water transport. For instance, exposing root systems to NPs could impair root activities like water transport and prevent water uptake ability of roots by blocking cell wall pores (Asli and Neumann, 2009), which would probably have an effect on plant growth and productivity. But there have only been a few evaluations of metal oxide NPs' "physical toxicity" on root functions (Asli and Neumann, 2009). Water movement efficiency in and through a root system is measured as root hydraulic conductivity, which calculates the root system's capacity to give water to a transpiring canopy. It is hence a key marker of plant performance and function, connected to photosynthetic and growth potential, and highly reactive to environmental changes (Melcher et al., 2012). Adams et al. (2017) and Wang et al. (2016) showed that wheat and maize plants exposed to CuO NP reduced root cell elongation, enhanced cell width, and abundance of cortical cell columns. It was suggested that the nanosize effect seen for CuO NPs could be causing a root physiological response adapted for clay-rich soils, given the evolutionary demands of soil conditions on plant root systems (Margenot et al., 2018). Because natural NPs are found in soils as clay minerals, metal (hydroxy) oxides, and organic matter, and because their concentrations are orders of magnitude higher than those found in most investigations on NPs, terrestrial plants' roots have developed in these environments that are naturally rich in NPs (Theng and Yuan, 2008).

Using different nanoparticles, promotes root growth and organizes the metabolic enzyme activities as well as gene expression and proteins in plants, affecting plant adaptation with the environment (Arruda et al., 2015). Several nanoparticles in low doses encourage root elongations, while high dose are toxic for plants and inhibit plant growth. Moreover, these did not occur in all nanoparticles due to several nanoparticles at high dose enhancing root growth. Overall, the impact of nanoparticles on root morphology relay on various factors like chemical formula, size, reactivity, and efficient doses. In this regard, Mukarram et al. (2021) reported that SiO_2-NPs enhanced root elongation in rice seedlings at 40 mg L^{-1} and lemongrass at 150 mg L^{-1}, on the other hand, high dose (250 and 1000 mg L^{-1}) inhibited growth in *Arabidopsis thaliana* (Slomberg and Schoenfisch, 2012). Also, Prasad et al. (2012) showed that ZnO_2-NPs at 1000 mg L^{-1} increased root growth of peanuts, while larger doses at 2000 mg L^{-1} had inhibitory effects. Hou et al. (2022) reported that L-aspartate nanocalcium at 100 mg L^{-1} increased total root surface area, average diameter, root volume, and apical number that enhanced uptake and translocation via vascular system, and reported that middle and fine roots had the maximum effect on nutrient uptake, thus it could help plants to cope with nutritional restrictions (Han et al., 2016; Liu et al., 2019). Lin and Xing (2007) studied the impact of different types of nanoparticles (multi-walled carbon nanotube, aluminum, alumina, zinc, and zinc oxide) on root growth of different plant species (radish, rape, ryegrass, lettuce, corn, and cucumber) and showed that the reduction of root growth greatly differed between nanoparticles and plants, moreover, nano-Zn or nano-ZnO suspensions at 2000 mg L^{-1} practically terminated root elongation of these plants. Moreover, Ma et al. (2010) studied the effect of nano-CeO_2, nano-La_2O_3, nano-Gd_2O_3, and nano-Yb_2O_3 on different plants (radish, rape, tomato, lettuce, wheat, cabbage, and cucumber) on root elongation, and showed that the impacts on root growth is extremely diverse between various nanoparticles and plant species, in this regard, nano-CeO_2 suspension at 2000 mg L^{-1} had no influence on root elongation of these plants, except lettuce. On the other hand, nano-La_2O_3, nano-Gd_2O_3, and nano-Yb_2O_3 suspensions at 2000 mg L^{-1} severely inhibited root elongation of all these plants. In addition, Yanik and Vardar (2015) reported that application with Al_2O_3 nanoparticles inhibited root elongation of wheat, callose formation, lignin deposition, cellular deformation, increased POD activity, reduced protein content, as well as DNA fragmentation, in addition, histochemical analyses and microscopic observations cleared that the toxicity caused by NPs only, not Al^{+3} ions. Several effects of various nanoparticles have been shown in Table 11.2.

TABLE 11.2
Some Effects of Various Nanoparticles on Plants

Nanofertilizers	Plant Species	Effect	References
Selenium and silicon nanoparticles	Rice	Increase root length, root volume, and root thickness	Badawy et al. (2021)
N.P.K nanofertilizer	*Pinus brutia* Ten	Increase root length and diameter	Shareef et al. (2021)
ZnO nanoparticle	Clusterbean	Increase of root length	Raliya and Tarafdar (2013)
Bioorganic nanofertilizer	Sugar beet	Increase root biomass by 42.6% and root yield by 12.6%	Jakiene et al. (2015)
nano nitrogen fertilizer and mixed micro elements (boron, zinc and manganese)	Sugar beet	Increase root fresh weight	Hassnein et al.(2019)
L-aspartate nano-calcium	Rapeseed	Increase root growth especially the growth of coarse roots by increasing root biomass and root dry weight, root length ratio, root fineness, root mass ratio and root tissue density	Hou et al. (2022)
Nano-Calcium	Cauliflower	Increase weight, length and root diameter	Al-Zubaidi and Tabban (2023)
nano-microelement mixtures (Zn, Fe, B, Mn)	Sugar beet	Increase root length, root diameter, root fresh weight and top fresh weight traits	Dewdar et al. (2018)
Nano-Fe_2O_3	Soybean	Increase root length	Alidoust and Isoda (2013)
Nano-SiO_2	Changbai larch (*Larix olgensis*)	Increase height, root collar diameter, main root length, and the number of lateral roots of seedlings	Bao-shan et al. (2004)
Nano Ag	Cabbage	Reduced root growth	Pokhrel and Dubey (2013)
ZnO MnO-NPs	Cabbage	Increased in root size by 10.3%	Murgueitio-Herrera et al. (2022)
Nano nitrogen	Lettuce	Increased main root weight	Sharaf Eldin et al. (2022)
Nano-selenium and copper	Strawberry	Increased root weight and length	El-Bialy et al. (2023)
ZnO nanoparticle	Mung bean	Increase of root length and root volume	(Raliya et al. 2016)

(continued)

TABLE 11.2 (Continued)
Some Effects of Various Nanoparticles on Plants

Nanofertilizers	Plant Species	Effect	References
Nano Ag	Maize	low concentration increased root length but high concentration decreased the length	Krishnaraj et al. (2012)
ZnO nanoparticle	Cucumber	Decreased root growth	de la Rosa et al. (2013)
Copper oxide nanoparticle	Lettuce and carrot	Increased root diameter for lettuce (+52%) and carrot (+26%), in addition, root hydraulic conductivity was similarly reduced by 16 and 45 nm CuO NPs for lettuce (-46%) but not for carrot	Margenot et al. (2018)
Silicon nanoparticle	Tomato	Decreased root volume	Haghighi and Pessarakli, (2013)
Silicon nanoparticle	Mango	Increased root length and width, number of lateral and secondary roots and root growth coefficient	El-Dengawy et al. (2021)

Generally, nanofertilizers have smaller particle sizes than the pores in leaves and roots that can increase the efficiency with which nutrients are taken up by the soil (Tarafder et al., 2020). By safeguarding domesticated plants and enhancing crop productivity (Guleria et al., 2023), nanofertilizers can aid in agro-food production for sustainable farming (Mahapatra et al., 2022).

11.6 NANOPARTICLES FOR NUTRIENT AVAILABILITY

There are many nanoparticles in the soil environment, and it is crucial to understand their behavior for a variety of soil activities related to plant nutrition and soil reclamation. Soil has a wide variety of nanoparticles due to the chemical weathering of silicates, oxides, and other minerals. Amorphous silica, hydrous aluminosilicates, clays, and oxides (magnetite and hematite) are the end products (Nowack and Bucheli, 2007). Additionally, NPs can be produced by microorganisms using pathways that use inorganic ions that take part in redox processes to provide metabolic energy. Iron oxide nanoparticles (NPs) are created when Fe (II) is oxidized. Similarly, a diversity of various manganese oxide NPs, ZnS NPs, and other oxidation-reduction activities in the soil are mediated by bacteria.

The majority of the organic NPs in soil are either present as coatings on mineral surfaces or are linked with their inorganic counterparts (Chorover et al., 2007). Numerous individual humic and fulvic molecules are in the nanometer size range, while humic and fulvic acids have molecular weights of hundreds to thousands of daltons or more. On mineral surfaces, natural organic matter aggregates in the nanometer to micron size range have been seen (Namjesnik-Dejanovic and Maurice, 2001). Due to their enzymatic activity, soil microorganisms maintain high yields of healthy crops by drawing energy and nutrients from the organic matter in the soil. For effective physical, chemical, and biological soil functioning, it must be preserved (Six et al., 2002).

When nanoparticles are released into the environment, the soil serves as a significant sink for them and is also a potential source of NPs for groundwater. To determine how NPs may affect the food chain and groundwater, it is crucial to comprehend how they move via natural soil systems. Research emphasis on the transport of NPs in porous medium and soils has grown (Lin et al., 2010). The mobility or relocation of NPs via soils is related to the interaction between particles and solid surfaces (Darlington et al., 2009) that might be affected by either environmental or particle physico-chemical characteristics (Tourinho et al., 2012). Both macropores and micropores can be found in the soil, making it a porous material. The soil structure's micropores, which are incredibly small pores, are made up of a web of humic substances and soil particles (Kretzschmar and Schafer, 2005). Since nanoparticles are small enough to fit into these micropore settings, they can clog the pores, which may impair the mobility of the particles. Aggregates of nanoparticles, on the other hand, can be too big to fit in the micropores, allowing them to stay in the macropores (Kumar et al., 2012). For greater use efficiency, the nanofertilizers can be sprayed onto the leaves or delivered promptly to a rhizospheric target. The emerging literature on nanotechnology has begun demonstrating the significance of nanoparticles in enhancing nutrient element bioavailability. Nanofertilizers may seem to be a highly effective alternative to mineral fertilizers. These nanoparticles of N fertilizers reduce N losses by leaching, emissions, and soil microbial immobilization (Liu et al., 2006). Subramanian et al. (2008) cleared that nanofertilizers can be used to regulate the release of nutrients from fertilizer granules, increasing the efficiency of nutrient utilization while preventing the loss of nutrient ions to the environment. Slow-release fertilizers are where nanoparticles have had the most success. They will firmly hold plant material due to their greater surface tension compared to ordinary surfaces. Additionally, nanocoatings can shield larger particles' surfaces from damage. Large amounts of fertilizer can be directly applied and may result in dangerously high local concentrations of ammonium salts, urea, nitrate, or phosphate compounds. Much of the fertilizer could dissolve in runoff water and have negative effects such as causing pollution while being unavailable to the target plants (Wilson et al., 2008). The nature of the coating material, the type of fertilizer, agronomic circumstances, and soil pH are only a few examples of the many variables that affect the controlled-release mechanism. Nanofertilizers may control the release of nutrients, give the right amount of nutrients to plants in the right proportions, and increase yield while protecting the environment (De-Rosa et al., 2010). Rahale (2010) cleared that nanofertilizer enhanced the NUE up to 45% over control, in addition releasing the nitrate from nanozeolite continued even 1176 h later, as well as slow and steady release of N from nanozeolite for more than 45 days while mineral fertilizers do it for 8 days. Moreover, ammonium-charged zeolites have demonstrated their ability to increase the solubilization of phosphate minerals, which will lead to improved agricultural plants' uptake and yield of phosphorus (Ramesh and Reddy, 2011).

Nanomaterials may cause tissue damage at the target site by a single or several pathways. The biological effects of nanomaterials depend critically on how they interact with diverse biological elements such as cells, bodily fluids, and proteins. The body may experience these resulting interaction consequences. When proteins are used as a substrate in a reaction with nanomaterials, complex molecules are created that are more mobile and can infiltrate tissue locations. There may be structural alterations on the nanoparticle surface that cause enzyme dysfunction (Vertegel et al., 2004). On the surface of the material, intramolecular or intermolecular bond splitting may be the cause of this rapid denaturation or degradation. NPs may come into contact with a variety of defenses during their absorption and movement throughout the body, which could destroy or dissolve those (Nel et al., 2006). Various studies clear that inorganic nanoparticles, such as TiO_2, SiO_2, and ZnO, had a toxic impact on bacteria and other organisms.

11.7 CONCLUSION

Nowadays, nanofertilizers are one of the most significant applications of nanotechnology in agriculture. As well as being easily absorbed, nanofertilizers are released slowly and are required in small amounts, minimizing the use of fertilizers and their unfavorable impact on the environment. Nanomaterials in small size assist plants to easily penetrate and regulate water channels and improve adsorption and allocation substances, which reflect on plant growth and productivity. The impact of nanoparticles on root morphology relay on various factors like chemical formula, size, reactivity, and efficient doses, in this regard, nanoparticles in low doses encourage root elongations, while high doses are toxic for plants and inhibit plant growth. Nanofertilizers increased root elongation, total root surface area, average diameter, root volume, and apical number that enhanced uptake and translocation via vascular system. More studies should be done on the plant cell level through gene expression, signal transduction, and anatomical studies of the root.

REFERENCES

Abbas, H. H.; Noufal, E.; Farid, I.; Attia, M.; Ahmed, R.; Abbas, M. (2021). Effect of traditional sources of Zn and ZnO-nano-particles foliar application on productivity and P-uptake of maize plants grown on sandy and clay loam soils. *Environ. Biodiv. Soil Security*, 5: 59–72.

Adams, J.; Wright, M.; Wagner, H.; Valiente, J.; Britt, D.; Anderson, A. (2017). Cu from dissolution of CuO nanoparticles signals changes in root morphology. *Plant Physiol. Biochem.*, 110: 108–117.

Afzal, S.; Sharma, D.; Singh, N. K. (2021). Eco-friendly synthesis of phytochemical-capped iron oxide nanoparticles as nano-priming agent for boosting seed germination in rice (*Oryza sativa* L.). *Environ. Sci. Pollut. Res.*, 28: 40275–40287.

Al-juthery; Al-Shami (2019). The effect of fertigation with nano npk fertilizers on some parameters of growth and yield of potato (*Solanum tuberosum* L.). *Al-Qadisiyah J. Agric. Sci.*, 9(2): 225–232.

Alidoust, D.; Isoda, A. (2013). Effect of Y-Fe_2O_3 nanoparticles on photosynthetic characteristic of soybean (*Glycine max* (L.) Merr.): foliar spray versus soil amendment. *Acta Physiol. Plant.*, 35(12): 3365–3375.

Almendros, P.; González, D.; Fernández, M. D.; García-Gomez, C.; Ana Obrador, A. (2022). Both Zn biofortification and nutrient distribution pattern in cherry tomato plants are influenced by the application of ZnO nanofertilizer. *Heliyon*, 8(3): e09130.

Al-Zubaidi, M. S.; Tabban, S. K. (2023). Effect of adding chemical fertilizer and spraing nano-calcium on some vegetative and root characteristics of cauliflower in calcareous soil. *J. of Survey in Fisheries Sci.*, 10(2S): 1013–1023.

Aminzare, M.; Eskandari, A.; Baroonian, M. H.; Berenov, A.; Razavi Hesabi, Z.; Taheri, M.; Sadrnezhaad, S. K. (2013). Hydroxyapatite nanocomposites: synthesis, sintering and mechanical properties. *Ceramics International*, 39(3): 2197–2206. https://doi.org/10.1016/J.CERAMINT.2012.09.023.

An, C.; Sun, C.; Li, N.; Huang, B.; Jiang, J.; Shen, Y.; Wang, C.; Zhao, X.; Cui, B.; Wang, C.; Li, X.; Zhan, S.; Gao, F.; Zeng, Z.; Cui, H.; Wang, Y. (2022). Nanomaterials and nanotechnology for the delivery of agrochemicals: strategies towards sustainable agriculture. *Journal of Nanobiotechnology*, 20(1). https://doi.org/10.1186/S12951-021-01214-7

Anjum, N. A.; Adam, V.; Kizek, R.; Duarte, A. C.; Pereira, E.; Iqbal, M.; Lukatkin, A. S.; Ahmad, I. (2015). Nanoscale copper in the soil–plant system – toxicity and underlying potential mechanisms. *Environ. Res.*, 138: 306–325.

Arruda, S. C. C.; Silva, A. L. D.; Galazzi, R. M.; Azevedo, R. A.; Arruda, M. A. Z. (2015). Nanoparticles applied to plant science: a review. *Talanta*, 131: 693–705.

Ashraf, H.; Anjum,T.; Riaz, S.; Ahmad, I.; Irudayaraj, J.; Javed, S.; Qaiser, U.; Naseem, S. (2021). Inhibition mechanism of green-synthesized copper oxide nanoparticles from Cassia fistula towards Fusarium oxysporum by boosting growth and defense response in tomatoes. *Environ. Sci.: Nano,* 2021, 8: 1729–1748.

Asli, S.; Neumann, P. M. (2009). Colloidal suspensions of clay or titanium dioxide nanoparticles can inhibit leaf growth and transpiration via physical effects on root water transport. *Plant Cell Environ.*, 32: 577–584.

Badawy, S. A.; Zayed, B. A.; Bassiouni, S. M. A.; Mahdi, A. H. A.; Majrashi, A.; Ali, E. F.; Seleiman, M. F. (2021). Influence of nano silicon and nano selenium on root characters, growth, ion selectivity, yield, and yield components of rice (*Oryza sativa* L.) under salinity conditions. *Plants*, 10: 1657.

Bahrulolum, H.; Nooraei, S.; Javanshir, N.; Tarrahimofrad, H.; Mirbagheri, V. S.; Easton, A. J.; Ahmadian, G. (2021). Green synthesis of metal nanoparticles using microorganisms and their application in the agrifood sector. *J. Nanobio., 2021* 19(1): 1–26. https://doi.org/10.1186/S12951-021-00834-3.

Bao-shan, L.; Shao-qi, D.; Chun-hui, L.; Li-jun, F.; Shu-chun, Q.; Min, Y. (2004). Effect of TMS (nanostructured silicon dioxide) on growth of Changbai larch seedlings. *J. Forest Res.*, 15: 138–140.

Basit, F.; Asghar, S.; Ahmed, T.; Ijaz, U.; Noman, M.; Hu, J.; Liang, X.; Guan, Y. (2022). Facile synthesis of nanomaterials as nano fertilizers: a novel way for sustainable crop production. *Environ. Sci. Pollution Res.,* 29(34): 51281–51297. https://doi.org/10.1007/S11356-022-20950-3.

Barrios, A. C.; Medina-Velo, I. A.; Zuverza-Mena, N.; Dominguez, O. E.; Peralta-Videa, J. R.; Gardea-Torresdey, J. L. (2016). Nutritional quality assessment of tomato fruits after exposure to uncoated and citric acid coated cerium oxide nanoparticles, bulk cerium oxide, cerium acetate and citric acid. *Plant Physiol. Biochem.* https://doi.org/10.1016/j.plaphy.2016.04.017.

Bindra, P.; Kaur, K.; Rawat, A.; De Sarkar, A.; Singh, M.; Shanmugam, V. (2019). Nano-hives for plant stimuli controlled targeted iron fertilizer application. *Chem. Eng. J.,* 375: 121995. https://doi.org/10.1016/J.CEJ.2019.121995.

Buzea, C.; Pacheco, I. (2017). Nanomaterial and nanoparticle: origin and activity. In: Ghorbanpour, M., Manika, K., Varma, A. (eds) *Nanoscience and Plant–Soil Systems. Soil Biology*. Springer, Cham, pp. 71–112. https://doi.org/10.1007/978-3-319-46835-8_3.

Chorover, J.; Kretzschmar, R.; Garcia-Pichel, F.; Sparks, D. L. (2007). Soil biogeochemical processes within the critical zone. *Elements*, 3: 321–326.

Darlington, T. K.; Neigh, A. M.; Spencer, M. T.; Nguyen, O. T.; Oldenburg, S. J. (2009). Nanoparticle characteristics affecting environmental fate and transport through soil. *Environ. Toxicol. Chem.*, 28: 1191–1199.

de la Rosa, G. ; López-Moreno, M. L.; de Haro, D.; Botez, C. E.; Peralta-Videa, J. R. ; Gardea-Torresdey, J. L. (2013). Effects of ZnO nanoparticles in alfalfa, tomato, and cucumber at the germination stage: root development and X-ray absorption spectroscopy studies. *Pure. Appl. Chem.,* 85(12): 2161–2174.

De-Rosa, M. C.; Monreal, C.; Schnitzer, M.; Walsh, R.; Sultan, Y. (2010). Nanotechnology in fertilizers. *Nat. Nanotechnol.*, 5: 91.

Dewdar, M. D.; Abbas, M. S.; El-Hassanin, A. S.; Abd El-Aleem, H. A. (2018). Effect of nano micronutrients and nitrogen foliar applications on sugar beet (*Beta vulgaris* L.) of quantity and quality traits in marginal soils in egypt. *Int. J. Curr. Microbiol. App. Sci.*, 7(8): 4490–4498.

Dimkpa, C. O.; Bindraban, P. S. (2018). Nanofertilizers: new products for the industry? *J. Agricultural Food Chem.*, 66(26): 6462–6473.

Ditta, A.; Arshad, M.; Ibrahim, M. (2015). Nanoparticles in sustainable agricultural crop production: applications and perspectives. In: Siddiqui, M.H. et al (eds) *Nanotechnology and Plant Sciences*. Springer, Cham, pp. 55–75.

Ditta, A.; Mehmood, S.; Imtiaz, M.; Rizwan, M. S.; Islam, I. (2020b). Soil fertility and nutrient management with the help of nanotechnology. In: Husen, A., Jawaid, M. (eds) *Nanomaterials for Agriculture and Forestry Applications*. Elsevier, Netherlands, pp. 273–287. https://doi.org/10.1016/B978-0-12-817852-2.00011-1.

Du, W.; Sun, Y.; Ji, R.; Zhu, J.; Wu, J.; Guo, H. (2011). TiO_2 and ZnO nanoparticles negatively affect wheat growth and soil enzyme activities in agricultural soil. *J. Env. Monitor.*, 13(4): 822–828.

Du, W.; Tan, W.; Peralta-Videa, J. R.; Gardea-Torresdey, J. L.; Ji, R.; Yin, Y.; Guo, H. (2017). Interaction of metal oxide nanoparticles with higher terrestrial plants: physiological and biochemical aspects. *Plant Physiol. Biochem.*, 11: 210–225.

El-Bialy, S. M.; El-Mahrouk, M. E.; Elesawy, T.; Omara, A. E.-D.; Elbehiry, F.; El-Ramady, H.; Áron, B.; Prokisch, J.; Brevik, E. C.; Solberg, S. Ø. (2023). Biological nanofertilizers to enhance growth potential of strawberry seedlings by boosting photosynthetic pigments, plant enzymatic antioxidants, and nutritional status. *Plants*, 12: 302.

El-Dengawy, E. F.; EL-Abbasy, U. K.; El-Gobba, M. H. (2021). Influence of nano-silicon treatment on growth behavior of 'Sukkary' and 'Gahrawy' mango root-stocks under salinity stress. *J. of Plant Production, Mansoura Univ.*, 12(1): 49–61.

El-Ramady, H.; Abdalla, N.; Taha, H. S.; Alshaal, T.; El-Henawy, A.; Faizy, S. E. D. A.; Shams, M. S.; Youssef, S. M.; Shalaby, T.; Bayoumi, Y.; Elhawat, N.; Shehata, S.; Sztrik, A.; Prokisch, J.; Fári, M.; Domokos-Szabolcsy, É.; Pilon-Smits, E. A.; Selmar, D.; Haneklaus, S.; Schnug, E. (2015). Selenium and nano-selenium in plant nutrition. *Environ. Chem. Lett.,* 14(1): 123–147. https://doi.org/10.1007/S10311-015-0535-1.

Elsayed, A. A.; EL-Gohary, A.; Taha, Z. K.; Farag, H. M.; Hussein, M. S.; AbouAitah, K. (2022). Hydroxyapatite nanoparticles as novel nano-fertilizer for production of rosemary plants. *Scientia Horticulturae*, 295: 110851.
Fernando, M. S.; De Silva, R. M.; De Silva, K. M. N. (2015). Synthesis, characterization, and application of nano hydroxyapatite and nanocomposite of hydroxyapatite with granular activated carbon for the removal of Pb2+ from aqueous solutions. *Appl. Surface Sci.,* 351: 95–103. https://doi.org/10.1016/J.APSUSC.2015.05.092
Gámez-Espinosa, E.; Barberia-Roque, L.; Obidi, O. F.; Deyá, C.; Bellotti, N. (2020). Antifungal applications for nano-additives synthesized with a bio-based approach. *Adv. Nat. Sci.: Nanosci. Nanotechnol.*, 11(1): 015019. https://doi.org/10.1088/2043-6254/AB790F.
Gómez-Merino, F. C.; Trejo-Téllez, L. I. (2018). The role of beneficial elements in triggering adaptive responses to environmental stressors and improving plant performance. *In:* Vats, S., (ed.) *Biotic and Abiotic Stress Tolerance in Plants*. Springer, Singapore, pp. 137–172.
Guleria, G.; Thakur, S.; Shandilya, M.; Sharma, S.; Thakur, S.; Kalia, S. (2023). Nanotechnology for sustainable agro-food systems: the need and role of nanoparticles in protecting plants and improving crop productivity. *Plant Physiol. Biochem.*, 194: 533–549.
Hafeez, A.; Razzaq, A.; Mahmood, T.; Jhanzab, HM (2015). Potential of copper nanoparticles to increase growth and yield of wheat. *J. Nanosci Adv. Technol.*, 1(1): 6–11.
Haghighi, M.; Pessarakli, M. (2013). Influence of silicon and nano-silicon on salinity tolerance of cherry tomatoes (*Solanum lycopersicum* L.) at early growth stage. *Sci Hortic.*, 161: 111–117.
Hajihashemi, S.; Kazemi, S. (2022). The potential of foliar application of nano-chitosan-encapsulated nano-silicon donor in amelioration the adverse effect of salinity in the wheat plant. *BMC Plant Bio.*, 22(1): 1–15. https://doi.org/10.1186/S12870-022-03531-X/FIGURES/6.
Han, P.; Qin, L.; Li, Y.; Liao, X.; Xu, Z.; Yu, C.; Hu, X.; Xie, L.; Liao, H.; Liao, X. (2016). Effects of different nutrient deficiencies on growth and root morphological changes of rapeseed seedlings (*Brassica napus* L.). *Chinese J. Oil Crop Sci.*, 38: 88–97.
Hassnein, A. M.; Azab, M. A.; El-Hawary, M. A.; Darwish, N. N. (2019). Effect of nano fertilization on sugar beet. *Al-Azhar J. of Agric. Res.*, 44(2): 194–201.
He, S.; Feng, Y.; Ni, J.; Sun, Y.; Xue, L.; Feng, Y.; Yu, Y.; Lin, X.; Yang, L. (2016). Different responses of soil microbial metabolic activity to silver and iron oxide nanoparticles. *Chemosphere,* 147: 195–202.
Hernandez-Viezcas, J. A.; Castillo-Michel, H.; Andrews, J. C.; Cotte, M.; Rico, C.; Peralta-Videa, J. R.; Ge, Y.; Priester, J. H.; Holden, P. A.; Gardea-Torresdey, J. L. (2013). In situ synchrotron X-ray fluorescence mapping and speciation of CeO_2 and ZnO nanoparticles in soil cultivated soybean (*Glycine max*). *ACS Nano*, 7(2): 1415–1423.
Hou, J.; Riaz, M.; Yan, L.; Lu, K.; Jiang, C. (2022). Effect of exogenous L-aspartate nano-calcium on root growth, calcium forms and cell wall metabolism of *Brassica napus* L. *NanoImpact*, 27: 100415.
Ibrahim, R. K.; Hayyan, M.; AlSaadi, M. A.; Hayyan, A.; Ibrahim, S. (2016). Environmental application of nanotechnology: air, soil, and water. *Environ. Sci. Pollut. Res. Int.*, 23: 13754–13788.
Iavicoli, I.; Leso, V.; Beezhold, D. H.; Shvedova, A. A. (2017). Nanotechnology in agriculture: opportunities, toxicological implications, and occupational risks. *Toxicology Appl. Pharmacol.,* 329: 96–111. https://doi.org/10.1016/J.TAAP.2017.05.025.
Jakienė, E.; Spruogis, V.; Romaneckas, K.; Dautartė, A.; avižienytė, D. (2015). The bioorganic nano fertilizer improves sugar beet photosynthesis process and productivity. *Zemdirbyst.*, 102: 141–146.
Jampílek, J.; Kráľová, K. (2017). Nanopesticides: preparation, targeting, and controlled release. In: Grumezescu, A. M. (ed) *New Pesticides and Soil Sensors*, pp. 81–127. London, Academic Press. https://doi.org/10.1016/B978-0-12-804299-1.00004-7.
Kalia, A.; Sharma, S. P.; Kaur, H.; Kaur, H. (2020). Novel nanocomposite-based controlled-release fertilizer and pesticide formulations: prospects and challenges. In: Abd-Elsalam, K. (ed) *Multifunctional Hybrid Nanomaterials for Sustainable Agri-Food and Ecosystems*. Elsevier, Netherlands, pp. 99–134. https://doi.org/10.1016/B978-0-12-821354-4.00005-4.
Kalwani, M.; Chakdar, H.; Srivastava, A.; Pabbi, S.; Shukla, P. (2022). Effects of nanofertilizers on soil and plant-associated microbial communities: emerging trends and perspectives. *Chemosphere*, 287: 132107.
Khan, M. N.; Mobin, M.; Abbas, Z. K.; AlMutairi, K. A.; Siddiqui, Z. H. (2017). Role of nanomaterials in plants under challenging environments. *Plant Physiol. Biochem.*, 110: 194–209.

Kottegoda, N.; Munaweera, I.; Madusanka, N.; Karunaratne, V. (2011). A green slow-release fertilizer composition based on urea-modified hydroxyapatite nanoparticles encapsulated wood. *Curr. Sci.*, 101(1): 73–78.

Kretzschmar, R.; Schafer, T. (2005). Metal retention and transport on colloidal particles in the environment. *Elements*, 1: 205–210.

Krishnaraj, C.; Ramachandran, R.; Mohan, K.; Kalaichelvan, P. T. (2012). Optimization for rapid synthesis of silver nanoparticles and its effect on phytopathogenic fungi. *Spectrochim. Acta A. Mol. Biomol. Spectrosc.*, 93: 95–99.

Kumar, R.; Rawat, K. S.; Mishra, A. K. (2012). Nanoparticles in the soil environment and their behaviour: an overview. *J. App. Natural Sci.* 4(2): 310–324.

Lalarukh, I.; Zahra, N.; Al Huqail, A. A.; Amjad, S. F.; Al-Dhumri, S. A.; Ghoneim, A. M.; Alshahri, A. H.; Almutari, M. M.; Alhusayni, F. S.; Al-Shammari, W. B.; Poczai, P.; Mansoora, N.; Ayman, M.; Abbas, M. H.; Abdelhafez, A. A. (2022). Exogenously applied ZnO nanoparticles induced salt tolerance in potentially high yielding modern wheat (*Triticum aestivum* L.) cultivars, *Environ. Technol. Innov.*, 27: 102799.

Lin, D.; Xing, B. (2007). Phytotoxicity of nanoparticles: inhibition of seed germination and root growth. *Environ. Pollut.*, 150(2): 243–250.

Lin, D.; Xing, X. (2008). Root uptake and phytotoxicity of ZnO nanoparticles. *Environ. Sci. Technol.*, 42: 5580–5585.

Lin, D.; Tian, X.; Wu, F.; Xing, B. (2010). Fate and transport of engineered nanomaterials in the environment. *J. Environ. Qual.*, 39: 1–13.

Liu, R.; Zhang, H.; Lal, R. (2016). Effects of stabilized nanoparticles of copper, zinc, manganese, and iron oxides in low concentrations on Lettuce (*Lactuca sativa*) seed germination: nanotoxicants or nanonutrients? *Water Air Soil Pollut.*, 227: 42.

Liu, X.; Feng, Z.; Zhang, S.; Zhang, J.; Xiao, Q.; Wang, Y. (2006). Preparation and testing of cementing nano-subnano composites of slower controlled release of fertilizers. *Sci. Agri. Sin. J.*, 39: 1598–1604.

Liu, Y. L.; Riaz, M.; Wu, X. W.; Yan, L.; Du, C. Q.; Jiang, C. C. (2019). Boron and calcium homeostasis affects trifoliate rootstock (*Poncirus Trifoliata*) root configuration and nutrient utilization. *Int. J. Agric. Biol.*, 21: 359–366.

Lu, J.; Cheng, M.; Zhao, C.; Li, B.; Peng, H.; Zhang, Y.; Shao, Q.; Hassan, M. (2022). Application of lignin in preparation of slow-release fertilizer: current status and future perspectives. *Indust. Crops Products,* 176: 114267. https://doi.org/10.1016/J.INDCROP.2021.114267.

López-Moreno, M. L.; de la Rosa, G.; Hernández-Viezcas, J. A.; Peralta-Videa, J. R.; Gardea-Torresdey, J. L. (2010). XAS corroboration of the uptake and storage of CeO_2 nanoparticles and assessment of their differential toxicity in four edible plant species. *J. Agric. Food. Chem.,* 58(6): 3689–3693.

Ma, C.; White, J. C.; Zhao, J.; Zhao, Q.; Xing, B. (2018). Uptake of engineered nanoparticles by food crops: characterization, mechanisms, and implications. *Ann. Rev. Food Sci. Technol.,* 9: 129–153. https://doi.org/10.1146/ANNUREV-FOOD-030117-012657.

Ma, C. X.; White, J. C.; Dhankher, O. P.; Xing, B. (2015). Metal-based nanotoxicity and detoxification pathways in higher plants. *Environ. Sci. Technol.*, 49(12): 7109–7122.

Ma, Y.; Kuang,L.; He, X.; Bai, W.; Ding, Y.; Zhang, Z.; Zhao, Y.; Chai, Z. (2010). Effects of rare earth oxide nanoparticles on root elongation of plants. *Chemosphere*,78(3): 273–279.

Madusanka, N.; Sandaruwan, C.; Kottegoda, N.; Sirisena, D.; Munaweera, I.; De Alwis, A.; Karunaratne, V.; Amaratunga, G. A. J. (2017). Urea–hydroxyapatite-montmorillonite nanohybrid composites as slow release nitrogen compositions. *Appl. Clay Sci.,* 150: 303–308. https://doi.org/10.1016/J.CLAY.2017.09.039.

Marchiol, L.; Iafisco, M.; Fellet, G.; Adamiano, A. (2020). Nanotechnology support the next agricultural revolution: perspectives to enhancement of nutrient use efficiency. *Adv. Agronomy,* 161: 27–116. https://doi.org/10.1016/BS.AGRON.2019.12.001.

Mahapatra, D. M.; Satapathy, K. C.; Panda, B. (2022). Biofertilizers and nanofertilizers for sustainable agriculture: phycoprospects and challenges. *Sci. Total Environ.*, 803: 149990.

Margenot, A. J.; Rippner, D. A.; Dumlao, M. R.; Nezami, S.; Green, P. G.; Parikh, S. J.; McElrone, A. J. (2018). Copper oxide nanoparticle effects on root growth and hydraulic conductivity of two vegetable crops. Plant Soil, https://doi.org/10.1007/s11104-018-3741-3.

Medici, M.; Mattetti, M.; Argyropoulos, D.; Poudel, A.; Singh, S. K.; Jiménez-Ballesta, R.; Singh Jatav, S.; Patra, A.; Pandey, A. (2023). Effect of nano-phosphorus formulation on growth, yield and nutritional quality of wheat under semi-arid climate. *Agronomy*, 13(3): 768. https://doi.org/10.3390/AGRONOMY13030768.

Mejías, J. H.; Salazar, F. J.; Pérez, L.; Hube, S.; Rodriguez, M.; Alfaro, M. A. (2021). Nanofertilizers: a cutting-edge approach to increase nitrogen use efficiency in grasslands. *Frontiers Environ. Sci.*, 9: 52.

Melcher, P. J.; Michele Holbrook, N.; Burns, M. J.; Zwieniecki, M. A.; Cobb, A. R.; Brodribb, T. J.; Choat, B.; Sack, L. (2012). Measurements of stem xylem hydraulic conductivity in the laboratory and field. *Methods Ecol. Evol.*, 3: 685–694.

Meredith, A. N.; Harper, B.; Harper, S. L. (2016). The influence of size on the toxicity of an encapsulated pesticide: a comparison of micron- and nano-sized capsules. *Environ. Int.,* 86: 68–74.

Mukarram, M.; Khan, M. M. A.; Corpas, F. J. (2021). Silicon nanoparticles elicit an increase in lemongrass (*Cymbopogon flexuosus* (Steud.) wats) agronomic parameters with a higher essential oil yield. *J. Hazard. Mater.*, 412, 125254.

Mukhopadhyay, S. S.; Kaur, N. (2016). Nanotechnology in soil-plant system. In: Kole, C., Kumar, D. S., Khodakovskaya, M. V. (eds) *Plant Nanotechnology: Principles and Practices.* Springer, Cham, pp. 329–348. https://doi.org/10.1007/978-3-319-42154-4_13/TABLES/1.

Mukhopadhyay, S. S.; Brar, M. S. (2006). *Mineralogy and management of soils rich in potassium containing minerals. Proceedings of the International Symposium on Balanced Fertilization, Ludhiana, India, during 22–25 November 2005, vol 1.* International Potash Institute, Berne, Switzerland, pp. 95–114.

Murgueitio-Herrera, E.; Falconí, C. E.; Cumbal, L.; Gómez, J.; Yanchatipán, K.; Tapia, A.; Martínez, K.; Sinde-Gonzalez, I.; Toulkeridis, T. (2022). Synthesis of iron, zinc, and manganese nanofertilizers, using andean blueberry extract, and their effect in the growth of cabbage and lupin plants. *Nanomaterials,* 12: 1921.

Nagar, N.; Devra, V. (2018). Green synthesis and characterization of copper nanoparticles using *Azadirachta indica* leaves. *Mater. Chem. Phys.,* 213: 44–51. https://doi.org/10.1016/J.MATCHEMPHYS.2018.04.007.

Najafi, S.; Razavi, S. M.; Khoshkam, M.; Asadi, A. (2020). Effects of green synthesis of sulfur nanoparticles from *Cinnamomum zeylanicum* barks on physiological and biochemical factors of Lettuce (*Lactuca sativa*). *Physiology Mol. Bio. Plants*, 26(5): 1055–1066. https://doi.org/10.1007/S12298-020-00793-3/FIGURES/10.

Namjesnik-Dejanovic, K.; Maurice, P. A. (2001). Conformations and aggregate structures of sorbed natural organic matter on muscovite and hematite. *Geochim Cosmochim Acta*, 65: 1047–1057.

Nel, A.; Xia, T.; Madler, L.; Li, N. (2006). Toxic potential of materials at the nanolevel. *Sci.*, 311: 622–627.

Nowack, B.; Bucheli, T. D. (2007). Occurrence, behavior and effects of nanoparticles in the environment. *Environ. Pollut.*, 150: 5–22.

Panpatte, D. G.; Jhala, Y. K.; Shelat, H. N.; Vyas, R. V. (2016). Nanoparticles: the next generation technology for sustainable agriculture. In: Singh, D.P. et al. (eds) *Microbial Inoculants in Sustainable Agricultural Productivity.* Springer, Delhi, India, pp. 289–300.

Peralta-Videa, J. R.; Hernandez-Viezcas, J. A.; Zhao, L.; Diaz, B. C.; Ge, Y.; Priester, J. H.; Holden, P. A.; Gardea-Torresdey, J. L. (2014). Cerium dioxide and zinc oxide nanoparticles alter the nutritional value of soil cultivated soybean plants. *Plant Physiol. Biochem.*, 80: 128.

Philippe, G.; Sørensen, I.; Jiao, C.; Sun, X.; Fei, Z.; Domozych, D. S.; Rose, J. K. (2020). Cutin and suberin: assembly and origins of specialized lipidic cell wall scaffolds. *Curr. Opin. Plant Bio.,* 55: 11–20. https://doi.org/10.1016/J.PBI.2020.01.008.

Pitambara, A.; Shukla, Y. M. (2019). Nanofertilizers: a recent approach in crop production. In: Panpatte, D. G., Jhala, Y. K. (eds) *Nanotechnology for Agriculture: Crop Production and Protection.* Springer, Singapore, *pp.* 25–58. https://doi.org/10.1007/978-981-32-9374-8.

Pokhrel, L. R.; Dubey, B. (2013). Evaluation of developmental responses of two crop plants exposed to silver and zinc oxide nanoparticles. *Sci. Total Environ.*, 45: 321–332.

Prasad, T. N. V.; Sudhakar, P.; Sreenivasulu, Y.; Latha, P.; Munaswamy, V.; Reddy, K. R.; Sreeprasad, T. S.; Sajanlal, P. R.; Pradeep, T. (2012). Effect of nanoscale zinc oxide particles on the germination, growth and yield of peanut. *J. Plant Nutr.*, 35: 905–927.

Pradhan, S.; Mailapalli, D. R. (2017). Interaction of engineered nanoparticles with the agri-environment. *J. Agri. Food Chem.*, 65(38): 8279–8294.

Rahale, C. S. (2010). *Nutrient Release Pattern of Nano-fertilizer Formulations.* PhD Thesis, Tamil Nadu Agricultural University, Coimbatore, Tamil Nadu, India.

Rai, V.; Acharya, S.; Dey, N. (2012). Implications of nanobiosensors in agriculture. *J. Biomat Nanobiotechnol.*, 3: 315–324.

Raliya, R.; Tarafdar, C. J.; Pratim, B. P. (2016). Enhancing the mobilization of native phosphorous in mung bean rhizosphere using ZnO NPs synthesized by soil fungi. *J. Agric. Food Chem.*, 64: 3111–3118.

Raliya, R.; Tarafdar, J. (2013). ZnO nanoparticle biosynthesis and its effect on phosphorous-mobilizing enzyme secretion and gum contents in Clusterbean (Cyamopsistetr agonoloba L.). *Agric. Res.*, 2: 48–57.

Raliya, R.; Saharan, V.; Dimkpa, C.; Biswas, P. (2018). Nanofertilizer for precision and sustainable agriculture: current state and future perspectives. *J. Agri. Food Chem.*, 66(26): 6487–6503.

Ramesh, K.; Reddy, D. D. (2011). Zeolites and their potential uses in agriculture. *Adv. Agron.*, 113: 219–241.

Reeve, P. J.; Fallowfield, H. J. (2018). Natural and surfactant modified zeolites: a review of their applications for water remediation with a focus on surfactant desorption and toxicity towards microorganisms. *J. Environ. Management,* 205: 253–261. https://doi.org/10.1016/J.JENVMAN.2017.09.077

Reddy, P. V. L.; Hernandez-Viezcas, J. A.; Peralta-Videa, J. R.; Gardea-Torresdey, J. L. (2016). Lessons learned: are engineered nanomaterials toxic to terrestrial plants? *Sci. Total Environ.* 568: 470–479.

Rico, C. M.; Barrios, A. C.; Tan, W.; Rubenecia, R.; Lee, S. C.; Varela-Ramirez, A.; Peralta-Videa, J. R.; Gardea-Torresdey, J. L. (2015). Physiological and biochemical response of soil-grown barley (*Hordeum vulgare* L.) to cerium oxide nanoparticles. *Environ. Sci. Pollut. Res., 22: 10551–10558.* doi:10.1007/s11356-015-4243.

Rico, C. M.; Lee, S. C.; Rubenecia, R.; Mukherjee, A.; Hong, J.; Peralta-Videa, J. R.; Gardea-Torresdey, J. L. (2014). Cerium oxide nanoparticles impact yield and modify nutritional parameters in wheat (*Triticum aestivum* L.). *J. Agric. Food Chem.*, 62: 9669–9675.

Rizwan, M.; Ali, S.; Qayyum, M. F.; Ok, Y. S.; Adrees, M. Ibrahim, M.; Zia-ur-Rehman, M. Farid, M.; Abbas, F. (2017). Effect of metal and metal oxide nanoparticles on growth and physiology of globally important food crops: a critical review. *J. Hazard Mater.*, 322: 2–16.

Rop, K.; Karuku, G. N.; Mbui, D.; Michira, I.; Njomo, N. (2018). Formulation of slow release NPK fertilizer (cellulose-graft-poly(acrylamide)/nano-hydroxyapatite/soluble fertilizer) composite and evaluating its N mineralization potential. *Ann. Agricultur. Sci.*, 63(2): 163–172. https://doi.org/10.1016/J.AOAS.2018.11.00.

Salih, A. M.; Al-Qurainy, F.; Khan, S.; Tarroum, M.; Nadeem, M.; Shaikhaldein, H. O.; Gaafar, A. R. Z.; Alfarraj, N. S. (2021). Biosynthesis of zinc oxide nanoparticles using Phoenix dactylifera and their effect on biomass and phytochemical compounds in Juniperus procera. *Sci. Rep.,* 11(1): 1–12. https://doi.org/10.1038/s41598-021-98607-3.

Sarmast, M. K.; Salehi, H. (2016). Silver nanoparticles: an influential element in plant nanobiotechnology. *Mol. Biotechnol.*, 58(7): 441–449.

Sayed, E. G.; Mahmoud, A. W. M.; Mohamed M. El-Mogy, Mahmoud A. A. Ali, Mahmoud A. M. Fahmy, G. A. T. (2022). *The effective role of nano-silicon application in improving productivity and quality of grafted tomato grown under salinity stress. Horticulturae*, ISSN 2311-7524.

Servin, A. D.; White, J. C. (2016). Nanotechnology in agriculture: next steps for understanding engineered nanoparticle exposure and risk. *NanoImpact*, 1: 9–12.

Shafiq, M.; Anjum, S.; Hano, C.; Anjum, I.; Abbasi, B. H. (2020). An overview of the applications of nanomaterials and nanodevices in the food industry. *Foods,* 9(2): 148. https://doi.org/10.3390/FOODS9020148.

Shalaby, T. A.; Abd-Alkarim, E.; El-Aidy, F.; Hamed, E.; Sharaf-Eldin, M.; Taha, N.; El-Ramady, H.; Bayoumi, Y.; dos Reis, A. R. (2021). Nano-selenium, silicon and H_2O_2 boost growth and productivity of cucumber under combined salinity and heat stress. *Ecotoxicol. Environ. Safety*, 212: 111962.

Shalaby, T. A.; Bayoumi, Y.; Abdalla, N.; Taha, H.; Alshaal, T.; Shehata, S.; Amer, M.; Domokos-Szabolcsy, É.; El-Ramady, H.; Shalaby, T. A.; Bayoumi, Y.; Abdalla, N.; Taha, H.; Alshaal, T.; El-Ramady, H.; Shehata, S. (2016). Nanoparticles, soils, plants and sustainable agriculture. In: Ranjan, S., Dasgupta, N. (eds) *Sustainable Agriculture Reviews.* Springer, Switzerland, pp. 283–312. https://doi.org/10.1007/978-3-319-39303-2_10.

Sharaf-Eldin, M. A.; Elsawy, M. B.; Eisa, M. Y.; El-Ramady, H.; Usman, M.; Rehman, M. Z. (2022). Application of nano-nitrogen fertilizers to enhance nitrogen efficiency for lettuce growth under different irrigation regimes. *Pak. J. Agri. Sci.*, 59(3): 367–379.

Shareef, S. S.; Qasim, H. A.; Omar, O. M. (2021). Effect of (N.P.K) nano and mineral fertilizer on some growth characteristics of *Pinus Brutia* Ten. seedlings by foliar application. *IOP Conf. Series: Earth and Environmental Science,* 910: 012012.

Sharma, G.; Kumar, A.; Devi, K. A.; Prajapati, D.; Bhagat, D.; Pal, A.; Raliya, R.; Biswas, P.; Saharan, V. (2020). Chitosan nanofertilizer to foster source activity in maize. *Int. J. Biol. Macromol.*, 145: 226–234.

Sharma, P., Urfan, M., Anand, R. et al. (2022) Green synthesis of zinc oxide nanoparticles using Eucalyptus lanceolata leaf litter: characterization, antimicrobial and agricultural efficacy in maize. *Physiol Mol. Biol. Plants*, 28: 363–381. https://doi.org/10.1007/s12298-022-01136-0

Shebl, A.; Hassan, A.; Salama, D.; El-Aziz, M. E. A.; Elwahed, M. A. (2019). Green synthesis of manganese zinc ferrite nanoparticles and their application as nanofertilizers for Cucurbita pepo L. *Beilstein Archives,* 2019(1): 45. https://doi.org/10.3762/BXIV.2019.45.V1.

Sheoran, P.; Goel, S.; Boora, R.; Kumari, S.; Yashveer, S.; Grewal, S. (2021). Biogenic synthesis of potassium nanoparticles and their evaluation as a growth promoter in wheat. *Plant Gene,* 27: 100310. https://doi.org/10.1016/J.PLGENE.2021.100310.

Singh, J.; Lee, B. K. (2016). Influence of nano-TiO_2 particles on the bioaccumulation of Cd in soybean plants (*Glycine max*): a possible mechanism for the removal of Cd from the contaminated soil. *J. Environ. Manag.*, 170: 88–96.

Six, J.; Feller, C.; Denef, K.; Ogle, S. M.; Moraes, J. C.; Albrecht, A. (2002). Soil organic matter, biota and aggregation in temperate and tropical soils-effects of no-tillage. *Agronomie*, 22: 755–775.

Slomberg, D. L.; Schoenfisch, M. H. (2012). Silica nanoparticle phytotoxicity to *Arabidopsis thaliana. Environ. Sci. Technol.*, 46: 10247–10254.

Solanki, P.; Bhargava, A.; Chhipa, H.; Jain, N.; Panwar, J. (2015). Nano-fertilizers and their smart delivery system. In: Rai, M., Ribeiro, C., Mattoso, L., Duran, N. (eds) *Nanotechnologies in Food and Agriculture*. Springer, Switzerland, pp. 81–101. https://doi.org/10.1007/978-3-319-14024-7_4.

Su, Y.; Ashworth, V.; Kim, C.; Adeleye, A. S.; Rolshausen, P.; Roper, C.; White, J.; Jassby, D. (2019). Delivery, uptake, fate, and transport of engineered nanoparticles in plants: a critical review and data analysis. *Environ. Sci.: Nano*, 6(8): 2311–2331. https://doi.org/10.1039/C9EN00461K.

Subramanian, K. S.; Manikandan, A.; Thirunavukkarasu, M.; Rahale, C. S. (2015). Nano-fertilizers for balanced crop nutrition. In: Rai, M. et al. (eds) *Nanotechnologies in Food and Agriculture*. Springer, Cham, pp. 69–80.

Subramanian, K. S.; Muniraj, I.; Uthandi, S. (2016). Role of actinomycete-mediated nanosystem in agriculture. In: Subramaniam, G. et al. (eds) *Plant Growth Promoting Actinobacteria*. Springer, Singapore, 233–247.

Subramanian, K. S.; Paulraj, C.; Natarajan, S. (2008). Nanotechnological approaches in nutrient management. In: Chinnamuthu, C.R., Chandrasekaran, B., Ramasamy, C. (eds) *Nanotechnology Applications in Agriculture. TNAU Technical Bulletin, Coimbatore*, India, pp. 37–42.

Suman, P. R.; Jain, V. K.; Varma, A. (2010). Role of nanomaterials in symbiotic fungus growth enhancement. *Curr. Sci.*, 99: 1189–1191.

Syu, Y.-yu, Hung, J. H.; Chen, J. C.; Chuang, H.-wen. (2014). Impacts of size and shape of silver nanoparticles on Arabidopsis plant growth and gene expression. *Plant Physiology Biochem.,* 83: 57–64. https://doi.org/10.1016/J.PLAPHY.2014.07.010.

Tahmasbi, D.; Zarghami, R.; Azghandi, A. V.; Chaichi, M. (2011). Effects of nanosilver and nitroxin biofertilizer on yield and yield components of potato minitubers. *Int. J. Agric. Biol.*, 13: 986–990.

Talgar, S.; Jianxiu, G. U.; Changshan, X. U.; Zhikun, Y.; Qing, Z.; Yuxue, L.; Yichun, L. (2011). Phytotoxic and genotoxic effects of ZnO nanoparticles on garlic (Allium sativum L.): a morphological study. *Nanotoxicology*, 1–8.

Tarafdar, J. C.; Sharma, S.; Raliya, R. (2013). Nanotechnology: interdisciplinary science of applications. *Afr. J. Biotechnol.*, 12(3): 219–226.

Tarafder, C.; Daizy, M.; Alam, M. M.; Ali, M. R.; Islam, M. J.; Islam, R.; Ahommed, M. S.; Aly, M. A. S.; Khan, M. Z. H. (2020). Formulation of a hybrid nanofertilizer for slow and sustainable release of micronutrients. *ACS Omega*, 5: 23960–23966.

Theng, B. K. G.; Yuan, G. (2008). Nanoparticles in the soil environment. *Elements*, 4: 395–399.

Tourinho, P. S.; Van Gestel, C. A. M.; Lofts, S.; Svendsen, C.; Soares, A. M.; Loureiro, S. (2012). Metal based nanoparticles in soil: fate, behavior, and effects on soil invertebrates. *Environ. Toxico. Chem.,* 31(8): 1679–1692.

Toksha, B.; Sonawale, V. A. M.; Vanarase, A.; Bornare, D.; Tonde, S.; Hazra, C.; Kundu, D.; Satdive, A.; Tayde, S.; Chatterjee, A. (2021). Nanofertilizers: a review on synthesis and impact of their use on crop yield and environment. *Environ. Technol. Innov.,* 24: 101986. https://doi.org/10.1016/J.ETI.2021.101986.

Tymoszuk, A. (2021). Silver nanoparticles effects on in vitro germination, growth, and biochemical activity of tomato, radish, and kale seedlings. *Materials*, 14(18): 5340. https://doi.org/10.3390/MA14185340.

Vejan, P.; Khadiran, T.; Abdullah, R.; Ahmad, N. (2021). Controlled release fertilizer: a review on developments, applications and potential in agriculture. *J. Controll. Release,* 339: 321–334. https://doi.org/10.1016/J.JCONREL.2021.10.003.

Vertegel, A. A.; Siegel, R. W.; Dordick, J. S. (2004). Silica nanoparticle size influences the structure and enzymatic activity of adsorbed lysozyme. *Langmuir*, 20: 6800–6807.

Wang, F.; Liu, X.; Shi, Z.; Tong, R.; Adams, C. A.; Shi, X. (2016a). Arbuscular mycorrhizae alleviate negative effects of zinc oxide nanoparticle and zinc accumulation in maize plants—a soil microcosm experiment. *Chemosphere*, 147: 88–97.

Wang, J.; Fang, Z.; Cheng, W.; Tsang, P. E.; Zhao, D. (2016b). Ageing decreases the phytotoxicity of zerovalent iron nanoparticles in soil cultivated with *Oryza sativa. Ecotoxicology*, 25: 1202–1210.

Wang, P.; Menzies, N. W.; Lombi, E.; McKenna, B. A.; Johannessen, B.; Glover, C. J.; Kappen, P.; Kopittke, P. M. (2013). Fate of ZnO nanoparticles in soil and cowpea (*Vigna unguiculata*). *Environ. Sci. Technol.*, 47(23): 13822–13830.

Wang, S.; Wang, F.; Gao, S.; Wang, X. (2016c). The role of nano-silicon and its foliar application on heavy metals accumulation in different rice cultivars. *Water Air Soil Pollut.*, 227: 228.

Wang, Z.; Xu, L.; Zhao, J.; Wang, X.; White, J. C.; Xing, B. (2016). CuO nanoparticle interaction with arabidopsis thaliana: toxicity, parent-progeny transfer, and gene expression. *Environ. Sci. Technol.*, 50: 6008–6016.

Wilson, M. A.; Tran, N. H.; Milev, A. S.; Kannangara, G.; Volk, H.; Lu, G. (2008). Nanomaterials in soils. *Geoderma*, 146(1): 291–302.

Yadav, J.; Jasrotia, P.; Kashyap, P. L.; Bhardwaj, A. K.; Kumar, S.; Singh, M.; Singh, G. P. (2022). Nanopesticides: current status and scope for their application in agriculture. *Plant Protect. Sci.,* 58(1): 1–17. https://doi.org/10.17221/102/2020-PPS.

Yanik, F.; Vardar, F. (2015). Toxic effects of aluminum oxide (Al_2O_3) nanoparticles on root growth and development in *Triticum aestivum. Water Air Soil Pollut.*, 226: 296.

Yuvaraj, M.; Subramanian, K. S. (2015). Controlled-release fertilizer of zinc encapsulated by a manganese hollow core shell. *Soil Sci. Plant Nutr.*, 61(2): 319–326.

Zhao, L.; Peralta-Videa, J. R.; Rico, C. M.; Hernandez-Viezcas, J. A.; Sun, Y.; Niu, G.; Servin, A.; Nunez, J. E.; Duarte-Gardea, M.; Gardea-Torresdey, J. L. (2014). CeO_2 and ZnO nanoparticles change the nutritional qualities of cucumber (*Cucumis sativus*). *J. Agric. Food Chem.*, 62: 2752–2759.

Zulfiqar, F.; Navarro, M.; Ashraf, M.; Akram, N. A.; Munné-Bosch, S. (2019). Nanofertilizer use for sustainable agriculture: advantages and limitations. *Plant Science,* 289: 110270. https://doi.org/10.1016/J.PLANTSCI.2019.110270.

Zuverza-Mena, N.; Martı´nez-Fernandez, D.; Du, W.; Hernandez-Viezcas, J. A.; Bonilla-Bird, N.; Lopez-Moreno, M. L.; Komarek, M.; Peralta-Videa, J. R.; Gardea-Torresdey, J. L. (2017). Exposure of engineered nanomaterials to plants: insights into the physiological and biochemical responses-a review. *Plant Physiol Biochem.,* 110: 236–264.

12 Nanofertilizers
The Smart and Precise Regulators of Root Growth

Tripti Sharma, Vandana Singh, Subash Chudali, Kumari Prabha Singh, Surbhi, Aarshi Kongbrailatpam, Taring Yakum, Meghna Laiphangbam, Prasun Kumar Singh, Surendra Pratap Singh and Astha Gupta

12.1 INTRODUCTION

The global population has expanded exponentially over the decades and is expected to reach a mark of 9.7 billion by 2050 (FAO, 2018). The agriculture sector may face the challenge of feeding the growing population and evading hunger and malnutrition, simultaneously with the threat posed by dynamics of climate change. So as to synchronize with the growing demand, the grain yield is required to be elevated by nearly 70% (Hunter et al., 2017). However, major agricultural resources such as arable land and water are limited, therefore yield can be optimized in sync with the growing demand by augmenting the use of fertilizers in agricultural practices.

Green revolution introduced the use of chemical fertilizers to achieve higher productivity to fulfil the demands of the growing population. This led countries facing food scarcity to become self-sufficient rather than produce surplus food (Nongbet et al., 2022). Application of fertilizers in optimum or recommended quantities to the soil or plant tissues (a process also known as fertilization) is important to supply nutrients that are critical in determining the growth, yield, and nutrient status of the crop. In soil, it helps improve the nutrient status, hence fertility and quality by enriching with deficient nutrients.

Fertilizer may be organic, inorganic, or biological (biofertilizer) in origin and contains a minimum 5% of at least one or more of the three primary elements viz. nitrogen (N), phosphorus (P), and potassium (K), commonly referred to as NPK (Kumar et al., 2019). Fertilizers promote growth and development of plants either by increasing the availability of nutrient elements present in soil or by altering the structural configuration of soil particles. Similarly, chemical fertilizers are synthetic inorganic compounds, such as urea, ammonium chloride, ammonium sulphate, etc., that are added to soil to improve the availability of nutrients to growing plants for better growth, yield, and nutrient quality (Mahil & Kumar, 2019). On the other hand, organic fertilizers are derivatives of living or once living matter, including plant or animal detritus, treated sludge and compost, generally regarded as biodegradable material. Application of industrial sludge to agricultural land by pre-treatment with earthworms – vermicomposting – also produces stabilized and enriched organic fertilizer or vermicompost, in comparatively less time (Lee et al., 2018).

Furthermore, biofertilizers are compositions of live or latent microbial cell cultures that enhance soil fertility, improving the nutrient availability in soil and its uptake by plants (Fasusi et al., 2021). Consequently, this promotes the biomass, yield, content of essential pigments like chlorophyll, anthocyanins, and carotenoids along with other nutritive parameters like phenolic compounds and

 DOI: 10.1201/9781003354109-12

antioxidants (Verma et al., 2019; Arora et al., 2018; Khalid et al., 2017). Recently, the use of nanotechnology in agriculture has showed the potential to revolutionize the way we grow crops by enhancing nutrient uptake, increasing crop yield, and reducing fertilizer application rates. Nanofertilizers (NFs) are nanoparticles (NPs) that are designed to deliver essential nutrients (both micro and macro nutrients) to crops more efficiently and sustainably than traditional fertilizers (Shang et al., 2019). Their particles range in size from 1 to 100 nanometers (nm). These particles can be made from a variety of materials, including metals, ceramics, quantum dots, emulsions, and polymers (both natural and synthetic) (Ruiz-Cañas et al., 2020).

12.2 COMPARISON OF TRADITIONAL FERTILIZERS WITH NANOFERTILIZERS

Chemical fertilizers have low nutrient-use efficiency (NUE), with even less than 50% of the applied fertilizer being utilized (Nongbet et al., 2022). Their exploited use beyond permissible limits has deleterious effects on the environment (soil, air, water) as well as living beings (plants, animals, human beings, and microbes). Heavy metals – a component of chemical fertilizers (for example, cadmium, arsenic etc.) – accumulate in soil and aquatic bodies via mineralization and leaching, leading them to enter the food chain and affecting human health directly or indirectly.

On the other hand, organic fertilizers are natural nutrient additives of soil supplying essential nutrients in available forms to plants. As decomposition of debris is a relatively slow process, the nutrients and minerals are released into the soil at a slow pace making it convenient for plants to process them in a suitable manner as per the requirement (Shaji et al., 2021). This balances out the risk of intoxication of any particular nutrient in the plant. Further, organic fertilizers are beneficiary for soil microbiota, that may contribute to enhanced drainage capacity together with circulation of air within the soil and consequently plant growth (Assefa & Tadesse, 2019; Shaji et al., 2021). This suggests that organic fertilizers are better and alternatives to chemical or inorganic fertilizers for sustainable agricultural practices.

Rhizosphere studies have revealed that plant growth-promoting rhizobacteria (PGPR) and mycorrhizal association forming fungi have the potential to be applied as biofertilizers. However, blue green algae and ferns like azolla are also a promising source of biofertilizers. These microbes act by serving as biocontrol agents, solubilizing or mobilizing available phosphate, zinc and potassium, fixing nitrogen and oxidizing sulfur (Kumar et al., 2022). Increase in nutrient levels may be achieved in one of the following ways: (a) Affecting plant metabolism and altering the nature of root exudates. (b) Affecting nutrient solubility and hence availability. (c) Enhancing interaction with other rhizospheric microbes. Biofertilizers are considered as better alternatives due to their cost-effective and environment sustainable attributes and ability to produce yield that is toxin-free and good for human and animal consumption (El-Ghamry et al., 2018). Simultaneously it also contributes towards enhancing production of phytohormones, protection of pests and pathogens, and ability to ameliorate abiotic stresses.

Contrarily, application of nanofertilizers may not only help to increase food production but can also reduce environmental impacts by decreasing the amount of excess nutrients that leach into waterways and cause pollution. The surface of these nanoparticles is engineered to interact with plant roots and soil in a way that enhances nutrient uptake and retention. One of the most significant advantages of nanofertilizers is their ability to release nutrients in a controlled manner, which improves nutrient uptake by plants and reduces waste (Kumar et al., 2018). Whereas, traditional fertilizers can release nutrients too quickly, which can lead to atmospheric gas emissions leaching and runoff, causing environmental problems (Kah et al., 2018). The controlled release of nutrients also reduces the frequency and quantity of fertilizer application, thereby reducing the cost and environmental impact of fertilization.

The uptake of nanofertilizers by plants is mainly governed by their size, coatings, and functional groups present on their surface, and their chemical compostion (Mittal et al., 2020). However,

TABLE 12.1
Comparison Between Nanofertilizers and Traditional Fertilizers

Properties	Nanofertilizers	Traditional/Conventional Fertilizers	References
Rate of nutrient loss	Minimal loss of nutrients from fertilizers	High loss rate from runoff, leaching, and drifting.	(Kalwani et al., 2022)
Nutrient uptake effectiveness	Improved uptake ratio and reduced use of fertilizer resources	The efficiency of nutrient intake is low, and it is not accessible to roots.	(Rizwan et al., 2021)
Soil adsorption and fixation	Lower	Higher	(Seleiman et al., 2020)
The effective time of release	Long-lasting and efficient	Used by the plant at the location and time of application; the remaining is changed into an insoluble form.	(Thavaseelan & Priyadarshana, 2021)
Mineral micronutrients' dispersion	Increased insoluble nutrient dispersion	Larger particle size results in lower solubility.	(Chhipa, 2017)
Solubility	Higher	Lower	(Chhipa, 2017)
Release techniques	Controlled release rate and pattern	High toxicity and soil imbalance are caused by the overuse of fertilizers.	(Avila-Quezada et al., 2022)
Bioavailability	Higher	Lower	(Seleiman et al., 2020)

their effect on the plant system can be manifested on seed germination, seed or seedling vigor, plant growth and physiology together with protection and amelioration under stress or unfavorable conditions. For example, in vitro studies in *Coriandrum sativum* have reported the drought stress-alleviating effect of proline-coated zinc oxide nanoparticles (Hanif et al., 2023). Similarly, nanofertilizers can regulate root growth by affecting multiple mechanisms, including hormonal regulation, ion transport, enzymatic activities, gene expression, and ROS signaling. These mechanisms are interdependent and can interact with each other to regulate root growth and development. Further research is needed to optimize the use of nanofertilizers for sustainable agriculture and to ensure their safety and efficacy.

Therefore, comparative analysis of traditional fertilizers versus nanofertilizers with some important properties have been explained in Table 12.1.

12.3 NANOTECHNOLOGY IN AGRICULTURE: NEED AND CURRENT STATUS

In recent times, nanotechnology is being applied in many areas like medicine (nanomedicine), catalysis, energy materials, engineering-bio and optical, agriculture, textile and food industry, etc. The variations in size and surface area properties of nanoparticles allow for its potential application in the above-mentioned sectors. Nanoparticles (NPs) can be synthesized by different methods like physical, chemical, and biological methods (Goutam et al., 2020). Nowadays, green pathways (i.e., biological methods) utilizing plant- and microbe- (including fungi, bacteria, etc.) derived extracts are largely employed for producing nanoparticles. These methods are preferred over conventional physical and chemical methods as they are comparatively less-toxic, convenient, cheaper, and eco-friendly. Several metallic as well as non-metallic nanoparticles have been synthesized with the application of green methods. For example, zinc oxide nanoparticles have been produced from leaf extracts of *Salvia officinalis, Brassica oleracea* var. *botrytis* and stem extract of *Tinospora cordifolia.* Similarly, extracts derived from sea-weeds like *Pterocladia capillacea* and bacteria such as *Bacillus subtilis* ZBP4 have also been utilized for synthesizing ZnO nanoparticles (Alrajhi & Ahmed, 2023; Manoj kumar et al., 2023; Sharma et al., 2023; Mansour et al., 2022; Hamk et al., 2023). Plant- and microbe-derived extracts have high amounts of reducing and bio-active agents that enable synthesis of particles at nanodimensions. Therefore, the ease of scaling up, and limited environmental impact has also escalated the synthesis of NPs using this technology.

Nanomaterials (NMs) developed by eco-friendly methods could improve the fertilization (i.e., addition of fertilizers) process, plant growth regulators, and pesticide application (Shang et al., 2019). They can reduce the amount of harmful chemicals that pollute the environment. Hence, this technology helps in reducing the environmental pollutants. In agriculture, nanotechnology has gained momentum since the last two decades by virtue of the enhanced efficiency of inputs and simultaneously decreasing related losses. Application of nanotechnology in agriculture has revolutionized a novel frontier in agriculture known as "nanofarming" (Haris et al., 2023). Because of the unique nature of farm production, like as an open system whereby energy and matter are exchanged freely, many methods have come under the umbrella of agriculture. It improves crop production and reduces pesticide applications as it provides precise and targeted application of agrochemicals, thereby making agriculture more sustainable.

The primary nanotechnological interventions in agriculture include formulation of pesticides and fertilizers derived from nanoformations. Nanoparticle derived "smart pesticides" are applied for protection of crops against identified pests and pathogens, on the other hand, nanofertilizers aid by enhancing seed germination, followed by plant growth, thus increasing crop yield (Kumar et al., 2019; Verma et al., 2018). Similarly, nanosensors for precision farming have been developed for adequate monitoring, analyzing, and management of agricultural practices (quality of soil, crop growth, etc.) (Bhagat et al., 2023). These sensors are also capable of detecting environmental changes (or stresses) facilitating plants' potential to withstand all such changes.

Further, nanobiotechnology has also found its utilization in the field of agriculture. Its applicability has been widely explored for designing and construction of carriers of genetic material – "nanocarriers" (Mat Jalaluddin et al., 2023). Nanocarriers are employed in genetic engineering of plants and producing transgenic plants with the ability to combat both biotic as well as abiotic stresses. Nanoparticle-mediated gene or DNA/RNA transfer in plants for the development of insect/pest-resistant varieties (Ma et al., 2023). Importance of nanoparticle-engineering for insecticide development is gaining importance as this would lead to understanding mechanisms operating in real time and because of its economic importance.

As an extension of its agricultural implications, nanotechnology has also proven beneficial for food processing and post-harvest management of agriculture produce. It may be used for storage with shelf-life extension preventing post-harvest associated losses (Haris et al., 2023). Nanopackaging is one of the recent advancements in this sector (Stuparu-Cretu et al., 2023). Additionally, nanomaterials may also be applied to soil for an improvement in the soil attributes, which may directly or indirectly affect crop production. In recent years, agricultural waste products are increasing so they are gaining attention to be used as renewable materials, which can be processed and use as a raw material for nanomaterial production (Uddin & Haque, 2023). It is believed that the nanotechnology for agriculture and allied sectors need to be balanced against issues related to the soil, water and environment, and the occupational health of workers. There is a large potential for nanotechnology in agriculture, along with some limitations, which also need to be addressed as the risk assessment. Therefore, nanotechnology will enable transformation of agriculture and the allied sectors, such as in the case of farming techniques, increasing the ability of plants to absorb nutrients, to detect disease, and control pests.

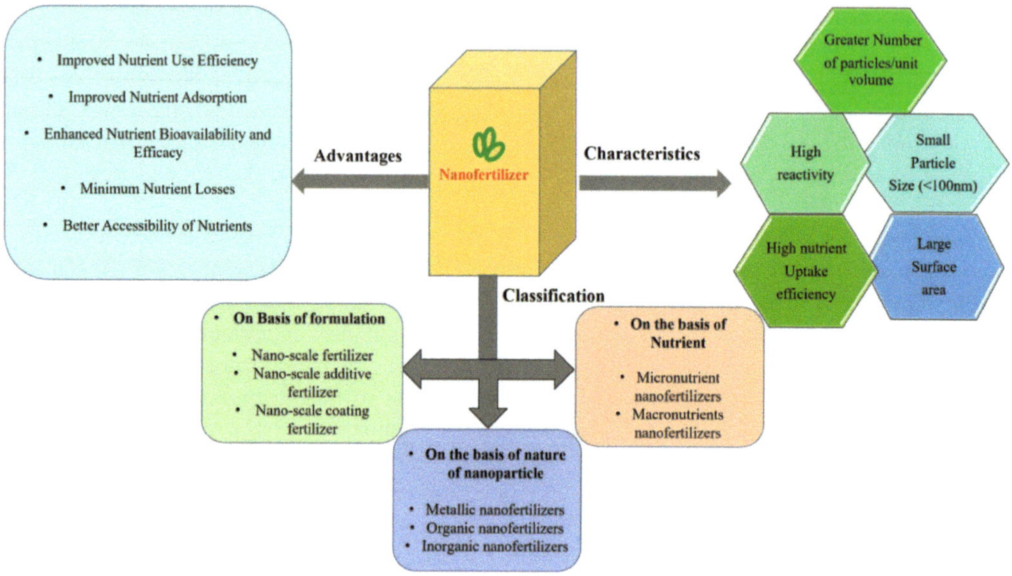

FIGURE 12.1 Summarizes the characteristics, classification, and advantages of nanofertilizers.

12.4 CHARACTERISTICS OF DIFFERENT TYPES OF NANOFERTILIZERS (NFS)

As discussed in previous sections, nanofertilizers have distinct qualities that make them superior to synthetic fertilizers (Singh & Kumar, 2017). They can penetrate deeper into plants due to their small particle size (<100 nm). Because nanofertilizers have bigger surface areas their reactivity is higher than synthetic fertilizers. They also offer a greater number of particles in unit volume thereby raising the number of nutrients available as well as entering the plant, increasing plant nutrient-use efficiency, bioavailability, and efficacy (Figure 12.1). In addition, incorporation of sugars and amino acids while nanofertilizer application improves its absorption (Gangwar et al., 2023). Nanofertilizers readily solublize and diffuse in soil, which contributes to increased accessibility of nutrients to the plants directly. Fertilizers are condensed in the nanoparticles to increase the nutrient uptake by the plant and decrease the density of fertilizer application (Chhipa, 2017). For example, using zeolite-based nanofertilizers makes it possible to provide plants with high nutrient levels throughout the entire growth phase. Furthermore, nanofertilizers gradually and precisely supply nutrients to plants, which thus reduces the salt build-up in the soil, and the loss of nutrients is reduced as major proportions are utilized by the plant.
Advantages of nanofertilizers

- Minimum nutrient loss.
- Slow and efficient use of fertilizers.
- Promote early seed germination.
- Ability to withstand abiotic stresses, for example, salinity and drought stress.
- Increase yield.
- Improve the whole health of plants.
- Increase nutrients in the produce.
- Help eradicate weeds.

12.4.1 Classification of Nanofertilizers

Nanofertilizers may be broadly classified on the following basis:

12.4.1.1 On the Basis of Formulation

Based on the formulation of nanofertilizers, they are of the following three types: (a) Nanoscale fertilizer – a traditional fertilizer whose size has been tailored in the range of 1–100 nm. For example, zinc oxide nanoparticle. (b) Nanoscale additive fertilizer – is a type of fertilizer, which mainly comprises of a traditional fertilizer containing a nanomaterial as an additive to the formulation. For example, carbon nanotube urea. (c) Nanoscale coating fertilizer – comprises nutrients encased within a nanofilm or interposed inside nanosized pores of host matter. For example, polymer coated urea (Jain et al., 2021). Such types of fertilizers are usually held in a carrier (for example, clay) resulting in the formation of nanocomposites that enables the controlled release of nutrients (Borges et al., 2019).

12.4.1.2 On the Basis of the Nature of Nanoparticles

Nanofertilizers can be classified into three major categories based on the nature of their constituent nanoparticle, namely metallic, organic, and inorganic nanofertilizers. Each category has its own unique properties (Ndaba et al., 2022). Metallic nanofertilizers are composed of metal nanoparticles, such as silver, gold, and copper. One example of metallic nanofertilizer is silver nanoparticles (Ag NPs), which have been found to improve the growth of wheat plants by increasing the uptake of nitrogen and phosphorus. Ag NPs have also been shown to reduce soil-borne pathogens, such as *Fusarium oxysporum* and *Rhizoctonia solani*, which cause root rot in plants. Organic nanofertilizers are composed of organic, i.e., carbon molecules – natural or synthetic – that have been synthesized into nanoparticles. These may be formulated into vesicles, dendrimers, liposomes, and polymersomes. They are biodegradable, non-toxic, and bio-compatible and are susceptible to degradation at higher temperatures. One example of organic nanofertilizer is chitosan nanoparticles, which have been found to enhance the growth of tomato plants by increasing the uptake of nutrients, such as nitrogen and phosphorus. Chitosan nanoparticles have also been shown to improve soil structure by reducing soil compaction, which enhances water and nutrient availability to plants. Inorganic nanofertilizers are composed of inorganic compounds, such as metal oxides, non-metals, hydroxides, and phosphates. One example of inorganic nanofertilizer is calcium phosphate nanoparticles, which have been found to enhance the growth of maize plants by increasing the uptake of phosphorus (Rane et al., 2015). Calcium phosphate nanoparticles have also been shown to improve soil structure by reducing soil compaction, which enhances water and nutrient availability to plants.

12.4.1.3 On the Basis of the Nutrient Delivered

The nutrient requirements of a plant are divided into two types – micro and macronutrients. Consequently, a nanofertilizer may be developed to deliver either of these nutrients and thus is referred to as a macronutrient or micronutrient nanofertilizer (Nongbet et al., 2022). A macronutrient nanofertilizer is an amalgamation of one or a combination of macronutrient(s) like nitrogen, magnesium, phosphorus, calcium with nanomaterials designed such that these are delivered accurately to the plant. Similarly, a micronutrient nanofertilizer comprises one or a combination of micronutrient(s) like iron, manganese, copper, zinc encapsulated inside a specific nanomaterial. However, hydroxyapatite nanoparticles (HA NPs) are widely employed to supply both macro and micronutrients to plants.

12.4.2 Different Types of Nanofertilizers Explored in Agriculture

12.4.2.1 Non-Porous Zeolite

Non-porous zeolite nanofertilizers are a type of fertilizer made from zeolite, a natural mineral with a porous structure. The zeolite particles are reduced to nanosize, which increases their surface area.

Non-porous zeolites are made by treating the zeolite with chemicals to remove the pores from the structure. This makes the zeolite less able to retain water and nutrients, but it may increase the stability and reduce the cation exchange capacity of the zeolite. Non-porous zeolite nanofertilizers are used in agriculture to improve soil structure, increase water retention, and reduce nutrient leaching (Mahesh et al., 2018).

They aid in the fertilizer's delayed release onto the plant, which encourages the plant to absorb all of the nutrients from the fertilizer. The greater surface area allows it to retain more molecules, which may be released anytime the plant requires them. The nanoparticle known as ZEOLITE (Clinoptilolite) was produced using a ball mill. Although nitrogen fertilizers are crucial for plants, they also seriously harm the environment and the plants because of their high solubility. As a result, urea was combined with nanoporous zeolite, and a notable improvement in the uptake of nitrogen-efficient urea with the gradual release was observed (Bajpai et al., 2020). Due to their high porosity and capacity for soil absorption, aluminium zeolites can also be employed. Moreover, it also helps dry soil to conserve its moisture level and helps to absorb the nutrients from the soil (Cataldo et al., 2021).

12.4.2.2 Nitrogen Nanofertilizers

Nitrogen is an essential macronutrient for plant growth, and it is often the limiting factor in agricultural systems in spite of atmospheric nitrogen accounting for nearly 78%. Nitrogen is utilized by plants mainly as ammonium and nitrates (Hansen & Elberling, 2023). The conventional source of nitrogen, i.e., urea has poor nitrogen-use efficiency (NUE), due to its conversion to undesirable forms and quick dissolution. This limitation has raised the need for development of N nanofertilizers, particularly with controlled release. These fertilizers are encapsulated within nanoparticles and may have surface coatings of polymers or sulfur so as to enhance NUE and slow release (Mali et al., 2020).

12.4.2.3 Potash Nanofertilizers

Fertilizers that contain potassium in the form of K^+ are referred to as potash fertilizers. Together with phosphorus and nitrogen, potassium is one of the three pivotal macronutrients necessary for plant growth and development. In order to give plants the potassium they require to improve crop yield and quality, potash fertilizers are frequently utilized in agriculture. In addition to boosting photosynthesis, regulates stomatal opening in the leaves, controls water storage, increasing disease resistance, and it is essential for many physiological functions. The gradual release of potash fertilizer was achieved by coating cylindrical pellets with a polyacrylamide-based coating (Subbarao et al., 2013). Clay and potash were combined, then dried for one hour before being layered with the toothpaste to safeguard proper attachment and then getting coated with the polyacrylamide polymer. The research demonstrated the variation in dissolution with the absence or presence of coating, and it was seen that when the potash was used in a lesser amount, then potash release was also slowed; less water can also be used to maintain fertilizer release (Zhang & Wang, 2005). Thus, these studies were done on rice as it is the main cash crop in India and some other countries. With the application of less quantity nanopotash, it was seen that it induces a greater number of grains per panicle in rice and ultimately, the yield gets increased (Subbarao et al., 2013).

12.4.2.4 Zinc Nanofertilizers

In addition to being a critical element for plant growth, zinc is also essential for many physiological activities. However, zinc shortage is a widespread issue particularly in alkaline soils, which can restrict plant development and agricultural yield. In order to address this shortage and increase crop productivity, zinc fertilizers can be used. Fertilizers using zinc nanoparticles are one form of micronutrient nanofertilizer. In comparison to conventional fertilizers, these fertilizers are made to supply plants with vital nutrients more effectively. By shrinking the size of the particles to the nanoscale range, zinc nanofertilizers are produced and higher plant growth and yield are possible as a result of increased surface area and higher nutrient uptake by plants (Singh et al., 2018). To obtain diffused and soluble zinc, the nanoparticles can be used to coat the zinc (Siddiqi & Husen, 2018). Further, to maintain the zinc's solubility, the nanoparticles' surface area and size should be equal (Malik & Kumar, 2014). However, as the pH increases, solubility of zinc decreases (Rutkowska et al., 2015).

12.4.2.5 Nanobiofertilizers

Formulation of nanobiofertilizers is a recent advancement in the field of agriculture. A general review of biofertilizers has been discussed in Section 12.2. So, nanobiofertilizers are an amalgamation of microorganisms into nanostructures or along with nanomaterials. These fertilizers combine together the life span of biofertilizers, facilitated transport of nanoparticles and the interaction between these two components (Kumari & Singh, 2020). Potential of selenium and copper nanobiofertilizers on strawberry seedlings has been studied by assessing various growth parameters, nutrients, and antioxidant status (El-Bialy et al., 2023)

12.5 DEVELOPMENT OF NANOFERTILIZER THROUGH SYNTHESIS OF NOVEL NANOPARTICLES

There are generally three main techniques used for synthesizing nanoparticles – physical, chemical, and green/biological. These can be categorized under two main approaches namely top-bottom and bottom-top approaches. Both approaches have distinct applications, advantages, and limitations, which are discussed below.

12.5.1 Top-Bottom/Destructive Approach

In this approach, the complex material is broken down into simpler forms by physical processes and therefore also called destructive approach (Figure 12.2). These processes, deployed to reduce particle size, cause certain imperfections on the particle's surface, which are very important for surface chemistry and physical attributes of the resultant product (Bernela et al., 2021). Although it is an easy and cost-effective approach, it does not offer much control over the size of nanoparticles and the end product that is drawbacks of this approach (Yadi et al., 2018).

12.5.1.1 Physical Methods

As mentioned above the top-bottom approach involves reduction of particle size to nanodimensions using physical methods/processes. Some of these physical processes include pearl or ball milling, pulse laser ablation, high pressure homogenization, sputter and electric arc deposition, ultrasonication, and spray pyrolysis (Raj et al., 2021). Several metal nanoparticles such as silver, gold, lead, and fullerene have been synthesized by employing physical processes. These methods are generally devoid of solvent-derived impurities and produce homogenous uni-disperse. However, the plethora of waste generated makes it economically less compatible.

12.5.2 Bottom-Top/Build-Up Process

In this approach, the atoms/molecules are combined to form a cluster, which then finally concludes in the formation of nanosized particles (Figure 12.2) and that is why it is also known as the build-up

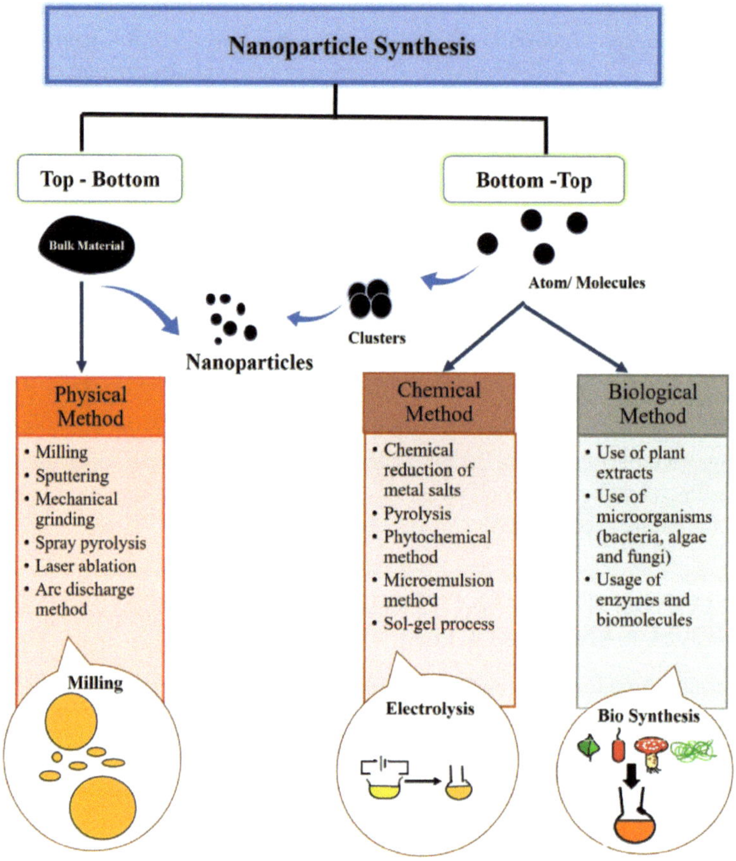

FIGURE 12.2 Depicts synthesis of nanoparticles (NPs): top-bottom and bottom-top approaches. Top-bottom and bottom-top approaches represents physical, chemical, and biological methods, respectively.

process (Heikal et al., 2020). This approach is harnessed in chemical and biological methods of NP synthesis (Yadi et al., 2018). It is feasible to produce NPs having uniform composition and less impurities/defects.

12.5.2.1 Chemical Methods

The chemical methods involved in synthesis of NPs (via the bottom-up approach) use organic/inorganic chemicals as agents for reduction, capping, and stabilization of synthesized particles (Raj et al., 2021). Some chemical methods are microemulsion methods, chemical vapor deposition, pyrolysis, sol-gel, electrochemical synthesis, sonochemical method, and co-precipitation. However, these are regarded as the easiest methods for production of metal NPs, the hazards, associated costs, and infidelity of reducing agents pose several limitations (Jamkhande et al., 2019).

12.5.2.2 Green or Biological Synthesis of Nanoparticles in Agriculture

The synthesis process is an essential aspect of nanoparticle production, and green synthesis methods have evolved as an eco-friendly and sustainable approach. Green synthesis of nanoparticles in agriculture involves the application of sustainable practices that reduce the consumption of harmful chemicals and energy while maximizing the use of renewable resources. These techniques frequently use natural extracts, materials derived from plants, microorganisms including algae, bacteria, fungi (microbial enzymes), or other environmentally benign sources as reducing agents, stabilizers, or templates for the creation of nanoparticles (Bordiwala, 2023). A brief summary of nanoparticles synthesized using plant- and microorganism-derived extracts is tabulated in Tables 12.2 and 12.3, respectively.

TABLE 12.2
Nanoparticles Synthesized Using Plant-Derived Extracts

			Morphology			
Nanoparticles	Plants Used	Characterization Techniques	Shape	Size	Application	References
Ag	Aloe Vera (*Aloe barbadensis miller)*	XRD, Scanning Electron Microscopy (SEM), UV-vis spectroscopy	Spherical	70.70 ± 22-192.02 ± 53 nm	Anti-bacterial, biomedical and biotechnological activity.	(Tippayawat et al., 2016)
	Azadirachta indica	DLS, photoluminescence, TEM and UV–Visible spectrophotometer	Spherical	34 nm	Anti-microbial and biomedical application	(Ahmed et al., 2016)
	Litsea cubeba	Fourier-transform infrared (FTIR) spectroscopy, Powder X-ray diffraction (XRD), Transmission electron microscope (TEM), Field-emission scanning electron microscopy (FE-SEM), Thermogravimetric analysis (TG and DTA), Energy-dispersive X-ray (EDX) spectroscopy, Dynamic light scattering (DLS) technology and UV-Vis spectroscopy	Spherical	8–18 nm	Catalytic reduction of 4-nitrophenol	(Doan et al., 2020)
ZnO	*Syzygium Cumini*	XRD, SEM, TEM, FTIR spectrum and UV–Vis spectrophotometer	Hexagonal	11.35 nm	Photocatalytic degradation of toxic dyes	(Sadiq et al., 2021)
	Thyme Plant (*Thymus)*	UV-Visible Spectroscopy(Cary 100 UV-Vis), FTIR, EDX, FESEM, HR-XRD	Spherical	39.4–51.86 nm	Biomedical, electronics, and optics activity	(Karam & Abdulrahman, 2022)
	Calotropis gigantea	XRD, SEM, EDAX, HR-TEM with SAED, FT-IR and UV-DRS. XRD	Hexagonal and pyramidal	31 nm	Photocatalysis, antibacterial, sensing nitrite and optoelectronic	(Kumar et al., 2020)

(*continued*)

TABLE 12.2 (Continued)
Nanoparticles Synthesized Using Plant-Derived Extracts

Nanoparticles	Plants Used	Characterization Techniques	Morphology		Application	References
			Shape	Size		
TiO2	*Mentha arvensis*	X-ray diffraction, UV–Visible absorbance spectroscopy, Fourier transform infrared spectroscopy and scanning electron microscopy methods.	Spherical	20–70 nm	Antibacterial and antifungal.	(Ahmad et al., 2020)
	Terminalia catappa and Carissa carandas	X-ray Diffractometer (XRD), Scanning Electron Microscopy (SEM), UV–Visible Spectroscopy (UV–Vis) and Fourier Transform Infrared Spectroscopy analysis	Spherical	10–21 nm	Production of metal nanoparticles	(Rajendhiran et al., 2021)
Pd	*Syzygium aqueum*	FTIR, BET, HR-TEM, ICP-MS, TGA, XRD, and FE-SEM/EDX.	Irregular	5–20 nm	Catalysis in Suzuki–Miyaura coupling reaction	(Manjare & Chaudhari, 2020)
	Urtica dioica	UV-vis, XPS, FT-IR, TEM, and XRD analyses	-	7.44 ± 1.94 nm	Anticancer, antibacterial, and antioxidant activity	(Gulbagca et al., 2021)
	Rosmarinus ofcinalis	FT-IR, XRD, FESEM, TEM and UV/Vis spectroscopy	Semi-spherical	15–90	Mizoroki–Heck catalytic, antibacterial and anti-fungal applications	(Rabiee et al., 2020)
Alumina (Al_2O_3)	Tea, coffee and triphala	Scanning electron microscope (SEM), UV–Visible (UV–Vis) spectroscopy and Fourier transform infrared spectroscopy (FTIR).	Spherical in case of tea and coffee Oval in case of triphala	50–200 nm 200–400 nm	Biomedical applications	(Sutradhar et al., 2013)
Cu	Walnut shells	TEM, EDX, TGA	–	15–22 nm	Antibacterial, anticancer and antioxidant	(Mehdizadeh et al., 2020)
	Cissus vitiginea	Ultraviolet-visible (UV-Vis) spectroscopy, Scanning Electron Microscopy (SEM), X-ray Diffraction (XRD), X-ray Photoelectron Spectroscopy (XPS),Transmission Electron Microscopy (TEM), Atomic Force Microscopy (AFM), Fourier Transform Infrared Spectroscopy (FT-IR)	Spherical	20 nm	Antibacterial and antioxidant activity	(Wu et al., 2020)

TABLE 12.2 (Continued)
Nanoparticles Synthesized Using Plant-Derived Extracts

Nanoparticles	Plants Used	Characterization Techniques	Morphology Shape	Morphology Size	Application	References
Mn_3O_4	*Helianthus annuus*	XRD, FTIR, SEM, HR-TEM, UV–visible spectroscopic	Spherical	10–70 nm	Applicable in agriculture field.	(Ramesh et al., 2021)
Au	*Hibiscus sabdarifa*	Fourier-transform infrared spectroscopy (FT-IR) spectroscopy, ultraviolet–visible spectroscopy (UV–Vis.), X-ray diffraction (XRD), field emission scanning electron microscopy (FE-SEM), and transmission electron microscopy (TEM) analysis	Spherical	15–45 nm	Production of Au nano particles containing anti-acute myeloid leukemia effect	(Zangeneh & Zangeneh, 2020)
	Nyctanthes arbortristis	TEM, XRD, FTIR, and UV–Vis spectroscopic analysis	Spherical	19.8 ± 5.0 nm	Biomedical application	(Das et al., 2011)
Pt	*Prosopis farcta*	X-ray diffraction, Field Emission Scanning Electron Microscopy (FESEM), Transmission Electron Microscopy (TEM), ImageJ, and ZETASIZER instrument.	Irregular	3.5 nm	–	(Jameel et al., 2020)
	Azadirachta indica	UV–visible spectrophotometer, TEM, and FT-IR.	Small sphere to large spheres	5–50 nm	Medicinal and catalytic activity.	(Thirumurugan et al., 2016)
Fe_2O_3	*Sida cordifolia*	X-ray diffraction, scanning, and transmission electronic microscopy, TG-DTA, FTIR, and UV spectroscopy	–	20 nm	Antibacterial activity	(Pallela et al., 2019)

TABLE 12.3
Nanoparticles Synthesized Using Microorganism-Derived Extracts

Nanoparticles	Microorganism	Location	Morphology		Optimum temperature for culturing (◦C)	Applications	References
			Shape	Size(nm)			
				Bacteria			
Au	*Shewanella oneidensis*	Extracellular	Spherical	12 ± 5	30	Biomedical and engineering applications	Suresh et al.(2011)
Ag	Bacillus subtilis	Extracellular	Spherical and occasionally triangular	5-60	37	-	Saifuddin et al. (2009)
CdS	Bacillus licheniformis	Extracellular	Triangle	20-40	Room temperature	Antimicrobial	Shivashankarappa et al.(2015)
				Fungi			
Au,Ag	*Neurospora crassa*	Intracellular, extracellular	Spherical	11 for Ag 32 for Au	28	-	Castro-Longoria et al. (2011)
Ag	Penicillium oxalicum	Extracellular	Spherical	60-80	37	Bactericidal	Feroze et al. (2020)
				Actinomycetes			
ZnO	Streptomyces sp.	-	Soherical	16-25	Room temperature	Antibacterials and anti-biofilms	Rajivgandhi et al.(2022)
				Algae			
Pd	Spirulina platensis	-	Spherical	10-20nm	-	Adsorbent	Sayadi et al. (2018)

12.6 ROOT: A POTENTIAL FUNCTIONAL PART OF PLANTS TO UPTAKE NUTRITION FROM THE SOIL

Roots are an important organ of a vascular plant and are the descending portion of a plant body that provides support to the plant and helps them grow tall and fast by taking in water and nutrients from the soil (Macdonald & Stevens, 2019). The fragment of a plant that connects it to the ground, usually underground, feeds and hydrates the rest of the plants by a variety of bough and filaments. Plants receive their water, nutrients, hormones, and anchoring via their roots, which has an impact on their economic outputs (Sundararajan et al., 2023). Plant root growth is genetically controlled, but environmental surroundings can have an impact. Rhizosphere and rhizodeposition, which support microbial activity, would not exist without roots (Damo et al., 2023). Plant hormones control the start, elongation, and creation of new root axes, which constitute root growth and development. The formation and growth of roots requires a sufficient amount of carbohydrates, either from current photosynthesis or from reserves.

12.6.1 Root Morphology, Root Morphogenesis and Biological Pathways

Root shape and density are crucial for obtaining nutrients with poor mobility in the soil (Marschner & Rengel, 2023). The initial root in plants arising from the seed is called radicle, which emerges from the embryo following seed germination. The root cap, apical meristem, elongation zone, and hair are the four zones that make up root morphology (Chouhan et al., 2023). New roots have a root coat that aids in soil encrustation. These root caps are shed as the root grows deeper, leaving a slimy surface that acts as a lubricant. New, elongating root cells are produced by the root cap's apical meristem. Then, the surrounding rhizosphere gives the growing root hairs water and mineral nutrients. However, root morphology is typically viewed as being less significant for more mobile nutrients such as NO^{3-} and NH^{4+} but is considerably significant for the acquisition of nutrients' lesser mobility (Lyzenga et al., 2023). For measuring the root morphological characters, a scanner-type image system was used (Barone et al., 2018). Instead of being actively absorbed, water diffuses through the root in response to a gradient in the water potential created by transpiration. When water travels through many tissues, symplastic, parallel apoplastic and transcellular pathways are crucial.

12.6.2 Growth of Root in Aquaponics

Hydroponics/aquaponics provide an environment that is extremely aerobic, which helps plants grow healthily. If given the right care, hydroponic roots grow and thrive far more swiftly and energetically than their equivalents in the soil. The optimal pH range for roots grown in hydroponic systems lies between 5.5 and 6.5 (Sharma et al., 2018). Roots that are produced hydroponically thrive in the dark. They can be exposed to some light, but prolonged exposure to even dim light and direct light are not recommended. Another crucial element that has a big impact on the health of hydroponic plant roots is the level of aeration. A steady stream of pure oxygen is necessary for the roots. If that airflow is even slightly altered, problems will unavoidably occur. Strong root systems unquestionably require water, just as they do light and air. Liquid mixes of microbial and other organic components are called root inoculants. These aid in healthy growth of plants and root system. A greater degree of nutritional content, pH, and growing environment control is permissible in hydroculture (Dhawi, 2023). It is possible to recycle water and nutrients, hence lowering the costs related. More oxygen availability in the root region leads to faster growth. Decrease or eradication of soil-related fungus, bacteria, and insects is an added benefit for root and overall plant growth.

12.7 APPLICATION OF NANOFERTILIZERS FOR ROOT GROWTH REGULATION

There have been several studies on the use of nanofertilizers for root growth regulation. These studies have shown that nanofertilizers can enhance root growth by improving nutrient uptake, increasing water-holding capacity, and regulating root growth hormones (Figure 12.3). Further, unique properties of metallic, organic, and inorganic nanofertilizers influence their effectiveness in root growth regulation (Ndaba et al., 2022).

Metallic nanofertilizers have been shown to enhance root growth by improving nutrient uptake, increasing soil microbial activity, and reducing soil pathogens. One example of metallic nanofertilizer is silver nanoparticles (Ag NPs), which have been found to improve the growth of wheat plants by increasing the uptake of nitrogen and phosphorus. Ag NPs have also been shown to reduce soil-borne pathogens, such as *Fusarium oxysporum* and *Rhizoctonia solani*, which cause root rot in plants.

Organic nanofertilizers have shown to improve root growth by increasing the availability of nutrients, improving soil structure, and reducing soil compaction. One example of organic nanofertilizer is chitosan nanoparticles, which have been found to enhance the growth of tomato plants by increasing the uptake of nutrients, such as nitrogen and phosphorus. Chitosan nanoparticles have also been shown to improve soil structure by reducing soil compaction, which enhances water and nutrient availability to plants.

Inorganic nanofertilizers have been shown to improve root growth by increasing the availability of nutrients, improving soil structure, and reducing soil acidity. One example of inorganic nanofertilizer is calcium phosphate nanoparticles, which have been found to enhance the growth of maize plants by increasing the uptake of phosphorus. Calcium phosphate nanoparticles have also been shown to improve soil structure by reducing soil compaction, which enhances water and nutrient availability to plants.

Some of the recent studies are discussed below.

1. Zinc oxide nanoparticles (ZnO NPs) – ZnO NPs have been shown to enhance root growth and increase nutrient uptake in plants. A study by Mahmoudi et al. (2020) showed that the application of ZnO NPs increased the root length, surface area, and volume of wheat plants. The study

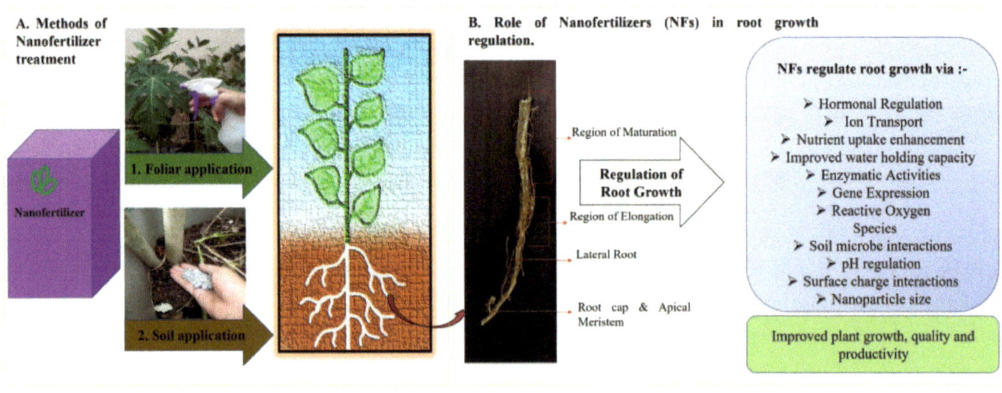

FIGURE 12.3 Application of nanofertilizers (NFs) in root development that promotes growth, quality, and productivity of plant. A. Two methods of nanofertilizers (NFs) treatment: 1. Foliar application 2. Soil application. B. Role of nanofertilizers (NFs) in root growth regulation.

also found that ZnO NPs increased the uptake of nitrogen, phosphorus, and potassium in wheat plants.

2. Silicon nanoparticles (Si NPs) – Si NPs have been shown to improve root growth and increase the resistance of plants to environmental stresses. Si NPs regulates defense related genes, protects ultra-cellular structures, increase plant growth, photosynthetic performance, antioxidants defence machinery and reduce Cr accumulation, H_2O_2, and MDA through in *Brassica napus* L. (Huang et al. 2024). The study also found that Si NPs increased the activity of antioxidant enzymes and reduced oxidative stress in tomato plants.
3. Carbon nanotubes (CNTs) – CNTs have been shown to improve root growth and increase the uptake of nutrients in plants. A study by Khodakovskaya et al. (2013) showed that the application of CNTs increased the root length, surface area, and volume of corn plants. The study also found that CNTs increased the uptake of nitrogen, phosphorus, and potassium in corn plants.
4. Silver nanoparticles (Ag NPs): Wang et al. (2020) studied the effect of silver nanoparticles (Ag NPs) on root growth in *Arabidopsis thaliana*. They found that the Ag NPs have an impact on root growth by regulating cell division. After uptake, the Ag NPs get concentrated in the cell wall and spaces between the cells, thus influencing the growth of root via dependence on dosage of these NPs. Further, it was also observed that these NPs induce formation of reactive oxygen species (ROS), which get accumulated in the root tips.
5. Copper nanoparticles (Cu NPs): Bakshi and Kumar (2021) investigated the effect of copper nanoparticles (Cu NPs) on plants exposed to either foliar or root exposure. They found that the application of Cu NPs improved root length, biomass, chlorophyll content, and photosynthetic activity, indicating the potential of Cu NPs to enhance plant growth and development.
6. Titanium dioxide nanoparticles (TiO_2 NPs): Suriyaprabha et al. (2012) investigated the effect of titanium dioxide nanoparticles (TiO_2 NPs) on the growth and root morphology of maize. They found that the application of TiO_2 NPs improved root length, surface area, and volume, as well as the number and length of lateral roots, indicating the potential of TiO_2 NPs to enhance root growth and improve plant development.
7. Iron oxide nanoparticles (Fe_3O_4 NPs): Gurav et al. (2020) studied the effect of iron oxide nanoparticles (Fe_3O_4 NPs) on the growth and yield of chickpea plants. They found that the application of Fe_3O_4 NPs improved root length, biomass, photosynthetic pigments, and yield of chickpea plants, indicating the potential of Fe_3O_4 NPs to enhance plant growth and improve crop productivity.
8. Chitosan nanoparticles (CS NPs): Zhang et al. (2018) investigated the effect of chitosan nanoparticles (CS NPs) on the root morphology and phosphorus uptake of soybean plants. They found that the application of CS NPs improved root length, surface area, and volume, as well as the phosphorus uptake and utilization efficiency of soybean plants, indicating the potential of CS NPs to enhance root growth and improve nutrient uptake.
9. Magnesium oxide nanoparticles (MgO NPs): Abdullah et al. (2019) studied the effect of magnesium oxide nanoparticles on the growth and nutrient uptake of wheat plants. They found that the application of MgO NPs improved root length, biomass, photosynthetic pigments, and nutrient uptake of wheat plants, indicating the potential of MgO NPs to enhance plant growth and improve nutrient use efficiency.
10. Gold nanoparticles (Au NPs): Kamran et al. (2019) investigated the effect of gold nanoparticles (Au NPs) on the root morphology and heavy metal uptake of lettuce plants. They found that the application of Au NPs improved root length, surface area, and volume, as well as the uptake and translocation of heavy metals, indicating the potential of Au NPs to enhance root growth and improve plant tolerance to heavy metal stress.

12.7.1 Mechanisms of Root Growth Regulation by Nanofertilizers

1. Hormonal regulation: nanofertilizers can regulate root growth by affecting the synthesis and transport of plant hormones. For instance, some nanofertilizers can inhibit the synthesis of the auxin hormone, which can reduce root elongation and promote lateral root development. On the other hand, some nanofertilizers can enhance the synthesis and transport of cytokinins, which can promote cell division and elongation, leading to improved root growth. Additionally, nanofertilizers can also influence the balance of gibberellins and abscisic acid, which are essential for root growth and stress tolerance. (Singh et al., 2017).
2. Ion transport: nanofertilizers can regulate root growth by improving the uptake and transport of essential ions such as nitrogen, phosphorus, and potassium. The nanoparticles can penetrate the root cell walls and membranes and release the nutrients directly into the cytoplasm, leading to enhanced nutrient availability and uptake. Additionally, some nanofertilizers can also chelate essential ions such as iron, zinc, and copper, which can enhance their mobility and availability for plant uptake (Soliman et al., 2015).
3. Enzymatic activities: nanofertilizers can regulate root growth by affecting the activities of enzymes responsible for nutrient uptake and assimilation. For instance, some nanofertilizers can enhance the activities of nitrate reductase and urease enzymes, which are essential for nitrogen assimilation and utilization. Additionally, some nanofertilizers can also enhance the activity of phosphatase enzymes, which can promote the release of phosphate ions from the soil and improve their availability for plant uptake. (Singh et al., 2017).
4. Gene expression: nanofertilizers can regulate root growth by influencing the expression of genes involved in root development and nutrient uptake. For example, some nanofertilizers can induce the expression of genes responsible for lateral root development and branching, leading to improved root architecture and nutrient uptake. Additionally, some nanofertilizers can also enhance the expression of genes involved in stress response and tolerance, leading to improved plant growth and productivity under stressful conditions. (Kottegoda et al., 2018).
5. Reactive oxygen species (ROS): nanofertilizers can regulate root growth by influencing the generation and scavenging of ROS. ROS are essential signaling molecules that regulate root growth and development, and their levels need to be tightly regulated for optimal growth and stress tolerance. Some nanofertilizers can enhance the levels of ROS, leading to improved antioxidant defense systems and stress tolerance. On the other hand, some nanofertilizers can scavenge ROS, leading to reduced stress and improved root growth. (Wang et al., 2016).
6. Soil microbe interactions: nanofertilizers can also regulate soil microbe interactions, leading to improved root growth. They can enhance the growth of beneficial microbes, such as mycorrhizal fungi, which can improve nutrient uptake by plants. (Mirzapour et al., 2019).
7. pH regulation: nanofertilizers can also regulate soil pH, which can affect root growth. By adjusting the pH, nanofertilizers can improve nutrient availability and reduce the toxicity of certain elements in the soil, promoting healthy root growth (Mirzapour et al., 2019).
8. Surface charge interactions: nanofertilizers have unique surface charges, which can interact with plant roots and regulate their growth. Studies have shown that positively charged nanofertilizers can enhance root growth, while negatively charged ones can inhibit it (Mirzapour et al., 2019).
9. Nanoparticle size: the size of nanofertilizers can also affect their ability to regulate root growth. Studies have shown that smaller particles are more effective at enhancing root growth, as they can penetrate the root system more easily and deliver nutrients and other compounds more efficiently (Mirzapour et al., 2019).

12.7.2 Advantages and Disadvantages of Using Nanofertilizers for Root Growth Regulation

12.7.2.1 Advantages

There are several advantages of nanofertilizers for root growth regulation as mentioned below (Zulfiqar et al., 2019).

1. Improved nutrient uptake: nanofertilizers are more efficient than traditional fertilizers at delivering nutrients to plants. They can be precisely targeted to specific areas of the root system, allowing for more efficient nutrient uptake. The nanofertilizers' small size and surface area allow for better absorption by plant roots, enabling plants to utilize the nutrients effectively. This results in healthier plants with improved growth and yield.
2. Reduced environmental impact: the use of traditional fertilizers has been associated with negative environmental impacts. Nanofertilizers, on the other hand, are more environmentally friendly. They are designed to be more efficient at delivering nutrients, which means less fertilizer is needed overall. This can reduce the amount of fertilizer that ends up in the environment, reducing the risk of pollution. Moreover, nanofertilizers can also be engineered to release nutrients slowly, providing a more sustained and controlled release of nutrients into the soil.
3. Increased yield: nanofertilizers can improve plant growth and yield. They are more efficient at delivering nutrients, which can result in higher yields and improved crop quality. Nanofertilizers can also be engineered to release nutrients slowly over time, providing a continuous source of nutrients to plants. This leads to more significant and sustained growth, improving the overall yield and quality of crops.
4. Improved root growth: nanofertilizers can be designed to regulate root growth, which is particularly beneficial in areas with poor soil quality or container gardening. By regulating root growth, nanofertilizers can ensure that plants have access to the nutrients they need, improving their overall health and growth. Moreover, nanofertilizers can help to develop stronger root systems, enabling plants to better withstand adverse environmental conditions like drought and flooding.
5. Enhanced disease resistance: nanofertilizers can be engineered to enhance plant resistance to diseases. By delivering specific nutrients to plants, nanofertilizers can help improve their overall health and resistance to disease. This can reduce the need for chemical treatments, making it a more sustainable approach to agriculture. Furthermore, nanofertilizers can also be designed to promote the growth of beneficial microorganisms in the soil, which can further help to prevent plant diseases.

12.7.2.2 Disadvantages

While nanofertilizers have shown promising role in improving plant growth, there are several disadvantages associated with their use. These include potential toxicity issues, environmental impacts, health risks, cost, and lack of regulation. Before widespread adoption of nanofertilizers, more research is needed to assess their safety and efficacy in the long term. It is important to carefully consider the potential risks and benefits of nanofertilizers before deciding to use them for root growth regulation. (Nongbet et al., 2022).

1. Toxicity issues

   One of the biggest concerns associated with the use of nanofertilizers is their potential toxicity. Nanoparticles have unique properties that can make them more toxic than larger particles of the same material. This increased toxicity is due to their small size, large surface area, and ability to penetrate biological membranes.

When nanofertilizers are applied to plants, the nanoparticles can be taken up by the plant and accumulate in the tissues. This can interfere with the plant's metabolic pathways, causing cellular damage or even death. The accumulation of nanoparticles in the soil can also lead to toxicity issues for other plants and organisms in the ecosystem.

In addition to toxicity concerns, the use of nanofertilizers may also lead to an accumulation of toxic metals in the soil. Some nanofertilizers are made with metal nanoparticles, such as silver or copper. These metals can accumulate in the soil and have negative impacts on soil quality and ecosystem health.

2. Environmental impacts

   Another significant disadvantage of nanofertilizers is their potential impact on the environment. When nanofertilizers are applied to plants, the nanoparticles can enter the groundwater and contaminate the water supply. This can have negative impacts on aquatic ecosystems and the health of people who depend on the water supply.

   Furthermore, nanoparticles can persist in the soil for long periods of time and accumulate over time. This can lead to potential long-term environmental impacts that are not yet fully understood.

3. Health risks

   There are also concerns about the potential health risks associated with the use of nanofertilizers. Workers who handle these materials may be exposed to nanoparticles, which can have negative impacts on their health. Nanoparticles can penetrate biological membranes and accumulate in the tissues, potentially causing damage or disease. Inhalation or ingestion of nanoparticles can also lead to potential health impacts.

   Consumers who consume plants grown with nanofertilizers may also be exposed to nanoparticles. While there is currently limited data on the health impacts of consuming plants grown with nanofertilizers, there are concerns that long-term exposure could lead to negative health impacts.

4. Cost

   Nanofertilizers are more expensive than conventional fertilizers. This cost may be prohibitive for farmers who are operating on a tight budget. Furthermore, the high cost of nanofertilizers may make it difficult for small-scale farmers to compete in the market.

5. Lack of regulation

   Currently, there are no regulations in place to govern the use of nanofertilizers. This lack of regulation may lead to potential risks for both the environment and human health. Furthermore, there is a lack of data on the long-term impacts of nanofertilizers, which makes it difficult to assess their safety and efficacy.

12.7.3 Applications of Nanofertilizers for Root Growth Regulation

Nanofertilizers, which are essentially fertilizers in nanoparticle form, have the potential to revolutionize the agricultural industry by enhancing plant growth, increasing yield, and reducing the use of conventional fertilizers. The potential future applications of nanofertilizers are wide-ranging and could have a significant impact on the agricultural industry. From improving nutrient uptake and crop resilience to reducing environmental impacts and improving soil health, nanofertilizers have the potential to help meet the global demand for food while reducing the reliance on conventional fertilizers. However, further research is needed to fully understand the safety and efficacy of these materials, as well as their potential impacts on human health and the environment.

1. Improved nutrient uptake: one of the most promising applications of nanofertilizers is their ability to improve the uptake of nutrients by plants. Nanoparticles can penetrate plant cells more easily than conventional fertilizers, allowing for more efficient absorption of nutrients such as

nitrogen, phosphorus, and potassium. This can lead to improved plant growth and increased yields.

2. Reduced environmental impact: conventional fertilizers can have negative environmental impacts, such as leaching into groundwater or releasing harmful gases into the atmosphere. Nanofertilizers have the potential to reduce these impacts by allowing for more targeted application of nutrients to plants. For example, nanoparticles can be engineered to release nutrients slowly over time, reducing the need for frequent applications and minimizing the risk of runoff.
3. Enhanced crop resilience: nanofertilizers can also be engineered to enhance crop resilience to environmental stressors such as drought, heat, or pests. For example, nanoparticles can be coated with molecules that protect plants from UV radiation or that enhance their ability to tolerate drought conditions. This can lead to higher crop yields and improved food security in regions with challenging growing conditions.
4. Improved soil health: nanofertilizers can also help improve soil health by providing nutrients in a form that is more readily available to plants. This can lead to improved soil structure, increased microbial activity, and reduced soil erosion. Additionally, some nanoparticles have been shown to have antimicrobial properties, which could help control soil-borne pathogens.
5. Reduced cost: finally, nanofertilizers have the potential to reduce the cost of fertilizer application by requiring less frequent applications and reducing the amount of fertilizer needed per application. This could lead to improved profitability for farmers and increased adoption of sustainable farming practices.

12.8 FUTURE DIRECTIONS

Nanotechnology has the potential to revolutionize the field of agriculture by providing a new generation of fertilizers that are more efficient and effective than traditional fertilizers. Nanofertilizers are designed to deliver nutrients directly to the root system of plants, promoting growth and enhancing the plant's ability to absorb nutrients. In recent years, research has been focused on developing nanofertilizers that can regulate root growth and improve plant health. One of the future directions in nanofertilizers for root growth regulation is the development of nanoparticles that can improve the availability of nutrients to plants. For example, iron oxide nanoparticles can be used to deliver iron to plants in a more efficient way, increasing the uptake of this vital nutrient. This can be particularly important in soils that are deficient in iron, which is essential for the production of chlorophyll and other plant pigments. Another area of research is the use of nanofertilizers to enhance plant resistance to biotic and abiotic stresses. For example, silver nanoparticles have been shown to have antimicrobial properties and can be used to control plant diseases. Similarly, nanoparticles of calcium carbonate and silica can be used to improve the tolerance of plants to drought and salinity stress.

In addition to improving nutrient availability and stress tolerance, nanofertilizers can also be used to regulate root growth and improve plant architecture. For example, nanoparticles of zinc oxide can be used to promote lateral root development, which can improve nutrient uptake and overall plant growth. Similarly, nanoparticles of titanium dioxide can be used to regulate the size and shape of plant cells, which can have a significant impact on plant growth and development. Hence, there is growing interest in the use of nanofertilizers for sustainable agriculture. One of the key challenges in agriculture is reducing the environmental impact of fertilizer use. Traditional fertilizers can contribute to soil and water pollution, which can have negative impacts on human health and the environment. Nanofertilizers have the potential to reduce the amount of fertilizer needed to achieve the same level of plant growth, which can help to reduce the environmental impact of agriculture.

ACKNOWLEDGEMENTS

The authors gratefully acknowledge the Department of Biotechnology, Sharda School of Engineering and Technology, Sharda University, Greater Noida, Uttar Pradesh, India, Sharda School of Agricultural Sciences and Sharda School of Allied Health Sciences, Sharda University, Greater Noida, Uttar Pradesh, India, Plant Molecular Biology Laboratory, Department of Botany, Dayanand Anglo-Vedic (PG) College, Chhatrapati Shahu Ji Maharaj University, Kanpur, India.

REFERENCES

Abdullah, R., Naveed, M., & Javaid, A. (2019). Magnesium oxide nanoparticles improve growth, photosynthetic pigments and nutrient uptake of wheat. *Environmental Science and Pollution Research, 26*, pp.10714–10722.

Ahmad, W., Jaiswal, K.K., & Soni, S. (2020). Green synthesis of titanium dioxide (TiO_2) nanoparticles by using *Mentha arvensis* leaves extract and its antimicrobial properties. *Inorganic and Nano-Metal Chemistry, 50*, pp.1032–1038.

Ahmed, S., Ahmad, M., Swami, B.L., & Ikram, S. (2016). Green synthesis of silver nanoparticles using *Azadirachta indica* aqueous leaf extract. *Journal Radiation Research and Applied Sciences, 9*, pp.1–7.

Alrajhi, A.H., & Ahmed, N.M. (2023). Green synthesis of Zinc Oxide nanoparticles using *Salvia officinalis* Extract. In Shanker, U., Hussain, C.M., Rani, M. (eds) *Handbook of Green and Sustainable Nanotechnology: Fundamentals, Developments and Applications* (pp. 1–21). Springer International Publishing.

Arora, M., Saxena, P., Abdin, M.Z., & Varma, A. (2018). Interaction between *Piriformospora indica* and *Azotobacter chroococcum* governs better plant physiological and biochemical parameters in *Artemisia annua* L. plants grown under in vitro conditions. *Symbiosis, 75*, pp.103–112.

Assefa, S., & Tadesse, S. (2019). The principal role of organic fertilizer on soil properties and agricultural productivity-a review. *Agricultural Research and Technology: Open Access Journal, 22*, p.556192.

Avila-Quezada, G.D., Ingle, A.P., Golińska, P., & Rai, M. (2022). Strategic applications of nano-fertilizers for sustainable agriculture: Benefits and bottlenecks. *Nanotechnology Reviews, 11*, pp.2123–2140.

Bajpai, A., Jadhav, K., Muthukumar, M., Kumar, S., & Srivatava, G. (2020). Use of nanotechnology in quality improvement of economically important agricultural crops. In Ghorbanpour, M., Bhargava, P., Varma, A., & Choudhary, D.K. (eds) *Biogenic Nano-Particles and their Use in Agro-ecosystems* (pp. 39–57). Springer.

Bakshi, M., & Kumar, A. (2021). Copper-based nanoparticles in the soil-plant environment: Assessing their applications, interactions, fate and toxicity. *Chemosphere, 281*, p.130940.

Barone, V., Baglieri, A., Stevanato, P., Broccanello, C., Bertoldo, G., Bertaggia, M., Cagnin, M., Pizzeghello, D., Moliterni, V.M.C., Mandolino, G., & Fornasier, F. (2018). Root morphological and molecular responses induced by microalgae extracts in sugar beet (*Beta vulgaris* L.). *Journal of Applied Phycology, 30*, pp.1061–1071.

Bernela, M., Rani, R., Malik, P., & Mukherjee, T.K. (2021). Nanofertilizers: Applications and future prospects. In Sindhu, R.K., Chitkara, M., & Sandhu, I.S. (eds) *Nanotechnology* (pp. 289–332). Jenny Stanford Publishing.

Bhagat, R., Ingle, A.P., & Chen, H. (2023). Nanosensors and nanobiosensors for sustainable agriculture. In Ingle, A.P. (ed) *Nanotechnology in Agriculture and Agroecosystems* (pp. 93–112). Elsevier.

Bordiwala, R.V. (2023). Green synthesis and applications of metal nanoparticles.-A Review Article. *Results in Chemistry, 5*, p.100832.

Borges, R., Wypych, F., Petit, E., Forano, C., & Prevot, V. (2019). Potential sustainable slow-release fertilizers obtained by mechanochemical activation of MgAl and MgFe layered double hydroxides and K_2HPO_4. *Nanomaterials, 9*, p.183.

Castro-Longoria, E., Vilchis-Nestor, A.R., & Avalos-Borja, M. (2011). Biosynthesis of silver, gold and bimetallic nanoparticles using the filamentous fungus *Neurospora crassa*. *Colloids and Surfaces B: Biointerfaces, 83*, pp.42–48.

Cataldo, E., Salvi, L., Paoli, F., Fucile, M., Masciandaro, G., Manzi, D., Masini, C.M., & Mattii, G.B. (2021). Application of zeolites in agriculture and other potential uses: A review. *Agronomy, 11*, p.1547.

Chhipa, H. (2017). Nanofertilizers and nanopesticides for agriculture. *Environmental Chemistry Letters, 15*, pp.15–22.

Chouhan, R., Nautiyal, A.K., Sharma, N., & Gandhi, S.G. (2023). Plant transcription factors and root development. In Srivastava, V., Mishra, S., Mehrotra, S., & Upadhyay, S.K. (eds) *Plant Transcription Factors* (pp. 63–76). Academic Press.

Damo, J.L.C., Shimizu, T., Sugiura, H., Yamamoto, S., Agake, S.I., Anarna, J., Tanaka, H., Sugihara, S., Okazaki, S., Yokoyama, T., & Yasuda, M. (2023). The application of sulfur influences microbiome of soybean rhizosphere and nutrient-mobilizing bacteria in andosol. *Microorganisms, 11*, p.1193.

Das, R.K., Gogoi, N., & Bora, U. (2011). Green synthesis of gold nanoparticles using *Nyctanthes arbortristis* flower extract. *Bioprocess and Biosystems Engineering, 34*, pp.615–619.

Dhawi, F. (2023). The role of plant growth-promoting microorganisms (PGPMs) and their feasibility in hydroponics and vertical farming. *Metabolites, 13*, p.247.

Doan, V.D., Thieu, A.T., Nguyen, T.D., Nguyen, V.C., Cao, X.T., Nguyen, T.L.H., & Le, V.T. (2020). Biosynthesis of gold nanoparticles using *Litsea cubeba* fruit extract for catalytic reduction of 4-nitrophenol. *Journal of Nanomaterials, 2020*, pp.1–10.

El-Bialy, S.M., El-Mahrouk, M.E., Elesawy, T., Omara, A.E.D., Elbehiry, F., El-Ramady, H., Áron, B., Prokisch, J., Brevik, E.C., & Solberg, S.Ø. (2023). Biological nanofertilizers to enhance growth potential of strawberry seedlings by boosting photosynthetic pigments, plant enzymatic antioxidants, and nutritional status. *Plants, 12*, p.302.

El-Ghamry, A., Mosa, A.A., Alshaal, T., & El-Ramady, H. (2018). Nanofertilizers vs. biofertilizers: New insights. *Environment, Biodiversity and Soil Security, 2*, pp.51–72.

FAO. (2018). *The future of food and agriculture -Alternative pathways to 2050. Summary version.* Rome, Italy, 60. Licence:CC BY-NC-SA 3.0 IGO. www.fao.org/3/CA1553EN/ca1553en.pdf.

Fasusi, O.A., Cruz, C., & Babalola, O.O. (2021). Agricultural sustainability: Microbial biofertilizers in rhizosphere management. *Agriculture, 11*, p.163.

Feroze, N., Arshad, B., Younas, M., Afridi, M.I., Saqib, S., & Ayaz, A. (2020). Fungal mediated synthesis of silver nanoparticles and evaluation of antibacterial activity. *Microscopy Research and Technique, 83*, pp.72–80.

Gangwar, J., Kadanthottu Sebastian, J., Puthukulangara Jaison, J., & Kurian, J.T. (2023). Nano-technological interventions in crop production—a review. *Physiology and Molecular Biology of Plants, 29(1)*, pp. 93–107.

Goutam, S.P., Saxena, G., Roy, D., Yadav, A.K., & Bharagava, R.N. (2020). Green synthesis of nanoparticles and their applications in water and wastewater treatment. In Saxena, G., & Bharagava, R. (eds) *Bioremediation of Industrial Waste for Environmental Safety: Volume I: Industrial Waste and Its Management* (pp. 349–379). Springer.

Gulbagca, F., Aygün, A., Gülcan, M., Ozdemir, S., Gonca, S., & Şen, F. (2021). Green synthesis of palladium nanoparticles: Preparation, characterization, and investigation of antioxidant, antimicrobial, anticancer, and DNA cleavage activities. *Applied Organometallic Chemistry, 35*, p.e6272.

Gurav, S.S., Jadhav, S.A., & Kadam, A.A. (2020). Iron oxide nanoparticles promote growth and yield of chickpea (*Cicer arietinum* L.) through enhanced photosynthetic pigments and improved soil characteristics. *Ecotoxicology and Environmental Safety, 202*, p.110886.

Hamk, M., Akçay, F.A., & Avcı, A. (2023). Green synthesis of zinc oxide nanoparticles using *Bacillus subtilis* ZBP4 and their antibacterial potential against foodborne pathogens. *Preparative Biochemistry & Biotechnology, 53*, pp.255–264.

Hanif, S., Sajjad, A., & Zia, M. (2023). Proline coated ZnO NPs as nanofertilizer against drought stress: An in vitro study to *Coriandrum sativum. Plant Cell, Tissue and Organ Culture (PCTOC), 155(2)*, pp. 493–504.

Hansen, H.F.E., & Elberling, B. (2023). Spatial distribution of bioavailable inorganic nitrogen from thawing permafrost. *Global Biogeochemical Cycles, 37*, p.e2022GB007589.

Haris, M., Hussain, T., Mohamed, H.I., Khan, A., Ansari, M.S., Tauseef, A., Khan, A.A., & Akhtar, N. (2023). Nanotechnology–A new frontier of nano-farming in agricultural and food production and its development. *Science of The Total Environment, 857*, p.159639.

Heikal, Y.M., & Abdel-Aziz, H.M. (2020). Biogenic nanomaterials and their applications in agriculture. In Ghorbanpour, M., Bhargava, P., Varma, A., & Choudhary, D. (eds) *Biogenic Nano-Particles and their Use in Agro-ecosystems* (pp. 489–514). Springer.

Huang, Q., Ayyaz, A., Farooq, M.A., Zhang, K., Chen, W., Hannan, F., Sun, Y., Shahzad, K., Ali, B. & Zhou, W. (2024). Silicon dioxide nanoparticles enhance plant growth, photosynthetic performance, and antioxidants defence machinery through suppressing chromium uptake in *Brassica napus L. Environmental Pollution*, *342*, p. 123013.

Hunter, M.C., Smith, R.G., Schipanski, M.E., Atwood, L.W., & Mortensen, D.A. (2017). Agriculture in 2050: Recalibrating targets for sustainable intensification. *Bioscience, 67*, pp.386–391.

Jain, D., Sanadhya, S., Saheewala, H., Joshi, A., Bhojiya, A.A., & Mohanty, S.R. (2021). The role of nanofertilizers in smart agriculture: An effective approach to increase nutrient use efficiency. In Kumar Srivastava, D., Kumar Thakur, A., & Kumar, P. (eds) *Agricultural Biotechnology: Latest Research and Trends* (pp.493–510). Springer

Jameel, M.S., Aziz, A.A., & Dheyab, M.A. (2020). Comparative analysis of platinum nanoparticles synthesized using sonochemical-assisted and conventional green methods. *Nano-Structures & Nano-Objects, 23*, p.100484.

Jamkhande, P.G., Ghule, N.W., Bamer, A.H., & Kalaskar, M.G. (2019). Metal nanoparticles synthesis: An overview on methods of preparation, advantages and disadvantages, and applications. *Journal of Drug Delivery Science and Technology, 53*, p.101174.

Kah, M., Kookana, R.S., Gogos, A., & Bucheli, T.D. (2018). A critical evaluation of nanopesticides and nanofertilizers against their conventional analogues. *Nature Nanotechnology, 13*, pp.677–684.

Kalwani, M., Chakdar, H., Srivastava, A., Pabbi, S., & Shukla, P. (2022). Effects of nanofertilizers on soil and plant-associated microbial communities: Emerging trends and perspectives. *Chemosphere, 287*, p.132107.

Kamran, M.A., Alharby, H.F., Shakeel, M., Rizwan, M., Xu, J., & Liu, J. (2019). Gold nanoparticles promote growth, root morphology and heavy metal uptake in lettuce plants: Implications for phytoremediation of heavy metal contaminated soils. *Ecotoxicology and Environmental Safety, 183*, p.109499.

Karam, S.T., & Abdulrahman, A.F. (2022). Green synthesis and characterization of ZnO Nanoparticles by using thyme plant leaf extract. *Photonics, 9*(8), p. 594. MDPI.

Khalid, M., Hassani, D., Bilal, M., Asad, F., & Huang, D. (2017). Influence of bio-fertilizer containing beneficial fungi and rhizospheric bacteria on health promoting compounds and antioxidant activity of *Spinacia oleracea* L. *Botanical Studies, 58*, pp.1–9.

Khodakovskaya, M.V., Kim, B.S., Kim, J.N., Alimohammadi, M., Dervishi, E., Mustafa, T., & Cernigla, C.E. (2013). Carbon nanotubes as plant growth regulators: Effects on tomato growth, reproductive system, and soil microbial community. *Small, 9*, 115–123. https://doi.org/10.1002/smll.201201225

Kottegoda, N., Munaweera, I., Siriwardhana, N., Attanayake, A., Karunaratne, V., & Rajapakse, A. (2018). Effects of zinc oxide nanoparticles on root growth and gene expression in *Arabidopsis thaliana. Journal of Environmental Science and Health, Part A, 53*, 137–146.

Kumar, C.R., Betageri, V.S., Nagaraju, G., Suma, B.P., Kiran, M.S., Pujar, G.H., & Latha, M.S. (2020). One-pot synthesis of ZnO nanoparticles for nitrite sensing, photocatalytic and antibacterial studies. *Journal of Inorganic and Organometallic Polymers and Materials, 30*, pp.3476–3486.

Kumar, R., Ashfaq, M., & Verma, N. (2018). Synthesis of novel PVA–starch formulation-supported Cu–Zn nanoparticle carrying carbon nanofibers as a nanofertilizer: Controlled release of micronutrients. *Journal of Materials Science, 53*, pp.7150–7164.

Kumar, R., Kumar, R., & Prakash, O. (2019). The impact of chemical fertilizers on our environment and ecosystem. *Chief Ed, 35*, p.69.

Kumar, S., Nehra, M., Dilbaghi, N., Marrazza, G., Hassan, A.A., & Kim, K.H. (2019). Nano-based smart pesticide formulations: Emerging opportunities for agriculture. *Journal of Controlled Release, 294*, pp.131–153.

Kumar, S., Sindhu, S.S., & Kumar, R. (2022). Biofertilizers: An ecofriendly technology for nutrient recycling and environmental sustainability. *Current Research in Microbial Sciences, 3*, p.100094.

Kumari, R., & Singh, D.P. (2020). Nano-biofertilizer: An emerging eco-friendly approach for sustainable agriculture. *Proceedings of the National Academy of Sciences, India Section B: Biological Sciences, 90*, pp.733–741.

Lee, L.H., Wu, T.Y., Shak, K.P.Y., Lim, S.L., Ng, K.Y., Nguyen, M.N., & Teoh, W.H. (2018). Sustainable approach to biotransform industrial sludge into organic fertilizer via vermicomposting: A mini-review. *Journal of Chemical Technology & Biotechnology, 93*, pp.925–935.

Lyzenga, W.J., Liu, Z., Olukayode, T., Zhao, Y., Kochian, L.V., & Ham, B.K. (2023). Getting to the roots of N, P, and K uptake. *Journal of Experimental Botany, 74*, pp.1784–1805.

Ma, Z., Zhang, Y., Li, M., Chao, Z., Du, X., Yan, S., & Shen, J. (2023). A first greenhouse application of bacteria-expressed and nanocarrier-delivered RNA pesticide for *Myzus persicae* control. *Journal of Pest Science, 96*, pp.181–193.

Macdonald, H., & Stevens, D. (2019). *Biotechnology and Plant Biology*. Scientific e-Resources.

Mahesh, M., Thomas, J., Kumar, K.A., Bhople, B.S., Saresh, N.V., Vaid, S.K., & Sahu, S.K. (2018). Zeolite farming: A sustainable agricultural prospective. *International Journal of Current Microbiology and Applied Sciences, 7*, pp.2912–2924.

Mahil, E.I.T., & Kumar, B.A. (2019). Foliar application of nanofertilizers in agricultural crops–A review. *Journal of Farm Sciences, 32*, pp.239–249.

Mahmoudi, H., Rostami, F., Tabatabaei, S., & Sharifi, M. (2020). Zinc oxide nanoparticles affect growth, physiology and nutrient uptake of wheat (*Triticum aestivum* L.) under drought stress. *Ecotoxicology and Environmental Safety, 203*, p.111023. https://doi.org/10.1016/j.ecoenv.2020.111023

Mali, S.C., Raj, S., & Trivedi, R. (2020). Nanotechnology a novel approach to enhance crop productivity. *Biochemistry and Biophysics Reports, 24*, p.100821.

Malik, S., & Kumar, A. (2014). Approach for nano-particle synthesis: Using as nano-fertilizer. *International Journal of Pharmaceutical Research and Bioscience, 3*, pp.2277–8713.

Manjare, S.B., & Chaudhari, R.A. (2020). Palladium nanoparticle-bentonite hybrid using leaves of *Syzygium aqueum* plant from India: Design and assessment in the catalysis of–C–C–coupling reaction. *Chemistry Africa, 3*, pp.329–341.

Manoj kumar, U., Kaliannan, D., Srinivasan, V., Balasubramanian, B., Kamyab, H., Mussa, Z.H., Palaniyappan, J., Mesbah, M., Chelliapan, S., & Palaninaicker, S. (2023). Green synthesis of zinc oxide nanoparticles using *Brassica oleracea* var. *botrytis* leaf extract: Photocatalytic, antimicrobial and larvicidal activity. *Chemosphere, 323*, p.138263.

Mansour, A.T., Alprol, A.E., Khedawy, M., Abualnaja, K.M., Shalaby, T.A., Rayan, G., Ramadan, K.M., & Ashour, M. (2022). Green synthesis of zinc oxide nanoparticles using red seaweed for the elimination of organic toxic dye from an aqueous solution. *Materials, 15*, p.5169.

Marschner, P., & Rengel, Z. (2023). Nutrient availability in soils. *In* Rengel, Z., Cakmak, I., & Whit, P.J. (eds) *Marschner's Mineral Nutrition of Plants* (pp. 499–522). Academic press.

Mat Jalaluddin, N.S., Asem, M., Harikrishna, J.A., & Ahmad Fuaad, A.A.H. (2023). Recent progress on nanocarriers for topical-mediated rnai strategies for crop protection—A Review. *Molecules, 28*, p.2700.

Mehdizadeh, T., Zamani, A., & Froushani, S.M.A. (2020). Preparation of Cu nanoparticles fixed on cellulosic walnut shell material and investigation of its antibacterial, antioxidant and anticancer effects. *Heliyon, 6*, p.e03528.

Mirzapour, M.H., Habibi, H., Pessarakli, M., & Bayat, H. (2019). Effect of foliar application of iron nanoparticles on growth, physiological attributes and iron concentration of strawberry under water-deficit stress. *Scientia Horticulturae, 250*, pp.1–7.

Mittal, D., Kaur, G., Singh, P., Yadav, K., & Ali, S.A. (2020). Nanoparticle-based sustainable agriculture and food science: Recent advances and future outlook. *Frontiers in Nanotechnology, 2*, p.579954.

Ndaba, B., Roopnarain, A., Haripriya, R.A.M.A., & Maaza, M. (2022). Biosynthesized metallic nanoparticles as fertilizers: An emerging precision agriculture strategy. *Journal of Integrative Agriculture, 21*, pp.1225–1242.

Nongbet, A., Mishra, A.K., Mohanta, Y.K., Mahanta, S., Ray, M.K., Khan, M., Baek, K.H., & Chakrabartty, I. (2022). Nanofertilizers: A Smart and Sustainable attribute to modern agriculture. *Plants, 11*, p.2587.

Pallela, P.N.V.K., Ummey, S., Ruddaraju, L.K., Gadi, S., Cherukuri, C.S., Barla, S., & Pammi, S.V.N. (2019). Antibacterial efficacy of green synthesized α-Fe_2O_3 nanoparticles using *Sida cordifolia* plant extract. *Heliyon, 5*, p.e02765.

Rabiee, N., Bagherzadeh, M., Kiani, M., & Ghadiri, A.M. (2020). *Rosmarinus officinalis* directed palladium nanoparticle synthesis: Investigation of potential anti-bacterial, anti-fungal and Mizoroki-Heck catalytic activities. *Advanced Powder Technology, 31*, pp.1402–1411.

Raj, S., Trivedi, R., & Soni, V. (2021). Biogenic synthesis of silver nanoparticles, characterization and their applications—a review. *Surfaces, 5*, pp.67–90.

Rajendhiran, R., Deivasigamani, V., Palanisamy, J., Masan, S., & Pitchaiya, S. (2021). *Terminalia catappa* and *Carissa carandas* assisted synthesis of TiO_2 nanoparticles–A green synthesis approach. *Materials Today: Proceedings, 45*, pp.2232–2238.

Rajivgandhi, G., Gnanamangai, B.M., Prabha, T.H., Poornima, S., Maruthupandy, M., Alharbi, N.S., Kadaikunnan, S., & Li, W.J. (2022). Biosynthesized zinc oxide nanoparticles (ZnO NPs) using actinomycetes enhance the anti-bacterial efficacy against *K. Pneumoniae. Journal of King Saud University-Science, 34*, p.101731.

Ramesh, R., Catherine, G., Sundaram, S.J., Khan, F.L.A., & Kaviyarasu, K. (2021). Synthesis of Mn_3O_4 nano complex using aqueous extract of *Helianthus annuus* seed cake and its effect on biological growth of *Vigna radiata. Materials Today: Proceedings, 36*, pp.184–191.

Rane, M., Bawskar, M., Rathod, D., Nagaonkar, D., & Rai, M. (2015). Influence of calcium phosphate nanoparticles, *Piriformospora indica* and *Glomus mosseae* on growth of *Zea mays. Advances in Natural Sciences: Nanoscience and Nanotechnology, 6*, p.045014.

Rizwan, M., Ali, S., ur Rehman, M.Z., Riaz, M., Adrees, M., Hussain, A., Zahir, Z.A., & Rinklebe, J. (2021). Effects of nanoparticles on trace element uptake and toxicity in plants: A review. *Ecotoxicology and Environmental Safety, 221*, p.112437.

Ruiz-Cañas, M.C., Quintero, H.I., Corredor, L.M., Manrique, E., & Romero Bohorquez, A.R. (2020). New nanohybrid based on hydrolyzed polyacrylamide and silica nanoparticles: Morphological, structural and thermal properties. *Polymers, 12*, p.1152.

Rutkowska, B., Szulc, W., Bomze, K., Gozdowski, D., & Spychaj-Fabisiak, E. (2015). Soil factors affecting solubility and mobility of zinc in contaminated soils. *International Journal of Environmental Science and Technology, 12*, pp.1687–1694.

Sadiq, H., Sher, F., Sehar, S., Lima, E.C., Zhang, S., Iqbal, H.M., Zafar, F., & Nuhanović, M. (2021). Green synthesis of ZnO nanoparticles from *Syzygium Cumini* leaves extract with robust photocatalysis applications. *Journal of Molecular Liquids, 335*, p.116567.

Saifuddin, N., Wong, C.W., & Yasumira, A.A. (2009). Rapid biosynthesis of silver nanoparticles using culture supernatant of bacteria with microwave irradiation. *E-journal of Chemistry, 6*, pp.61–70.

Sayadi, M.H., Salmani, N., Heidari, A., & Rezaei, M.R. (2018). Bio-synthesis of palladium nanoparticle using *Spirulina platensis* alga extract and its application as adsorbent. *Surfaces and Interfaces, 10*, pp.136–143.

Seleiman, M.F., Almutairi, K.F., Alotaibi, M., Shami, A., Alhammad, B.A., & Battaglia, M.L. (2020). Nano-fertilization as an emerging fertilization technique: Why can modern agriculture benefit from its use? *Plants, 10*, p.2.

Shaji, H., Chandran, V., & Mathew, L. (2021). Organic fertilizers as a route to controlled release of nutrients. In Lewu, F.B., Volova, T., Thomas, S., & K.R., Rakhimol (eds) *Controlled Release Fertilizers for Sustainable Agriculture* (pp. 231–245). Academic Press.

Shang, Y., Hasan, M.K., Ahammed, G.J., Li, M., Yin, H., & Zhou, J. (2019). Applications of nanotechnology in plant growth and crop protection: A review. *Molecules, 24*, p.2558.

Sharma, N., Acharya, S., Kumar, K., Singh, N., & Chaurasia, O.P. (2018). Hydroponics as an advanced technique for vegetable production: An overview. *Journal of Soil and Water Conservation, 17*, pp.364–371.

Sharma, R., Sharma, R., Singh, R.R., & Kumari, A. (2023). Evaluation of biogenic zinc oxide nanoparticles from *Tinospora cordifolia* stem extract for photocatalytic, anti-microbial, and antifungal activities. *Materials Chemistry and Physics, 297*, p.127382.

Shivashankarappa, A., & Sanjay, K.R. (2015). Study on biological synthesis of cadmium sulfide nanoparticles by *Bacillus licheniformis* and its antimicrobial properties against food borne pathogens. *Nanoscience Nanotechnology Research, 3*, pp.6–15.

Siddiqi, K.S., & Husen, A. (2018). Properties of zinc oxide nanoparticles and their activity against microbes. *Nanoscale Research Letters, 13*, pp.1–13.

Singh, A., Singh, N.Á., Afzal, S., Singh, T., & Hussain, I. (2018). Zinc oxide nanoparticles: A review of their biological synthesis, antimicrobial activity, uptake, translocation and biotransformation in plants. *Journal of Materials Science, 53*, pp.185–201.

Singh, M.D., & Kumar, B.A. (2017). Bio efficacy of nano zinc sulphide (ZnS) on growth and yield of sunflower (*Helianthus annuus* L.) and nutrient status in the soil. *International Journal of Agriculture Sciences, 9*, pp.0975–3710.

Singh, S.K., Singh, A.K., & Pandey, A. (2017). Nanofertilizers: New products for the industry. *Journal of Nanoparticle Research, 19*, pp.1–15.

Soliman, M.H., El-Sheekh, M.M., & Younis, R.M. (2015). Role of zinc oxide nanoparticles in regulation of root growth and nutrient uptake in wheat (*Triticum aestivum* L.). *Environmental Science and Pollution Research, 22*, pp.15526–15534.

Stuparu-Cretu, M., Braniste, G., Necula, G.A., Stanciu, S., Stoica, D., & Stoica, M. (2023). Metal oxide nanoparticles in food packaging and their influence on human health. *Foods, 12*, p.1882.

Subbarao, C.V., Kartheek, G., & Sirisha, D. (2013). Slow release of potash fertilizer through polymer coating. *International Journal of Applied Science and Engineering, 11*, pp.25–30.

Sundararajan, S., Sivakumar, H.P., Rajendran, V., Kumariah, M., & Ramalingam, S. (2023). Adventitious roots in rice, the model cereal: Genetic factors and the influence of environmental cues—a mini review. *Plant Cell, Tissue and Organ Culture (PCTOC), 154*, pp.1–12.

Suresh, A.K., Pelletier, D.A., Wang, W., Broich, M.L., Moon, J.W., Gu, B., Allison, D.P., Joy, D.C., Phelps, T.J., & Doktycz, M.J. (2011). Biofabrication of discrete spherical gold nanoparticles using the metal-reducing bacterium *Shewanella oneidensis*. *Acta Biomaterialia, 7*, pp.2148–2152.

Suriyaprabha, R., Karunakaran, G., Yuvakkumar, R., Rajendran, V., Kannan, N., & Hong, S.I. (2012). Studies on the impact of nanosized titanium dioxide on the growth and root morphology of crop plants. *Journal of Nanoparticle Research, 14*, pp.1–14.

Sutradhar, P., Debnath, N., & Saha, M. (2013). Microwave-assisted rapid synthesis of alumina nanoparticles using tea, coffee and triphala extracts. *Advances in Manufacturing, 1*, pp.357–361.

Thavaseelan, D., & Priyadarshana, G. (2021). Nanofertilizer use for sustainable agriculture. *Journal of Research Technology and Engineering, 2*, pp.41–59.

Thirumurugan, A., Aswitha, P., Kiruthika, C., Nagarajan, S., & Christy, A.N. (2016). Green synthesis of platinum nanoparticles using *Azadirachta indica*–An eco-friendly approach. *Materials Letters, 170*, pp.175–178.

Tippayawat, P., Phromviyo, N., Boueroy, P., & Chompoosor, A. (2016). Green synthesis of silver nanoparticles in aloe vera plant extract prepared by a hydrothermal method and their synergistic antibacterial activity. *PeerJ, 4*, p.e2589.

Uddin, M.H., & Haque, M.M. (2023). Preparation and characterization of cellulose nanoparticles from agricultural wastes and their application in polymer composites. *Scholars International Journal of Chemistry and Material Sciences, 6*, pp.18–23.

Verma, D.K., Pandey, A.K., Mohapatra, B., Srivastava, S., Kumar, V., Talukdar, D., Yulianto, R., Zuan, A.T.K., Jobanputra, A.H., & Asthir, B. (2019). Plant growth-promoting rhizobacteria: An eco-friendly approach for sustainable agriculture and improved crop production. In Verma, D.K. (ed) *Microbiology for Sustainable Agriculture, Soil Health, and Environmental Protection* (pp. 3–80). Apple Academic Press.

Verma, S.K., Das, A.K., Patel, M.K., Shah, A., Kumar, V., & Gantait, S. (2018). Engineered nanomaterials for plant growth and development: A perspective analysis. *Science of the Total Environment, 630*, pp.1413–1435.

Wang, L., Sun, J., Lin, L., Fu, Y., Alenius, H., Lindsey, K., & Chen, C. (2020). Silver nanoparticles regulate *Arabidopsis* root growth by concentration-dependent modification of reactive oxygen species accumulation and cell division. *Ecotoxicology and Environmental Safety, 190*, p.110072.

Wang, P., Lombi, E., Zhao, F. J., Kopittke, P. M., & Naidu, R. (2016). Nanotechnology: A new opportunity in plant sciences. *Trends in Plant Science, 21*, 699–712.

Wu, S., Rajeshkumar, S., Madasamy, M., & Mahendran, V. (2020). Green synthesis of copper nanoparticles using *Cissus vitiginea* and its antioxidant and antibacterial activity against urinary tract infection pathogens. *Artificial Cells, Nanomedicine, and Biotechnology, 48*, pp.1153–1158.

Yadi, M., Mostafavi, E., Saleh, B., Davaran, S., Aliyeva, I., Khalilov, R., Nikzamir, M., Nikzamir, N., Akbarzadeh, A., Panahi, Y., & Milani, M. (2018). Current developments in green synthesis of metallic nanoparticles using plant extracts: A review. *Artificial Cells, Nanomedicine, and Biotechnology, 46*, pp.S336–S343.

Zangeneh, M.M., & Zangeneh, A. (2020). Novel green synthesis of *Hibiscus sabdariffa* flower extract conjugated gold nanoparticles with excellent anti-acute myeloid leukemia effect in comparison to daunorubicin in a leukemic rodent model. *Applied Organometallic Chemistry, 34*, p.e5271.

Zhang, H., Zhang, X., Wang, W., Zhang, X., Yang, H., & Yu, H. (2018). Chitos nanoparticles promote root growth and phosphorus uptake of soybean plants. *Frontiers in Plant Science, 9*, 1–9.
Zhang, Q.C., & Wang, G.H. (2005). Studies on nutrient uptake of rice and characteristics of soil microorganisms in a long-term fertilization experiments for irrigated rice. *Journal of Zhejiang University. Science. B, 6*, p.147.
Zulfiqar, F., Navarro, M., Ashraf, M., Akram, N.A., & Munné-Bosch, S. (2019). Nanofertilizer use for sustainable agriculture: Advantages and limitations. *Plant Science, 289*, p.110270.

13 Microbial Dynamics in the Plant Rhizosphere

A Comprehensive Metagenomic Perspective

Nidhi Selwal, Farida Rahayu, Nahid Akhtar, Ab Waheed Wani and Atif Khurshid Wani

13.1 INTRODUCTION

Increased soil fertility and crop productivity result from the plant rhizo-microbiome, which consists of various microbial communities, playing important roles in many processes central to ecosystem functioning, such as nutrient recycling, suppressing disease pathogens, secreting plant growth-promoting enzymes, and mineralizing organic matter (Mahmud et al., 2021). The metabolic compatibility and signaling events of root exudates secreted into the rhizosphere environment, the host-plant immune system, and interactions among different microbes within the plant rhizosphere all contribute to the recruitment of numerous microbes from the surrounding soils (Jain et al., 2020). The soil microbial population's activity and complexity are major factors in the health of the soil ecosystem (Fierer, 2017). The soil's biological, chemical, and physical qualities all have an impact on these bacteria. Soil ecosystem functioning can be affected by changes in microbial community composition, which can be influenced by land-use and management methods (Luo et al., 2020). The form and function of microbial communities are altered by agricultural management techniques, which in turn complicates the contributions of microbial communities to vital ecosystem processes including nutrient and carbon cycling (Schmidt et al., 2019; Wani et al., 2022a). The majority of research on how different land management techniques affect soil ecosystem functions has been on how such techniques change the make-up and activity of soil microorganisms. Researchers have looked at how the make-up of microbial communities in various ecosystems affects the health of those systems. The taxonomic makeup and abundance of soil microbial communities are only partially predictive of their functional potentials, as has been established in the literature. Soil microbial communities play an important role in ecosystems, and understanding how land use and management practices such as organic matter mineralization, nutrient cycling, organic pollutant degradation, and plant–microbe interactions affect these communities and the functioning of soil ecosystems is essential in understanding these roles (McKinley, 2019). Recent research has shown that analyzing the functional genes of soil microbial communities can help anticipate their roles. Understanding the dynamics of essential processes, such as nutrient and carbon cycling, and how they are influenced by agricultural management practices requires knowledge of the functional gene composition and diversity of microorganisms in the rhizosphere of agricultural soils (Garcia & Kao-Kniffin, 2018). The variety of microbial functional genes in the plant rhizosphere has been investigated using full metagenomic sequencing of soil samples from the rhizosphere. Metagenomics can help researchers learn more about the functional potentials of microbial communities in soils by providing data on a wide range of functional genes present in a sample (Mendes et al., 2017). Using this approach,

DOI: 10.1201/9781003354109-13

we can learn about the genes' functions and the roles they play in the rhizosphere soil ecosystem. Particular genes involved in phloroglucinol (phl) production and nitrogen fixation (nif) can directly impart useful features on plants, while genes involved in the creation of pyrroloquinoline quinone (pqq) contribute to a wide variety of indirect roles in plant cells. In combination, these genes promote robust plant development (Chukwuneme et al., 2021). Agricultural soils' functional gene diversity has been shown to be affected by land-use practices (Manoharan et al., 2017). However, there is a dearth of research on how land use and management practices have affected the variety of functional genes in the rhizosphere of plants in African soils. There is also a lack of research on the individual functional genes involved in critical processes, such as the biogeochemical cycling of nutrients and the encouragement of plant development in the maize rhizosphere (Upadhyay et al., 2022). Results from this study show that changes in land use (from grassland to cultivated land) and management strategies (tillage and no tillage) have a significant impact on the genetic variation of plant growth regulators and carbon-cycling enzymes in maize rhizosphere soils (Kumar et al., 2021; Wani et al., 2023). Shotgun metagenomic sequencing was performed to examine rhizosphere soil samples from formerly grassland soil and highly developed agricultural land in order to learn more about the variety of genes involved in plant growth promotion and carbon cycling (Chukwuneme et al., 2021). The rhizosphere soils of the reclaimed grassland were compared to those of the heavily farmed area, looking specifically at the content and variety of functional genes involved in plant growth promotion and carbon cycling. In addition, two hypotheses were developed based on prior research. Since the maize rhizosphere is home to beneficial bacteria, it must also be rich in a wide variety of microbial functional genes that play key roles in the soil ecosystem (Brahmaprakash et al., 2017). Secondly, the land-use and management histories of the two fields would result in a strikingly different variety of genes that contribute to plant growth and carbon cycling (Mishra et al., 2022).

13.2 FUNDAMENTALS OF METAGENOMICS

Metagenomics has significant potential in harnessing the abundant genetic reservoirs present in noncultivable microbes, through the bypassing of traditional cultivation methods and the direct extraction of genomic DNA from environmental samples (Arya & Ravindra, 2020a). There are two approaches that may be used to investigate the abundant genetic resources found in metagenomic libraries: sequence-driven screening and function-driven screening. The utilization of sequence-based screening methods is more prevalent compared to function-based techniques, mostly attributed to the convenient accessibility of metagenomics sequencing data and the availability of several software tools for data analysis (Sukhum et al., 2019). Nevertheless, sequence-based methodologies include intrinsic limitations, since their efficacy is heavily reliant on the precision of genome annotation and the comprehensiveness of the accessible data (Boolchandani et al., 2019; Wani et al., 2022b). These methodologies rely on the utilization of algorithms and the utilization of existing information inside databases to deduce the functionalities of recently identified genes. Therefore, these methods may exhibit limited efficacy in cases where the sequence similarity does not align with a functional association, or when the newly discovered gene bears only a tenuous resemblance to genes that have been subject to biochemical examination. Additionally, if a specific gene is capable of performing multiple functions within the cellular context, these methods may prove inadequate. Hence, the utilization of function-driven screening is the favored methodology for the identification of genes with unique activities or for the investigation of sequence variations within protein families associated with specific functions (Leis et al., 2013).

13.3 SHOTGUN METAGENOMICS

Upon elucidating the microbial diversity inside a metagenome, it is highly advantageous to get insights on its metabolic capabilities (Shu & Huang, 2021). The utilization of a comprehensive metagenome strategy involves the acquisition of entire DNA for the purpose of constructing full shotgun libraries (Méndez-García et al., 2018). The decision regarding the selection of a sequencing platform may be impacted by the computing resources and software capabilities necessary for handling and processing the sequencing output. It is important to acknowledge that the influence and promise of shotgun metagenomics may also be observed in the categorization of species at the taxonomy level (Quince et al., 2017). The diverse array of microorganisms acquired by comprehensive metagenome shotgun sequencing is likely to yield novel genes exhibiting unique functionalities (Franzosa et al., 2018).

13.3.1 The Evaluation of Taxonomy with Molecular Markers

In a theoretical context, the use of a metagenome shotgun sequencing technique enables the acquisition of a comprehensive depiction of all the genomes contained inside the given sample (Bharti & Grimm, 2021). This enables us to not only have a diverse selection of phylogenetic markers for taxonomic identification, but also allows us to acquire ribosomal markers or any other markers often used in amplicon sequencing (Yen & Johnson, 2021). One software solution for doing taxonomic annotation by extracting ribosomal marker genes from metagenomic sequences by multithreading is Parallel-meta (Suter, 2016). The software employs a reconstruction approach based on hidden Markov models (HMM) to gather ribosomal sequences from short reads (Wani et al., 2022c). Subsequently, the rebuilt sequences are aligned with other 16S gene databases using Megablast (Oliveira & Gruber 2021; Wani et al., 2022d). As elucidated in the part dedicated to metaprofiling analysis, the enhancement of taxonomical annotation might be achieved by the utilization of several phylogenetic markers. Hence, in the context of whole metagenome shotgun sequencing, it is possible to employ software tools for the purpose of querying single copy marker genes inside external databases. Two programs that employ these methodologies are MOCAT and AMPHORA. MOCAT utilizes the RefMG database, which consists of a compilation of 40 single copy marker genes (Shikha et al., 2021). On the other hand, AMPHORA incorporates a database comprising approximately 31 single copy universal markers. Following the discovery of the single copy marker, these processes proceed with an alignment of numerous sequences for operational taxonomic units (OTUs), computation of distances between them, and subsequent clustering. In many instances, the taxonomical annotation is conducted by utilizing reference genomes, resulting in the attainment of species resolution (Shikha et al., 2021).

13.4 NEXT-GENERATION SEQUENCING (NGS) METAGENOMICS

The utilization of Sanger sequencing technology significantly influenced the initial phase of research on microbial communities (Rastogi & Sani, 2011). In contemporary times, there have been significant advancements in the sequencing yield and sequence length as compared to the Sanger sequencing technique. At present, Sanger sequencing has the capability to get a maximum of 96 sequences each run, each with an average length of 650 base pairs (bp) (Shendure et al., 2017). This quantity and length of sequences may be sufficient for doing phylogenetic marker analysis. In various domains, NGS technologies have proven to be advantageous due to their ability to simultaneously sequence numerous DNA molecules with varying yields and sequence lengths (Figure 13.1) (Ari & Arikan, 2016). These low-cost platforms have had a beneficial influence in diverse fields.

One of the pioneering technologies that brought about a significant transformation in the fields of genomics and metagenomics was the 454-sequencing platform, often known as "pyrosequencing" (Meera Krishna et al., 2019). The underlying idea of this technique involves a sequential addition of nucleotides, one at a time. During the DNA polymerization procedure, the released pyrophosphate (PPi) is converted into a luminescent signal. The detection of light emission originating from a plate including several microwells housing a specific DNA fragment is accomplished by the machine (Liu

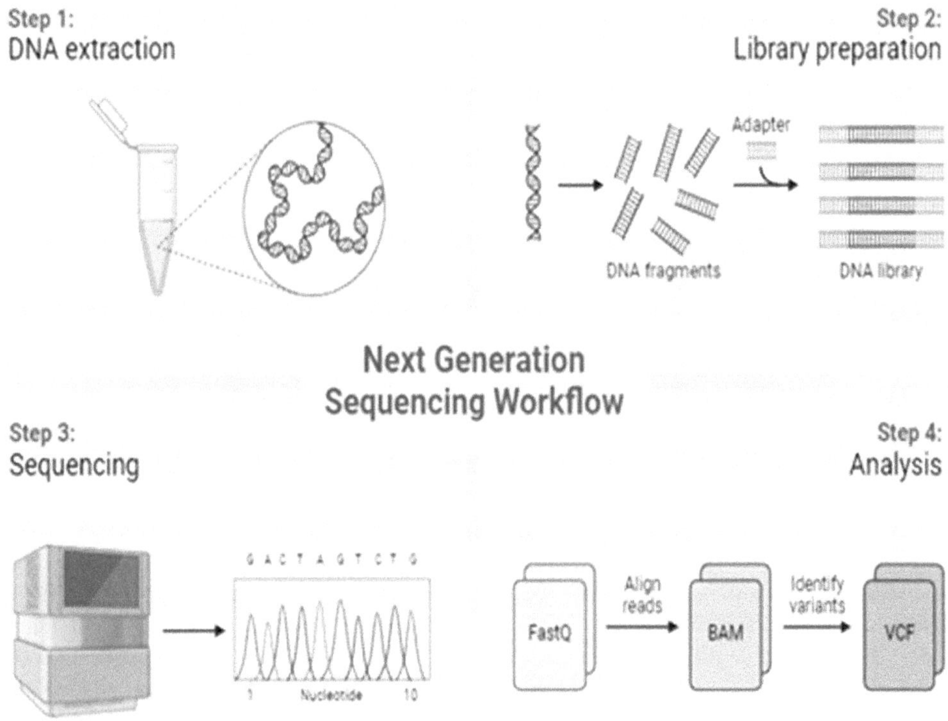

FIGURE 13.1 New generation sequencing metagenomics workflow.

& Lei, 2021). This light emission is subsequently converted into nucleotide sequences, accompanied by an associated base quality value. The aforementioned method provided a greater output in comparison to Sanger sequencing, while maintaining a reduced cost and shorter read lengths (Verma et al., 2017). One prominent limitation of this method is the occurrence of artificial insertions and deletions, mostly attributed to extended homopolymeric areas. Despite the numerous benefits that this technique has offered to the field of metagenomics, it has now become outdated. Roche, the current proprietor of the technology, has recently made public statements on the discontinuation of the 454 division. As a result, support for the platform will be terminated by mid-2016 (Escobar-Zepeda et al., 2015). However, it is possible to modify existing tools built for analyzing 454 data to also analyze data collected from other systems. The Ion Torrent platform may be seen as a technology that is comparable to the 454 platform, since it generates a comparable output and read length that is equivalent to what was achieved during the intermediate phase of its development (Goodwin et al., 2016). The Ion Torrent PGM is recognized as the most compact potentiometer available, capable of detecting the alteration in hydrogen potential resulting from the release of a proton following the addition of a nucleotide in the sequencing reaction taking place in many microwells (Escobar-Zepeda et al., 2015). The Ion Torrent sequencing platform has a maximum output of around 500 million reads, with a predominant read length of 400 base pairs. In this particular scenario, a notable advantage may be observed in relation to cost reduction, as the expense associated with Ion Torrent sequencing is just one-tenth of that incurred by pyrosequencing. One noted trade-off in some platforms is the reduction in read length, which is accompanied with increased yields and error rates. This trade-off is implemented as a means to decrease sequencing costs. A notable example of this is the Illumina technology, which has gained significant popularity because to its affordability and ability to produce large quantities of data (Hodkinson & Grice, 2015). The fundamental principle behind Illumina chemistry is the utilization of reversible-termination sequencing by synthesis, which employs nucleotides that are fluorescently tagged (Ambardar et al., 2016). In summary, the process involves the attachment and distribution of DNA fragments within a flow cell, whereby the sequencing reaction is initiated by the introduction of a tagged nucleotide. The machine detects the signal when the fluorescent molecule of the incorporated labeled nucleotide is stimulated by a laser. Subsequently, the fluorophore molecule is extracted, enabling the subsequent integration of the next nucleotide. The sequencing of DNA fragments can be performed from either one or both ends, resulting in single-end or pair-end sequencing, respectively. In this process, the maximum length of each read is limited to 300 base pairs (Chaudhary et al., 2021). Currently, this method has the greatest output among the second-generation sequencing technologies, rendering it well-suited for the multiplexing of hundreds of samples.

Presently, the aforementioned methods are widely employed in metagenome projects (Hu et al., 2021). However, during the past 5 years, efforts have been made to further advance sequencing techniques in order to address the inherent biases associated with these technologies and to provide an improved balance between output, cost, and read length. Currently, the third-generation sequencing technologies, such as the PacBio RS developed by Pacific Bioscience and the Oxford Nanopore, have addressed the issues of amplification bias and short read length (Akaçin et al., 2022). These technologies are characterized by their ability to sequence single molecules in real-time. The time and cost reduction provided by these technologies are also significant advantages. Nevertheless, it is worth noting that the error rate associated with this particular technology is comparatively greater when compared to alternative technologies. However, it is important to highlight that this issue may be rectified if the sequencing depth is increased to a sufficient level. In the realm of computational tools, the availability of software specifically designed for metagenomics analysis is quite limited (Akaçin et al., 2022). One notable advancement in second and third generation sequencing technology is the elimination of the need for DNA cloning vectors or bacterial hosts during library creation (Riley & Guss, 2021). This development has led to a simplification of the library preparation process and a decrease in DNA contamination originating from nonmetagenomic species. Despite the considerable capabilities of next-generation sequencing technologies in facilitating the discovery of previously unknown microbial realms and enabling exploration of uncharted settings, it is crucial to acknowledge and address the specific limits and biases associated with these methods. It is important to take into account that the processing of data acquired from second or third generation sequencing methods necessitates specific computing prerequisites (Bahassi & Stambrook, 2014). As the size of the dataset increases, there is a corresponding increase in the need for computer resources and the complexity of bioinformatics studies. Furthermore, it is imperative to have a substantial capacity for data storage in order to effectively preserve and analyze the voluminous amounts of data. In the context of bioinformatic analysis, proficiency in both high-performance servers and UNIX operating system is essential (Castrignanò et al., 2020). Proficiency in programming and scripting is highly advantageous for the successful execution and installation of the existing metagenomics software, which is utilized for the purpose of parsing and interpreting the obtained findings (Koutsandreas et al., 2019a). Hence, it is recommended that biologists or researchers in the field of biology acquire fundamental computing abilities to effectively leverage metagenomic data.

13.5 COMPUTATIONAL TOOLS FOR PLANT METAGENOME ANALYSIS

The aforementioned activities need significant computing resources and storage capabilities, as well as a comprehensive grasp of employing computational techniques from several disciplines (such as information theory, signal processing, and systems science) in tandem with at least one year of practical experience, in order to yield reliable and credible results. Hence, there is a significant demand for metagenomic analysis systems that incorporate automated workflows for diverse processing objectives (Navgire et al., 2022a). These systems integrate tools in the form of services

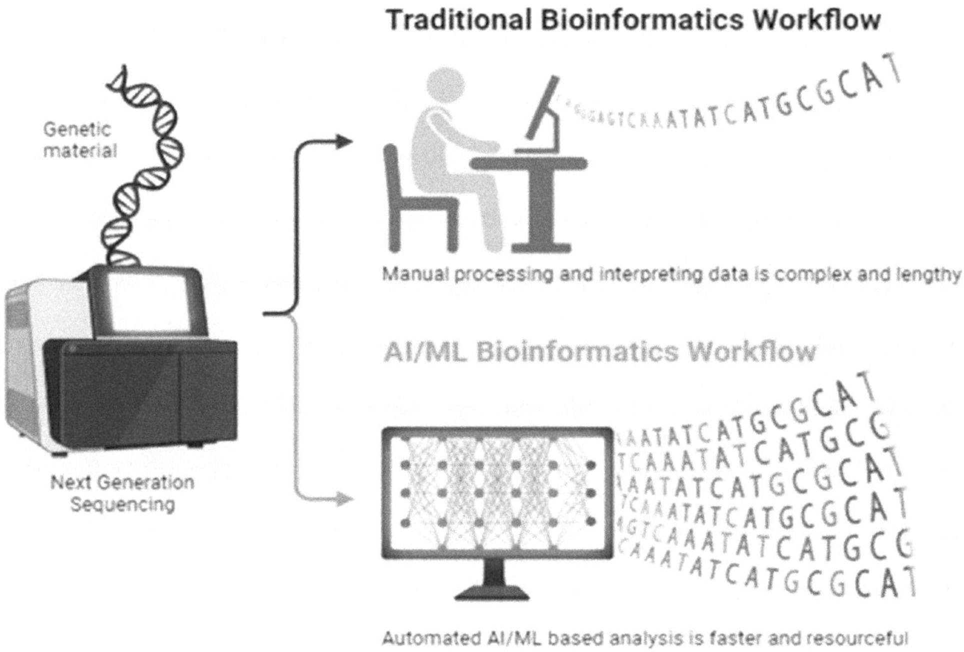

FIGURE 13.2 Next-generation sequencing: comparison between traditional and computational bioinformatics workflow.

into processing pipelines. A number of analytic processes have been developed for metagenomic research including a number of algorithms that have been devised for the analysis of genetic data pertaining to individual organisms. The application of NGS in metagenomic analysis has shown the limitations of existing approaches designed for analyzing data from single organisms in the context of metagenomic research (Figure 13.2) (Watts & Hurwitz, 2020). In order to analyze metagenomic sequencing data, bioinformatics software tools such as CloVR-metagenomics have been utilized (Jaenicke et al., 2018). The Galaxy platform, which is a metagenomics pipeline, has been documented in previous studies (Navgire et al., 2022a). The third item in the IMG/M dataset is represented by the values 71 and 72. The MetAMOS and MG-RAST platforms were utilized in the study (Vollmers et al., 2017). The resources RAMMCAP and Smash Community possess a high level of proficiency in accurately and comprehensively evaluating metagenomic data.

13.5.1 CloVR-Metagenomics

CloVR-metagenomics, a desktop software designed for automated sequence analysis, necessitates the utilization of two separate inputs in order to function well (Navgire et al., 2022a). To conduct a comparison analysis, it is necessary to get the raw sequencing data in fast format, as well as the metadata file containing sample-specific information. In the initial stage, the methodology utilizes UCLUST to group identical sequence reads together, followed by performing BLAST homology searches against the COG and RefSeq databases for the purpose of functional and taxonomic identification, respectively (Blakeley-Ruiz & Kleiner, 2022). In order to identify traits that exhibit differential abundance, the outcomes of both investigations are input into the integrated Meta stats program. The utilization of custom scripts in the R language, namely within the R Project for Statistical Computing version 2, is employed for the purpose of normalizing taxonomic or functional counts in order to facilitate grouping and display (Ladoukakis et al., 2014). One of the key features of CloVR's architecture is its ability to provide users with the option to use either local resources or leverage cloud computing capabilities (Langmead & Nellore, 2018). The platform heavily depends on third-party software for quality control and gene identification, which is limited to the single-genome and 16S rRNA software (Gao et al., 2021). This reliance on other software introduces a potential downside, since the platform's performance may be affected by the read length of sequencing datasets. The Galaxy platform, often known as the metagenomics pipeline, is a computational framework that facilitates the analysis of metagenomic data (Blank et al., 2018). Galaxy, an open-source framework, is currently being developed to facilitate the integration of computational tools and databases inside a unified workspace. This framework is specifically tailored for medical research endeavors that necessitate the handling of substantial volumes of data (Haefner et al., 2021). The novel Galaxy approach for metagenomic data employs automated analysis by using integrated specialized tools in conjunction with raw sequencing data (raw reads). According to a recent study, when applied to raw sequencing reads, a customized integrated toolset is employed to conduct a sequence of computational analyses. This is accomplished by a newly developed Galaxy procedure designed specifically for metagenomic datasets (Batut et al., 2018). The aforementioned studies encompass the following:

- The process of evaluating the quality and filtering of the readings using a proprietary tool.
- Thmm ,m,e process of modifying text and transforming data types using specialized tools.
- The NCBI-NT database was queried to identify homologous sequences.
- The utilization of specialized instruments in taxonomic study.
- Result visualization with the assistance of bespoke tools.

One of the most notable features of this platform is its ability to allow users to create workflow processes by integrating a wide range of custom tools, whether they are third-party or proprietary. These tools are capable of doing different analytical tasks, and the platform also provides a user interface that is extremely intuitive.

13.5.2 IMG/M

IMG/M encompasses a comprehensive collection of bacterial and archaeal species, together with a suite of analytical tools designed for the purpose of data exploration and comparative analysis (Paez-Espino et al., 2019). The sequencing data that has been compiled may be analyzed utilizing data exploration techniques to identify genes, contigs, and scaffolds, along with their corresponding functional characterizations (Arya & Ravindra, 2020b). The comparative data analysis suite encompasses many methodologies for analyzing metagenomic samples. These methodologies include the following: (i) the use of profile-based selection tools to ascertain the gene content and phylogenetic profile of a given sample (Wheeler et al., 2016), (ii) the application of gene neighborhood analysis tools, and (iii) the utilization of multiple sequence alignment tools. This platform has the capability to publish and maintain a user's (Meta) genome by utilizing its cloud infrastructure through the graphical user interface (GUI) of its web server (Yang, 2019). However, it is important to note that the user retains the responsibility for ensuring the quality control of raw readings and the process of assembly. The IMG/M software is specifically designed for the purpose of metagenome assembly, distinguishing it from other tools available for metagenomic analysis (Pal et al., 2018).

13.5.3 MetAMOS

MetAMOS software package is a widely used tool in the field of metagenomics. The pipeline is designed to accept either raw sequence reads or fully assembled contigs as input, and thereafter perform the assembly process to generate a metagenomic dataset (Navgire et al., 2022b).

The components of this pipeline provide a comprehensive analytical workflow encompassing the following: (i) quality control utilizing two distinct tools (FASTX-Toolkit 5, Babraham Bioinformatics - FastQC 6), (Koutsandreas et al., 2019b), (ii) sequence assembly into contigs employing eight diverse assembly approaches leveraging four distinct assembly tools (Kayani et al., 2021). Various tools may be integrated into MetAMOS workflows to do thorough analysis of metagenomic datasets, encompassing raw sequencing reads, contig, and scaffold data (Mukherjee et al., 2023).

13.5.4 MG-RAST (Metagenomics Rapid Annotation using Subsystem Technology)

Both unprocessed sequences from read datasets and pre-assembled contigs can be utilized as inputs in this procedure (Toshchakov et al., 2015). The system has four primary tasks that have been subdivided into components.

- Data normalization is a process in database management that involves organizing data in a structured and efficient manner.
- The identification of potential protein-coding genes and coding elements involves the screening of sequences against publicly available databases using predetermined search parameters.
- The input data is processed to compute functional annotations and taxonomic designations.
- The resulting data is then visualized using the SEED Viewer.

The pipeline installation ensures optimal data management using relational database technology by storing all job-relevant data in flat file and SQLite (SQLite 7) formats. A comparative metagenomic study of the primary dataset may be conducted by comparing the outcomes derived from the preceding phases with additional metagenomes or whole genomes acquired from the SEED environment. This platform, similar to IMG/M, offers a graphical user interface (GUI) that is designed to be user-friendly and is accessible through an HTTP server (Meyer et al., 2019). This design facilitates the ease of managing and analyzing data to the greatest extent possible. In addition to these features, the platform provides a diverse array of functional and comparative genomics capabilities, as well as the capacity to manage both assembled and unassembled data. Despite the lack of essential elements for ensuring the quality of reading and assembly processes, the pipeline provides a pragmatic and established taxonomic annotation system that effectively utilizes the capabilities of publicly available sequence databases (Roumpeka et al., 2017). The metagenomic platform known as RAMMCAP, which stands for rapid Analysis of Multiple Metagenomes with a Clustering and Annotation Pipeline, is designed to prioritize programmatic optimization in order to minimize the computational expenses associated with various processing tasks (Carlos & Da Costa, 2017). The CD-HIT approach is employed for the purpose of clustering sequences derived from one or several metagenomic samples (Zou et al., 2020). Subsequently, a secondary clustering of the protein sequences is conducted concurrently with the operation of identifying open reading frames (ORFs) in the unprocessed reads. This clustering is performed using a local method known as ORF finder (Zou et al., 2020).

13.5.5 NextFlow and MetaFlow | MICS pipelines

These pipelines are widely used in the academic community for the study of microbiome marker data (Arisdakessian et al., 2020). MetaFlow|mic is a novel framework designed for the analysis of microbiome data. It consolidates the functionalities of many tools such as DADA2, Mothur, VSEARCH, and others, into a user-friendly collection of pipelines. In addition to a basic sequence of instructions, our pipeline constitutes a comprehensive system founded on established norms, enabling seamless adaptability, versatility, and reproducibility. Furthermore, the system encompasses three jobs of significant complexity, a demultiplexing pipeline exclusive to the platform, as well as two comprehensive analysis pipelines. The first analysis pipeline is designed for the examination of bacterial data utilizing the 16S marker (Jovel et al., 2016), while the second pipeline is tailored for the study of fungal data employing the ITS marker (Pauvert et al., 2019). To improve the deployment of processes across diverse platforms, NextFlow was created (DI Tommaso et al., 2017). As a result, every analytic pipeline incorporates the essential tools required to perform quality assurance and quality control on the reads, as well as to estimate diversity at the operational taxonomic unit and precise sequence variation levels. In addition to Nextflow, the use of R, Python, Docker, and Singularly containers have been seen in the dissemination of analytic pipelines (Navgire et al., 2022b).

13.6 METAGENOMICS IN RHIZOSPHERE MANAGEMENT

When it comes to soil formation, pathogen pressure suppression, nutrient solubilization, and nutrient uptake, soil microorganisms play an essential role (Das et al., 2022). Therefore, rhizosphere engineering would benefit greatly from the use of a wide variety of biological instruments and methods aimed at altering the composition of the microbiome. As a result, our ability to influence and regulate the microbiome of the rhizosphere is severely constrained. Bio inoculation is the most reliable and efficient method for altering the microbiota. Many new products, including those made with PGPR (plant growth-promoting rhizobacteria) and AM (arbuscular mycorrhiza) fungus, have recently entered the global market (Pirttilä et al., 2021). To effectively mimic the highly organized crosslink seen in natural rhizosphere soils, bioinoculants need the ability to create relationships with another adjacent microbiota. The goal of this strategy is to increase the plant's tolerance to biotic and abiotic stress by the introduction of a diverse community of beneficial microorganisms (Majeed et al., 2018). The current progress in synthetic biological tools and gene editing methodologies presents a unique opportunity to manipulate the microbiome in order to achieve certain functionalities. Hence, the manipulation of plant characteristics and crop breeding are key strategies for engineering the rhizosphere of plants, as outlined in this concise study (Mashabela et al., 2022). The comprehensive comprehension of root architecture, host-specific root exudates, and other plant-related features that facilitate the selection of certain beneficial microorganisms holds the potential to modify crops with these qualities through the application of gene-editing techniques such as CRISPR (Kumar & Dubey, 2020). Therefore, this particular technique has more potential since it replicates the interconnected relationships that facilitate the identification of advantageous microorganisms, so contributing to the evolutionary progression of the holobiont. In the forthcoming years, there will be a growing capacity to deliberately manipulate the rhizosphere through advancements in engineering methodologies. A systematic strategy is necessary for the effective engineering of the rhizosphere microbiota (Kaul et al., 2021). By comprehending the fundamental processes involved in modulating the rhizosphere, there is potential to optimize the sustainability and efficacy of agricultural cultivation by emulation of advantageous symbiotic interactions among soil, microorganisms, and plants (Bhatt & Bhatt, 2021).

13.6.1 Monitoring Rhizosphere Microbiota

Recent findings in the literature provide compelling evidence on the significance of plant microbiota in facilitating the expansion of the plant genome for the purpose of agroecological transition. Numerous scholarly works have provided evidence to support the notion that plants engage in the recruitment of microorganisms, which subsequently establish colonies within their root tissues (Romano et al., 2020). This process gives rise to a distinct microbiota that is associated with the plants in the rhizosphere, defined as the soil volume influenced by root activities, as opposed to the larger surrounding soil environment. As a result, the performance of plants is improved by altering the ecological processes carried out by the rhizosphere and endophytic root bacteria. For example, the utilization of microbes has been found to have the potential to mitigate illnesses, improve the mineralization and delivery of nutrients, and stimulate plant development (Bargaz et al., 2018).

Environmental or ecological indicators are created with the purpose of evaluating the operational condition of an ecosystem. These indicators offer simplified data on a complex phenomenon of interest and possess the potential to be utilized for a diverse range of purposes in the advancement of rhizosphere engineering. The utilization of the rhizosphere effect indicator holds potential for researchers to effectively develop and evaluate cropping systems that rely on plant-rhizosphere microbiota (Mira et al., 2022; Wani et al., 2023). This approach aims to significantly reduce inputs, a crucial requirement for the advancement of pesticide-free agriculture in France and other European nations. Therefore, it is recommended that plant breeders use the rhizosphere effect indicator as a selection feature in order to release genotypes that effectively manifest their genetic potential in conjunction with the soil microbiome (Thakur et al., 2023). The potential of novel genotypes to engage in interactions with advantageous soil organisms, so triggering systemic resistance, holds considerable promise. The inclusion of new genotypes is crucial for the establishment of plant mixes that facilitate the coexistence of different genotypes (Semchenko et al., 2019). This coexistence is achieved through favorable interactions between plants, leading to a reduction in fertilizer usage by leveraging rhizosphere processes. The assessment of functional interactions between plant characteristics and microbiota is necessary in order to address these problems. The functioning of the rhizosphere may be effectively summarized by examining nutrient fluxes and enzyme activity. However, for a more precise understanding of plant–microbiota interactions, it is necessary to employ methodologies that utilize biomolecular information (Kaur et al., 2022). The enhancement of the rhizosphere effect indicator's functionality may facilitate a deeper comprehension of the rhizosphere's role in plant-soil feedbacks and ecosystem services (Kumawat et al., 2022).

13.6.2 Nutrient Cycling and Metagenomics

Phosphorus (P) is widely recognized as a crucial macronutrient for plants within agricultural ecosystems. It plays a significant role in determining agricultural productivity and the capacity of microbial communities to obtain carbon (C), nitrogen (N), phosphorus (P), and sulfur (S) resources. The primary productivity of crops in terrestrial ecosystems is significantly constrained by the limited supply of soil phosphorus (P) (Tang et al., 2018). While the use of phosphorus (P) fertilizer has the potential to mitigate P shortage and enhance the availability of soil nutrients to plants, it may also give rise to environmental degradation issues. In addition, it should be noted that plants have the ability to only uptake free orthophosphates. However, it is important to acknowledge that a significant portion of phosphorus in soil exists in forms that are not readily accessible to plants, as they are immobilized inside refractory inorganic and organic compounds. Soil microorganisms have the capacity to significantly influence the availability of phosphorus (P) through various mechanisms. These include the production of organic acids such as gluconic, oxalic, and citric acids, as well as the release of protons and chelating cations to facilitate the uptake of aluminum (Al), iron (Fe), and calcium (Ca) (Rawat et al., 2020). Additionally, soil microorganisms secrete enzymes such as phosphatases, phytases, and C-P lyases, which further contribute to the regulation of P availability. Previous research has demonstrated that the application of nanophos and plant growth-promoting rhizobacteria can yield substantial improvements in the abundance of soil microorganisms, soil enzyme activity, soil protein levels, as well as facilitate plant growth, enhance soil health, increase crop production, and improve soil nutrient utilization efficiency (Chaudhary et al., 2021). Hence, it is imperative to enhance our comprehension of the alterations occurring in soil microorganisms linked to the cycles of soil nutrients, such as carbon, nitrogen, phosphorus, and sulfur, within agricultural ecosystems. Additionally, there is a pressing need to enhance the efficacy of mineral fertilizers (Sahu et al., 2018). An increasing body of research indicates that the diversity of soil microorganisms exerts a prominent influence on the regulation of soil nutrient cycling, organic matter decomposition, pathogen control, and crop productivity, and health within agrosystems (Hartmann & Six, 2022). Prior research has demonstrated that diverse factors, including fertilization, soil type, soil physicochemical properties, and rhizosphere effects, exert an influence on functional soil microorganisms and genes associated with carbon (xylA, manA, and cdaR), nitrogen (nirK2 and nifH), and phosphorus (bpp) cycles (Wan et al., 2020). For instance, the application of long-term inorganic phosphorus (P) fertilizers at rates ranging from 26 to 70 kg P per hectare had a notable impact on the population and structure of soil functional microorganisms, as well as the genes associated with nutrient cycling (such as carbon, nitrogen, phosphorus, and sulfur) in both forest and arable soils (Bolo et al., 2021). Rhe introduction of inorganic phosphorus (P) into the soil, at levels ranging from 26 to 70 kg P ha^{-1}, has been shown to enhance soil respiration. However, it has also been seen that this increase in inorganic P input might hinder the activity of enzymes involved in carbon (C) cycling, leading to a decrease in microbial C usage efficiency. Consequently, these findings indicate that the presence of inorganic P in the soil can have detrimental impacts on soil organic C. In tropical soils deficient in phosphorus, there was a greater prevalence of genes associated with phosphorus acquisition (sph and DAD1), carbon substrate degradation (pel and yesW), and nitrogen transformations (nprE and npr) (Liu et al., 2024). Conversely, in phosphorus-rich soils, a higher abundance of genes involved in the degradation of aromatic compounds (hbaA, nagX, and mdlC), transformation of various nitrogenous compounds (pnpA, nirA, and eutB), and taurine degradation (tauD and cysJ) was observed. Furthermore, the acquisition of soil organic phosphorus (P) is linked to the breakdown of soil organic matter that comprises organic nitrogen (N), sulfur (S), and carbon (C). Hence, the soil phosphorus (P) turnover process is governed not only by the presence of P functional genes but also by the effect of carbon (C), nitrogen (N), and sulfur (S) functional genes (Liu et al., 2024). One example of a biological process that can contribute to the solubilization and mineralization of phosphorus in soil is the citric acid cycle. This cycle can generate energy and organic acids, which in turn can facilitate the release of phosphorus.

Another process that can influence soil pH and phosphorus availability is nitrification, which can lead to a reduction in soil pH and subsequent release of phosphorus (Vanzolini et al., 2017). Additionally, the presence of hydrogen sulfide (H_2S) can interact with ferric phosphate, resulting in the release of bound phosphorus (Flores-Alsina et al., 2016). This interaction is facilitated by acidophilic and sulfur-oxidizing bacteria. Hence, it is imperative to examine the alterations in functional genes and their interplay in response to prolonged phosphorus fertilizer application.

13.6.3 Disease Detection Via Metagenomics

Throughout history, humanity has consistently encountered perilous microbial illnesses, encompassing both bacterial and viral pathogens. Pathogens, such as viruses, undergo continuous evolution as a result of antigenic shift and drift mechanisms (Rouzine & Rozhnova, 2018). Viral evolution gives rise to novel kinds and serovars, which can have lethal consequences for human populations. Examples of such instances include influenza pandemics, severe acute respiratory syndrome (SARS), and Middle East respiratory syndrome (MERS). The recent occurrence of the new coronavirus SARS-CoV-2 pandemic serves as a notable illustration (Peiris & Poon, 2021). The global impact of this phenomenon has led to significant human casualties and has imposed a substantial economic burden. In order to address such circumstances, it is imperative to establish a comprehensive system that facilitates the timely identification of nascent viral infections. Implementing these measures will effectively mitigate the risk of potential epidemics and significantly contribute to the preservation of human lives (Ellwanger et al., 2021). The majority of new viruses and viral outbreaks with high mortality rates are believed to have originated from animal hosts. The implementation of routine surveillance in high-risk areas, such as zoological facilities and animal markets, utilizing metagenomics, has the potential to facilitate the evaluation of novel viruses and diseases (Bogaardt, 2020). This paper centers on the potential of viral metagenomics and presents a proposal for a monitoring system that utilizes this approach to enable early identification and subsequent prevention of developing viral diseases. The advent of NGS technology has initiated the age of precision medicine, revolutionizing the approach to cancer patient treatment and illness diagnosis (Morganti et al., 2019). Simultaneously, the emergence of these technologies has led to a significant increase in microbiome and metagenomic research in the past decade. A considerable portion of this research is dedicated to examining the interactions between host genes and microbes, which contribute to the onset and transmission of infectious diseases, as well as elucidating their crucial function in preserving overall well-being. As our understanding of the origins of infectious illnesses expands, it is more evident that metagenomic NGS technologies have significant promise for both the prompt detection and treatment of such diseases. In this study, we introduce a comprehensive methodology for the execution and utilization of "precision metagenomics" on diverse sequencing platforms in the context of clinical samples. This pipeline encompasses the integration of many stages, including DNA/RNA extraction, library preparation, sequencing, and bioinformatics studies (Afshinnekoo et al., 2017; Wani et al., 2022e). These stages collectively provide taxonomic categorization, screening for antimicrobial resistance (AMR) markers, and conducting functional analysis, specifically assessing the abundance of biochemical and metabolic pathways. In addition, the pipeline consists of three distinct tracks: STAT, Comprehensive, and Targeted. The STAT track provides data within a 24-hour timeframe. The Comprehensive track gives a more thorough analysis and takes between 5 and 7 days to complete, while also providing information on antibiotic resistance (Tacconelli et al., 2018). The Targeted track, also requiring 5–7 days, employs a more sensitive analysis specifically designed for identifying and analyzing particular bacteria (Rajapaksha et al., 2019). Lastly, the obstacles that must be confronted prior to achieving complete integration within the therapeutic environment will be examined.

The availability of clinical metagenomic next-generation sequencing (mNGS) testing for pathogen identification is increasing, including many sample types and justifications (Miller & Chiu, 2021a). Furthermore, there are anticipated future applications for this technology. The prevailing body of research indicates that mNGS has comparable sensitivity to targeted polymerase chain reaction (PCR) tests, while also detecting a greater number of possible infections compared to conventional diagnostic approaches (Duan et al., 2021). Although several organism detections align with the anticipated range of pathogens based on patient symptoms, there is a scarcity of formal research that verifies if these detections do indicate genuine positive infections and contribute to improved clinical outcomes. The presence of contamination and the discovery of clinically nonsignificant organisms lead to a decrease in specificity, which is a significant problem. This highlights the need for cautious interpretation of the pathogenicity of the organisms and their potential link with the clinical illness. Additional investigation is required in order to ascertain the potential enhancement in clinical outcomes and cost-effectiveness associated with the utilization of mNGS testing (Han et al., 2023). The diagnosis of infectious diseases is dependent on two main approaches: direct detection of bacteria in samples by the use of culture, antigen, or molecular markers, and indirect detection, which is based on the host's immune response and the presence of antibodies indicating infection (Miller & Chiu, 2021b). mNGS uses nucleic acid sequences (namely DNA and/or RNA) as biomarkers for the purpose of detecting harmful microorganisms in samples. The assays described in the literature can be designed with the purpose of selectively amplifying certain organism types, such as the 16S ribosomal and 18S-internal transcribed spacer gene amplification methods used for bacterial and fungal identification, respectively (Miller & Chiu, 2021b). Alternatively, these assays can also be designed to identify a broad spectrum of microbial species without any special targeting. The primary objective of this study is to examine the application of untargeted mNGS in the field of infectious disease diagnostics. This approach offers the most comprehensive pathogen identification capabilities over a wide spectrum.

In the majority of cases involving infection, culture is the preferred method for determining the appropriate antimicrobial treatment by isolating and testing the susceptibility of the pathogen.

However, there are instances where culture fails to cultivate a pathogen due to factors such as its inability to grow under laboratory conditions, its fastidious nature, or its suppression as a result of prior antibiotic use. This assertion holds special validity in the context of viral infections; however, it may also be extended to encompass bacterial, fungal, and parasitic illnesses (Fernández-García et al., 2018; Wani et al., 2023). The utilization of PCR testing has become the prevailing methodology for the identification of many organisms that pose challenges in cultivation. However, it is worth noting that the extensive scope required to encompass infrequent and newly appearing diseases often necessitates a substantial number of tests. Clinicians possess a significant inclination to exclude infection as the underlying cause for patient syndromes due to the resemblance in presentation between certain noninfectious conditions, such as autoimmune disease and vasculitis. Administering empirical treatment involving steroids and/or other immunosuppressants for these conditions would be unsuitable or potentially contraindicated if an infection were present. Hence, the utilization of mNGS can prove valuable in the identification of causative pathogens, particularly those that are infrequently encountered, and in offering reassurance regarding the absence of a severe infection prior to initiating immunosuppressive treatment (Megha et al., 2021). It is important to acknowledge, however, that mNGS may exhibit limitations in detecting infections of low magnitude or those localized in anatomical regions other than the site being tested. Similar to other diagnostic tests, the effectiveness of mNGS testing relies on several factors (Han et al., 2019). These include the symptoms, the severity of the symptoms, and the timing of when the sample is collected. Additionally, the quality of the sample, its location, and its type in relation to the source of infection are important considerations. Lastly, the performance characteristics of the mNGS assay, such as its analytical sensitivity, specificity, and detection range, also play a crucial role in determining the usefulness of the test (Miao et al., 2018). The present study highlights the need of considering certain factors while doing mNGS analysis and interpreting its clinical implications.

13.7 CHALLENGES AND FUTURE DIRECTIONS

In recent years, notable progress has been achieved with the emergence of metagenomics. In order to ascertain the interconnected biochemical mechanisms that govern microbial organization in the subsurface, it is necessary to conduct a comprehensive analysis of microbial structure, diversity, and dynamics. The reported findings demonstrate that metagenomic techniques have become a contemporary tool for managing many elements of microbial populations and hold promise for its use in the interaction between plants and microorganisms. The subterranean microbiome architecture plays a crucial role in facilitating plant growth and development. The microbial communities play an undeniable role in promoting soil fertility, facilitating nutrient uptake by plants, providing resistance against both biotic and abiotic stressors, and aiding in the phytoremediation of organic pollutants. In certain instances, it has been discovered that the composition of microbial communities can differ based on the depth of the soil. However, it has been noted that in deeper soil horizons, no discernible population variations have been detected. In the majority of investigations, it has been consistently reported that Proteobacteria exhibits a high abundance, suggesting its involvement in active processes such as nitrogen fixation, nutrient absorption capability, and polycyclic aromatic hydrocarbon degradation. In addition to Proteobacteria, Bacteroidetes, Firmicutes, and Acidobacteria have also been observed in greater abundance. Furthermore, the utilization of integrated omics technologies, including next-generation sequencing and metagenomics, has facilitated the exploration of the vast microbial resources present in the subsurface horizons. Further investigation is required to undertake comprehensive study, including other omics studies such as meta-transcriptomics and meta-proteomics, in order to uncover the in situ functions linked to microbial structures. Furthermore, the utilization of innovative methodologies will undoubtedly enhance the understanding of the unique and varied processes behind the activity of plant growth-promoting rhizobacteria (PGPRs), as well as reveal novel identities of these PGPRs. This will offer fresh perspectives on the potential applications of PGPRs in the field of biology. Additional investigation will yield innovative perspectives for enhanced comprehension of the metagenomic methodology pertaining to microorganisms. This will ideally lead to an inevitable enhancement of contemporary metagenomic methodologies in studying the interactions between microbial populations and commercially significant plant species.

Several practical challenges arise while doing a metagenomics investigation on rhizosphere soils. The primary consideration is the inherent quality of the initial genetic material. In order to conduct a successful metagenomics study, it is important to get a DNA sample of superior quality, possessing integrity and purity, extracted specifically from the rhizosphere soil. The presence of polyphenolics in DNA has been shown to significantly impede the PCR and subsequent sequencing processes. In order to tackle this matter, there exists a plethora of adapted methods and appropriate isolation kits that have been documented. The rhizobiome composition varies across various plants, necessitating distinct extraction processes, as demonstrated by the analysis of PCR-DGGE profiles. One of the primary challenges is the selection of an appropriate sequencing platform, given the multitude of NGS platforms available. One additional point of consideration is to the process of assembling metagenomes. The process of assembling a single genome is inherently intricate due to the existence of repeating parts in the reads. Consequently, the reconstruction of multiple genomes from a single sample is undoubtedly a challenging task that necessitates the utilization of specialized assemblers, algorithms, and software. The construction and categorization of the rhizosphere, which encompasses a diverse range of nonclonal microorganisms and is susceptible to horizontal gene transfers, presents additional challenges. Assemblers may hypothesize that repetitive changes might potentially be indicative of variability within a species. The subsequent obstacle is to the recognition of uncommon species inside the rhizosphere, which serves as a habitat for numerous uncharacterized and unculturable microorganisms. Conducting comprehensive sequencing to thoroughly cover a specific species in this context is not a viable undertaking. Extracting uncommon species from datasets of huge sizes can be challenging, unless the sample exclusively comprises a limited number of communities. The last concern is to inaccuracies in taxonomic categorization. When the reads are directly compared to reference databases, there exists a potential for matches to occur in noncoding areas and provide results from a distantly related organism. The assessment is contingent upon the caliber of the database. Moreover, it is worth noting that in the context of functional annotations, the occurrence of gene prediction errors tends to rise with the species richness in the sample. This is due to the fact that software tools may erroneously anticipate the presence of genes by relying on the detection of structure level homology, rather than considering the sequence level homology. However, there exist certain substitute options inside the program that may be utilized to address and resolve this particular problem.

13.8 CONCLUSION

Metagenomic samples originating from noncultivable microbes serve as a valuable repository for genes that possess unidentified activities. Numerous function-based metagenomic screening techniques have been devised to investigate the abundant genetic reservoirs, leading to the discovery of a considerable quantity of new enzymes exhibiting distinctive metabolic functionalities. Nevertheless, the identification of genes with unique activities is hindered by several obstacles, including inadequate expression of target genes within host cells and ineffective activity tests for the gene products. The conventional agar plate screening approach is plagued by its limited sensitivity and throughput capabilities. In order to address these challenges, many high-throughput techniques have been developed, including FACS-driven screening and microfluidics-driven screening (Wani et al., 2022f). These methodologies have significantly broadened the range of tools available for investigating the extensive sequence variability present in metagenomes. The utilization of microfluidics-based screening has demonstrated its capacity to screen more than 107 variations within a single day. When combined with fluorescence-activated cell sorting (FACS), microfluidic devices have the potential to significantly enhance the range of high-throughput metagenomic screening. The primary constraint of this methodology resides in the detection method, predominantly reliant on fluorescent signal. In subsequent periods, it is plausible that additional techniques for detection, such as mass spectrometry, nuclear magnetic resonance (NMR), and colorimetric assay, might be integrated with microfluidic devices to expedite the identification of new biocatalysts or other genes that have significant functional roles within the microbiota. However, the majority of function-based methodologies utilized for metagenomic screening encounter limitations due to the skewed and inadequate expression observed in *Escherichia coli*. There exists a pressing necessity to expand the repertoire of alternate hosts that exhibit proficient expression of exogenous genes derived from metagenomic sources. In order to achieve this objective, it is imperative to devise shuttle vectors that possess enhanced host compatibility, hence enabling the expression of the metagenomic library in diverse host organisms. In order to incorporate foreign genes, it is possible to modify the translation patterns in *E. coli* by the manipulation of ribosomal proteins or the components responsible for transcriptional/translational regulation.

REFERENCES

Afshinnekoo, E., Chou, C., Alexander, N., Ahsanuddin, S., Schuetz, A. N., & Mason, C. E. (2017). Precision metagenomics: Rapid metagenomic analyses for infectious disease diagnostics and public health surveillance. *Journal of Biomolecular Techniques: JBT, 28*(1), 40. https://doi.org/10.7171/JBT.17-2801-007

Akaçin, İ., Ersoy, Ş., Doluca, O., & Güngörmüşler, M. (2022). Comparing the significance of the utilization of next generation and third generation sequencing technologies in microbial metagenomics. *Microbiological Research, 264*, 127154. https://doi.org/10.1016/J.MICRES.2022.127154

Ambardar, S., Gupta, R., Trakroo, D., Lal, R., & Vakhlu, J. (2016). High throughput sequencing: An overview of sequencing chemistry. *Indian Journal of Microbiology*, *56*(4), 394–404. https://doi.org/10.1007/S12088-016-0606-4/TABLES/1

Ari, Ş., & Arikan, M. (2016). Next-generation sequencing: Advantages, disadvantages, and future. In *Plant Omics: Trends and Applications* (pp. 109–135). Springer. https://doi.org/10.1007/978-3-319-31703-8_5/COVER

Arisdakessian, C., Cleveland, S. B., & Belcaid, M. (2020). MetaFlowlmics: Scalable and reproducible nextflow pipelines for the analysis of microbiome marker data. In J. Wang (Ed.), *Practice and Experience in Advanced Research Computing* (pp. 120–124). https://doi.org/10.1145/3311790.3396664

Arya, P., & Ravindra. (2020a). Metagenomics based approach to reveal the secrets of unculturable microbial diversity from aquatic environment. In *Recent Advancements in Microbial Diversity* (pp. 537–559). Academic Press. https://doi.org/10.1016/B978-0-12-821265-3.00022-0

Bahassi, E. M., & Stambrook, P. J. (2014). Next-generation sequencing technologies: Breaking the sound barrier of human genetics. *Mutagenesis*, *29*(5), 303–310. https://doi.org/10.1093/MUTAGE/GEU031

Bargaz, A., Lyamlouli, K., Chtouki, M., Zeroual, Y., & Dhiba, D. (2018). Soil microbial resources for improving fertilizers efficiency in an integrated plant nutrient management system. *Frontiers in Microbiology, 9*, 364232. https://doi.org/10.3389/FMICB.2018.01606/BIBTEX

Batut, B., Gravouil, K., Defois, C., Hiltemann, S., Brugère, J. F., Peyretaillade, E., & Peyret, P. (2018). ASaiM: A Galaxy-based framework to analyze microbiota data. *GigaScience*, *7*(6), 1–7. https://doi.org/10.1093/GIGASCIENCE/GIY057

Bharti, R., & Grimm, D. G. (2021). Current challenges and best-practice protocols for microbiome analysis. *Briefings in Bioinformatics*, *22*(1), 178–193. https://doi.org/10.1093/BIB/BBZ155

Bhatt, K., & Bhatt, P. (2021). Rhizospheric biology. In A. Verma, J. K. Saini, & H. B. Singh (Eds.), *Phytomicrobiome Interactions and Sustainable Agriculture* (pp. 164–186). Springer. https://doi.org/10.1002/9781119644798.CH9

Blakeley-Ruiz, J. A., & Kleiner, M. (2022). Considerations for constructing a protein sequence database for metaproteomics. *Computational and Structural Biotechnology Journal, 20*, 937–952. https://doi.org/10.1016/J.CSBJ.2022.01.018

Blank, C., Easterly, C., Gruening, B., Johnson, J., Kolmeder, C. A., Kumar, P., May, D., Mehta, S., Mesuere, B., Brown, Z., Elias, J. E., Hervey, W. J., McGowan, T., Muth, T., Nunn, B. L., Rudney, J., Tanca, A., Griffin,

T. J., & Jagtap, P. D. (2018). Disseminating metaproteomic informatics capabilities and knowledge using the Galaxy-P framework. *Proteomes*, *6*(1), 7. https://doi.org/10.3390/PROTEOMES6010007

Bogaardt, C. (2020). *Metagenomic surveillance of viruses at the human-animal interface.* (Doctoral dissertation, University of Edinburgh). Edinburgh Research Archive. https://doi.org/10.7488/ERA/1329

Bolo, P., Kihara, J., Mucheru-Muna, M., Njeru, E. M., Kinyua, M., & Sommer, R. (2021). Application of residue, inorganic fertilizer and lime affect phosphorus solubilizing microorganisms and microbial biomass under different tillage and cropping systems in a Ferralsol. *Geoderma, 390*, 114962. https://doi.org/10.1016/J.GEODERMA.2021.114962

Boolchandani, M., D'Souza, A. W., & Dantas, G. (2019). Sequencing-based methods and resources to study antimicrobial resistance. *Nature Reviews Genetics*, 20(6), 356–370. https://doi.org/10.1038/s41576-019-0108-4

Brahmaprakash, G. P., Sahu, P. K., Lavanya, G., Nair, S. S., Gangaraddi, V. K., & Gupta, A. (2017). Microbial functions of the rhizosphere. *Plant-Microbe Interactions in Agro-Ecological Perspectives, 1*, 211–226. https://doi.org/10.1007/978-981-10-5813-4_10/COVER

Carlos, J., & Da Costa, S. (2017). *Development of an automated pipeline for meta-omics data analysis.* Master's thesis, Universidade do Minho (Portugal).

Castrignanò, T., Gioiosa, S., Flati, T., Cestari, M., Picardi, E., Chiara, M., Fratelli, M., Amente, S., Cirilli, M., Tangaro, M. A., Chillemi, G., Pesole, G., & Zambelli, F. (2020). ELIXIR-IT HPC@CINECA: High performance computing resources for the bioinformatics community. *BMC Bioinformatics*, *21*(10), 1–17. https://doi.org/10.1186/S12859-020-03565-8/FIGURES/5

Chaudhary, P., Chaudhary, A., Parveen, H., Rani, A., Kumar, G., Kumar, R., & Sharma, A. (2021). Impact of nanophos in agriculture to improve functional bacterial community and crop productivity. *BMC Plant Biology*, *21*(1), 1–12. https://doi.org/10.1186/S12870-021-03298-7/FIGURES/4

Chaudhary, P., Sharma, A., Chaudhary, A., Khati, P., Gangola, S., & Maithani, D. (2021). Illumina based high throughput analysis of microbial diversity of maize rhizosphere treated with nanocompounds and Bacillus sp. *Applied Soil Ecology, 159*, 103836. https://doi.org/10.1016/J.APSOIL.2020.103836

Chukwuneme, C. F., Ayangbenro, A. S., & Babalola, O. O. (2021). Metagenomic analyses of plant growth-promoting and carbon-cycling genes in maize rhizosphere soils with distinct land-use and management histories. *Genes*, *12*(9), 1431. https://doi.org/10.3390/GENES12091431/S1

Das, P. P., Singh, K. R., Nagpure, G., Mansoori, A., Singh, R. P., Ghazi, I. A., Kumar, A., & Singh, J. (2022). Plant-soil-microbes: A tripartite interaction for nutrient acquisition and better plant growth for sustainable agricultural practices. *Environmental Research, 214*, 113821. https://doi.org/10.1016/J.ENVRES.2022.113821

DI Tommaso, P., Chatzou, M., Floden, E. W., Barja, P. P., Palumbo, E., & Notredame, C. (2017). Nextflow enables reproducible computational workflows. *Nature Biotechnology*, *35*(4), 316–319. https://doi.org/10.1038/nbt.3820

Duan, H., Li, X., Mei, A., Li, P., Liu, Y., Li, X., Li, W., Wang, C., & Xie, S. (2021). The diagnostic value of metagenomic next-generation sequencing in infectious diseases. *BMC Infectious Diseases*, *21*(1), 1–13. https://doi.org/10.1186/S12879-020-05746-5/TABLES/7

Ellwanger, J. H., da Veiga, A. B. G., Kaminski, V. de L., Valverde-Villegas, J. M., de Freitas, A. W. Q., & Chies, J. A. B. (2021). Control and prevention of infectious diseases from a One Health perspective. *Genetics and Molecular Biology, 44*(1), e20200256. https://doi.org/10.1590/1678-4685-GMB-2020-0256

Escobar-Zepeda, A., De León, A. V. P., & Sanchez-Flores, A. (2015). The road to metagenomics: From microbiology to DNA sequencing technologies and bioinformatics. *Frontiers in Genetics, 6*(DEC), 155161. https://doi.org/10.3389/FGENE.2015.00348/BIBTEX

Fernández-García, M., Rojo, D., Rey-Stolle, F., García, A., & Barbas, C. (2018). Metabolomic-based methods in diagnosis and monitoring infection progression. *Experientia Supplementum (2012)*, *109*, 283–315. https://doi.org/10.1007/978-3-319-74932-7_7/FIGURES/3

Fierer, N. (2017). Embracing the unknown: Disentangling the complexities of the soil microbiome. *Nature Reviews Microbiology*, *15*(10), 579–590. https://doi.org/10.1038/nrmicro.2017.87

Flores-Alsina, X., Solon, K., Kazadi Mbamba, C., Tait, S., Gernaey, K. V., Jeppsson, U., & Batstone, D. J. (2016). Modelling phosphorus (P), sulfur (S) and iron (Fe) interactions for dynamic simulations of anaerobic digestion processes. *Water Research, 95*, 370–382. https://doi.org/10.1016/J.WATRES.2016.03.012

Franzosa, E. A., McIver, L. J., Rahnavard, G., Thompson, L. R., Schirmer, M., Weingart, G., Lipson, K. S., Knight, R., Caporaso, J. G., Segata, N., & Huttenhower, C. (2018). Species-level functional profiling of metagenomes and metatranscriptomes. *Nature Methods*, *15*(11), 962–968. https://doi.org/10.1038/s41592-018-0176-y

Gao, B., Chi, L., Zhu, Y., Shi, X., Tu, P., Li, B., Yin, J., Gao, N., Shen, W., & Schnabl, B. (2021). An introduction to next generation sequencing bioinformatic analysis in gut microbiome studies. *Biomolecules*, *11*(4), 530. https://doi.org/10.3390/BIOM11040530

Garcia, J., & Kao-Kniffin, J. (2018). Microbial group dynamics in plant rhizospheres and their implications on nutrient cycling. *Frontiers in Microbiology*, *9*(JUL), 370159. https://doi.org/10.3389/FMICB.2018.01516/BIBTEX

Goodwin, S., McPherson, J. D., & McCombie, W. R. (2016). Coming of age: Ten years of next-generation sequencing technologies. *Nature Reviews Genetics, 17*(6), 333–351. https://doi.org/10.1038/nrg.2016.49

Haefner, N., Wincent, J., Parida, V., & Gassmann, O. (2021). Artificial intelligence and innovation management: A review, framework, and research agenda☆. *Technological Forecasting and Social Change, 162*, 120392. https://doi.org/10.1016/J.TECHFORE.2020.120392

Han, D., Li, Z., Li, R., Tan, P., Zhang, R., & Li, J. (2019). mNGS in clinical microbiology laboratories: On the road to maturity. *Critical Reviews in Microbiology*, *45*(5–6), 668–685. https://doi.org/10.1080/1040841X.2019.1681933

Han, D., Yu, F., Zhang, D., Yang, Q., Xie, M., Yuan, L., Zheng, J., Wang, J., Zhou, J., Xiao, Y., Zheng, S., & Chen, Y. (2023). The real-world clinical impact of plasma mNGS testing: An observational study. *Microbiology Spectrum*, *11*(2). https://doi.org/10.1128/SPECTRUM.03983-22/SUPPL_FILE/SPECTRUM.03983-22-S0002.XLSX

Hartmann, M., & Six, J. (2022). Soil structure and microbiome functions in agroecosystems. *Nature Reviews Earth & Environment*, *4*(1), 4–18. https://doi.org/10.1038/s43017-022-00366-w

Hodkinson, B. P., & Grice, E. A. (2015). Next-generation sequencing: A review of technologies and tools for wound microbiome research. *Advances in Wound Care*, *4*(1), 50–58. https://doi.org/10.1089/WOUND.2014.0542

Hu, T., Chitnis, N., Monos, D., & Dinh, A. (2021). Next-generation sequencing technologies: An overview. *Human Immunology*, *82*(11), 801–811. https://doi.org/10.1016/J.HUMIMM.2021.02.012

Jaenicke, S., Albaum, S. P., Blumenkamp, P., Linke, B., Stoye, J., & Goesmann, A. (2018). Flexible metagenome analysis using the MGX framework. *Microbiome*, *6*(1), 76. https://doi.org/10.1186/S40168-018-0460-1/TABLES/2

Jain, A., Chakraborty, J., & Das, S. (2020). Underlying mechanism of plant–microbe crosstalk in shaping microbial ecology of the rhizosphere. *Acta Physiologiae Plantarum*, *42*(1), 1–13. https://doi.org/10.1007/S11738-019-3000-0/FIGURES/1

Jovel, J., Patterson, J., Wang, W., Hotte, N., O'Keefe, S., Mitchel, T., Perry, T., Kao, D., Mason, A. L., Madsen, K. L., & Wong, G. K. S. (2016). Characterization of the gut microbiome using 16S or shotgun metagenomics. *Frontiers in Microbiology*, *7*(APR), 180723. https://doi.org/10.3389/FMICB.2016.00459/BIBTEX

Kaul, S., Choudhary, M., Gupta, S., & Dhar, M. K. (2021). Engineering host microbiome for crop improvement and sustainable agriculture. *Frontiers in Microbiology, 12*, 635917. https://doi.org/10.3389/FMICB.2021.635917/BIBTEX

Kaur, M., Kaur, R., & Sodhi, H. S. (2022). Utilization of omics approaches for underpinning plant-microbe interaction. In S. S. M. R. Hussain, K. P. S. R. Rao, & P. K. S. R. K. Kothari (Eds.), *New and Future Developments in Microbial Biotechnology and Bioengineering: Sustainable Agriculture: Advances in Microbe-Based Biostimulants* (pp. 217–242). Elsevier. https://doi.org/10.1016/B978-0-323-85577-8.00009-3

Kayani, M. U. R., Huang, W., Feng, R., & Chen, L. (2021). Genome-resolved metagenomics using environmental and clinical samples. *Briefings in Bioinformatics*, *22*(5), 1–20. https://doi.org/10.1093/BIB/BBAB030

Koutsandreas, T., Ladoukakis, E., Pilalis, E., Zarafeta, D., Kolisis, F. N., Skretas, G., & Chatziioannou, A. A. (2019a). Anastasia: An automated metagenomic analysis pipeline for novel enzyme discovery exploiting next generation sequencing data. *Frontiers in Genetics*, *10*(MAY), 434571. https://doi.org/10.3389/FGENE.2019.00469/BIBTEX

Koutsandreas, T., Ladoukakis, E., Pilalis, E., Zarafeta, D., Kolisis, F. N., Skretas, G., & Chatziioannou, A. A. (2019b). Anastasia: An automated metagenomic analysis pipeline for novel enzyme discovery exploiting next generation sequencing data. *Frontiers in Genetics*, *10*(MAY), 434571. https://doi.org/10.3389/FGENE.2019.00469/BIBTEX

Kumar, A., & Dubey, A. (2020). Rhizosphere microbiome: Engineering bacterial competitiveness for enhancing crop production. *Journal of Advanced Research, 24*, 337–352. https://doi.org/10.1016/J.JARE.2020.04.014

Kumar, G. C., Dutta, D., Chaudhary, J., & Meena, A. L. (2021). Functional Diversity Management through Microbial Integrity for Sustainability. In A. Rakshit (Ed.), *Soil Science: Fundamentals to Recent Advances* (pp. 361–388). Springer. https://doi.org/10.1007/978-981-16-0917-6_18/COVER

Kumawat, K. C., Razdan, N., & Saharan, K. (2022). Rhizospheric microbiome: Bio-based emerging strategies for sustainable agriculture development and future perspectives. *Microbiological Research, 254*, 126901. https://doi.org/10.1016/J.MICRES.2021.126901

Ladoukakis, E., Kolisis, F. N., & Chatziioannou, A. A. (2014). Integrative workflows for metagenomic analysis. *Frontiers in Cell and Developmental Biology*, *2*(NOV), 117351. https://doi.org/10.3389/FCELL.2014.00070/BIBTEX

Langmead, B., & Nellore, A. (2018). Cloud computing for genomic data analysis and collaboration. *Nature Reviews Genetics,* 19(4), 208–219. https://doi.org/10.1038/nrg.2017.113

Leis, B., Angelov, A., & Liebl, W. (2013). Screening and expression of genes from metagenomes. *Advances in Applied Microbiology, 83*, 1–68. https://doi.org/10.1016/B978-0-12-407678-5.00001-5

Liu, H., & Lei, Y. (2021). A critical review: Recent advances in "digital" biomolecule detection with single copy sensitivity. *Biosensors and Bioelectronics, 177*, 112901. https://doi.org/10.1016/J.BIOS.2020.112901

Liu, L., Gao, Y., Yang, W., Liu, J., & Wang, Z. (2024). Community metagenomics reveals the processes of nutrient cycling regulated by microbial functions in soils with P fertilizer input. *Plant and Soil*, *499*(1), 1–16. https://doi.org/10.1007/S11104-023-05875-1/FIGURES/6

Luo, D., Cheng, R. M., Liu, S., Shi, Z. M., & Feng, Q. H. (2020). Responses of soil microbial community composition and enzyme activities to land-use change in the Eastern Tibetan Plateau, China. *Forests*, *11*(5), 483. https://doi.org/10.3390/F11050483

Mahmud, K., Missaoui, A., Lee, K., Ghimire, B., Presley, H. W., & Makaju, S. (2021). Rhizosphere microbiome manipulation for sustainable crop production. *Current Plant Biology, 27*, 100210. https://doi.org/10.1016/J.CPB.2021.100210

Majeed, A., Muhammad, Z., & Ahmad, H. (2018). Plant growth promoting bacteria: Role in soil improvement, abiotic and biotic stress management of crops. *Plant Cell Reports*, *37*(12), 1599–1609. https://doi.org/10.1007/S00299-018-2341-2/TABLES/2

Manoharan, L., Kushwaha, S. K., Ahrén, D., & Hedlund, K. (2017). Agricultural land use determines functional genetic diversity of soil microbial communities. *Soil Biology and Biochemistry, 115*, 423–432. https://doi.org/10.1016/J.SOILBIO.2017.09.011

Mashabela, M. D., Masamba, P., & Kappo, A. P. (2022). Metabolomics and chemoinformatics in agricultural biotechnology research: Complementary probes in unravelling new metabolites for crop improvement. *Biology*, *11*(8), 1156. https://doi.org/10.3390/BIOLOGY11081156

McKinley, V. L. (2019). Effects of Land Use and Restoration on Soil Microbial Communities. In E. A. Paul (Ed.), *Soil Microbiology, Ecology, and Biochemistry* (pp. 173–242). Springer. https://doi.org/10.1007/978-3-030-10777-2_7

Meera Krishna, B., Khan, M. A., & Khan, S. T. (2019). Next-generation sequencing (NGS) platforms: An exciting era of genome sequence analysis. *Microbial Genomics in Sustainable Agroecosystems, 2*, 89–109. https://doi.org/10.1007/978-981-32-9860-6_6/COVER

Megha, P., & Prasad, T. (2021). Metabolomics: A promising tool to study disease biomarkers and host-pathogen interactions. In S. Hameed & Z. Fatima (Eds.), *Integrated Omics Approaches to Infectious Diseases* (pp. 403–423). Springer. https://doi.org/10.1007/978-981-16-0691-5_22

Mendes, L. W., Braga, L. P. P., Navarrete, A. A., Souza, D. G. de, Silva, G. G. Z., & Tsai, S. M. (2017). Using metagenomics to connect microbial community biodiversity and functions. *Current Issues in Molecular Biology*, *24*(1), 103–118. https://doi.org/10.21775/CIMB.024.103

Méndez-García, C., Bargiela, R., Martínez-Martínez, M., & Ferrer, M. (2018). Metagenomic protocols and strategies. In M. Nagarajan (Ed.), *Metagenomics: Perspectives, Methods, and Applications* (pp. 15–54). Academic Press. https://doi.org/10.1016/B978-0-12-813403-0.00002-X

Meyer, F., Bagchi, S., Chaterji, S., Gerlach, W., Grama, A., Harrison, T., Paczian, T., Trimble, W. L., & Wilke, A. (2019). MG-RAST version 4—Lessons learned from a decade of low-budget ultra-high-throughput metagenome analysis. *Briefings in Bioinformatics*, *20*(4), 1151–1159. https://doi.org/10.1093/BIB/BBX105

Miao, Q., Ma, Y., Wang, Q., Pan, J., Zhang, Y., Jin, W., Yao, Y., Su, Y., Huang, Y., Wang, M., Li, B., Li, H., Zhou, C., Li, C., Ye, M., Xu, X., Li, Y., & Hu, B. (2018). Microbiological diagnostic performance of metagenomic next-generation sequencing when applied to clinical practice. *Clinical Infectious Diseases*, *67*(suppl_2), S231–S240. https://doi.org/10.1093/CID/CIY693

Miller, S., & Chiu, C. (2021a). The role of metagenomics and next-generation sequencing in infectious disease diagnosis. *Clinical Chemistry*, *68*(1), 115–124. https://doi.org/10.1093/CLINCHEM/HVAB173

Miller, S., & Chiu, C. (2021b). The role of metagenomics and next-generation sequencing in infectious disease diagnosis. *Clinical Chemistry*, *68*(1), 115–124. https://doi.org/10.1093/CLINCHEM/HVAB173

Mira, S., Emily, M., Mougel, C., Ourry, M., & Le Cadre, E. (2022). A field indicator for rhizosphere effect monitoring in arable soils. *Plant and Soil*, *478*(1), 325–346. https://doi.org/10.1007/S11104-021-05284-2

Mishra, A., Singh, L., & Singh, D. (2022). Unboxing the black box—One step forward to understand the soil microbiome: A systematic review. *Microbial Ecology*, *85*(2), 669–683. https://doi.org/10.1007/S00248-022-01962-5

Morganti, S., Tarantino, P., Ferraro, E., D'Amico, P., Viale, G., Trapani, D., Duso, B. A., & Curigliano, G. (2019). Complexity of genome sequencing and reporting: Next generation sequencing (NGS) technologies and implementation of precision medicine in real life. *Critical Reviews in Oncology/Hematology*, *133*, 171–182. https://doi.org/10.1016/J.CRITREVONC.2018.11.008

Mukherjee, A., D'Ugo, E., Giuseppetti, R., Magurano, F., & Cotter, P. D. (2023). Metagenomic approaches for understanding microbial communities in contaminated environments: Bioinformatic tools, case studies and future outlook. In A. Mukherjee & P. D. Cotter (Eds.), *Metagenomics to Bioremediation: Applications, Cutting Edge Tools, and Future Outlook (pp.* 103–156). Academic Press. https://doi.org/10.1016/B978-0-323-96113-4.00003-2

Navgire, G. S., Goel, N., Sawhney, G., Sharma, M., Kaushik, P., Mohanta, Y. K., Mohanta, T. K., & Al-Harrasi, A. (2022a). Analysis and interpretation of metagenomics data: An approach. *Biological Procedures Online*, *24*(1), 1–22. https://doi.org/10.1186/S12575-022-00179-7

Navgire, G. S., Goel, N., Sawhney, G., Sharma, M., Kaushik, P., Mohanta, Y. K., Mohanta, T. K., & Al-Harrasi, A. (2022b). Analysis and interpretation of metagenomics data: An approach. *Biological Procedures Online*, *24*(1), 1–22. https://doi.org/10.1186/S12575-022-00179-7

Oliveira, L. S., & Gruber, A. (2021). Rational design of profile hidden Markov models for viral classification and discovery. *Bioinformatics*, *37*(1),151–170. https://doi.org/10.36255/EXONPUBLICATIONS.BIOINFORMATICS.2021.CH9

Paez-Espino, D., Roux, S., Chen, I. M. A., Palaniappan, K., Ratner, A., Chu, K., Huntemann, M., Reddy, T. B. K., Pons, J. C., Llabrés, M., Eloe-Fadrosh, E. A., Ivanova, N. N., & Kyrpides, N. C. (2019). IMG/VR v.2.0: An integrated data management and analysis system for cultivated and environmental viral genomes. *Nucleic Acids Research*, *47*(D1), D678–D686. https://doi.org/10.1093/NAR/GKY1127

Pal, R. R., More, R. P., & Purohit, H. J. (2018). Bioinformatics tools for shotgun metagenomic data analysis. In H. J. Purohit, V. C. Kalia, & R. P. More (Eds.), *Soft Computing for Biological Systems* (pp. 91–110). Springer. https://doi.org/10.1007/978-981-10-7455-4_6/COVER

Pauvert, C., Buée, M., Laval, V., Edel-Hermann, V., Fauchery, L., Gautier, A., Lesur, I., Vallance, J., & Vacher, C. (2019). Bioinformatics matters: The accuracy of plant and soil fungal community data is highly dependent on the metabarcoding pipeline. *Fungal Ecology, 41*, 23–33. https://doi.org/10.1016/J.FUNECO.2019.03.005

Peiris, M., & Poon, L. L. M. (2021). Severe Acute Respiratory Syndrome (SARS) and Middle East Respiratory Syndrome (MERS) (Coronaviridae). *Encyclopedia of Virology, 1–5*, 814. https://doi.org/10.1016/B978-0-12-814515-9.00138-7

Pirttilä, A. M., Tabas, H. M. P., Baruah, N., & Koskimäki, J. J. (2021). Biofertilizers and biocontrol agents for agriculture: How to identify and develop new potent microbial strains and traits. *Microorganisms*, *9*(4), 817. https://doi.org/10.3390/MICROORGANISMS9040817/S1

Quince, C., Walker, A. W., Simpson, J. T., Loman, N. J., & Segata, N. (2017). Shotgun metagenomics, from sampling to analysis. *Nature Biotechnology*, *35*(9), 833–844. https://doi.org/10.1038/nbt.3935

Rajapaksha, P., Elbourne, A., Gangadoo, S., Brown, R., Cozzolino, D., & Chapman, J. (2019). A review of methods for the detection of pathogenic microorganisms. *Analyst*, *144*(2), 396–411. https://doi.org/10.1039/C8AN01488D

Rastogi, G., & Sani, R. K. (2011). Molecular techniques to assess microbial community structure, function, and dynamics in the environment. In I. Ahmad, F. Ahmad, & J. Pichtel (Eds.), *Microbes and Microbial Technology: Agricultural and Environmental Applications* (pp. 29–57). Springer. https://doi.org/10.1007/978-1-4419-7931-5_2/COVER

Rawat, P., Das, S., Shankhdhar, D., & Shankhdhar, S. C. (2020). Phosphate-solubilizing microorganisms: Mechanism and their role in phosphate solubilization and uptake. *Journal of Soil Science and Plant Nutrition*, *21*(1), 49–68. https://doi.org/10.1007/S42729-020-00342-7

Riley, L. A., & Guss, A. M. (2021). Approaches to genetic tool development for rapid domestication of non-model microorganisms. *Biotechnology for Biofuels*, *14*(1), 1–17. https://doi.org/10.1186/S13068-020-01872-Z

Romano, I., Ventorino, V., & Pepe, O. (2020). Effectiveness of plant beneficial microbes: Overview of the methodological approaches for the assessment of root colonization and persistence. *Frontiers in Plant Science, 11*, 487940. https://doi.org/10.3389/FPLS.2020.00006/BIBTEX

Roumpeka, D. D., Wallace, R. J., Escalettes, F., Fotheringham, I., & Watson, M. (2017). A review of bioinformatics tools for bio-prospecting from metagenomic sequence data. *Frontiers in Genetics*, *8*(MAR), 238989. https://doi.org/10.3389/FGENE.2017.00023/BIBTEX

Rouzine, I. M., & Rozhnova, G. (2018). Antigenic evolution of viruses in host populations. *PLOS Pathogens*, *14*(9), e1007291. https://doi.org/10.1371/JOURNAL.PPAT.1007291

Sahu, A., Bhattacharjya, S., Mandai, A., Thakur, J. K., Atoliya, N., Sahu, N., Manna, M. C., & Patra, A. K. (2018). Microbes: A sustainable approach for enhancing nutrient availability in agricultural soils. In V. Meena (Ed.), *Role of Rhizospheric Microbes in Soil: Volume 2: Nutrient Management and Crop Improvement* (pp. 47–75). Springer. https://doi.org/10.1007/978-981-13-0044-8_2/COVER

Schmidt, J. E., Kent, A. D., Brisson, V. L., & Gaudin, A. C. M. (2019). Agricultural management and plant selection interactively affect rhizosphere microbial community structure and nitrogen cycling. *Microbiome*, *7*(1), 1–18. https://doi.org/10.1186/S40168-019-0756-9/FIGURES/4

Semchenko, M., Nettan, S., Sepp, A., Zhang, Q., Abakumova, M., Davison, J., Kalamees, R., Lepik, A., Püssa, K., Saar, S., Saarma, M., Thetloff, M., & Zobel, K. (2019). Soil biota and chemical interactions promote co-existence in co-evolved grassland communities. *Journal of Ecology*, *107*(6), 2611–2622. https://doi.org/10.1111/1365-2745.13220

Shendure, J., Balasubramanian, S., Church, G. M., Gilbert, W., Rogers, J., Schloss, J. A., & Waterston, R. H. (2017). DNA sequencing at 40: Past, present and future. *Nature*, *550*(7676), 345–353. https://doi.org/10.1038/nature24286

Singh, S., & Shankar, S. (2021). Microbial metagenomics: Potential and challenges. In P. A. M. L. B. Mitra & N. K. J. Singh (Eds.), *Advances in Animal Genomics* (pp. 109–122). Elsevier. https://doi.org/10.1016/B978-0-12-820595-2.00008-4

Shu, W. S., & Huang, L. N. (2021). Microbial diversity in extreme environments. *Nature Reviews Microbiology*, *20*(4), 219–235. https://doi.org/10.1038/s41579-021-00648-y

Sukhum, K. V., Diorio-Toth, L., & Dantas, G. (2019). Genomic and metagenomic approaches for predictive surveillance of emerging pathogens and antibiotic resistance. *Clinical Pharmacology & Therapeutics*, *106*(3), 512–524. https://doi.org/10.1002/CPT.1535

Suter, E. A. (2016). *Aggregate-Associated Microbial Processes in the Cariaco Basin and Implications for Cycling of Carbon, Nitrogen, and Sulfur.* Doctoral dissertation, Stony Brook University). ProQuest Dissertations and Theses Global.

Tacconelli, E., Carrara, E., Savoldi, A., Harbarth, S., Mendelson, M., Monnet, D. L., Pulcini, C., Kahlmeter, G., Kluytmans, J., Carmeli, Y., Ouellette, M., Outterson, K., Patel, J., Cavaleri, M., Cox, E. M., Houchens, C. R., Grayson, M. L., Hansen, P., Singh, N., … Zorzet, A. (2018). Discovery, research, and development of new antibiotics: The WHO priority list of antibiotic-resistant bacteria and tuberculosis. *The Lancet Infectious Diseases*, *18*(3), 318–327. https://doi.org/10.1016/S1473-3099(17)30753-3

Tang, Z., Xu, W., Zhou, G., Bai, Y., Li, J., Tang, X., Chen, D., Liu, Q., Ma, W., Xiong, G., He, H., He, N., Guo, Y., Guo, Q., Zhu, J., Han, W., Hu, H., Fang, J., & Xie, Z. (2018). Patterns of plant carbon, nitrogen, and phosphorus concentration in relation to productivity in China's terrestrial ecosystems. *Proceedings of the National Academy of Sciences of the United States of America*, *115*(16), 4033–4038. https://doi.org/10.1073/PNAS.1700295114/SUPPL_FILE/PNAS.201700295SI.PDF

Thakur, N., Nigam, M., Mann, N. A., Gupta, S., Hussain, C. M., Shukla, S. K., Shah, A. A., Casini, R., Elansary, H. O., & Khan, S. A. (2023). Host-mediated gene engineering and microbiome-based technology optimization for sustainable agriculture and environment. *Functional and Integrative Genomics*, *23*(1), 1–26. https://doi.org/10.1007/S10142-023-00982-9/FIGURES/9

Toshchakov, S. V., Kublanov, I. V., Messina, E., Yakimov, M. M., & Golyshin, P. N. (2015). *Genomic Analysis of Pure Cultures and Communities*. 5–27. Springer. https://doi.org/10.1007/8623_2015_126

Upadhyay, S. K., Rajput, V. D., Kumari, A., Espinosa-Saiz, D., Menendez, E., Minkina, T., Dwivedi, P., & Mandzhieva, S. (2022). Plant growth-promoting rhizobacteria: A potential bio-asset for restoration of degraded soil and crop productivity with sustainable emerging techniques. *Environmental Geochemistry and Health, 2022*, 1–24. https://doi.org/10.1007/S10653-022-01433-3

Vanzolini, J. I., Galantini, J. A., Martínez, J. M., & Suñer, L. (2017). Changes in soil pH and phosphorus availability during decomposition of cover crop residues. *Archives of Agronomy and Soil Science*, *63*(13), 1864–1874. https://doi.org/10.1080/03650340.2017.1308493

Verma, M., Kulshrestha, S., & Puri, A. (2017). Genome sequencing. *Methods in Molecular Biology, 1525*, 3–33. https://doi.org/10.1007/978-1-4939-6622-6_1/COVER

Vollmers, J., Wiegand, S., & Kaster, A. K. (2017). Comparing and evaluating metagenome assembly tools from a microbiologist's perspective—not only size matters! *PLOS ONE*, *12*(1), e0169662. https://doi.org/10.1371/JOURNAL.PONE.0169662

Wan, W., Tan, J., Wang, Y., Qin, Y., He, H., Wu, H., Zuo, W., & He, D. (2020). Responses of the rhizosphere bacterial community in acidic crop soil to pH: Changes in diversity, composition, interaction, and function. *Science of The Total Environment, 700*, 134418. https://doi.org/10.1016/J.SCITOTENV.2019.134418

Wani, A. K., Akhtar, N., Naqash, N., Chopra, C., Singh, R., Kumar, V., Kumar, S., Mulla, S. I., & Américo-Pinheiro, J. H. P. (2022a). Bioprospecting culturable and unculturable microbial consortia through metagenomics for bioremediation. *Cleaner Chemical Engineering, 2*, 100017.

Wani, A. K., Akhtar, N., Naqash, N., Rahayu, F., Djajadi, D., Chopra, C., Singh, R., Mulla, S. I., Sher, F., & Américo-Pinheiro, J. H. P. (2023). Discovering untapped microbial communities through metagenomics for microplastic remediation: Recent advances, challenges, and way forward. In *Environmental Science and Pollution Research* (pp. 1–24). Springer.

Wani, A. K., Akhtar, N., Sher, F., Navarrete, A. A., & Américo-Pinheiro, J. H. P. (2022b). Microbial adaptation to different environmental conditions: Molecular perspective of evolved genetic and cellular systems. *Archives of Microbiology*, *204*(2), 144.

Wani, A. K., Akhtar, N., Singh, R., Chopra, C., Kakade, P., Borde, M., Al-Khayri, J. M., Suprasanna, P., & Zimare, S. B. (2022c). Prospects of advanced metagenomics and meta-omics in the investigation of phytomicrobiome to forecast beneficial and pathogenic response. *Molecular Biology Reports*, *49*(12), 12165–12179.

Wani, A. K., Chopra, C., Singh, R., Ahmad, S., & Américo-Pinheiro, J. H. P. (2023). Mining microbial tapestry using high-throughput sequencing and In silico analysis of Trehalose synthase (TreS) derived from hot spring metagenome. *Biocatalysis and Agricultural Biotechnology, 52*, 102829. https://doi.org/10.1016/j.bcab.2023.102829

Wani, A. K., Dhanjal, D. S., Akhtar, N., Chopra, C., Goyal, A., & Singh, R. (2022d). Role of genomics, metagenomics, and other meta-omics approaches for expunging the environmental contaminants by bioremediation. In T. G. A. & R. S. (Eds.), *Omics for Environmental Engineering and Microbiology Systems* (pp. 19–51). CRC Press.

Wani, A. K., Hashem, N. M., Akhtar, N., Singh, R., Madkour, M., & Prakash, A. (2022e). Understanding microbial networks of farm animals through genomics, metagenomics and other meta-omic approaches for livestock wellness and sustainability–A Review. *Annals of Animal Science*, *22*(3), 839–853.

Wani, A. K., Roy, P., & Kumar, V. (2022f). Metagenomics and artificial intelligence in the context of human health. *Infection, Genetics and Evolution, 100*, 105267.

Watts, G. S., & Hurwitz, B. L. (2020). Metagenomic next-generation sequencing in clinical microbiology. *Clinical Microbiology Newsletter*, *42*(7), 53–59. https://doi.org/10.1016/J.CLINMICNEWS.2020.03.004

Wheeler, N. E., Barquist, L., Kingsley, R. A., & Gardner, P. P. (2016). A profile-based method for identifying functional divergence of orthologous genes in bacterial genomes. *Bioinformatics*, *32*(23), 3566–3574. https://doi.org/10.1093/BIOINFORMATICS/BTW518

Yang, J. (2019). Cloud computing for storing and analyzing petabytes of genomic data. *Journal of Industrial Information Integration, 15*, 50–57. https://doi.org/10.1016/J.JII.2019.04.005
Yen, S., & Johnson, J. S. (2021). Metagenomics: A path to understanding the gut microbiome. *Mammalian Genome 2021 32:4*, *32*(4), 282–296. https://doi.org/10.1007/S00335-021-09889-X
Zou, Q., Lin, G., Jiang, X., Liu, X., & Zeng, X. (2020). Sequence clustering in bioinformatics: An empirical study. *Briefings in Bioinformatics*, *21*(1), 1–10. https://doi.org/10.1093/BIB/BBY090

14 Interactions Between Plants and Beneficial Microbes, Signaling Events, and Effective Cellular Networks for Inducing an Immune Response Against Diseases and Insects

Anshul Arya, Sujata Singh Yadav and Sapna Tiwari

14.1 INTRODUCTION

The development of plants is accompanied by the proliferation of a wide variety of microorganisms, each of which plays a significant part in the development and well-being of the plant. It is necessary to do research on novel microbes in order to overcome the existing difficulty of crop production. This is because the microbiota that are found in plants has a number of intriguing qualities that allow them to perform their functions. The cultivation of crops is confronted with a number of obstacles, including demographic development, production in a sustainable manner, climate change, and other similar issues. Because microorganisms have the potential to be utilized as biopesticides or as biofertilizers, there is a growing demand for the utilization of these microorganisms in agricultural operations as an alternative to the utilization of chemicals (Mitter et al. 2016). According to Hardoim et al. (2015), the recruitment of microorganisms through plants occurs from the surrounding reservoir of microbes. These reservoirs include the phyllosphere (the leaf surface in relation to the external environment), the soil/rhizosphere, the spermosphere (the exterior of the germinated seed), the anthosphere (the flowers' external environment), and the carposphere (the external fruit environment). The majority of the time, the microbiota of the root is transferred horizontally. This microbiota is comprised of a wide variety of microorganisms, the most of which are Verrucomicrobia, Acidobacteria, Proteobacteria, Actinobacteria, Planctomycetes, and Bacteroidetes. Seeds, on the other hand, have the potential to facilitate the vertical transmission of the bacterium. A major source of microorganisms is the seed, which is found to be multiplying in the roots of the plant that is growing (Liu et al. 2021). Microorganisms colonize the roots, rhizosphere, and the above-ground sections of plants are able to take advantage of the exclusive ecological niche that is provided by the root system of plants. According to Hiltner (1904), the rhizosphere is one of the ecosystems that is regarded to be the most complex, and it is also thought to be a prominent location for the activities of bacteria. Soil microorganisms have been shown to govern plant productivity, particularly in habitats that are deficient in nutrients. This is especially true in ecological systems. In addition, the mineralization process has been a significant factor in the robust regulation of plant production by

DOI: 10.1201/9781003354109-14

free-living microorganisms. Microorganisms that are present in the soil play a significant role in a variety of ecosystems and have an impact on a wide range of ecological processes. These processes include the acquisition of nutrients (Sprent, 2001), the cycling of carbon (Hogberg et al. 2001), the formation of soil (Rillig and Mummey, 2006), and the cycling of nitrogen (Kowalchuk and Stephen, 2001).

Observations indicate that only a limited number of plant signals may distinguish between harmful and helpful bacteria, hence triggering an immune response in plants. However, the specific methods via which plants distinguish between helpful and harmful microbes are still not understood (Naveed et al. 2014). Microbes can have commensal, pathogenic, or mutualistic associations with host plants. Plant-associated microbiomes encounter physical barriers composed of soil and air (Kaushal et al. 2016). The phyllospheric microbiome, exerts a substantial influence on the overall growth and development of plants. Moreover, these organisms contribute to the removal of pollutants from plants. Furthermore, they contribute to the preservation of plant health and impede the growth of plant diseases and are greatly influenced by the process of long-distance transfer. Subterranean ecosystems are abundant with advantageous and functionally important microorganisms. The rhizosphere serves as an ideal environment that facilitates the growth of both plants and microorganisms. The microbiomes in the rhizosphere are primarily influenced by plant-derived compounds (Begum et al. 2019).

Both biotic and abiotic stresses have a substantial influence on the metabolism of plants. Biotic factors include organisms include pathogenic fungus, bacteria, viruses, nematodes, and insects. In contrast, abiotic stress encompasses factors such as temperature, salinity, poisonous chemical compounds, dehydration, waterlogging, and metal ions. Moreover, scientific studies have shown that microbes found on or inside plant tissue produce many signals, including carbohydrates and proteins, volatile chemical compounds, hormones, and hormone mimics (Plett et al. 2018). Microorganisms have signals consisting of carbohydrates and proteins known as microbe- or pathogen-associated molecular patterns (MAMPs or PAMPs), which are essential for the survival of microorganisms (Boller, 2009). Plants possess a wide range of pattern recognition receptors (PRRs) that are located on the plasma membrane. These receptors play a crucial role in regulating the immunological responses of plants by binding to MAMPs and PAMPs. This is supported by the fact that these receptors are conserved and not synthesized inside plant cells. Similarly, the beneficial microbes operate as elicitors, triggering a signal that activates plant defence systems against a wide variety of pathogens. They utilize many methods of operation.

In order to activate the defence mechanism of the plants against disease and insects, this chapter discusses the aspects that are crucial to the interaction between beneficial microbes and the plants. The interaction between plants and beneficial microbes is comparable to the interaction between plants and pathogens or pests, with the exception of a few changes that culminate in a symbiotic or beneficial association between the plants and microbes. This association is responsible for the activation of defence mechanisms in plants against harmful pathogens or pests. Another topic that is covered in this chapter is the signaling events that are responsible for providing protection to particularly distant areas of the plants against disease and pests.

14.2 BENEFICIAL MICROORGANISM

The use of beneficial microorganisms, such as bacteria and fungus are basically utilized for the benefits of the plants either by providing nutrients or by providing the protection against abiotic and biotic stresses (Table 14.1). Pest and disease impacts are reduced by using natural opponents, which is known as biological control. Biological control agents (BCAs) comprise microorganisms, such as bacteria, fungi, and viruses, as well as predators and parasitoid insects. They are hired to oversee both living and non-living factors that cause stress and are considered ecologically friendly methods to improve the health and productivity of crops. Certain beneficial microorganisms have been acknowledged for their ability to enhance plant growth and act as biopesticides (Chandler et al. 2011) where "biopesticide" refers to a natural substance or agent derived from a living micro-organism used to manage plant pests. A comprehensive investigation has been carried out on the antagonistic characteristics of several microbial species, such as *Trichoderma, Pseudomonas, Bacillus* species, and other rhizosphere bacteria. In addition, mycoviruses and bacteriophages have the potential to be used as biological control agents (MBCAs) against plant diseases (van Lenteren et al. 2018). Currently, more than one hundred microbial biological control agents (MBCAs) have been officially approved and registered in many countries, such as in the United States, Europe, New Zealand, Japan, Australia, Canada, and Brazil (van Lenteren et al. 2018). Plant pathogens can be controlled and disease suppression can be aided by the use of MBCAs through direct antagonism, synergistic interactions, or indirect antagonism antibiosis, parasitism, predation (Ghorbanpour et al. 2018), enzyme synthesis, competition for resources and space (Spadaro and Droby, 2016). In addition, MBCAs have the ability to stimulate plant development and improve resilience to challenging conditions through induced systemic resistance (ISR), which is the process of enhancing the natural defence mechanisms of plants. MBCAs contribute to the suppression of diseases and control of phytopathogens through the production of antimicrobial compounds, antagonist action (including induced resistance, hyper-parasitism, and competitive inhibition), and by the production of antimicrobial compounds. The key factors that determine induced resistance in plants are the rapid activation of defence mechanisms and the effective identification of pathogens. The observed resistance phenomena can be ascribed to a complex sequence of events including salicylic acid (SA), an abundance of defensive proteins, or the induction of signaling pathways that depend on plant hormones. However, the main challenges associated with the deployment of MBCAs are the intricacies of directly administering bio-control agents to the soil and the subsequent cost load. Other notable issues regarding MBCAs that hinder their use and make them potentially dangerous organisms include their tendency to reproduce rapidly, adapt genetically, and potentially outcompete natural organisms, all of which could lead to negative environmental effects.

Plant growth-promoting microbes such as fungi and bacteria, have been found to significantly influence the growth and spread of plants (Table 14.1). Several studies have presented data on the growth-enhancing properties of plant growth-promoting rhizobacteria (PGPR), PGPB, and PGPF (Kakoi et al. 2014). PGPRs not only function as biocontrol agents, but they also act as phytostimulators and biofertilizers, taking use of the benefits provided by the rhizosphere and microenvironment. The effectiveness of PGPRs depends on their talents, method of communication, and surroundings. Bacteria impact plant development through both direct and indirect methods. The direct production of bacterial compounds enhances the uptake of micronutrients and vital nutrients from the soil. Moreover, these bacteria enhance the synthesis of plant-growth regulators such as IAA, deaminase, and ACC. The chemicals indicated above enhance plant growth and counteract the excessive inhibitory effects of stress ethylene on development (Bruto et al. 2014). In contrast, the indirect method counteracts the growth of harmful organisms in plants and promotes the production of substances that inhibit fungal growth. Bacterial metabolites elicit systemic resistance by acting as extracellular signals. This initiates a series of internal processes. The translocation and reception of a signal by remote plant cells triggers the initiation of defensive systems. Evidence suggests that the cotton plant affected by *Verticillium* saw an increase in plant pathogenic fungus and a decrease in helpful bacteria and arbuscular mycorrhizal fungi. The soil microbiota has been shown to affect the resilience of strawberry plants to *Verticillium dahlia* and *Macrophomina phaseolina*. Another experiment demonstrated that the presence of aphids on the pepper plant resulted in an alteration of the root exudate. The stated event leads to a decrease in the resistance of pepper plants to aphids due to the presence of rhizobacteria that affect plants. A supplementary study has confirmed that compost has a significant impact on the development of tomato plants and has the ability to alleviate the harmful effects of diseases caused by *Verticillium dahlia* and *Fusarium oxysporum*. The addition of compost further helped to the decrease in disease severity caused by these pathogens. Thus, it was inferred that the incorporation of compost might mitigate the detrimental effects generated by fungal infections on plant microbiomes as it has the plenty of beneficial microorganism.

14.3 SYNCOMS

In recent times, synthetic communities (SynComs) consisting of microorganisms have emerged as a potentially valuable tool for investigating the metabolic connections and dynamics of microbial communities (Finkel et al. 2020). For instance, a consortium consisting of *Chitinophaga* sp. and *Flavobacterium* sp. exhibits a higher effectiveness in inhibiting the growth of the fungal root pathogen *Rhizoctonia solani* as compared to their separate application (Carrion et al. 2019). Niu et al. (2017) conducted a study where they observed that a synthetic community consisting of seven bacterial species, mainly from the Proteobacteria, Firmicutes, Bacteroidetes, and Actinobacteria taxa, effectively prevented the colonization and growth of the *Fusarium verticillioides*, the pathogen of maize seedling blight.

14.4 INTERACTION BETWEEN BENEFICIAL MICROBES AND THE PLANTS

A multitude of microorganisms reside in the soil microenvironment around the roots, which contains a diverse range of nutrients, minerals, and metabolites (Figure 14.1). In the rhizosphere, plant root exudates exert a pulling force on microbial colonization. PGPR are effective in colonizing plant roots and, after a successful contact with the plant, they form microcolonies and/or create a biofilm (Table 14.1). Plant-associated biofilms have the ability to protect against external challenges, decrease competition among microorganisms, and enhance the development, production, and quality of crops in the host plant (Ramey et al. 2004). Biofilm production arises when a single microbial cell attaches to a non-living or living surface and multiplies to produce several microcolonies. The microcolonies are made up of linked cells that are surrounded by a matrix made of exopolysaccharides, which are

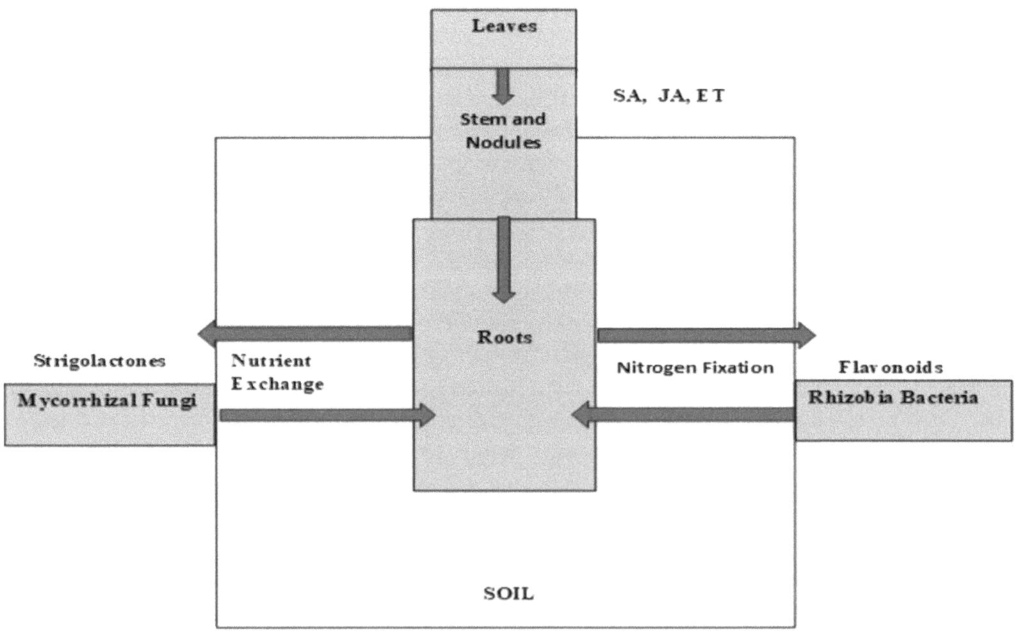

FIGURE 14.1 Interaction between plants and beneficial microorganism, signaling events.

TABLE 14.1
Interactions Between Plants, Beneficial Microbes, and Insects

Interaction Type	Components Involved	Mechanism	Outcome for Plant
		Beneficial Microbes	
Mycorrhizal Fungi	Plant roots, Mycorrhizal fungi	Plants release strigolactones to attract fungi. Fungi form arbuscules within root cells. Nutrient exchange (phosphate from fungi, carbohydrates from plant).	Enhanced nutrient uptake, especially phosphate.
Rhizobia Bacteria	Legume roots, Rhizobia bacteria	Plants release flavonoids to attract rhizobia. Rhizobia produce Nod factors, leading to nodule formation. Rhizobia fix atmospheric nitrogen	Improved nitrogen availability through nitrogen fixation.
Plant Growth-Promoting Rhizobacteria (PGPR)	Plant roots, PGPR	PGPR colonize roots and produce growth-promoting substances (e.g., auxins, gibberellins). Induce systemic resistance (ISR) against pathogens.	Enhanced plant growth and systemic resistance to pathogens.
		Pathogen Interaction	
Pathogen Attack	Plant leaves, Pathogens (bacteria, fungi, viruses)	- PRRs detect PAMPs, triggering PTI. Pathogen effectors may suppress PTI. R proteins detect effectors, triggering ETI.	Activation of immune responses (PTI and ETI).
Insect Interaction			
Insect Herbivory	Plant leaves, Herbivorous insects	- Insect feeding causes damage, releasing DAMPs. Activation of JA and ET signaling pathways. Production of anti-herbivory compounds	Defense against herbivory, reduced damage from insects.
		Systemic Responses	
Systemic Acquired Resistance (SAR)	Plant tissues, SA	- Local infection leads to SA production. - SA moves systemically to other parts of the plant. - Prepares plant for future pathogen attacks.	Enhanced broad-spectrum resistance to pathogens.
Induced Systemic Resistance (ISR)	Plant tissues, JA, ET	Beneficial microbes (e.g., PGPR) or insect damage induce JA and ET signaling. Activation of systemic defenses.	Increased resistance to pathogens and herbivorous insects.

extracellular polymeric substances. Furthermore, biofilms may contain extracellular DNA, proteins, and other substances. The phenotypic traits and gene regulation of the proliferating cells inside the biofilm are different from those of the planktonic cells of the same organism. Currently, there are several *Pseudomonas* spp. and *Bacillus* spp., which are being utilized as the beneficial microbes. It has been observed that these microorganisms may create biofilms and inhabit the surfaces of plant leaves and roots. Root exudates serve as the nourishment and carbon source for the microbial colonization of plant roots (Bais et al. 2006). Bacteria communicate chemically through quorum sensing during the process of biofilm development. This enables the microbial communities to promptly counteract competitor species, improve nutrient uptake, and adjust to changing environmental conditions. Furthermore, it controls the growth and well-being of bacterial populations. Diffusible signals, such as autoinducer-2, 2-heptyl-3-hydroxy-4-quinoline, and N-acyl-homoserine lactones (AHLs), are employed in bacterial intercellular communication to coordinate specific activities and attain greater operational cohesion. These signaling molecules are different for different bacterial spp. example AHLs, are found in Proteobacteria, gamma-butyrolactones in *Streptomyces*, cis-11-methyl-2-dodecanoic acid in *Xanthomonas*, and oligopeptides in Gram-positive microbes (Danhorn and Fuqua, 2007). Non-ribosomal peptide synthetases (NRPSs) are complicated enzymes that play a crucial role in the production of LPs, which have the ability to inhibit the growth of other microbes, including phytopathogens (Stein, 2005). Bacillomycin, bacilysin, lichenysin, surfactins, fengycins, iturins, mycobacillin, and bacillomycin are all categorized as lipopeptides (LPs). *Bacillus* also produces polyketides, which are a major group of secondary metabolites that can act antagonistically.

14.5 PRIMING WITH MICROBES

Priming refers to a specific state in which plants exhibit increased ability to withstand or defend stress as a result of a prior, appropriate stimulation. The colonization of plant roots by PGPR induces systemic resistance in specific areas of the plant, which is known to provide protection against infections (Lugtenberg and Kamilova, 2009). Research has proposed that the activation of disease resistance by beneficial microorganisms (BCA) occurs through a novel process of transmitting signals that result in ISR. This process is distinct from the conventional SA-dependent systemic acquired resistance (SAR). It has been observed that neither pathogenesis-related (PR) proteins nor SA play a role in ISR (Yan et al. 2002). Previous studies have shown that the ISR in plants relies on the signaling pathways of jasmonic acid (JA) and ethylene (ET). The link between the transcription factor MYB72 and EIL3 is of utmost importance, whereas the NPR1 gene is also required for ISR (Van der Ent et al. 2008). The expression of the JA gene during rhizobacteria-mediated ISR against *Pseudomonas syringae* may be regulated by an additional transcription factor called MYC2. The tomato variety Pst DC3000 was identified by Pozo et al. (2008). The activation of ISR is considered to have negligible fitness costs since defence mechanisms are only triggered in response to a stress challenge. In contrast, treatment of plants with SA or JA rapidly promotes defence and disease resistance, even in the absence of stress (van Hulten et al. 2006).

14.6 PLANT IMMUNITY

The plant immune system, as elucidated by Jones and Dangl (2006), may be understood using a four-phase "zigzag" paradigm. In the first phase, the plant immune system is activated by the detection of pathogen-associated molecular patterns (PAMPs) generated by microbes by employing cell surface localized pattern recognition receptors (PRRs). Plants are able to identify infections induced by PAMPs or microbe-associated molecular patterns (MAMPs) as a component of their first defensive mechanism. PRRs are capable of detecting several microbial components, including chitin in fungi, flagellin (flagellin/FLS2 and EF-Tu/EFR (Boller and Felix, 2009)) in bacteria, lipopolysaccharides, and fungal-oomycete cellulose-binding elicitor proteins (Zipfel, 2014). Several studies (Bakker et al. 2007) have investigated this process. The detection of PRR leads to the activation of PAMP-triggered immunity (PTI), which hinders the growth and establishment of the pathogen. Some pathogens produce effectors under the specific infection those enhance their virulence during phase 2, moreover, these effectors can interfere with PTI thereby result in effector-triggered susceptibility (ETS). Phase three, involves the activation of effector-triggered immunity (ETI) in response to the recognition of the effector by nucleotide binding-leucine rich repeat (NB-LRR) receptor proteins. Plant NBS-LRR proteins indirectly detect several virulence proteins by the changes they cause on host target proteins. However, certain NBS-LRR proteins have the ability to directly attach to pathogen proteins. Proteins from either the pathogen or the host that have been altered experience structural alterations in the amino-terminal and LRR domains when they interact with plant NBS-LRR proteins. It is theorized that these structural changes enable the NBS domain to interact with ADP instead of ATP, thereby promoting "downstream" signaling through an unknown mechanism that leads to resistance against pathogens. The majority of R genes encode proteins that have leucine-rich repeats (LRRs) and nucleotide-binding sites (NBSs). NBS-LRR proteins are involved in recognizing certain pathogen effectors, also called avirulence (Avr) proteins, which are thought to enhance the ability of pathogens to cause disease when the corresponding R gene is absent. NBS-LRR proteins, often referred to as NB-LRR or NB-ARC-LRR proteins, can be categorized as either TIR or non-TIR proteins based on the sequences that come before the NBS domain. The amino-terminal domain of plant NBS-LRR proteins belonging to the TIR class exhibits homology with the Toll and interleukin 1 receptors. The non-TIR class lacks a specific definition; nonetheless, the majority of NBS-LRR proteins in this group possess α-helical coiled-coil–like sequences in their amino-terminal domains. This acknowledgement can result in the formation of resistance to diseases, which might manifest as a hypersensitive reaction (HR) at the location of the infection. In the absence of plants evolving new receptors, it is possible that natural selection has forced pathogens to inhibit ETI during phase four by either getting rid of or producing more effectors that enhance disease development.

14.7 QUORUM SENSING

Plants release substances that imitate quorum sensing (QS) and interfere with the formation of the root microbiome. These compounds include molecules that imitate acyl-homoserine lactones (AHL), rosmarinic acid, and the flavanone naringenin. Proline and gamma-aminobutyric acid, both of which are synthesized by plants, modulate the concentration of QS molecules in *Agrobacterium* spp. "Quorum quenching" (QQ) or "quorum interfering" (QI) is a method by which the plant gains benefits by either inhibiting or stimulating the bacterial quorum sensing (QS) system. The study conducted by Vandeputte et al. (2015) revealed that two specific volatile compounds, namely 2,3-butanediol and acetoin, were initially detected in *Bacillus* isolates. These compounds were found to have a stimulating effect on plant development in Arabidopsis. In addition, the compounds dimethyl hexa decylamine (DMHDA) and indole from PGPRs enhance the growth of roots in Arabidopsis by suppressing auxin signaling. This leads to an increase in root length, as well as the quantity and density of root hairs (Bailly et al. 2014).

14.8 MECHANISM OF DEFENSE SIGNALING IN PLANTS BY MICROBES

The main effects of biocontrol activities are the production of antifungal metabolites, the stimulation of systemic resistance, and the availability of nutrients. A variety of rhizobacteria have been recognized as makers of compounds with antifungal characteristics, such as tensin, 2,4-diacetylphloroglucinol, pyoluteorin, pyrrolnitrin, and viscosinamide. The symbiotic relationship between rhizobacteria and plant roots provides immunity against harmful microbes, fungi, and viruses. ISR is the scientific term used to describe this phenomenon (Sgroy et al. 2009). Inflammatory response can be triggered by several bacterial components, such as siderophores, flagella, lipopolysaccharides, cyclic lipopeptides, homoserine lactones, and volatile chemicals such acetoin and 2,3-butanediol (Sziderics et al. 2007). AMFs often acquire essential nutrients, such as nitrogen (N), phosphorus (P), potassium (K), calcium (Ca), zinc (Zn), and several other trace elements, by forming a symbiotic relationship with the roots of the host plant. This illustrates that AMFs provide the plant with essential nutrients when it encounters abiotic stresses. In contrast, fungal structures produce arbuscules, which aid in the transport of inorganic chemicals and minerals to the host plants (Ghosh et al. 2012).

Induced resistance in crops can be triggered by many conditions, such as the presence of beneficial microbes in the root zone and insect herbivory (Romera et al. 2019; Meena et al. 2017a). Both kinds of induced resistance, namely sSAR and ISR are the outcomes of consecutive inoculations or challenges that trigger the activation of latent resistant mechanisms linked to a specific pathogen. The primary distinction between SAR and ISR is in the elicitor and regulatory processes involved, but there are several shared components in the signaling pathways that control both SAR and ISR. The non-affected regions of the plant that are exposed to pathogen invasion elicit immunological responses. SAR is a type of resistance that is produced across long distances and provides protection against a wide range of pathogens. The regulation of SAR is mainly controlled by the buildup of SA inside the plant itself. SAR is characterized by the activation of pathogenesis-related (PR) genes and antimicrobial proteins. Recent advancements in our understanding of the plant immune system have identified specific PRRs that are capable of detecting microbial substances like chitin, bacterial flagellin, or polysaccharides. These receptors are commonly referred to as pathogen- or microbe-associated molecular patterns (PAMPs or MAMPs) (Boller and Felix, 2009). PAMPs activate defence systems in crops, ultimately bolstering the host's immunity against infections. Plants respond to internal signals produced by the plant itself, called damage-associated molecular patterns (DAMPs), after being harmed by an invasion. This reaction has been studied by Boller and Felix in (2009), Zipfel in (2009), and Villena et al. in (2018). Pattern recognition mediates the transmission of PTI, which serves as an early defence mechanism capable of monitoring a wide range of possible intruders (Dodds and Rathjen, 2010). The pathogen effectively utilises virulence effector molecules to evade the first defensive mechanism (PTI) by suppressing or restricting host recognition through PTI signaling. The pathogen's ability to infect the plant host is facilitated by a process called ETI (Meena et al. 2020). ETI functions as an additional means of defence in addition to the elicitor-mediated defence. Most of the genes involved in the ETI pathway consist of cytoplasmic receptor proteins called intracellular nucleotide-binding sites and leucine-rich repeat domains (NBS-LRR or NLRs). Consequently, after plants have developed a second protective layer, the identification of attacker-specific effector chemicals is made easier by NBS-LRR receptor proteins. These proteins then transfer the signals to genes responsible for defence mechanisms (Zhang et al. 2020). The initiation of PTI and ETI often leads to the development of resistance in plant tissues located far from the site of infection. This process involves the transmission of several long-distance signals, which strengthen the defensive power of undamaged portions of the plant. The acronym used to refer to this distinct form of resistance triggered by infections is SAR. Plants have the ability to recognize herbivorous insects that use a similar communication method (Pieterse et al. 2014a). Beneficial bacteria make a conscious effort to prevent local host defence responses in the roots and trigger systemic defensive priming in order to maintain a balanced relationship with the host (Pieterse et al. 2014b). In contrast, beneficial bacteria that trigger ISR should produce elicitors that are responsible for the establishment of systemic immunity. This specific microRNA modulates the expression of genes involved in the defence response during both PTI and ETI, therefore conferring protection against several pathogens, including bacteria, viruses, and fungi (Li et al. 2010). ETI elicits more potent and expedited plant defence responses compared to PTI. ETI is commonly associated with the formation of a necrotic lesion, which can hinder the pathogen's dissemination from the infection site. As a result of these regional events, there is a possibility that ETI and PTI may provoke a response.

14.9 FACTORS THAT IMPACT RESISTANCE

The existence of SIR depends on a variety of circumstances, encompassing both living and non-living components. Abiotic variables encompass both chemical elements (such as heavy metal compounds, water, salt, and salicylic acid) and physical conditions (including mechanical forces, UV radiation, low temperature, and high temperature). Biotic factors encompass the biological compounds generated by microorganisms, fungi, bacteria, and viruses.

14.10 SIGNALING IN PLANTS

Molecular signals are chemical messages that are decoded by creatures with the ability to read them. The understanding of how plants detect and interpret various specific compounds produced by microbes, such as QS molecules and mVOCs, is mostly lacking (Schulz-Bohm et al. 2017).

The different components of molecular signaling's are described below:

14.10.1 Amino Acid Receptors

Multiple studies have presented data indicating that amino acids are involved in facilitating interactions between plants and microorganisms (Moe, 2013). There are some receptors present on the beneficial microbes, which has the ability to detect the amino acids released in environment to get stimulated for the movement towards the desirable host. Methyl-accepting chemotaxis proteins (MCPs) are bacterial receptors that specifically bind to amino acids. In addition to periplasmic binding proteins, MCPs are able to directly detect ligands from the external surroundings. The receptors are responsible for the regulation of the autophosphorylation activity of the histidine kinase CheA. They do this by transducing binding signals across the cellular membrane. CheA is made up of five different domains, numbered P1 through P5, each of which performs a different function. Following activation, CheA is responsible for initiating an intracellular phosphorelay, which eventually regulates the rotation sense of the flagella. The function of HISTIDINE KINASE (CHEMOTAXIS PROTEIN A; CheA), which adds a phosphate group to the CHE-MOTAXIS PROTEIN Y (CheY) response regulator, is influenced by the detection of amino acids by these chemoreceptors. The phosphorylated CheY subsequently engages with the flagellar motor, resulting in modifications to the rotation and/or speed of the flagella in order to control bacterial behavior. Gaining a more extensive knowledge of the receptors responsible for detecting amino acids, along with the similar proteins of CheA and CheY present in beneficial bacteria, could offer additional insights into the perception of amino acids and possibly other substances released by plants in the rhizosphere. While several MCPs have been identified in *Bacillus subtillis*, only McpB has exhibited a specific affinity for amino acid binding (Allard-Massicotte et al. 2016). Recent study suggests that McpB plays a role in the recognition of *B. subtillis.*

The work undertaken by Kørner et al. (2013) investigated the possible involvement of intracellular receptor-like kinases (RLKs) in interactions with plant viruses. RLKs are known to have a role in recognizing PAMPs through PRRs in plants. The extracellular domain (ECD), the transmembrane domain (TMD), and the intracellular kinase domain (KD) are the three components that come together to produce the receptor kinases, which are proteins that are found at the surface of the cell. These proteins are commonly referred to as RLKs in plants. They are known to play a crucial role in a wide range of cellular activities, ranging from the regulation of developmental processes to the physiological responses to environmental stimuli. The activation of mitogen-protein kinases (MAPKs) triggers the production of antimicrobial compounds.

14.10.2 G Protein

The participation of beneficial microbes in the stimulation of heterotrimeric G proteins in plants to counteract biotrophic and necrotrophic diseases is essential. Varying infections elicit distinct activation of G-protein subunits (α, β, and γ), regulating the entrance of the pathogen into a plant cell. Plants utilize G-protein-mediated defence mechanisms to provide resistance against a diverse range of diseases. The activation of certain G-protein subunits depends on the pathogen's method of obtaining nutrients. Recent study (Escudero et al. 2019) has shown that G proteins have a direct role in both biotic and abiotic stress responses in plants. This includes activating phospholipases, NADPH oxidases, and ion channels, as well as producing reactive oxygen species (ROS). G proteins consist of three subunits: Gα, Gβ, and Gγ. The activation of these receptors is attributed to seven transmembrane G-protein coupled receptors (GPCRs). Upon coupling to GTP, the activated Gα subunit separates from the Gβγ complex, showcasing the subunit's inherent GTPase activity. The Gα subunit functions autonomously from GTP, whereas the Gβ and Gγ subunits come together to create a heterodimer. The signaling cascades that follow are controlled by the Gβγ complex and the Gαγ component, which have been isolated from each other (Oldham and Hamm, 2008). Thylakoid formation 1 (THF1), the Arabidopsis cupin domain protein (AtPirin1/PRN1), prephenate dehydrogenase 1 (PD1), and phospholipase are the four effector molecules that interact with G protein (Holland and Jez, 2018). Based on accounts, a solitary Arabidopsis G proteins receive external signals through a minimum of three different categories of GPCR-like molecules and convey the information using at least four different types of effector molecules (Urano and Jones, 2014). The genes associated with JA signaling pathways in Arabidopsis are controlled by heteromeric G proteins, mainly Gα (Li et al. 2019). Patel et al. (2016) investigated the role of G protein signaling, particularly the Gα subunit, in the relationship between the pea powdery mildew pathogen and its host pea when *Trichoderma* and *Pseudomonas* were present either separately or in combination. Furthermore, the research results revealed that the activation of the Gα subunit during the contact with bioagents leads to the initiation of JA-mediated signaling, especially the activation of COI1 and LOX1. Following the activation of the Gα subunit, the suppression of pathogen growth by ROS production was detected. Stomatal closure is linked to the activation of ABA (abscisic acid) signaling following the creation of ROS, namely H_2O_2 production with both single and combination treatments involving beneficial microorganisms. The Gα component not only helps in closing stomata when pathogens invade, but it also blocks K+ channels through PAMPs (Melotto et al. 2017). The dimer of Gβγ directly interacts with extra-large G proteins (XLGs) to activate plant immunity.

14.10.3 Chemicals and Hormones

IAA is a growth hormone that regulates both biotic and abiotic stress, whereas ABA is a stress hormone specifically involved in stress regulation. Salicylic acid, also known as SA or 2-hydroxybenzoic acid, is a crucial stress hormone. The hormone that governs several elements of plant development, either via direct or indirect means. Intracellular calcium concentrations play a crucial role in regulating various biological processes such as development, thermogenesis, drought resistance, and disease resistance. Additionally, the accumulation of SA, Strigolactones (SL), a type of phytohormones derived from carotenoids, serves as dual-purpose regulators of plant development and signaling molecules in the rhizosphere. Recent studies by Bouwmeester et al. (2019) and Banasiak et al. (2020) have shown that the secretion of SLs into the rhizosphere enhances the colonization of beneficial AM fungi in *Medicago truncatulata*, suggesting that plants employ signaling mechanisms to recruit helpful microbes and alleviate phosphate starvation stress. Benzoxazinoids (BXs) are chemical compounds that serve as chemo-attractants for certain beneficial microorganisms (Neal et al. 2012; Cotton et al. 2019). BXs are subjected to many abiotic and environmental challenges, including as photoperiod, water shortage, light intensity, and jasmonic acid, in addition to their role in promoting iron absorption and defence against insect herbivory (Hu et al. 2018). Stringlis et al.

(2018) discovered that scopoletin, a coumarin with antibacterial properties, selectively hinders the development of pathogenic *Verticillium dahlia* and *Fusarium oxysporum*. Hormones such as JA, ET, auxin, and NO are mostly linked to ISR, which is a defensive response in plants. These hormones also play a crucial role in the plant's defence mechanism.

Jasmonates, a group of plant hormones produced by the breakdown of fatty acids in cell membranes, have attracted significant attention in the last 20 years. Jasmonates, such as methyl jasmonate (MeJA), 12-oxophytodienoic acid (12-OPDA), JA, and jasmonic acid conjugated with certain amino acids like leucine (JA-leucine) and isoleucine (JA-isoleucine), are widely present in the plant world. Following tri trophic interactions, senescence, fruit development, and direct and indirect defence responses, these compounds play a crucial role in essential processes that are necessary for the survival and growth of plants. MeJA has attracted considerable interest as a potential airborne signal molecule due to its volatile nature and ability to diffuse through membranes. It has been suggested that MeJA could play a role in facilitating communication between plants, regulating reproductive processes, and modulating defence responses. JA can undergo several metabolic transformations, leading to the production of unique jasmonates. MeJA is synthesized by methylation processes catalysed by JA methyltransferases, which are believed to occur in the cytoplasm (Seo et al. 2001). The role of jasmonates in regulating the production of many secondary metabolites, including terpenoids, alkaloids, phenylpropanoids, and antioxidants, has been extensively studied. Currently, there is a significant amount of data that supports the idea that jasmonates, when they bind to receptors, start a signal-transduction pathway that may either suppress or activate genes controlled by JA (Kazan and Manners, 2008). There are two important mechanisms at play here: (1) In response to injuries, leaf chemicals are produced that are poisonous and harmful to nutrition; and (2) the production and release of volatile compounds that attract parasites and herbivore predators, therefore serving as an indirect defensive mechanism. Plants have the capacity to carry out a wide variety of activities, including strengthening their cell walls, producing signaling compounds, and eventually undergoing cellular death (Scheel, 1998). Jasmonates and their derivatives are of major relevance in activating genes that are implicated in defensive responses (Rosahl and Feussner, 2005). These defence mechanisms are activated in response to both living and non-living stresses, such as insect invasion, microbial infections, drought, and UV irradiation. It is widely recognized that when necrotrophic organisms, which need to destroy host cells to get nourishment, enter the plant, defensive systems that depend on JA are triggered. The initial thorough studies on Solanaceae involved the examination of JA's role in signaling pathways triggered by wound-induced responses and insect damage. The crucial discovery of wound-induced proteinases expressed in potatoes and tomatoes, along with the confirmation that external treatment with JA can trigger the expression of these genes, played a central role in establishing the initial research framework for wound-induced signaling in plants. Moreover, it was discovered that MeJA induces the expression of defense-related genes in grapes (Belhadj et al. 2006). It has been suggested that (pro)systemin works as the main regulatory molecule that interacts with wound-induced systemic defence response pathways (Schilmiller and Howe, 2005). An alternate hypothesis is that systemin exerts its actions in a localised manner, specifically in or in close proximity to the site of the injury, by facilitating the buildup of JA. Consequently, the systemic reaction is strengthened in magnitude. Hence, model substantiates the idea that systemin and JA interact to transmit distal signals through a positive feedback mechanism. The cyclopentenone precursor of JA, 12-oxo-phytodienoic acid (OPDA) also controls gene expression and enhances resistance to insect and fungal attack in the absence of JA. Currently, more than one thousand volatile chemicals, such as alcohols, terpenoids, sulfides, and sesquiterpenes, have been found. The development of methods to enhance plant growth by modifying hormone signaling might greatly benefit from the use of elicitors produced from beneficial microbes. The term "elicitor" refers to a stimulus or agent that triggers a certain response or reaction in an organism or system. The compound Reticine A, which is a newly discovered substance that triggers a response in plants, has been found to induce SAR in tobacco plants against the tobacco mosaic virus (TMV)

(Wang et al. 2021). The outcome is an elevation in the production of PR proteins and an augmentation in the activity of defence enzymes through the activation of the SA biosynthesis genes ICS and PAL.

Salicylic acid is often associated with defensive responses against biotrophic diseases, while JA is engaged in defence responses against necrotrophic pathogens and insects (Beckers and Spoel, 2006) without activating PR genes (Bari and Jones, 2009). To summarize, there are several important situations related to defence response mechanisms in plants: (1) Various genetic studies confirm that jasmonate-signaling pathways are crucial for defence responses triggered by wounds; (2) When tissue is damaged, along with systemin, it can activate the production of proteinase inhibitors through a signaling pathway that requires JA; (3) There is a complex interaction between the signaling pathways of different plant hormones.

14.10.4 Role of Genes in Signaling

Although both ISR and SAR employ NON-EXPRESSOR OF PATHOGENESIS-RELATED GENES 1 (NPR1) as a regulator, their roles in those pathways are different (Dong, 2004). The PR and NPR1 genes play a crucial role in the plant's immune response to pathogens. NPR1 does not operate by direct stimulation of the PR gene. DNA-binding transcription factors (TFs) govern this process, with MYC2 serving as a key regulator of the JA signaling pathway and functioning as a helix-loop-helix transcription factor (Zhai and Li, 2020). The proposed connection between PTI and MYB72, a transcription factor is also discovered in order to induce ISR by the systemic translocation of a long-distance molecular or electric signal.

14.11 INDUCED SYSTEMIC RESISTANCE

The autoregulation of symbionts' colonization density is mostly attributed to beneficial symbiotic microbial interactions, such as mycorrhizal and rhizobial symbiosis (Meena et al. 2017). This process might perhaps be the underlying factor responsible for the occurrence of ISR. Multiple studies (Mou et al. 2020) have shown the importance of JA and ET signaling in the control of rhizobacterial-mediated ISR. When plants are infected by a pathogen, they identify MAMPs/PAMPs, which are small molecular patterns that are shared by different microbes with the help of PRRs, located on the plasma membrane of plants. Furthermore, it leads to the induction of PTI. The proposed connection between PTI and MYB72, a transcription factor, induces ISR by the systemic translocation of a long-distance molecular or electric signal. Consequently, the degree of resistance increases from the local level to the systemic level (Pieterse et al. 2014b). While both ISR and SAR employ NON-EXPRESSOR OF PATHOGENESIS-RELATED GENES 1 (NPR1) as a regulator, they have different purposes. Arabidopsis exhibits ISR when exposed to *P. simiae* WCS417 inoculation in the roots, which also triggers the expression of MYB72 (Zamioudis et al. 2015). The induction of MYB72 is mediated by separate signaling pathways in cases of iron shortage and ISR, which is highly fascinating. In addition, rhizobacteria that stimulate ISR enhance the synthesis of coumarin, even under conditions of sufficient iron levels. These transcription factors further drive the manufacture of other chemical components or ROS, which are essential for defense-related genes or the activation of JA/ET-dependent defence genes, hence promoting resistance response. According to Meena et al. (2020), first studies on MAMPs and other PGPR-induced ISR emitters discovered the participation of supplementary compounds such as lipopolysaccharides (LPS), pyoverdin, salicylic acid, and iron-regulated metabolite. Plants expressing SAR showed an accumulation of PR genes, while plants exhibiting ISR had gene expression sensitive to JA and ET. Therefore, ISR is not influenced by SA genes and relies on the activation of JA/ET signaling pathways, as evidenced by the study conducted by Van Wees et al. (1999). Moreover, it has been confirmed that miR825 and miR825* have a substantial impact on the control of ISR by suppressing the expression of genes associated with resistance (Niu et al. 2016a, 2016b). miR825 and miR825* are two small RNAs that are generated from a single miRNA gene. They function as negative regulators in plant defense responses to pathogens. *Trichoderma harzianum* can trigger ISR in maize leaves by activating the C6 zinc finger protein-like elicitor Thph1 and Thph2 proteins (Saravanakumar et al. 2016). The pathogen's lifestyle and mechanism of infection influence the timing, content, and quantity of these molecules (De Vos et al. 2005). Pathogenesis-related proteins (PRPs) are the characteristic genes of the JA and SA pathways in many agricultural plants. Increased expression of the PR1, PR2, and PR5 genes leads to activation of the SA signaling pathway (Ali et al. 2017). Expression of the PR3, PR4, and PR12 genes in Arabidopsis leads to the activation of the JA pathway.

14.12 CROSS-KINGDOM INTERFERENCE WITH RNA

Cross-kingdom interference with RNA (CKRI) refers to the phenomenon of RNA molecules from one biological kingdom interfering with the gene expression of another biological kingdom is also utilized as one type of defense in plants against several pathogens. Exosome-mediated CKRI is a mechanism that utilizes vesicle trafficking of short RNAs and microRNAs to suppress the expression of virulence genes in microbes. It serves as the third line of defence against pathogens (Li et al. 2010). In the past 15 years, there has been a notable advancement in the study of plant defence mechanisms. It has been discovered that dsRNAs obtained from plant transgenes can trigger gene silencing in invading pathogens and pests through a process called RNA interference (RNAi) (Iqbal et al. 2020). The term used to describe this event is host-induced gene silencing (HIGS). Nowara et al. (2010) demonstrated the feasibility of using HIGS to control a fungal disease. In order to accomplish this, scientists inserted a set of double-stranded RNAs (both in a permanent and temporary manner) into genetically modified plants. The objective was to specifically target the genes of *Blumeria graminis* that encode the effector Avra10 or 1,3-β-glucanosyltransferases, which are involved in the development of haustoria. HIGS demonstrates the potential for RNA molecules to be transported from a plant to a fungus and trigger RNAi. Therefore, HIGS is an example of cross-kingdom RNA interference (CKRNAi) by its successful execution. Further studies have confirmed that plant endogenous short RNAs (sRNAs) may be transported into parasitic organisms, where they trigger gene silencing in the invaders through RNAi. This event demonstrated the natural phenomenon of CKRNAi, and many research suggested that it may serve as an effective technique for host defence. The original occurrence took place during the interaction between cotton and *Verticillium dahliae*, whereby infection with *V. dahliae* led to an augmentation in the synthesis of two microRNAs (miRNAs) in cotton. Subsequently, these miRNAs were transported to the fungal hyphae, where they suppressed certain genes of *V. dahliae* (Zhang et al. 2016). Cai et al. (2018) presented proof that exosome-like vesicles carrying sRNAs may be released by Arabidopsis. Subsequently, *Botrytis cinerea* hyphae internalize these vesicles at the infection site. This work defined two small interfering RNAs (siRNAs) that specifically concentrate in fungal cells and are anticipated to target gene associated with *B. cinerea* that control vesicle trafficking.

14.13 NEMATODE PROTECTION

The beneficial microbes employ several methods to combat nematodes, including as parasitization, toxin generation, antibiotic release, enzyme secretion, disruption of nematode-plant recognition, nutrition competition, and induction. Siddiqui and Mahmood (1999) found that plants may be improved in terms of their health and ability to withstand diseases by boosting their systemic resistance. The bacteria in question possess a wide range of inhibitory characteristics against different kinds of nematodes, including both predatory and free-living nematodes, as well as plant and animal parasitic nematodes (Siddiqui and Mahmood, 1999). The categorization of organisms in the genus *Pasteuria* is mostly based on pathological and morphological characteristics, including the size and shape of sporangia and endospores, the structures at a microscopic level, the life cycles, and the distribution among hosts (Atibalentja et al. 2000). *Pasteuria* is a kind of bacteria that forms endospores and is mycelial. It is an obligate parasite that lives in water fleas and plant-parasitic nematodes (Bekal et al. 2001). Infestation of the root-knot nematode *Meloidogyne* spp. is done by the microorganism *Pasteuria penetrans*. *Pasteuria* spores germinate when second-stage juveniles attach to the cuticles, enter the roots, and start feeding. After the germ tubes penetrate the cuticle, vegetative microcolonies rapidly multiply throughout the female's body during development. Afterwards, fully developed endospores are released into the soil when the female nematode's reproductive system deteriorates. Preston et al. (2003) suggested that the existence of an adhesin-associated epitope, N-acetyglucosamine, which is located on the surface of the spores, enhances the adherence of polypeptides produced from different *Pasteuria* strains. This specific antigenic determinant provides an immunochemical technique for differentiating species and biotypes that exhibit diverse preferences for hosts. Collagen may aid in the identification process by interacting with a receptor on the nematodes cuticle. Specifically, this occurs when the nematode is exposed to the Persidis opportunistic parasite bacteria *Brevibacillus laterosporus* strain G4 and *Bacillus* sp. B16. Hydrolytic enzymes have a role in breaking down the various components of the nematode cuticle that surround the openings during bacterial infection (Huang et al. 2005).

14.14 ISR AGAINST VIRAL DISEASE

Induced systemic resistance involves the inoculation of susceptible plants with a virus to bestow resistance and prevent further infection by the same or another susceptible virus. The absence of virus introduction during the vulnerable stages of the crop or the plant's partial resistance to the virus inoculum, even at the same concentration, shields the plant from viral infection. The term cross protection is also utilized for such resistance. The development of resistance can occur when the mobility of the virus within the infected cells is limited, hindering its capacity to propagate from one cell to another. The virus has the potential to spread throughout the plant as a result of its resistance to replication or accumulation. However, this spread may happen at low levels and may not manifest noticeable symptoms. Two possibilities were proposed about induced resistance. The depletion theory, first proposed by Silberschmidt and Caner in (1967), suggests that neighboring uninfected tissues undergo a reduction in metabolites as a result of the increased metabolic activity of infected cells. As a result, these tissues experience a decrease in their ability to support the production of viruses and acquire resistance to the infection. (2) Bozarth and Ross (1964) proposed in their resistance-induced chemicals idea that the infection does generate some compounds that confer resistance in unaffected tissues, the researchers were unable to elucidate the depletion theory, whereby metabolites are depleted after removal of the lower inoculated leaves. An evaluation of this hypothesis was carried out by Van Loon (1997). Loebenstein (2009) suggests that acquired/induced resistance in plants might be attributed to an interferon-like mechanism, besides, this resistance is non-specific and has been reported to be triggered by many agents, such as viruses, fungi, TMV protein, and yeast RNA and susceptible to dactinomycin, which might cause the lesion to increase in size. Most of the prevailing resistance (R) genes were discovered through interactions between plants and viruses such type of resistance is called dominant resistance. These genes belong to the NBS-LRR class, which identifies viral avirulence (avr) gene products through "gene-for-gene" contact. The concept of the "guard hypothesis," proposing a direct physical connection between the avr and R gene products, was initially proposed by Soosaar et al. (2005). Subsequently, a multitude of NBS-LRR proteins that confer resistance to viruses have been discovered and classified based on their N-terminal structure. This consists of either a coiled-coil (CC) domain or a toll-interleukin-1 receptor (TIR) domain (De Ronde et al. 2014). The resistance to potato virus X (PVX) is facilitated by the CC-NBS-LRR protein, which is produced by the potato gene and operates by detecting the PVX capsid protein (CP). Activation is frequently accompanied by the entry of calcium ions, signaling mediated by MAPK, generation of ROS, buildup of SA, and significant transcriptional reprogramming. The activation of the R gene is usually followed by a HR, which restricts the pathogen to the hypersensitive lesion and stops it from spreading further into the plant. Furthermore, the functional module responsible for mediating HR to viruses requires the interaction between the protein senescence-associated gene 101 (SAG101) and two lipases, namely increased disease susceptibility 1 (EDS1) and phytoalexin deficient 4 (PAD4) (Wagner et al. 2013). Local defence mediated by R-genes triggers defensive signaling in distant tissues, known as "systemic acquired resistance" or "SIR." The N and Rx1 genes were employed to illustrate this process (Liu et al. 2010). Phloem-resident multiprotein complexes (RTMs) consisting of RTM members may play a role in resistance mechanisms that hinder the long-distance migration of potyviruses. Ishibashi and Ishikawa (2013) provided evidence that the tomato Tm-1 gene, which produces a protein with a structure similar to a triose phosphate isomerase barrel, directly interacts with the viral replicase and effectively inhibits the replication of the viral genome. This interaction leads to resistance against TMV, and it is not dependent on the NBS-LRR mechanism.

14.15 RECESSIVE RESISTANCE

The inheritance of over 200 resistance genes that specifically target plant viruses is predominantly recessive, making it the most common kind of resistance. These genes are being used in breeding strategies to control plant pathogenic viruses. Recessive resistance arises due to the absence of crucial host components required for the virus to effectively complete its life cycle. Several eukaryotic initiation factors (eIF4E and eIF4G) have been discovered as recessive resistance genes that have been cloned in several crop species (Revers and Nicaise, 2014). Initially, these parameters were confirmed to be valid in potyvirus infection and then utilised in the study of additional plant viruses such as Bymovirus, Cucumovirus, Ipomovirus, Sobemovirus, Carmovirus, and Waikivirus. The presence of a large amount of "resistant eIF4E" makes "susceptible eIF4E" inaccessible from viruses' recruitment. The translational apparatus, including translation initiation factor 4B (eIF4B), translation elongation factors 1A and 1B (eEF1A and eEF1B), and poly(A)-binding proteins (PABPs), was found crucial for the replication of viruses in Arabidopsis and tobacco. The presence of these translation elements may facilitate the identification of novel sources of resistance, which are crucial for conducting future research on the management of viral diseases (Li et al. 2014b). The documentation of recessive resistance genes in potatoes has included genetic characterization. The resistance genes rlm1 and rpv1, which confer protection against LMV and PPV, are situated in a chromosomal area that lacks genes encoding translation factors (Decroocq et al. 2006).

14.16 RNA INTERFERENCE

RNAi, also known as gene silencing, is triggered by double-stranded RNAs (dsRNAs). Double-stranded RNA triggers a reaction in the host that activates the RNA-silencing process. These molecules act as MAMPs or PAMPs, and RNAi can be seen as a pattern-triggered immune response (PTI). An important difference between viral MAMPs or PAMPs and those of pathogens such as fungi and bacteria is that the former is identified within cells (Ding and Voinnet, 2007). Silencing may be classified into two main types: host-gene encoded miRNAs play a role in regulating genes, whereas siRNAs are a sign of antiviral RNA interference. Viral double-stranded RNA (dsRNA) molecules, which are produced during viral replication as intermediates (RIs), replicative forms (RFs), or secondary RNA structures, trigger the antiviral RNAi response. The dicer-like (DCL) protein, which is an RNase type III-like enzyme in the host, identifies these formations and divides them into siRNA molecules consisting of 21 to 24 nucleotides (Sharma et al. 2013). The RNA-induced silencing complex (RISC) is a functional protein complex that incorporates siRNAs. Upon activation, this intricate system breaks down viral RNA molecules that have sequences that are complementary to the guide strand. The slicer, a crucial component of RISC, enables the destruction of the target RNA. The protein is a member of the Argonaute (AGO) family, as described by Vaucheret (2008) and Sharma et al. (2013). Subsequently, the plant system synthesizes and amplifies the main siRNAs, which is crucial for triggering an RNA interference (RNAi) response that effectively combats viral infections both locally and throughout the entire plant. The amplification process involves the participation of RNA-dependent RNA polymerases (RdRps), which have the ability to convert viral messenger RNA (mRNA) into dsRNA through both siRNA-dependent and siRNA-independent mechanisms (Csorba et al. 2009). Antiviral RNAi is a highly successful method for combating both RNA and DNA viruses (Incarbone and Dunoyer, 2013). Viruses also face a secondary barrier consisting of resistance genes. These genes are specifically triggered by a particular virus and provide resistance only against that virus; certain genes even exhibit functionality against related viruses (for example, the Ty-1 resistance gene from tomato against TYLCV). This gene contains an RNA-dependent RNA polymerase (RdRp) and, as stated by Verlaan et al. (2013), provides resistance to TYLCV by enhancing the signal of RNA interference. Within plant systems, three essential processes have been identified in the pathways that lead to silencing: (1) cytoplasmic RNA silencing, also known as PTGS, which is facilitated by small interfering RNAs (siRNAs); (2) silencing that is facilitated by microRNAs (miRNAs) encoded by plants themselves; and (3) transcriptional gene silencing (TGS), which occurs through siRNA-directed methylation of DNA and histone proteins. The three RNA silencing mechanisms outlined before have been shown to have a protective role against viral infections that enter plants (Bologna and Voinnet, 2014).

14.17 DUAL RESISTANCE

Regarding systemically induced viruses, the generation of vsiRNA from transgenes can trigger the gene silencing in the transgenic line after virus infection. This is the first description of the creation of transgenic plants that possess resistance to the most common viruses.

14.18 INTERACTION BETWEEN PLANT, MICROBES, AND INSECTS

Various mechanisms like morphological, biochemical and molecular are responsible to elicit defence response in host-plants against their insect-pests (Figure 14.2). The signaling compounds involved in defense response are produced in response to previous plant damage. Plant pathogens can affect interactions between host-plants and their insect-pests. In nature, plants are attacked by many organisms, including insects and plant pathogens. In response, plants have developed defence strategies that allow them to recognize herbivore-insects and pathogens and thus reduce damage in many cases. The induction of the plant defence response leading to the production of secondary metabolites is thought to have an important role as it has antagonistic and/or synergistic interactions.

Several recent studies have shown that in addition to the independent effects of plant pathogens and insect-pests attack on the host plant, the complicated interactions too exist between these two pests feeding on the same plant. In response to this interaction a range of physical or chemical defence elicit in the host plant. In addition, plant-pathogen infection alters the emission of volatile organic compounds or visual cues. This in return can negatively affect the performance of insect-pests (Ajlan and Potter, 1991; Cardoza et al. 2003). Similarly, damage caused by insect-pests can elicit chemical and physical changes in plants that can affect the development of disease. For example, in cotton due to previous infestation of spider mite some physiological changes occur, which reduces the degree of infection by wilt fungi *Verticillium dahliae*. In rice infestation of white-backed planthopper (*Sogatella furcifera*) elicit resistance against *Magnaporthe grisea* (Karban et al. 1987; Simon and Hilker, 2003). Disease infected plants can also elicit the resistance in host plant to insect-pests by preventing egg laying, suppressing reproductive development of insect, reducing performance and causing larval and pupal mortality (Myers and Cory, 2016; Queiroz et al. 2016).

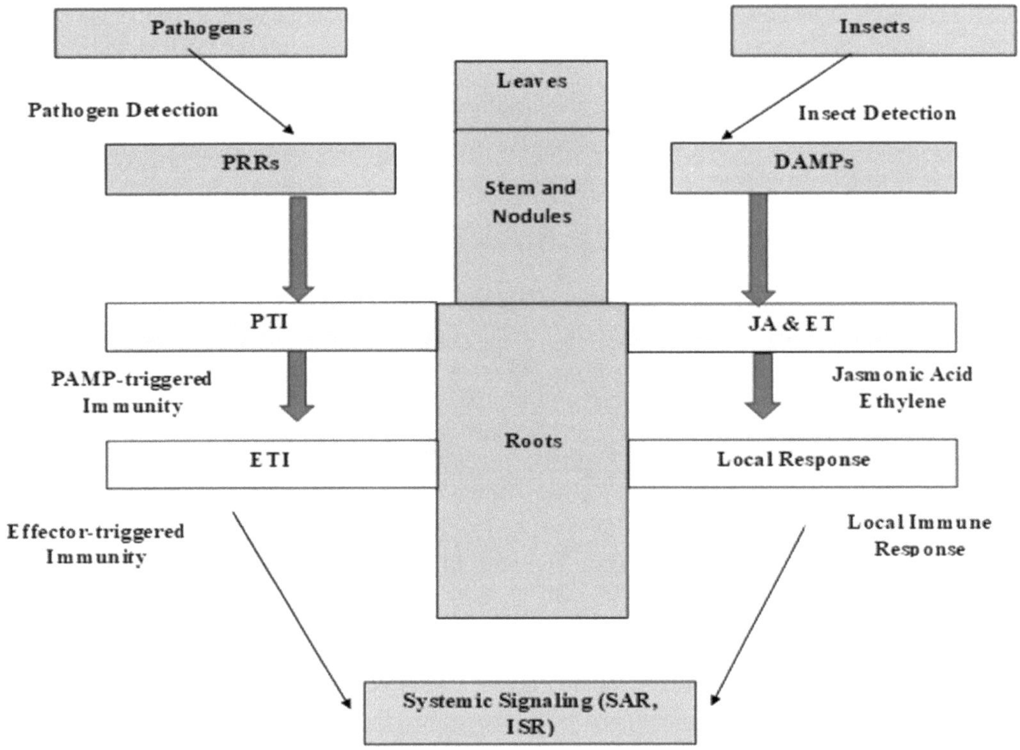

FIGURE 14.2 Systemic acquired resistance and induced systemic resistance.

14.19 STEPS INVOLVED IN HOST PLANT-DEFENCE RESPONSE DUE TO PATHOGEN INFECTION

1. In the plants infested by bacterial and fungal pathogen, the pathways of defence system are triggered by the elicitor–receptor interaction.
2. First of all, ion permeability of the plasma membrane is modified to initiate defence signaling.
3. Ion channels across the plasma membrane get opened due to the receptor molecule activation. This leads to an influx of calcium ions into the plant cells, results in stimulation of ROS production in the cell.
4. Around the location of proliferation of pathogen in plants, ROS accumulation is reported (Nadarajah et al. 2015).
5. This ROS accumulation causes an oxidative eruption that elicit a defence mechanism or provide signal to the activation of other defences (Nurnberger and Scheel, 2001). Signaling pathways involved in host plant due to plant-microbial interaction against insect-pests. There are two major signaling pathways that are responsible for defence in plants effective against distinct types of insect-pests (Koornneef and Pieterse, 2008). The pathways are as follows:
   1. The SA-dependent pathways: it is responsible for SAR, which leads to an increase in local phytohormone production. SA is an endogenous growth regulator responsible for various metabolic and physiological effects during plant resistance, growth, and development

(Aljbory and Chen, 2018; Khan et al. 2015). SA mainly induces defence to manage piercing and sucking insects as compared to the biting and chewing insect-pests (Zhao et al. 2009). Following are the ways by which defence is elicited by SA signaling in the host-plant:

a. ROS production is responsible for the induction of plant defence to insect pests, e.g., interaction of tomato and cotton bollworm (*Lycopersicon esculentum* Mill. and *Helicoverpa armigera* Hubner).
b. Salicylic acid induces production of H_2O_2, which protects plants against pathogens and insect-pests feeding because H_2O_2 is detrimental for the alimentary canal of insects, (Peng et al. 2004).
c. The emission of plant volatiles are triggered by SA which acts as lures for the predators of insects (Maffei et al. 2007a).
d. Methyl salicylate (MeSA) is produced once the cell has been infected by the pathogen. This volatile molecule travels to other parts of the plant through the plasmodesmata (Viswanathan et al. 2008) and attracts insect predators to increase indirect resistance in host-plant (Aljbory and Chen, 2018).

2. The JA-dependent pathways: it is responsible for induced systemic resistance (ISR). JA (Jasmonic acid) triggers both direct as well as indirect defences (Guo et al. 2018). JA is a derivative of linolenic acid and accumulates in plant tissue upon wounding due to insect infestation or pathogen infection (Zhang et al. 2008). Following are the ways by which defence is elicited by JA signaling in the host-plant:
   a. Various defence reactions induced by jasmonates are antioxidant activity, emission of volatiles and alkaloids (Guo et al. 2018).
   b. JA induces different genes during defence against insects (Aljbory and Chen, 2018).
   c. Glucosinolate concentration is also induced by jasmonates.
   d. Herbivore associated elicitors (HAEs) are chemical cues produced by insect herbivores. HAEs activate particular plant defence responses to fight insect pests. These structurally diverse compounds are not common elicitors of responses against insects in all plant species and normally limited to specific plant–arthropod interaction.
   e. JA also induces antioxidant enzymes (POD and PPO) (War et al. 2012).

Examples:

- In tomato plants ISR is induced due to enhancement in the expression of production of defense-related genes lipoxygenase D and F (LOXD, LOXF), which leads to the increased production of lipopeptides when colonized by *B. amyloliquefaciens* against fall armyworms (Li et al. 2015).
- PGPR accumulates Phenolic compounds which produce a HR in plants (Kiprovski et al. 2016). In rice plants, when roots are colonized by *Pseudomonas fluorescens*, phenolic compounds are accumulated which induces ISR (De Vleesschauwer et al. 2008).
- In rice plants due to *P. fluorescens* ISR is stimulated against the larvae of leaf-folder due to the activation of chitinases and trypsin inhibitors (Saravanakumar et al. 2008). Root colonization by PGPR in cotton plants induces production of higher levels of JA, which may elicit resistance against *Spodoptera exigua* (Zebelo et al. 2016).
- In tomato, *Bacillus subtilis* induces resistance against whitefly by increasing the expression of both JA-dependent genes including proteinase and proteases inhibitor genes and JA-independent genes including terpenoid biosynthetic pathways genes, photosynthetic genes and phenylpropanoid (Valenzuela- Soto et al. 2010).
- Roots of Arabidopsis when colonized by *Pseudomonas fluorescens* resulted in the increase of susceptibility to *Myzus persicae* (Pineda et al. 2012).

14.20 CONCLUSION

Managing plant diseases should be accomplished by the focused modification of naturally occurring defence systems of host plants. This should be the most effective and sustainable technique for managing plant diseases. The use of microorganisms that are capable of antagonistic activity against pathogens in order to establish systemic resistance in the plant is an effective strategy for the treatment of crop diseases. When it comes to controlling pathogens and abiotic stresses, another strategy that may be utilized is the administration of elicitors to plants. On the other hand, in order to accomplish complete protection of plants against diseases pest and abiotic stresses, an integrated strategy to disease management and control that makes use of microorganisms and the metabolites they produce is required. In the recent past, there have been extraordinary efforts undertaken for the purpose of identifying metabolites that provide disease resistance in a variety of crops. Several metabolites, including enzymes, amino acids, lipids, and organic acids, have been recognized as various metabolic indicators for disease resistance in a variety of crops. Recently, the growing method of metabolome network research (exploring the connections between the various metabolites) has opened the door to the possibility of investigating the ways in which various metabolites interact with one another in order to protect the plant against pest and diseases. The use of resistance-related constitutive and induced metabolites in the process of imparting disease resistance to plants would be an intriguing research topic to investigate in the not-too-distant future.

REFERENCES

Ajlan AM, Potter DA. (1991). Does immunization of cucumber against anthracnose by *Colletotrichum lagenarium* affect host suitability for arthropods. *Entomol Exp Appl* 58: 83–91.

Ali S, Mir ZA, Bhat JA, Chandrashekar N, Papolu PK, Rawat S, Grover A. (2017). Identification and comparative analysis of *Brassica juncea* pathogenesis-related genes in response to hormonal, biotic and abiotic stresses. *Acta Physiol Plant* 39: 268.

Aljbory Z. Chen MS. (2018). Indirect plant defence against insect herbivores: a review. *Insect Science* 25:2–23.

Allard-Massicotte R, Tessier L, Le´ cuyer F, Lakshmanan V, Lucier J-F, Garneau D, Caudwell L, Vlamakis H, Bais HP, Beauregard PB. (2016). *Bacillus subtilis* early colonization of *Arabidopsis thaliana* roots involve multiple chemotaxis receptors. *mBio* 7:e01664–16.

Atibalentja N, Noel GR, Domier LL. (2000). Phylogenetic position of the North American isolate of Pasteuria that parasitizes the soybean cyst nematode, Heterodera glycines, as inferred from 16S rDNA sequence analysis. *Int J Syst Evol Microbiol* 50:605–613.

Bailly A, Groenhagen U, Schulz S, Geisler M, Eberl L, Weisskopf L. (2014). The inter-kingdom volatile signal indole promotes root development by interfering with auxin signalling. *Plant J* 80:758–771.

Bais HP, Weir TL, Perry LG, et al. (2006). The role of root exudates in rhizosphere interactions with pl ants and other organisms. *Annu Rev Plant Biol* 57:234–266.

Bakker PAHM, Pieterse CMJ, Van Loon LC. (2007). Induced systemic resistance by fluorescent *Pseudomonas spp. Phytopathology* 97(2): 239–243.

Banasiak J, Borghi L, Stec N, Martinoia E, Jasinski M. (2020). The full-size ABCG transporter of *Medicago truncatula* is involved in strigolactone secretion, affecting arbuscular mycorrhiza. *Front Plant Sci* 11:18.

Bari R, Jones JD. (2009). Role of plant hormones in plant defence responses. *Plant Mol Biol* 69(4):473–88.

Beckers GJ, Spoel SH. (2006). Fine-tuning plant defence signalling: Salicylate versus Jasmonate. *Plant Biol (Stuttg)* 8(1):1–10.

Begum N, Qin C, Ahanger MA, Raza S, Khan MI, Ashraf M, Ahmed N, Zhang L. (2019). Role of arbuscular mycorrhizal fungi in plant growth regulation: implications in abiotic stress tolerance. *Front Plant Sci* 10:1068.

Bekal S, Borneman J, Springer J, Giblin-Davis RM, Becker JO. (2001). Phenotypic and molecular analysis of a Pasteuria strain parasitic to the sting nematode. *J Nematol* 33(2-3):110–115.

Belhadj A, Saigne C, Telef N, Cluzet S, et al. (2006). Methyl jasmonate induces defense responses in grapevine and triggers protection against *Erysiphe necator*. *J Agric Food Chem* 54:9119–9125.

Boller T, Felix G. (2009). A renaissance of elicitors: perception of microbe-associated molecular patterns and danger signals by pattern-recognition receptors. *Ann Rev Plant Biol* 60:379–407.

Bologna NG, Voinnet O. (2014). The diversity, biogenesis, and activities of endogenous silencing small RNAs in *Arabidopsis*. *Annu Rev Plant Biol* 65:473–503.

Bouwmeester HJ, Fonne-Pfister R, Screpanti C, De Mesmaeker A. (2019). Strigolactones: plant hormones with promising features. *Angew Chem Int Ed* 58:12778–12786.

Bozarth RF, Ross AF. (1964). Systemic resistance induced by localized virus infections: extent of changes in uninfected plant parts. *Virology* 24:446–455.

Bruto M, Prigent-Combaret C, Muller D, Moenne-Loccoz, Y. (2014). Analysis of genes contributing to plant-beneficial functions in plant growth-promoting rhizobacteria and related proteobacteria. *Scientific Reports* 4:6261. 10.1038/srep06261.

Cai Z, Cao Y, Luo Y, Hu H, Ling H. (2018). Signalling mechanism(s) of epithelial mesenchymal transition and cancer stem cells in tumour therapeutic resistance. *Cli Chim Acta* 483:156–163.

Cardoza YJ, Lait CG, Schmelz EA, Huang J, Tumlinson JH. (2003). Fungus induced biochemical changes in peanut plants and their effect on development of beet armyworm, *Spodoptera exigua* Hubner (Lepidoptera: Noctuidae) larvae. *Environ Entomol* 32:220–228.

Carrion VJ, Perez-Jaramillo J, Cordovez V, Tracanna V, De Hollander M, Ruiz-Buck D, Mendes LW, Van Ijcken WFJ, Gomez-Exposito R, Elsayed SS. (2019). Pathogen-induced activation of disease-suppressive functions in the endophytic root microbiome. *Science* 366:606–612.

Chandler D, Bailey AS, Tatchell GM, Davidson G, Greaves J, Grant WP. (2011). The development, regulation and use of biopesticides for integrated pest management. *Philos Trans R Soc Lond B Biol Sci* 366:1987–1998.

Cotton TEA, Petriacq P, Cameron DD, Al Meselmani M, Schwarzenbacher R, Rolfe SA, Ton J. (2019). Metabolic regulation of the maize rhizobiome by benzoxazinoids. *ISME J* 13:1647–1658.

Csorba T, Pantaleo V, Burgyan J. (2009). RNA silencing: an antiviral mechanism. *Adv Virus Res* 75:35–71.

Danhorn T, Fuqua C (2007). Biofilm formation by plant-associated bacteria. *Annual Review of Microbiology* 61:401–422.

De Ronde D, Butterbach P, Kormelink R. (2014). Dominant resistance against plant viruses. *Front Plant Sci* 5:307.

De Vleesschauwer D, Djavaheri M, Bakker PA, Hofte M. (2008). *Pseudomonas fluorescens* WCS374r-induced systemic resistance in rice against *Magnaporthe oryzae* is based on pseudobactin-mediated priming for a salicylic acid-repressible multifaceted defence response. *Plant Physiol* 148:1996–2012.

De Vos M, Van Oosten VR, Van Poecke RMP, Van Pelt JA, Pozo MJ, Mueller M, Decroocq V, Sicard O, Alamillo JM, Lansac M, Eyquard JP, Garcıa JA. (2006). Multiple resistance traits control plum pox virus infection in *Arabidopsis thaliana*. *Mol Plant-Microbe Interact* 19:541–549.

Decroocq V, Ion L, Lansac M, Eyquard JP and Schurdi-Levraud V. (2006). Unravelling the Prunus/Plum pox virus interactions. *EPPO Bulletin*, 36: 346–349. https://doi.org/10.1111/j.1365-2338.2006.01013.x

Ding SW, Voinnet O. (2007). Antiviral immunity directed by small RNAs. *Cell* 130:413–426.

Dodds PN, Rathjen JP. (2010). Plant immunity: towards an integrated view of plant–pathogen interactions. *Nat Rev Genet* 11 (8):539–548.

Dong X. (2004). NPR1, all things considered. *Curr Opin Plant Biol* Oct;7(5):547–552.

Escudero V, Torres MA, Delgado M, Sopena-Torres S, Swami S, Morales J, Molina A. (2019). Mitogen-activated protein kinase phosphatase 1 (MKP1) negatively regulates the production of reactive oxygen species during Arabidopsis immune responses. *Mol Plant Microbe Interact* 32(4):464–478.

Finkel OM, Salas-Gonzalez I, Castrillo G, Conway JM, Law TF, Teixeira PJPL, Wilson ED, Fitzpatrick CR, Jones CD, Dangl JL. (2020). A single bacterial genus maintains root growth in a complex microbiome. *Nature* 587:103–108.

Ghorbanpour M, Omidvari M, Abbaszadeh-Dahaji P, Omidvar R, Kariman K. (2018). Mechanisms underlying the protective effects of beneficial fungi against plant diseases. *Biol Control* 117:147–157.

Ghosh R, Sengupta S, Bhattacharyya S. (2012) Arbuscular mycorrhizal fungi associated with some fern species collected from Kumaon region of western Himalaya. *Indian Phytopathol.* 65:282–285.

Guo Q, Major IT Howe GA. (2018). Resolution of growth–defence conflict: mechanistic insights from jasmonate signaling. *Curr Opin Plant Biol* 44:72–81.

Hardoim PR, Van Overbeek LS, Berg G, et al. (2015). The hiddenworld within plants: ecological and evolutionary considerations for defining functioning of microbialendophytes. *Microbiol Mol Biol Rev* 79:293–320.

Hiltner L. (1904). Uber neue erfahrungen und probleme auf dem gebiete der bodenbakteriologie. *Arbeiten der Deutschen Land- wirtschaft Gesellschaft* 98:59–78.

Hogberg P, Nordgren A, Buchmann N, Taylor AFS, Ekblad A, Hogberg MN, Nyberg G, Ottosson-Lofvenius M, Read DJ. (2001). Large-scale forest girdling shows that current photosynthesis drives soil respiration. *Nature* 411:789–792.

Holland CK, Jez JM. (2018). Reaction mechanism of prephenate dehydrogenase from the alternative tyrosine biosynthesis pathway in plants. *Chem BioChem* 19(11):1132–1136.

Hu L, Mateo P, Ye M, Zhang X, Berset JD, Handrick V, Radisch D, Grabe V, Kollner TG, Gershenzon J. (2018). Plant iron acquisition strategy exploited by an insect herbivore. *Science* 361:694–697.

Huang G, Dong R, Allen R, Davis EL, Baum TJ, Hussey RS. (2005). Two chorismate mutase genes from the root-knot nematode Meloidogyne incognita. *Mol Plant Pathol* 6(1):23–30.

Incarbone M, Dunoyer P. (2013). RNA silencing and its suppression: novel insights from in planta analyses. *Trends Plant Sci* 18:382–392.

Iqbal A, Qiang D, Alamzeb M, Xiangru W, Huiping G, Hengheng Z, Nianchang P, Xiling Z, Meizhen S. (2020). Untangling the molecular mechanisms and functions of nitrate to improve nitrogen use efficiency. *J Sci Food Agric* 100:904–914.

Ishibashi K, Ishikawa M. (2013). The resistance protein Tm-1 inhibits formation of a tomato mosaic virus replication protein-host mem- brane protein complex. *J Virol* 87:7933–7939.

Jones J, Dangl J. (2006). The plant immune system. *Nature* 444:323–329.

Kakoi K, Yamaura M, Kamiharai T, Tamari D, Abe M, Uchiumi T, Kucho K. (2014). Isolation of mutants of the nitrogen-fixing actinomycetes frankia. *Microbes Environ* 29:31–37.

Karban R, Adamchak R, Schnathorst WC. (1987). Induced resistance and interspecific competition between spider-mites and a vascular wilt fungus. *Science* 235:678–680.

Kaushal M, Wani SP. (2016). Plant-growth-promoting rhizobacteria: drought stress alleviators to ameliorate crop production in drylands. *Ann Microbiol* 66:35–42.

Kazan K, Manners JM. (2008). Jasmonate signaling: toward an integrated view. *Plant Physiol* 146:1459–1468.

Khan MIR, Fatma M, Per TS, Anjum NA, Khan NA. (2015). Salicylic acid- induced abiotic stress tolerance and underlying mechanisms in plants. *Front Plant Sci* 6:462.

Kiprovski B, Malenčić S, Bursać M, Cvejić J, Sikora V. (2016). Isoflavone content and antioxidant activity of soybean inoculated with plant-growth promoting rhizobacteria. *J Serb Chem Soc* 81:1239–1249.

Koornneef A, Pieterse CMJ. (2008). Cross talk in defense signaling. *Plant Physiol* 146:839–844.

Kørner CJ, Klauser D, Niehl A, Domínguez-Ferreras A, Chinchilla D, Boller T, Heinlein M, Hann DR. (2013). The immunity regulator BAK1 contributes to resistance against diverse RNA viruses. *Mol Plant Microbe Interact* 26:1271–1280.

Kowalchuk GA, Stephen JR. (2001). Ammonia-oxidizing bacteria: a model for molecular microbial ecology. *Annu Rev Microbiol* 55:485–529.

Li H, Soares MA, Torres MS, Bergen M, White JFJ. (2015). Endophytic bacterium, *Bacillus amyloliquefaciens,* enhances ornamental host a resistance to diseases and insect pests. *J Plant Interact* 10: 224–229.

Li L, Su B, Qi X, Zhang X, Song S, Shan X. (2019). JA-induced endocytosis of AtRGS1 is involved in G-protein mediated JA responses. *Int J Mol Sci* 20(15):3779.

Li Y, Zhang QQ, Zhang J, Wu L, Qi Y, Zhou JM. (2010). Identification of microRNAs involved in pathogen-associated molecular pattern-triggered plant innate immunity. *Plant Physiol* 152(4):2222–2231.

Li Z, Gonzalez PA, Sasvari Z, Kinzy TG, Nagy PD. (2014b). Methylation of translation elongation factor 1A by the METTL10- like See1 methyltransferase facilitates tombusvirus replication in yeast and plants. *Virology* 448:43–54.

Liu J, Wang X. (2021). Plant diseases and pests detection based on deep learning: a review. *Plant Methods* *17*(1):1–19.

Liu PP, Yang Y, Pichersky E, Klessig DF. (2010). Altering expression of benzoic acid/salicylic acid carboxyl methyl transferase 1 compromises systemic acquired resistance and PAMP-triggered immunity in *Arabidopsis. Mol Plant-Microbe Interact* 23:82–90.

Loebenstein G. (2009). Local lesions and induced resistance. *Adv Virus Res* 75:73–117.

Lugtenberg B, Kamilova F. (2009). Plant-growth-promoting rhizobacteria. *Annu Rev Microbiol* 63:541–556.
Maffei ME, Mithofer A, Boland W. (2007a). Before gene expression: early events in plant–insect interaction. *Trends Plant Sci* 12:310–316.
Meena KK, Sorty AM, Bitla UM, Choudhary K, Gupta P, Pareek A, Singh DP, Prabha R, , Sahu PK, Gupta VK, Singh HB, Krishanani KK, Minhas PS (2017) Abiotic stress responses and microbe-mediated mitigation in plants: The omics strategies. *Frontiers in Plant Science*, 8:1664–462X
Meena M, Prasad V, Upadhyay RS. (2017a). Evaluation of biochemical changes in leaves of tomato infected with *Alternaria alternata* and its metabolites. *Vegetos* 30:2.
Meena M, Swapnil P, Divyanshu K, Kumar S, Harish, Tripathi YN, Zehra A, Marwal A, Upadhyay RS. (2020). PGPR-mediated induction of systemic resistance and physiochemical alterations in plants against the pathogens: current perspectives. *J Basic Microbiol* 60 (8):1–34.
Melotto M, Zhang L, Oblessuc PR, He SY. (2017). Stomatal defence a decade later. *Plant Physiol* 174(2):561–571.
Mitter B, Pfaffenbichler N, Sessitsch A. (2016). Plant–microbe partnerships in 2020. *Microb Biotechnol* 9:635–640.
Moe LA. (2013). Amino acids in the rhizosphere: from plants to microbes. *Am J Bot* 100:1692–1705.
Mou W, Kao Y, Michard E, Simon AA, Li D, Wudick MM, Lizzio MA, Feijo JA, Chang C. (2020). Ethylene-independent signaling by the ethylene precursor ACC in *Arabidopsis* ovular pollen tube attraction. *Nat Commun* 11(1):4082.
Myers JH, Cory JS. (2016). Ecology and evolution of pathogens in natural populations of Lepidoptera. *Evol Appl* 9:231–247.
Nadarajah K, Ali HZ, Muni NM. (2015). The host pathogen interactions observed between *Fusarium spp.* F2 and rice. *Plant Omics* 8:572–580.
Naveed M, Hussain MB, Zahir ZA, Mitter B, Sessitsch A. (2014). Drought stress amelioration in wheat through inoculation with *Burkholderia phytofirmans* strain PsJN. *Plant Growth Regul* 73:121–131.
Neal AL, Ahmad S, Gordon-Weeks R, Ton J. (2012). Benzoxazinoids in root exudates of maize attract *Pseudomonas putida* to the rhizosphere. *PLoS One* 7:e35498.
Niu B, Paulson JN, Zheng X, Kolter R. (2017). Simplified and representative bacterial community of maize roots. *Proc Natl Acad Sci U S A* 114: E2450–E2459.
Niu D, Wang X, Wang Y, Song X, Wang J, Guo J, Zhao H. (2016a). *Bacillus cereus* AR156 activates PAMP-triggered immunity and induces a systemic acquired resistance through a NPR1-and SA-dependent signaling pathway. *Biochem Biophys Res Commun* 469(1):120–125.
Niu D, Xia J, Jiang C, Qi B, Ling X, Lin S, Zhang W, Guo J, Jin H, Zhao H. (2016b). *Bacillus cereus* AR156 primes induced systemic resistance by suppressing miR825/825* and activating defence-related genes in *Arabidopsis*. *J Integr Plant Biol* 58(4):426–439.
Nowara D, Gay A, Lacomme C, Shaw J, Ridout C, Douchkov D, Hensel G, Kumlehn J, Schweizer P. (2010). HIGS: host-induced gene silencing in the obligate biotrophic fungal pathogen *Blumeria graminis*. *Plant Cell* 22(9):3130–41.
Nurnberger T, Scheel D. (2001). Signal transmission in the plant immune response. *Trends Plant Sci* 6:372–379.
Oldham WM, Hamm HE. (2008). Heterotrimeric G protein activation by G-protein-coupled receptors. *Nat Rev Mol Cell Biol* 9(1):60.
Patel JS, Sharma BK, Singh HB, Upadhyay RS, Kharwar RN, Ahmed M. (2016). *Pseudomonas fluorescens* and *Trichoderma asperellum* enhance expression of Gα subunits of the pea heterotrimeric G-protein during *Erysiphe pisi* infection. *Front Plant Sci* 6:1206.
Peng J, Deng X, Jia S, Huang J, Miao X, Huang Y. (2004). Role of salicylic acid in tomato defence against cotton bollworm, Helicoverpa armigera Hubner. *Z Naturforsch C Biosci* 59:856–862.
Pieterse CMJ, Pierik R, Van Wees SCM. (2014a). Different shades of JAZ during plant growth and defense. *New Phytol* 204 (2):261–264.
Pieterse CMJ, Zamioudis C, Berendsen RL, Weller DM, Van Wees SCM, Bakker PAHM. (2014b). Induced systemic resistance by beneficial microbes. *Ann Rev Phytopathol* 52:347–375.
Pineda A, Zheng SJ, Van Loon JJA, Dicke M. (2012). Rhizobacteria modify plant-aphid interactions: a case of induced systemic susceptibility. *Plant Biol* 14:83–90.
Plett JM, Martin FM. (2018). Know your enemy, embrace your friend: using omics to understand how plants respond differently to pathogenic and mutualistic micro-organisms. *Plant J* 93:729–746.

Pozo MJ, Van der Ent S, Van Loonm LC, Pieterse CM. (2008). Transcription factor MYC2 is involved in priming for enhanced defense during rhizobacteria-induced systemic resistance in Arabidopsis thaliana. *New Phytol* 180:511–523.

Preston JF, Dickson DW, Muruniak JE, Nong G, Brito JA, Schmidt LM, Giblin-Davis RM. (2003). *Pasteuria spp.*: systematics and phylogeny of these bacterial parasites of phytopathogenic nematodes. *J Nematol* 35:198–207.

Queiroz RB, Donkersley P, Silva FN, Al-Mahmmoli IH, Al-Sadi AM, Carvalho CM, Elliot SL. (2016). Invasive mutualisms between a plant pathogen and insect vectors in the Middle East and Brazil. *R Soc Open Sci* 3:160557.

Ramey BE, Koutsoudis M, Bodman BV, Fuqua C. (2004). Biofilm formation in plant–microbe associations. *Curr Opin Microbiol* 7:602–609.

Revers F, Nicaise V. (2014). Plant resistance to infection by viruses. In: *Encyclopedia of Life Sciences*. Wiley-Blackwell, John Wiley & Sons Ltd, Chichester, 1–10. DOI: 10.1002/9780470015902.a0000757.pub3

Rillig MC, Mummey DL. (2006). Mycorrhizas and soil structure. *New Phytol* 171:41–53.

Romera FJ, García MJ, Lucena C, Martínez-Medina A, Aparicio MA, Ramos J, Alcántara, E, Angulo M, Pérez Vicente R. (2019). Induced Systemic Resistance (ISR) and Fe Deficiency Responses in Dicot Plants. *Front Plant Sci* 10:287.

Rosahl S, Feussner I. (2005). Oxylipins. In: *Plant Lipids: Biology, Utilization and Manipulation* (Murphy DJ, ed.). Blackwell Publishing Ltd./CRC Press, Oxford and Boca Raton, 329–354.

Saravanakumar D, Lavanya N, Muthumeena B, Raguchander T, Suresh S, Samiyappan R. (2008). Pseudomonas fluorescens enhances resistance and natural enemy population in rice plants against leaf folder pest. *J Appl Entomol* 132:469–479.

Saravanakumar K, Fan L, Fu K, Yu C, Wang M, Xia H, Sun J, Li Y, Chen J. (2016). Cellulase from *Trichoderma harzianum* interacts with roots and triggers induced systemic resistance to foliar disease in maize. *Sci Rep* 6(1):35543.

Scheel D. (1998). Resistance response physiology and signal transduction. *Curr Opin Plant Biol* 1:305–310.

Schilmiller AL, Howe GA. (2005). Systemic signalling in the wound response. *Curr Opin Plant Biol* 8:369–377.

Schulz-Bohm K, Martınsa nchez L, Garbeva P. (2017). Microbial volatiles: small molecules with an important role in intra-and inter-kingdom interactions. *Front Microbiol* 8:2484.

Seo HS, Song JT, Cheong JJ, Lee YH, et al. (2001). Jasmonic acid carboxyl methyltransferase: a key enzyme for jasmonate-regulated plant responses. *Proc Natl Acad Sci USA* 98:4788–4793.

Sgroy V, Cassán F, Masciarelli O, Del Papa MF, Lagares A, Luna V. (2009). Isolation and characterization of endophytic plant growth-promoting (pgpb) or stress homeostasis-regulating (pshb) bacteria associated to the halophyte *Prosopis strombulifera. Appl Microbiol Biotechnol* 85:371–381.

Sharma N, Sahu PP, Puranik S, Prasad M. (2013). Recent advances in plant-virus interaction with emphasis on small interfering RNAs (siRNAs). *Mol Biotechnol* 55:63–77.

Silberschmidt K, Caner J. (1967). Physiological factors responsible for a specific form of resistance against plant virus infection. *Arq Inst Biol S Paulo* 34:127–133.

Soosaar JLM, Burch-Smith TM, Dinesh-Kumar SP. (2005). Mechanisms of plant resistance to viruses. *Nat Rev Microbiol* 3:789–798.

Simon M, Hilker M. (2003). Herbivores and pathogens on willow: do they affect each Other. *Agric For Entomol* 5:275–284.

Spadaro D, Droby S. (2016). Development of biocontrol products for postharvest diseases of fruit: the importance of elucidating the mechanisms of action of yeast antagonists. *Trends Food Sci Technol* 47:39–49.

Sprent JI. (2001). *Nodulation in Legumes*. Gardens, Kew, UK: Royal Bot.

Stein T. (2005). *Bacillus subtilis* antibiotics: structure, syntheses and specific functions. *Mol Microbiol* 56:845–857.

Stringlis IA, Yu K, Feussner K, De Jonge R, Van Bentum S, Van Verk MC, Berendsen RL, Bakker PAHM, Feussner I, Pieterse CMJ. (2018). MYB72-dependent coumarin exudation shapes root microbiome assembly to promote plant health. *Proc Natl Acad Sci* 115:5213–5222.

Sziderics A, Rasche F, Trognitz F, Sessitsch A, Wilhelm E. (2007). Bacterial endophytes contribute to abiotic stress adaptation in pepper plants (*Capsicum annuum* L.). *Can J Microbiol* 53:1195–1202.

Urano D, Jones AM. (2014). Heterotrimeric G protein-coupled signalling in plants. *Annu Rev Plant Biol* 65:365–384.

Valenzuela-Soto JH, Estrada-Hernandez MG, Ibarra-Laclette E, Delano-Frier JP. (2010). Inoculation of tomato plants (*Solanum lycopersicum*) with growth-promoting *Bacillus subtilis* retards whitefly *Bemisia tabaci* development. *Planta* 231:397–410.

Van der Ent S, Verhagen BWM, Van Doorn R, Bakker D, Verlaan, MG, Pel MJC, Joosten RG, Proveniers MCG, Van Loon LC, Ton J, Pieterse CMJ. (2008). MYB72 is required in early signaling steps of rhizobacteria-induced systemic resistance in arabidopsis. *Plant Physiology*, 146(3):1293–1304.

van Hulten M, Pelser M, Van Loon LC, Pieterse CM, Ton J. (2006). Costs and benefits of priming for defense in *Arabidopsis*. *Proc Natl Acad Sci USA* 103:5602–5607.

van Lenteren JC, Bolckmans K, Köhl J, Ravensberg WJ, Urbaneja A. (2018). Biological control using invertebrates and microorganisms: plenty of new opportunities. *BioControl* 63: 39–59.

Van Loon LC. (1997). Induced resistance in plants and the role of pathogenesis-related proteins. *Eur J Plant Pathol* 103:753–765.

Van Wees SCM, Luijendijk M, Smoorenburg I, Van Loon LC, Pieterse CMJ. (1999). Rhizobacteria-mediated induced systemic resistance (ISR) *Arabidopsis* is not associated with a direct effect on expression of known defense-related genes but stimulates the expression of the jasmonate-inducible gene ATVSP upon challenge. *Plant Mol Biol* 41(4):537–549.

Vandeputte OM, Kiendrebeogo M, Rasamiravaka T, Stevigny C, Duez P, Rajaonson S, Diallo B, Mol A, Baucher M, El Jaziri M. (2015). The flavanone naringenin reduces the production of quorum sensing-controlled virulence factors in *Pseudomonas aeruginosa* PAO1. *Microbiology* 157:2120–2132.

Vaucheret H. (2008). Plant Argonautes. *Trends Plant Sci* 13:350–358.

Verlaan MG, Hutton SF, Ibrahem RM, Kormelink R, Visser RGF, Scott JW. (2013). The tomato yellow leaf curl virus resistance genes Ty-1 and Ty-3 are allelic and code for DFDGD-class RNA- dependent RNA polymerases. *PLoS Genet* 9:e1003399.

Villena J, Kitazawa H, Van Wees SCM, Pieterse CMJ, Takahashi H. (2018) Receptors and signalling pathways for recognition of bacteria in livestock and crops: prospects for beneficial microbes in healthy growth strategies. *Front Immunol* 9:2223.

Viswanathan DV, McNickle G, Thaler JS. (2008). Heterogeneity of plant phenotypes caused by herbivore-specific induced responses influences the spatial distribution of herbivores. *Ecol Entomol* 33:86–94.

Wagner S, Stuttmann J, Rietz S, Guerois R, Brunstein E, Bautor J, Niefind K, Parker JE. (2013). Structural basis for signaling by exclusive EDS1 hetero- meric complexes with SAG101 or PAD4 in plant innate immunity. *Cell Host Microbe* 14:619–630.

Wang Y, Li X, Fan B, Zhu C, Chen Z. (2021). Regulation and function of defense-related callose deposition in plants. *Int J Mol Sci* 22:2393.

War AR, Paulraj MG, Ahmad T, Buhroo AA, Hussain B, Ignacimuthu S, Sharma HC. (2012). Mechanisms of plant defence against insect herbivores. *Plant Signal Behav* 7:1306–1320.

Yan Z, Reddy MS, Ryu CM, McInroy JA, Wilson M, Kloepper JW. (2002). Induced systemic resistance against tomato late blight elicited by plant growth-promoting rhizobacteria. *Phytopathology* 92:1329–1333.

Siddiqui ZA, Mahmood I. (1999). Role of bacteria in the management of plant parasitic nematodes: a review. *Bioresour Technol* 69(2):167–179.

Zamioudis C, Korteland J, Van Pelt JA, Van Hamersveld M, Dombrowski N, Bai Y, Hanson J, Van Verk MC, Ling H, Schulze-Lefert P. (2015). Rhizobacterial volatiles and photosynthesis-related signals coordinate MYB 72 expressions in *Arabidopsis* roots during onset of induced systemic resistance and iron-deficiency responses. *Plant J* 84:309–322.

Zebelo S, Song Y, Kloepper JW, Fadamiro H. (2016). Rhizobacteria activates (+)-δ-cadinene synthase genes and induces systemic resistance in cotton against beet armyworm (Spodoptera exigua). *Plant Cell Environ* 39:935–943.

Zhai Q, Deng L, Li C. (2020). Mediator subunit MED25: at the nexus of jasmonate signaling. *Curr Opin Plant Biol* 57:78–86.

Zhang CY, Liu HC, Zhang XS, Guo QX, Bian SM. Wang JY, Zhai LL (2020) VcMYB4a, an R2R3-MYB transcription factor from Vaccinium corymbosum, negatively regulates salt, drought, and temperature stress. *Gene*, 757 (2020), Article 144935.

Zhang J, Wang X, Vikash V, Ye Q, Wu D, Liu Y, Dong W. (2016). ROS and ROS-mediated cellular signaling. *Oxid Med Cell Longev* 2016:4350965.

Zhang SZ, Hua BZ, Zhang F. (2008). Induction of the activities of antioxidative enzymes and the levels of malondialdehyde in cucumber seedlings as a consequence of *Bemisia tabaci* (Hemiptera: aleyrodidae) infestation. *Arthropod-Plant Inte* 2:209–213.

Zhao L, Chen J, Cheng D, Sun J, Liu Y, Tian Z. (2009). Biochemical and molecular characterizations of *Sitobion avenae*-induced wheat defence responses. *Crop Protect* 28:435–442.

Zipfel C. (2009). Early molecular events in PAMP-triggered immunity. *Curr Opin Plant Biol* 12(4):414–420.

Zipfel C (2014) Plant pattern-recognition receptors. *Trends Immunol.* Jul;35(7):345–51